AF807267

Recent Developments in Multivariate and Random Matrix Analysis

Thomas Holgersson • Martin Singull
Editors

Recent Developments in Multivariate and Random Matrix Analysis

Festschrift in Honour of Dietrich von Rosen

Springer

Editors
Thomas Holgersson
Department of Economics and Statistics
Linnaeus University
Växjö, Sweden

Martin Singull
Department of Mathematics
Linköping University
Linköping, Sweden

ISBN 978-3-030-56772-9 ISBN 978-3-030-56773-6 (eBook)
https://doi.org/10.1007/978-3-030-56773-6

© Springer Nature Switzerland AG 2020
This work is subject to copyright. All rights are reserved by the Publisher, whether the whole or part of the material is concerned, specifically the rights of translation, reprinting, reuse of illustrations, recitation, broadcasting, reproduction on microfilms or in any other physical way, and transmission or information storage and retrieval, electronic adaptation, computer software, or by similar or dissimilar methodology now known or hereafter developed.
The use of general descriptive names, registered names, trademarks, service marks, etc. in this publication does not imply, even in the absence of a specific statement, that such names are exempt from the relevant protective laws and regulations and therefore free for general use.
The publisher, the authors, and the editors are safe to assume that the advice and information in this book are believed to be true and accurate at the date of publication. Neither the publisher nor the authors or the editors give a warranty, expressed or implied, with respect to the material contained herein or for any errors or omissions that may have been made. The publisher remains neutral with regard to jurisdictional claims in published maps and institutional affiliations.

This Springer imprint is published by the registered company Springer Nature Switzerland AG.
The registered company address is: Gewerbestrasse 11, 6330 Cham, Switzerland

To Dietrich

Preface

Dietrich von Rosen received his PhD in Mathematical Statistics from the Stockholm University in 1985 with his dissertation "Multivariate Linear Normal Models with Special References to the Growth Curve Model." He stayed at the same university as assistant professor until 1990, after which he worked as a senior lecturer in Mathematical Statistics at the Uppsala University. In 1998, he obtained a full professorship in Statistics at the Swedish University of Agricultural Sciences. Since 2009, he has been also active as an adjoined professor in Mathematical Statistics at the Linköping University. Furthermore, von Rosen has been working at the medical university in Stockholm, Karolinska Institutet, for 6 years and is an Honorary Doctor at the University of Tartu, Estonia, since 2014. Professor von Rosen has written more than 120 peer-reviewed articles and 2 books (1 together with Professor Tõnu Kollo, University of Tartu) and has been a supervisor for more than 20 PhD theses in areas such as statistics, mathematical statistics, and biometry. He is a frequently invited speaker at international conferences, the editor-in-chief of Journal of Multivariate Analysis, and the associate editor of four other journals.

This volume is a tribute to Professor von Rosen on the occasion of his 65th birthday. It contains a collection of 20 papers (ordered alphabetically by the first author) by 45 colleagues and researchers, some of whom are his former students, from 16 countries. The contents of the papers evolve around multivariate analysis and random matrices with topics such as high-dimensional analysis, goodness-of-fit measures, variable selection and information criteria, inference of covariance structures, the Wishart distribution, and, of course, growth curve models. Although a scientific book perhaps is rarely read in its full length, we hope and believe that Professor von Rosen cannot resist reading this one from front to back. Indeed, each of the contributions is founded on his own work in one way or another.

It would not be right to characterize von Rosen only by his academic and scientific merits. Although any successful scientist will receive a certain amount of attention, much of von Rosen's popularity among colleagues and friends is due to his kind and relaxed attitude. He is always eager to interact with his audience but yet very polite and respectful with everyone.

When taking on the task of launching a project like this, one never knows how it will progress. In our case, it happened that a few invited contributors were unable to send in a manuscript before deadline. Because of this, we decided to contribute with more than a single paper each. Our sincere thanks go to all authors of the papers for their contributions and commitment to this Festschrift. We would also like to convey our apologies to those we may have overlooked or failed to contact and who would also have liked to be a part of this volume. Just drop an email to us, and we will include you in the forthcoming Festschrift to von Rosen's 75th birthday!

In closing, we wish Dietrich a joyful 65th birthday and are looking forward to many more years of cooperation and friendship.

Växjö, Sweden
Linköping, Sweden
July 2020

Thomas Holgersson
Martin Singull

Contents

1 **Spectral Analysis of Large Reflexive Generalized Inverse and Moore-Penrose Inverse Matrices** 1
Taras Bodnar and Nestor Parolya

2 **Testing for Double Complete Symmetry** 17
Carlos A. Coelho and Martin Singull

3 **Convexity of Sets Under Normal Distribution in the Structural Alloy Steel Standard** 41
Kai-Tai Fang, Zhen Luo, and Yung Liang Tong

4 **Comments on Maximum Likelihood Estimation and Projections Under Multivariate Statistical Models** 51
Katarzyna Filipiak, Mateusz John, and Augustyn Markiewicz

5 **Growth Curve Model with Orthogonal Covariance Structure** 67
Miguel Fonseca and Martin Singull

6 **Holonomic Gradient Method for the Cumulative Distribution Function of the Largest Eigenvalue of a Complex Wishart Matrix with Noncentrality Matrix of Rank One** 83
Yuta Fukasawa and Akimichi Takemura

7 **Some Tests for the Extended Growth Curve Model and Applications in the Analysis of Clustered Longitudinal Data** 103
Jemila S. Hamid and Sayantee Jana

8 **Properties of BLUEs and BLUPs in Full vs. Small Linear Models with New Observations** 123
Stephen J. Haslett, Augustyn Markiewicz, and Simo Puntanen

9 **A Collection of Moments of the Wishart Distribution** 147
Thomas Holgersson and Jolanta Pielaszkiewicz

10 **Risk and Bias in Portfolio Optimization** 163
Thomas Holgersson and Martin Singull

11 **Approximating Noncentral Chi-Squared to the Moments and Distribution of the Likelihood Ratio Statistic for Multinomial Goodness of Fit** 175
Björn Holmquist, Anna Sjöström, and Sultana Nasrin

12 **Covariance Structure Tests for t-distribution** 199
Tõnu Kollo and Marju Valge

13 **Variable Selection in Joint Mean and Covariance Models** 219
Chaofeng Kou and Jianxin Pan

14 **On Shrinkage Estimators and "Effective Degrees of Freedom"** 245
Lynn R. LaMotte, Julia Volaufova, and Simo Puntanen

15 **On Explicit Estimation of the Growth Curve Model with a Block Circular Covariance Structure** 255
Yuli Liang and Deliang Dai

16 **Space Decomposition and Estimation in Multivariate Linear Models** 267
Joseph Nzabanita and Innocent Ngaruye

17 **Detection of Sparse and Weak Effects in High-Dimensional Feature Space, with an Application to Microbiome Data Analysis** ... 287
Tatjana Pavlenko, Annika Tillander, Justine Debelius, and Fredrik Boulund

18 **Exploring Consistencies of Information Criterion and Test-Based Criterion for High-Dimensional Multivariate Regression Models Under Three Covariance Structures** 313
Tetsuro Sakurai and Yasunori Fujikoshi

19 **Mean Value Test for Three-Level Multivariate Observations with Doubly Exchangeable Covariance Structure** 335
Ivan Žežula, Daniel Klein, and Anuradha Roy

20 **Estimation of the Common Mean of Two Multivariate Normal Distributions Under Symmetrical and Asymmetrical Loss Functions** 351
Dan Zhuang, S. Ejaz Ahmed, Shuangzhe Liu, and Tiefeng Ma

Contributors

S. Ejaz Ahmed Department of Mathematics and Statistics, Brock University, St. Catharines, ON, Canada

Taras Bodnar Department of Mathematics, Stockholm University, Stockholm, Sweden

Fredrik Boulund The Centre for Translational Microbiome Research (CTMR), Department of Microbiology, Tumor, and Cell Biology, Karolinska Institutet, Stockholm, Sweden

Carlos A. Coelho Mathematics Department (DM) and Centro de Matemática e Aplicações (CMA), Faculdade de Ciências e Tecnologia, Universidade Nova de Lisboa, Caparica, Portugal

Deliang Dai Department of Economics and Statistics, Linnaeus University, Växjö, Sweden

Justine Debelius The Centre for Translational Microbiome Research (CTMR), Department of Microbiology, Tumor, and Cell Biology, Karolinska Institutet, Stockholm, Sweden

Kai-Tai Fang Division of Science and Technology, BNU-HKBU United International College Zhuhai, Zhuhai, China

Katarzyna Filipiak Institute of Mathematics, Poznań University of Technology, Poznań, Poland

Miguel Fonseca Centro de Matemática e Aplicações (CMA), FCT, UNL, Almada, Portugal
Departamento de Matemática, FCT, UNL, Almada, Portugal

Yasunori Fujikoshi Department of Mathematics, Graduate School of Science, Hiroshima University, Hiroshima, Japan

Yuta Fukasawa The Risk Data Bank of Japan, Limited, Tokyo, Japan

Jemila S. Hamid University of Ottawa, Ottawa, ON, Canada

Stephen J. Haslett School of Fundamental Sciences, Massey University, Palmerston North, New Zealand

Thomas Holgersson Department of Economics and Statistics, Linnaeus University, Växjö, Sweden

Björn Holmquist Department of Statistics, Lund University, Lund, Sweden

Sayantee Jana Indian Institute of Management, Nagpur, India

Mateusz John Institute of Mathematics, Poznań University of Technology, Poznań, Poland

Daniel Klein Institute of Mathematics, Faculty of Science, P. J. Šafárik University, Košice, Slovakia

Tõnu Kollo Institute of Mathematics and Statistics, University of Tartu, Tartu, Estonia

Chaofeng Kou Department of Mathematics, The University of Manchester, Manchester, UK

Lynn R. LaMotte Biostatistics Program, LSU Health-NO, School of Public Health, New Orleans, LA, USA

Yuli Liang Department of Statistics, Örebro Univeristy School of Business, Örebro, Sweden

Shuangzhe Liu University of Canberra, Canberra, ACT, Australia

Zhen Luo Pfizer China, Pudong, Shanghai, China

Tiefeng Ma Southwestern University of Finance and Economics, Chengdu, China

Augustyn Markiewicz Department of Mathematical and Statistical Methods, Poznań University of Life Sciences, Poznań, Poland

Sultana Nasrin Department of Statistics, Lund University, Lund, Sweden

Innocent Ngaruye Department of Mathematics, College of Science and Technology, University of Rwanda, Butare, Rwanda

Joseph Nzabanita Department of Mathematics, College of Science and Technology, University of Rwanda, Butare, Rwanda

Jianxin Pan Department of Mathematics, The University of Manchester, Manchester, UK

Nestor Parolya Delft Institute of Applied Mathematics, Delft University of Technology, Delft, The Netherlands

Tatjana Pavlenko Department of Mathematics, KTH Royal Institute of Technology, Stockholm, Sweden

Jolanta Pielaszkiewicz Department of Computer Science, Linköping University, Linköping, Sweden

Simo Puntanen Faculty of Information Technology and Communication Sciences, Tampere University, Tampere, Finland

Anuradha Roy Department of Management Science and Statistics, The University of Texas at San Antonio, San Antonio, TX, USA

Tetsuro Sakurai School of General and Management Studies, Suwa University of Science, Nagano, Japan

Martin Singull Department of Mathematics, Linköping University, Linköping, Sweden

Anna Sjöström Department of Statistics, Lund University, Lund, Sweden

Akimichi Takemura Faculty of Data Science, Shiga University, Shiga, Japan

Annika Tillander Department of Statistics and Machine Learning, Linköping University, Linköping, Sweden

Yung Liang Tong Department of Mathematics, Georgia Institute of Technology, Atlanta, GA, USA

Marju Valge Institute of Mathematics and Statistics, University of Tartu, Tartu, Estonia

Julia Volaufova Biostatistics Program, LSU Health-NO, School of Public Health, New Orleans, LA, USA

Ivan Žežula Institute of Mathematics, Faculty of Science, P. J. Šafárik University, Košice, Slovakia

Dan Zhuang Fujian Normal University, Fuzhou, China

Chapter 1
Spectral Analysis of Large Reflexive Generalized Inverse and Moore-Penrose Inverse Matrices

Taras Bodnar and Nestor Parolya

Abstract A reflexive generalized inverse and the Moore-Penrose inverse are often confused in statistical literature but in fact they have completely different behaviour in case the population covariance matrix is not a multiple of identity. In this paper, we study the spectral properties of a reflexive generalized inverse and of the Moore-Penrose inverse of the sample covariance matrix. The obtained results are used to assess the difference in the asymptotic behaviour of their eigenvalues.

1.1 Introduction

Let $\mathbf{Y}_n = (\mathbf{y}_1, \mathbf{y}_2, \ldots, \mathbf{y}_n)$ be the $p \times n$ data matrix which consists of n column vectors of dimension p with $E(\mathbf{y}_i) = \mathbf{0}$ and $Cov(\mathbf{y}_i) = \boldsymbol{\Sigma}$ for $i \in 1, \ldots, n$. We assume that $p/n \to c \in (1, +\infty)$ as $n \to \infty$. This type of limiting behavior is also referred to a "large dimensional asymptotics" or "the Kolmogorov asymptotics". In this case, the traditional estimators perform very poorly and tend to over/underestimate the unknown parameters of the asset returns, i.e., the mean vector and the covariance matrix.

Throughout this paper it is assumed that there exists a $p \times n$ random matrix $\mathbf{X}_n$ which consists of independent and identically distributed (i.i.d.) real random variables with zero mean and unit variance such that

$$\mathbf{Y}_n = \boldsymbol{\Sigma}^{1/2}\mathbf{X}_n \, , \tag{1.1}$$

where the matrix $\boldsymbol{\Sigma}^{1/2}$ denotes the symmetric square root of the population covariance matrix $\boldsymbol{\Sigma}$. Other square roots of $\boldsymbol{\Sigma}$, like the lower triangular matrix

T. Bodnar (✉)
Department of Mathematics, Stockholm University, Stockholm, Sweden
e-mail: taras.bodnar@math.su.se

N. Parolya
Delft Institute of Applied Mathematics, Delft University of Technology, Delft, The Netherlands
e-mail: n.parolya@tudelft.nl

© Springer Nature Switzerland AG 2020

T. Holgersson, M. Singull (eds.), *Recent Developments in Multivariate and Random Matrix Analysis*, https://doi.org/10.1007/978-3-030-56773-6_1

of the Cholesky decomposition, can also be used. Note that the observation matrix $\mathbf{Y}_n$ consists of dependent rows although its columns are independent. It is worth mentioning that although the assumption of time independence looks quite restrictive in real-life applications, the model can be extended to dependent variables (for instance, causal AR(1) models)[1] by imposing more complicated conditions on the elements of $\mathbf{Y}_n$ (see, [2] for details) or by assuming an m-dependent structure (e.g., [11] for a finite number of dependent entries, and [9] and [16] for a possibly increasing number). Nevertheless, this will not change the main results of our paper and would only make the proofs more technical. That is why we assume independence for the sake of brevity and transparency.

The sample covariance matrix is given by (e.g., [13])

$$\mathbf{S}_n = \frac{1}{n}\mathbf{Y}_n\mathbf{Y}_n' = \frac{1}{n}\boldsymbol{\Sigma}^{1/2}\mathbf{X}_n\mathbf{X}_n'\boldsymbol{\Sigma}^{1/2}. \tag{1.2}$$

Throughout the paper it is assumed that the sample size n is smaller than the dimension p of random vectors $\mathbf{y}_i$, $i = 1, \ldots, n$, that is $p/n \to c > 1$ as $n \to \infty$. In this case the sample covariance matrix is singular and its inverse does not exist (cf., [4, 6, 12]). On the other side, the inverse of the population covariance matrix $\boldsymbol{\Sigma}$ is present in many applications from finance, signal processing, biostatistics, environmentrics, etc. (see, e.g., [7, 8, 10]). In practice, the Moore-Penrose inverse is usually employed (cf., [5, 12]), while the other types of the generalize inverse (see, [15]) can also be used.

In this paper we compare the spectral properties of two generalized inverses of the singular sample covariance matrix given by:

- Moore-Penrose inverse:

$$\mathbf{S}_n^+ = \left(\frac{1}{n}\mathbf{Y}_n\mathbf{Y}_n'\right)^+ = \frac{1}{n}\mathbf{Y}_n\left(\frac{1}{n}\mathbf{Y}_n'\mathbf{Y}_n\right)^{-2}\mathbf{Y}_n' = \frac{1}{n}\boldsymbol{\Sigma}^{1/2}\mathbf{X}_n\left(\frac{1}{n}\mathbf{X}_n'\boldsymbol{\Sigma}\mathbf{X}_n\right)^{-2}\mathbf{X}_n'\boldsymbol{\Sigma}^{1/2} \tag{1.3}$$

- Reflexive inverse

$$\mathbf{S}_n^- = \boldsymbol{\Sigma}^{-1/2}\left(\frac{1}{n}\mathbf{X}_n\mathbf{X}_n'\right)^+\boldsymbol{\Sigma}^{-1/2} = \frac{1}{n}\boldsymbol{\Sigma}^{-1/2}\mathbf{X}_n\left(\frac{1}{n}\mathbf{X}_n'\mathbf{X}_n\right)^{-2}\mathbf{X}_n'\boldsymbol{\Sigma}^{-1/2} \tag{1.4}$$

Although the Moore-Penrose inverse $\mathbf{S}^+$ can directly be computed from the observation matrix $\mathbf{Y}_n$, the derivation of its stochastic properties might be challenging in some practical problems. On the other side, the computation of the considered reflexive inverse $\mathbf{S}^-$ is not possible in practice, but the derivation of

---

[1] Bai and Zhou [2] define a general dependence structure in the following way: for all k, $E(Y_{jk}Y_{lk}) = \sigma_{lj}$, and for any non-random matrix $\mathbf{B}$ with bounded norm, it holds that $E\left|\mathbf{y}_k^\top\mathbf{B}\mathbf{y}_k - \mathrm{tr}(\mathbf{B}\boldsymbol{\Sigma})\right|^2 = o\left(n^2\right)$ where $\boldsymbol{\Sigma} = (\sigma_{lj})$.

the stochastic properties is considerably simplified. The goal of this paper is to quantify the difference between the two generalized inverse matrices with the aim to develop a statistical methodology to assess the estimation errors when the Moore-Penrose inverse is used. To this end, we note that a reflexive inverse is not uniquely defined and that the stochastic properties of the reflexive inverse will be derived in the special case of (1.4).

The rest of the paper is structured as follows. In the next section, we study the asymptotic properties of $\mathbf{S}^+$ and $\mathbf{S}^-$ by deriving two integral equations whose solutions are the Stieltjes transforms of the limiting spectral distributions of $\mathbf{S}^+$ and $\mathbf{S}^-$. These findings are used in Sect. 1.3, where the asymptotic behaviour of the Frobenius norm of the difference between $\mathbf{S}^+$ and $\mathbf{S}^-$ is investigated. Section 1.4 presents the results of a numerical illustration, while concluding remarks are provided in Sect. 1.5.

1.2 Asymptotic Properties of $\mathbf{S}^+$ and $\mathbf{S}^-$

For a symmetric matrix $\mathbf{A}$ we denote by $\lambda_1(\mathbf{A}) \geq \ldots \geq \lambda_p(\mathbf{A})$ its ordered eigenvalues and by $F^{\mathbf{A}}(t)$ the corresponding empirical distribution function (e.d.f.), that is

$$F^{\mathbf{A}}(t) = \frac{1}{p} \sum_{i=1}^{p} \mathbb{1}\{\lambda_i(\mathbf{A}) \leq t\},$$

where $\mathbb{1}\{\cdot\}$ is the indicator function. Furthermore, for a function $G : \mathbb{R} \to \mathbb{R}$ of bounded variation the Stieltjes transform is introduced by

$$m_G(z) = \int_{-\infty}^{+\infty} \frac{1}{\lambda - z} dG(\lambda); \quad z \in \mathbb{C}^+ \equiv \{z \in \mathbb{C} : \Im z > 0\}.$$

Note that there is a direct connection between $m_G(z)$ and moment generating function of G, $\Psi_G(z)$ given by

$$\Psi_G(z) = -\frac{1}{z} m_G\left(\frac{1}{z}\right) - 1.$$

In Theorem 1.1 we present the expressions of the Stieltjes transform for $\mathbf{S}^+$ and $\mathbf{S}^-$.

Theorem 1.1 *Let the $p \times n$ noise matrix $\mathbf{X}_n$ consist of i.i.d. real random variables with zero mean and unit variance. Assume that Σ is nonrandom symmetric and* *positive definite with a bounded spectral norm and the e.d.f.* $H_n = F^{\Sigma}$ *converges weakly to a nonrandom distribution function* H. *Then*

(i) *the e.d.f.* $F^{\mathbf{S}_n^+}$ *converges weakly almost surely to some deterministic c.d.f.* P^+ *whose Stieltjes transformation* m_P *satisfies the following equation:*

$$m_{P^+}(z) = -\frac{1}{z}\left(2 - c^{-1} + \int_{-\infty}^{+\infty} \frac{dH(\tau)}{z\tau c(zm_{P^+}(z)+1) - 1}\right);$$

(ii) *the e.d.f.* $F^{\mathbf{S}_n^-}$ *converges weakly almost surely to some deterministic c.d.f.* P^- *whose Stieltjes transformation* m_{P^-} *satisfies the following equation:*

$$m_{P^-}(z) = -\frac{1}{z} - \frac{1}{z}\int_{-\infty}^{+\infty} \frac{dH(\tau)}{\tau c z^2 m_{P^-}(z)\left(1 - \frac{c}{1-c-czm_{P^-}(z)}\right) - 1}.$$

Proof

(i) The result of part (i) of the theorem was derived in Theorem 2.1 of [5].

(ii) In order to prove the asymptotic result in the case of $\mathbf{S}_n^-$, whose eigenvalues are the same as those of the product of two matrices $\mathbf{\Sigma}^{-1}$ and $\tilde{\mathbf{S}}_n^+ = 1/n\mathbf{X}_n\left(\frac{1}{n}\mathbf{X}_n'\mathbf{X}_n\right)^{-2}\mathbf{X}_n'$, we use subordination results of free probability (see, e.g., [3]). Namely for two Borel probability measures μ and ν on $[0, +\infty)$ (both are limits of $F^{\tilde{\mathbf{S}}_n^+}$ and $F^{\mathbf{\Sigma}^{-1}}$, respectively), there exist two unique analytic functions $F_1, F_2 : \mathbb{C} \setminus [0, +\infty) \to \mathbb{C} \setminus [0, +\infty)$ such that

$$\frac{F_1(z)F_2(z)}{z} = \eta_\mu(F_1(z)) = \eta_\nu(F_2(z)) = \eta(z)\,, \tag{1.5}$$

where $\eta(z) = 1 - \frac{z}{g_{P^-}(1/z)}$ is the eta transform and g_{P^-} is the Cauchy transform (the negative Stieltjes transform), i.e., $g_{P^-}(z) = -m_{P^-}(z)$, for the limit of $F^{\mathbf{S}_n^-}$ denoted by P^-. In particular, from the first and the second equalities in (1.5) we get that for any $z \in \mathbb{C}^+ = \{z : \Im(z) > 0\}$ such that F_2 is analytic at z, the function $F_2(z)$ satisfies the following equation:[2]

$$\eta_\nu(F_2(z)) = \eta_\mu\left(\frac{\eta_\nu(F_2(z))z}{F_2(z)}\right). \tag{1.6}$$

[2]In fact, a similar equation also holds for F_1 when one replaces μ and ν in (1.6) but for our purposes it is enough to know one of them.

On the other hand, from the last equality in (1.5) we have that $g_{P^-}(z)$ satisfies the equality

$$g_{P^-}(z) = \frac{1}{zF_2(1/z)} g_\nu\left(\frac{1}{F_2(1/z)}\right). \tag{1.7}$$

Thus, we need first to find the so called subordination function $F_2(z)$ from (1.6) and plug it into (1.7). For simplicity we further suppress the subindex of g_{P^-}.

Let

$$\Theta(z) = \frac{F_2(z)}{z\eta(z)} \tag{1.8}$$

and rewrite (1.6) using $\eta(z) = \eta_\nu(F_2(z))$ in the following way

$$\eta(z) = \eta_\mu\left(\frac{\eta(z)z}{F_2(z)}\right) = \eta_\mu\left(\frac{1}{\Theta(z)}\right). \tag{1.9}$$

Using the definition of the eta transform we get

$$1 - \frac{1}{\frac{1}{z}g(1/z)} = 1 - \frac{1}{\Theta(z)}\frac{1}{g_\mu(\Theta(z))}$$

and, hence,

$$\frac{1}{z}g(1/z) = \Theta(z)g_\mu(\Theta(z)). \tag{1.10}$$

From [5] we get that

$$g_\mu(z) = -m_{P^-}(z) = \frac{1}{z}\left(2 - c^{-1} + \frac{m_{MP}(1/z)}{z}\right), \tag{1.11}$$

where $m_{MP}(z)$ is the Stieltjes transformation of the Marchenko-Pastur law given by (see, e.g., [1])

$$m_{MP}(z) = \frac{1}{2cz}\left(1 - c - z + \sqrt{(1+c-z)^2 - 4c}\right). \tag{1.12}$$

Thus, Eq. (1.10) becomes

$$
\begin{aligned}
\frac{1}{z}g(1/z) &= \Theta(z)\frac{1}{\Theta(z)}\left(2 - c^{-1} + \frac{m_{MP}(1/\Theta(z))}{\Theta(z)}\right)\\
&= 2 - c^{-1} + \frac{1 - c - \frac{1}{\Theta(z)} + \sqrt{(1 + c - \frac{1}{\Theta(z)})^2 - 4c}}{\Theta(z)2c\frac{1}{\Theta(z)}}\\
&= \frac{1}{2c}\left(2c - 2 + (1 + c - \frac{1}{\Theta(z)}) + \sqrt{(1 + c - \frac{1}{\Theta(z)})^2 - 4c}\right),
\end{aligned}
$$

or, equivalently, by rearranging terms we obtain

$$
2\left(c(\frac{1}{z}g(1/z) - 1) + 1\right) - (1 + c - \frac{1}{\Theta(z)}) = \sqrt{(1 + c - \frac{1}{\Theta(z)})^2 - 4c}\,, \tag{1.13}
$$

where by squaring of both sides, we get

$$
\left(c(\frac{1}{z}g(1/z) - 1) + 1\right)^2 - \left(c(\frac{1}{z}g(1/z) - 1) + 1\right)(1 + c - \frac{1}{\Theta(z)}) + c = 0
$$

or,

$$
\begin{aligned}
&1 + c^2(\frac{1}{z}g(1/z) - 1)^2 + 2c(\frac{1}{z}g(1/z) - 1) - 1 + \frac{1}{\Theta(z)}\\
&- c(\frac{1}{z}g(1/z) - 1)(1 + c - \frac{1}{\Theta(z)}) = 0,
\end{aligned}
$$

which yields

$$
c(\frac{1}{z}g(1/z) - 1) + (1 - c) + \frac{1}{\Theta(z)c(\frac{1}{z}g(1/z) - 1)} + \frac{1}{\Theta(z)} = 0\,. \tag{1.14}
$$

From (1.14) we find $\Theta(z)$ as a function of $g(1/z)$ expressed as

$$
\Theta(z) = \frac{1 + c(\frac{1}{z}g(1/z) - 1)}{c(\frac{1}{z}g(1/z) - 1)(c - 1 - c(\frac{1}{z}g(1/z) - 1))} \tag{1.15}
$$

or, in terms of $F_2(z)$ given by

$$
\begin{aligned}
F_2(z) &= \frac{z - cz + cg(1/z)}{c(\frac{1}{z}g(1/z) - 1)(c - 1 - c(\frac{1}{z}g(1/z) - 1))} \frac{\frac{1}{z}g(1/z) - 1}{\frac{1}{z}g(1/z)} \\
&= \frac{z - cz + cg(1/z)}{c(-z + 2cz - cg(1/z))} \frac{z^2}{g(1/z)} .
\end{aligned} \tag{1.16}
$$

At last, we use the property $g_\nu(z) = 1/z - 1/z^2 g_H(1/z)$ and plug $F_2(z)$ in (1.7). This leads to

$$
\begin{aligned}
g(z) &= \frac{1}{zF_2(1/z)}\left(F_2(1/z) - F_2^2(1/z)g_H(F_2(1/z))\right) \\
&= \frac{1}{z} - \frac{F_2(1/z)}{z} g_H(F_2(1/z)) = \frac{1}{z} - \frac{F_2(1/z)}{z}\int_{-\infty}^{+\infty} \frac{dH(\tau)}{F_2(1/z) - \tau} \\
&= \frac{1}{z} + \int_{-\infty}^{+\infty} \frac{dH(\tau)}{\tau z/F_2(1/z) - z} = \frac{1}{z} + \int_{-\infty}^{+\infty} \frac{dH(\tau)}{-\tau c z^3 g(z)\left(1 - \frac{c}{1 - c + czg(z)}\right) - z} .
\end{aligned}
$$

The latter can be rewritten in terms of Stieltjes transform $m_{P^-}(z)$ by

$$
m_{P^-}(z) = -\frac{1}{z} - \frac{1}{z}\int_{-\infty}^{+\infty} \frac{dH(\tau)}{\tau c z^2 m_{P^-}(z)\left(1 - \frac{c}{1 - c - czm_{P^-}(z)}\right) - 1} \tag{1.17}
$$

and the second part of theorem is proved.

Although the Stieltjes transforms of two generalize inverse matrices are quite different, they have one in common: besides the fact that they are equal for $\mathbf{\Sigma} = \mathbf{I}$, we also observe that they become close to each other in case c tends to one from the right. Thus, if p/n is close to one from the right, then there should be no large difference in using the Moore-Penrose inverse $\mathbf{S}^+$ or the generalized reflexive inverse $\mathbf{S}^-$ asymptotically.

1.3 Quantification the Difference Between the Asymptotic Behaviour of $\mathbf{S}^+$ and $\mathbf{S}^-$

Using the Stieltjes transforms computed in the case of $\mathbf{S}^+$ and $\mathbf{S}^-$, we are able to compare their moments. In order to quantify this difference more carefully we consider the quadratic or the so-called Frobenius norm of the difference between

$\mathbf{S}^+$ and $\mathbf{S}^-$ given by

$$\begin{aligned}\|\mathbf{S}^- - \mathbf{S}^+\|_F^2 &= \mathrm{tr}\left((\mathbf{S}^- - \mathbf{S}^+)(\mathbf{S}^- - \mathbf{S}^+)^\top\right)\\ &= \mathrm{tr}\left(\mathbf{S}^-\mathbf{S}^-\right) - 2\mathrm{tr}\left(\mathbf{S}^+\mathbf{S}^-\right) + \mathrm{tr}\left(\mathbf{S}^+\mathbf{S}^+\right)\\ &= \mathrm{tr}\left(\mathbf{S}^-\mathbf{S}^-\right) - \mathrm{tr}\left(\mathbf{S}^+\mathbf{S}^+\right)\\ &= \|\mathbf{S}^-\|_F^2 - \|\mathbf{S}^+\|_F^2 .\end{aligned}$$

In Theorem 1.2 the asymptotic equivalents for both Frobenius norms are provided.

Theorem 1.2 *Under assumptions of Theorem 1.1 it holds almost surely as $p/n \to c > 1$*

$$1/p\|\mathbf{S}^+\|_F^2 \longrightarrow c^{-1}\left(\frac{1}{m_{\underline{F}}^2(0)} - c\int_{-\infty}^{+\infty}\frac{\tau^2 dH(\tau)}{(1+\tau m_{\underline{F}}(0))^2}\right)^{-1}, \tag{1.18}$$

$$1/p\|\mathbf{S}^-\|_F^2 \longrightarrow \frac{(1+c(c-1))}{c^2(c-1)^3}\left(\int_{-\infty}^{+\infty}\frac{dH(\tau)}{\tau}\right)^2 + \frac{1}{(c(c-1))^2}\int_{-\infty}^{+\infty}\frac{dH(\tau)}{\tau^2}, \tag{1.19}$$

where $m_{\underline{F}}(0)$ satisfies the following equation

$$\frac{1}{m_{\underline{F}}(0)} = c\int_{-\infty}^{+\infty}\frac{\tau dH(\tau)}{1+\tau m_{\underline{F}}(0)} .$$

Proof In order to prove the statement of the theorem, we rewrite the Frobenius norm of random matrix $\mathbf{A}$ for all $z \in \mathbb{C}^+$ in the following way

$$\begin{aligned}1/p\|\mathbf{A}\|_F^2 = 1/p\mathrm{tr}(\mathbf{A}^2) &= -\frac{1}{2}\frac{\partial^2}{\partial z^2}\frac{1}{p}\frac{\mathrm{tr}(\mathbf{A}-1/z\mathbf{I})^{-1}}{z}\Bigg|_{z=0}\\ &= -\frac{1}{2}\frac{\partial^2}{\partial z^2}\frac{m_{F^{\mathbf{A}}}(1/z)}{z}\Bigg|_{z=0} .\end{aligned} \tag{1.20}$$

We start with $\mathbf{A} = \mathbf{S}^+$. Let

$$\Gamma_n(z) = \frac{m_{F^{\mathbf{S}^+}}(1/z)}{z}.$$

Due to Theorem 1.1.i we get that $\Gamma_n^+(z)$ converges almost surely to deterministic $\Gamma^+(z)$, which satisfies the following asymptotic equation

$$\Gamma^+(z) = -\left(2 - c^{-1} + \int_{-\infty}^{+\infty} \frac{zdH(\tau)}{\tau c(\Gamma^+(z)+1) - z}\right). \tag{1.21}$$

For the computation of $1/p\|\mathbf{S}^+\|_F^2$, the quantities $\Gamma^+(z)$, $\frac{\partial}{\partial z}\Gamma^+(z)$ and $\frac{\partial^2}{\partial z^2}\Gamma^+(z)$ should be evaluated at zero. We rewrite (1.21) in an equivalent way

$$\Gamma^+(z) = -1 - c^{-1} z m_{\underline{F}}(z), \tag{1.22}$$

where $m_{\underline{F}}(z)$ is the limiting Stieltjes transform of $m_{F^{1/n\mathbf{Y}^\top\mathbf{Y}}}$, which satisfies the following equation (see, e.g., [14])

$$m_{\underline{F}}(z) = \left(c\int_{-\infty}^{+\infty} \frac{\tau dH(\tau)}{1+\tau m_{\underline{F}}(z)} - z\right)^{-1}. \tag{1.23}$$

The advantage of (1.22) over (1.21) lies in the calculation of the derivatives, which can be done easier by the former. Since the quantity $m_{\underline{F}}(z)$ is bounded at zero for $c > 1$, we immediately obtain

$$\Gamma^+(0) \equiv \lim_{z\to 0^+} \Gamma^+(z) = -1\,. \tag{1.24}$$

The first derivative of $\Gamma^+(z)$ is given by

$$\frac{\partial}{\partial z}\Gamma^+(z) = -c^{-1} m_{\underline{F}}(z) - c^{-1} z \frac{\partial}{\partial z} m_{\underline{F}}(z) \tag{1.25}$$

and, thus, taking the limit $z \to 0^+$ and using that $\frac{\partial}{\partial z} m_{\underline{F}}(z) = O(1)$ as $z \to 0^+$, we get

$$\Gamma'^+(0) \equiv \lim_{z\to 0^+} \frac{\partial}{\partial z}\Gamma^+(z) = -c^{-1} m_{\underline{F}}(0)\,, \tag{1.26}$$

where $m_{\underline{F}}(0)$ satisfies the equation

$$\frac{1}{m_{\underline{F}}(0)} = c\int_{-\infty}^{+\infty} \frac{\tau dH(\tau)}{1+\tau m_{\underline{F}}(0)}\,. \tag{1.27}$$

The second derivative of $\Gamma^+(z)$ is equal to

$$\frac{\partial^2}{\partial z^2}\Gamma^+(z) = -2c^{-1}\frac{\partial}{\partial z}m_{\underline{F}}(z) - c^{-1}z\frac{\partial^2}{\partial z^2}m_{\underline{F}}(z). \tag{1.28}$$

Denoting $m'_{\underline{F}}(z) = \frac{\partial}{\partial z}m_{\underline{F}}(z)$ and using (1.23) together with (1.27), we obtain

$$m'_{\underline{F}}(0) \equiv \lim_{z\to 0^+} m'_{\underline{F}}(z) = \lim_{z\to 0^+} \frac{c\int\limits_{-\infty}^{+\infty}\frac{\tau^2 m'_{\underline{F}}(z)dH(\tau)}{(1+\tau m_{\underline{F}}(z))^2}+1}{\left(c\int\limits_{-\infty}^{+\infty}\frac{\tau dH(\tau)}{1+\tau m_{\underline{F}}(z)}\right)^2} = \frac{m'_{\underline{F}}(0)c\int\limits_{-\infty}^{+\infty}\frac{\tau^2 dH(\tau)}{(1+\tau m_{\underline{F}}(0))^2}+1}{\frac{1}{m^2_{\underline{F}}(0)}}$$

or, equivalently,

$$m'_{\underline{F}}(0) = \left(\frac{1}{m^2_{\underline{F}}(0)} - c\int\limits_{-\infty}^{+\infty}\frac{\tau^2 dH(\tau)}{(1+\tau m_{\underline{F}}(0))^2}\right)^{-1}. \tag{1.29}$$

Finally, the application of (1.28) and (1.29) together with $\frac{\partial^2}{\partial z^2}m_{\underline{F}}(z) = O(1)$ as $z \to 0^+$ leads to

$$\Gamma''^{+}(0) \equiv \lim_{z\to 0^+}\frac{\partial^2}{\partial z^2}\Gamma^+(z) = -2\frac{c^{-1}}{\frac{1}{m^2_{\underline{F}}(0)} - c\int\limits_{-\infty}^{+\infty}\frac{\tau^2 dH(\tau)}{(1+\tau m_{\underline{F}}(0))^2}}.$$

Now, the first result of the theorem follows from (1.20).

Similarly, for the second identity of Theorem 1.2 we denote $\Gamma^-(z)$ as a limit of

$$\Gamma_n^-(z) = \frac{m_{F_{\mathbf{S}^-}}(1/z)}{z}. \tag{1.30}$$

Then using Theorem 1.1, the limiting function $\Gamma^-(z)$ satisfies the following asymptotic equation

$$\Gamma^-(z) = -1 - \int\limits_{-\infty}^{+\infty}\frac{z dH(\tau)}{\tau c\Gamma^-(z)\left(1 - \frac{c}{1-c-c\Gamma^-(z)}\right) - z}. \tag{1.31}$$

Here we immediately get that $\Gamma^{-}(z)$ is bounded at zero and it holds that

$$\lim_{z\to 0^+} \Gamma^{-}(z) = -1, \tag{1.32}$$

$$\lim_{z\to 0^+} \frac{\partial}{\partial z}\Gamma^{-}(z) = -\frac{1}{c(c-1)} \int_{-\infty}^{+\infty} \frac{dH(\tau)}{\tau}, \tag{1.33}$$

where, the result for the first derivative of $\Gamma^{-}(z)$ follows from the identity

$$\left(1 - z\int_{-\infty}^{+\infty} \frac{\tau c\left(1 - \frac{c(1-c)}{(1-c-c\Gamma^{-}(z))^2}\right) dH(\tau)}{\left(\tau c\Gamma^{-}(z)\left(1 - \frac{c}{1-c-c\Gamma^{-}(z)}\right) - z\right)^2}\right)\frac{\partial}{\partial z}\Gamma^{-}(z)$$

$$= -\int_{-\infty}^{+\infty} \frac{\tau c\Gamma^{-}(z)\left(1 - \frac{c}{1-c-c\Gamma^{-}(z)}\right) dH(\tau)}{\left(\tau c\Gamma^{-}(z)\left(1 - \frac{c}{1-c-c\Gamma^{-}(z)}\right) - z\right)^2}$$

$$= -\int_{-\infty}^{+\infty} \frac{dH(\tau)}{\left(\tau c\Gamma^{-}(z)\left(1 - \frac{c}{1-c-c\Gamma^{-}(z)}\right) - z\right)} - \int_{-\infty}^{+\infty} \frac{z\,dH(\tau)}{\left(\tau c\Gamma^{-}(z)\left(1 - \frac{c}{1-c-c\Gamma^{-}(z)}\right) - z\right)^2}.$$

For the second derivative we get the following equality

$$(1 + O(z))\frac{\partial^2}{\partial z^2}\Gamma^{-}(z) = 2\frac{\partial}{\partial z}\Gamma^{-}(z) \int_{-\infty}^{+\infty} \frac{\tau c\left(1 - \frac{c(1-c)}{(1-c-c\Gamma^{-}(z))^2}\right) dH(\tau)}{\left(\tau c\Gamma^{-}(z)\left(1 - \frac{c}{1-c-c\Gamma^{-}(z)}\right) - z\right)^2}$$

$$- 2\int_{-\infty}^{+\infty} \frac{dH(\tau)}{\left(\tau c\Gamma^{-}(z)\left(1 - \frac{c}{1-c-c\Gamma^{-}(z)}\right) - z\right)^2} + O(z)$$

and taking the limit $z \to 0^+$ from both sides leads to

$$\lim_{z\to 0^+} \frac{\partial^2}{\partial z^2}\Gamma^{-}(z) = 2\frac{\partial}{\partial z}\Gamma^{-}(z)\frac{c(1-c(1-c))}{c^2(c-1)^2} \int_{-\infty}^{+\infty} \frac{dH(\tau)}{\tau} - 2\frac{1}{(c(c-1))^2} \int_{-\infty}^{+\infty} \frac{dH(\tau)}{\tau^2}$$

$$= -2\frac{(1+c(c-1))}{c^2(c-1)^3}\left(\int_{-\infty}^{+\infty} \frac{dH(\tau)}{\tau}\right)^2 - 2\frac{1}{(c(c-1))^2} \int_{-\infty}^{+\infty} \frac{dH(\tau)}{\tau^2}. \tag{1.34}$$

The second statement of the theorem follows now from (1.34) and (1.20).

For a better visualization of the results of Theorem 1.2 from the statistical point of view, we present its empirical counterpart. Here, the almost sure convergence of the asymptotic equivalents are summarize in Corollary 1.1.

Corollary 1.1 *Under assumptions of Theorem 1.1 for* $p/n \to c > 1$ *holds almost surely*

$$1/p\left|\|\mathbf{S}^{+}\|_F^2 - c^{-1}\left(\frac{p}{m_{\underline{F}}^2(0)} - c\|(\boldsymbol{\Sigma} + m_{\underline{F}}(0)\mathbf{I})^{-1}\|_F^2\right)^{-1}\right| \to 0, \tag{1.35}$$

$$1/p\left|\|\mathbf{S}^{-}\|_F^2 - \frac{(1+c(c-1))}{c^2(c-1)^3}\frac{1}{p}\left(tr(\boldsymbol{\Sigma}^{-1})\right)^2 - \frac{1}{(c(c-1))^2}\|\boldsymbol{\Sigma}^{-1}\|_F^2\right| \to 0, \tag{1.36}$$

where $m_{\underline{F}}(0)$ *satisfies asymptotically (approximately) the following equation*

$$\frac{p}{m_{\underline{F}}(0)} = c \cdot tr(\boldsymbol{\Sigma} + m_{\underline{F}}(0)\mathbf{I})^{-1}.$$

The results of Corollary 1.1 show how the Frobenius norms of both inverses are connected with their population counterpart $\|\boldsymbol{\Sigma}^{-1}\|_F^2$. Namely, looking on the asymptotic behavior of $\|\mathbf{S}^{+}\|_F^2$ one can easily deduce that it is not possible to estimate $\|\boldsymbol{\Sigma}^{-1}\|_F^2$ consistently using the Moore-Penrose inverse because of the nonlinearity which is present in $(\boldsymbol{\Sigma} + m_{\underline{F}}(0)\mathbf{I})^{-1}$. On the other side, it is doable by reflexive generalized inverse $\mathbf{S}_n^{-}$. Indeed, using the proof of Theorem 1.2 one can find that

$$1/p\left|\mathrm{tr}(\mathbf{S}_n^{-}) - \frac{1}{c(c-1)}\mathrm{tr}(\boldsymbol{\Sigma}^{-1})\right| \to 0,$$

which together with (1.36) implies that

$$1/p\left|(c(c-1))^2\left[\|\mathbf{S}^{-}\|_F^2 - \left(\frac{1}{(c-1)} + c\right)\frac{1}{p}\left(\mathrm{tr}(\mathbf{S}^{-})\right)^2\right] - \|\boldsymbol{\Sigma}^{-1}\|_F^2\right| \to 0, \tag{1.37}$$

almost surely.

1.4 Numerical Illustration

Using the results of the previous section, we present several numerical results to quantify the difference $\|\mathbf{S}^- - \mathbf{S}^+\|_F^2 = \|\mathbf{S}^-\|_F^2 - \|\mathbf{S}^+\|_F^2$ for some $\boldsymbol{\Sigma}$. In order to avoid the normalization $1/p$ we will use the normalized Frobenius loss (NFL) expressed as

$$NFL = \frac{\|\mathbf{S}^- - \mathbf{S}^+\|_F^2}{\|\mathbf{S}^+\|_F^2} = \frac{\|\mathbf{S}^-\|_F^2}{\|\mathbf{S}^+\|_F^2} - 1, \tag{1.38}$$

which measures the difference between $\mathbf{S}^-$ and $\mathbf{S}^+$ normalized by the Frobenius norm of the Moore-Penrose inverse. The application of (1.38) with $\mathbf{S}^+$ and $\mathbf{S}^-$ as in (1.3) and (1.4) leads to the so called empirical NFL, while the usage of the results of Theorem 1.2 corresponds to the asymptotic NFL. The latter can be interpreted how much both Frobenius norms differ asymptotically.

In Fig. 1.1 we present the results of a simulation study where the normalized Frobenius losses are computed for several values of the concentration ratio $c > 1$ as a function of dimension p. For the sake of illustration, we provide the results obtained for the samples generated from the multivariate normal distribution, while similar values were also obtained for other multivariate distributions. We set the mean vector to zero and, without loss of generality, use the diagonal covariance matrix $\boldsymbol{\Sigma}$ with 20% of eigenvalues equal to one, 40% equal to three and the rest equal to ten. The results for $c = 1.07$ confirm our expectations discussed after the proof of Theorem 1.1, namely there is no large difference between both norms: NFL is small and the empirical NFL converges quite fast to its asymptotic counterpart. By increasing c we observe that the NFL increases indicating that both norms deviate from each other. For example, in case $c = 2$ the asymptotic and empirical NFLs are close to 1.2 what means that the Frobenius norm for reflexive inverse is more than a double of the Frobenius norm of the Moore-Penrose inverse. This observation becomes more severe for larger values of c. Moreover, for $c = 10$ we observe a bit slower convergence of the sample NFL to its asymptotic value.

The obtained findings indicate that the usage of $\mathbf{S}_n^+$ instead of $\mathbf{S}_n^-$ must be done with much care and are only reliable if the concentration ratio p/n is close to one. Otherwise, one can not expect neither a good approximation of functionals depending on the inverse covariance matrix nor consistent estimators for them.

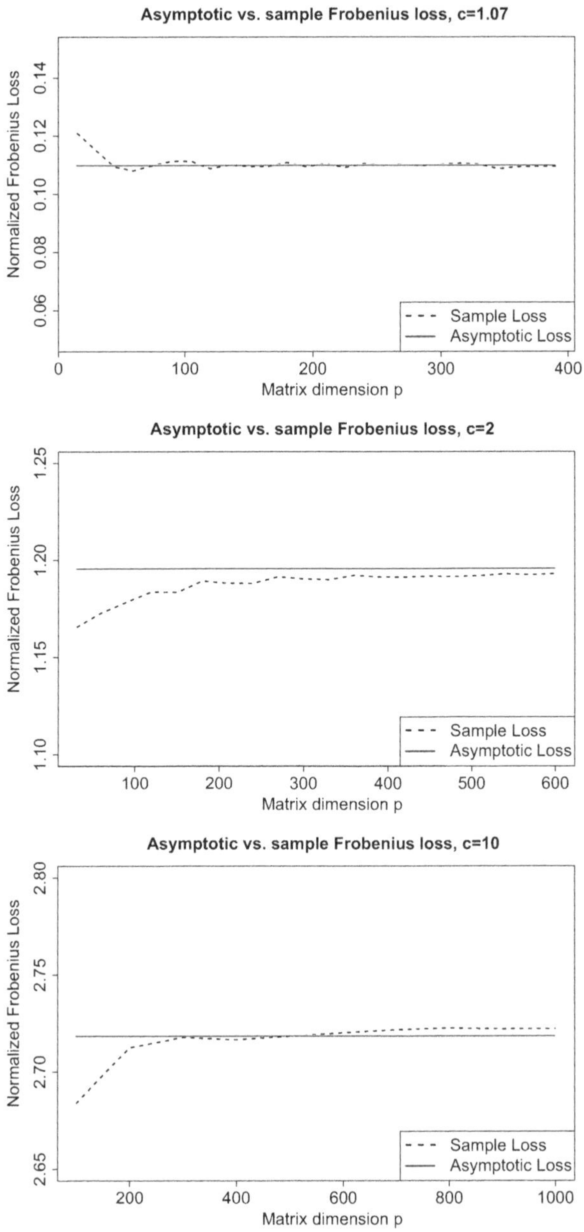

Fig. 1.1 Normalized Frobenius losses $\frac{\|\mathbf{S}^- - \mathbf{S}^+\|_F^2}{\|\mathbf{S}^+\|_F^2}$ for several values of $c > 1$ as function of p

1.5 Summary

In many statistical applications the dimension of the data-generating process is larger than the sample size. However, one still needs to compute the inverse of the sample covariance matrix, which is present in many expressions. There are a plenty of ways how to define a generalized inverse with the Moore-Penrose inverse and reflexive generalized inverse matrices be the mostly used ones. While the former is easily computable and is unique, the latter can not be calculated from the data. On the other side, the behavior of the Moore-Penrose inverse in many high-dimensional applications is far away from a satisfactory one, whereas the reflexive inverse obeys very convenient statistical and asymptotic properties. Namely, the application of the reflexive inverse allows to estimate functionals of the precision matrix (the inverse of population covariance matrix) consistently in high-dimension.

In this work we study spectral properties of both inverses in detail. The almost sure limits of the Stieltjes transforms of their empirical distribution functions of eigenvalues (so-called limiting spectral distributions) is provided in the case the dimension p and the sample size n increase to infinity simultaneously such that their ratio p/n tends to a positive constant c greater than one. We discover that both limiting spectral distributions differ considerably and only coincide when c tends to one. The results are illustrated via the calculation of asymptotic normalized Frobenius loss of both inverse matrices. Finally, we justify the obtained theoretical findings via a simulation study.

Acknowledgments The authors would like to thank Professor Thomas Holgersson, Professor Martin Singull, and the anonymous Reviewer for their helpful suggestions which improved the presentation of the paper.

References

1. Bai, Z.D., Silverstein, J.W.: Spectral Analysis of Large Dimensional Random Matrices. Springer, New York (2010)
2. Bai, Z.D., Zhou, W.: Large sample covariance matrices without independence structures in columns. Stat. Sin. **18**, 425–442 (2008)
3. Belinschi, S.T., Bercovici, H.: A new approach to subordination results in free probability. J. Anal. Math. **101**, 357–365 (2007)
4. Bodnar, T., Okhrin, Y.: Properties of the singular, inverse and generalized inverse partitioned Wishart distributions. J. Multivar. Anal. **99**, 2389–2405 (2008)
5. Bodnar, T., Dette, H., Parolya, N.: Spectral analysis of the Moore-Penrose inverse of a large dimensional sample covariance matrix. J. Multivar. Anal. **148**, 160–172 (2016)
6. Bodnar, T., Gupta, A.K., Parolya, N.: Direct shrinkage estimation of large dimensional precision matrix. J. Multivar. Anal. **146**, 223–236 (2016)
7. Bodnar, T., Parolya, N. Schmid, W.: Estimation of the global minimum variance portfolio in high dimensions. Eur. J. Oper. Res. **266**, 371–390 (2018)
8. Bodnar, T., Dmytriv, S., Parolya, N., Schmid, W.: Tests for the weights of the global minimum variance portfolio in a high-dimensional setting. IEEE Trans. Signal Process. **67**, 4479–4493 (2019)

9. Friesen, O., Löwe, M., Stolz, M.: Gaussian fluctuations for sample covariance matrices with dependent data. J. Multivar. Anal. **114**, 270–287 (2013)
10. Holgersson, T., Karlsson, P., Stephan, A.: A risk perspective of estimating portfolio weights of the global minimum-variance portfolio. AStA Adv. Stat. Anal. **104**, 59–80 (2020)
11. Hui, J., Pan, G.: Limiting spectral distribution for large sample covariance matrices with m-dependent elements. Commun. Stat. Theory Methods **39**, 935–941 (2010)
12. Imori, S., von Rosen, D.: On the mean and dispersion of the Moore-Penrose generalized inverse of a Wishart matrix. Electron. J. Linear Algebra **36**, 124–133 (2020)
13. Kollo, T., von Rosen, D.: Advanced Multivariate Statistics with Matrices. Springer, Dordrecht (2005)
14. Silverstein, J.W., Bai, Z.D.: On the empirical distribution of eigenvalues of a class of large dimensional random matrices. J. Multivar. Anal. **54**, 175–192 (1995)
15. Wang, G., Wei, Y., Qiao, S., Lin, P., Chen, Y.: Generalized Inverses: Theory and Computations. Springer, Singapore (2018)
16. Wei, M., Yang, G., Yang, L.: The limiting spectral distribution for large sample covariance matrices with unbounded m-dependent entries. Commun. Stat. Theory Methods **45**, 6651–6662 (2016)

Chapter 2
Testing for Double Complete Symmetry

Carlos A. Coelho and Martin Singull

Abstract In this paper the authors take an approach over the likelihood ratio test for double complete symmetry, which enables a particularly simple implementation of the test as well as a particularly adequate way towards obtaining very sharp approximations for the exact distribution of the test statistic. Numerical studies show the very good performance of the near-exact distributions derived, namely their very good performance for very small sample sizes and their asymptotic behavior for increasing numbers of variables involved.

2.1 Introduction

Patterned covariance matrices arise in a variety of contexts and have been studied by many authors. In a seminal paper Wilks [18] considered patterned structures when dealing with measurements on k equivalent psychological tests. This led to a covariance matrix with equal diagonal elements and equal off-diagonal elements, i.e., a covariance matrix given by

$$\Sigma = \sigma^2 \big((1-\rho)I_p + \rho J_p\big) : p{\times}p,$$

with $-\frac{1}{p-1} < \rho < 1$ and where I_p denotes the identity matrix of order p and J_p a $p{\times}p$ matrix of ones. This structure is called *complete symmetry*, *uniform* or *intraclass* covariance structure. In [18] Wilks developed statistical test criteria for testing equality in means, equality in variances and equality in covariances. The structure implies that both the inverse and the determinant have closed form

C. A. Coelho (✉)
Mathematics Department (DM) and Centro de Matemática e Aplicações (CMA), Faculdade de Ciências e Tecnologia, Universidade Nova de Lisboa, Caparica, Portugal
e-mail: cmac@fct.unl.pt

M. Singull
Department of Mathematics, Linköping University, Linköping, Sweden
e-mail: martin.singull@liu.se

© Springer Nature Switzerland AG 2020

T. Holgersson, M. Singull (eds.), *Recent Developments in Multivariate and Random Matrix Analysis*, https://doi.org/10.1007/978-3-030-56773-6_2

expressions and the maximum likelihood estimators can easily be calculated [10, p. 114].

Votaw [17] extended the complete symmetry model to a model with blocks called *compound symmetry, type I* and *type II*. The compound symmetry covariance matrices are, for the $p = 4$ case, given as

$$\Sigma_I = \begin{bmatrix} \alpha & \beta & \beta & \beta \\ \beta & \gamma & \delta & \delta \\ \beta & \delta & \gamma & \delta \\ \beta & \delta & \delta & \gamma \end{bmatrix} \quad \text{and} \quad \Sigma_{II} = \begin{bmatrix} \alpha & \beta & \kappa & \sigma \\ \beta & \alpha & \sigma & \kappa \\ \kappa & \sigma & \gamma & \delta \\ \sigma & \kappa & \delta & \gamma \end{bmatrix}.$$

In [17] Votaw considered different psychometric and medical research problems where the compound symmetry is applicable. In [14] and [15] Szatrowski discusses block compound symmetry and the model is applied to the analysis of an educational testing problem.

A *circular stationary model*, where variables are thought of as being equally spaced around a circle was considered by Olkin and Press [12]. The covariance between two variables in the circular stationary model depends on the distance between the variables and the covariance matrices for $p = 4$ and $p = 5$ have the structures

$$\Sigma = \sigma^2 \begin{bmatrix} 1 & \rho_1 & \rho_2 & \rho_1 \\ \rho_1 & 1 & \rho_1 & \rho_2 \\ \rho_2 & \rho_1 & 1 & \rho_1 \\ \rho_1 & \rho_2 & \rho_1 & 1 \end{bmatrix} \quad \text{and} \quad \Sigma = \sigma^2 \begin{bmatrix} 1 & \rho_1 & \rho_2 & \rho_2 & \rho_1 \\ \rho_1 & 1 & \rho_1 & \rho_2 & \rho_2 \\ \rho_2 & \rho_1 & 1 & \rho_1 & \rho_2 \\ \rho_2 & \rho_2 & \rho_1 & 1 & \rho_1 \\ \rho_1 & \rho_2 & \rho_2 & \rho_1 & 1 \end{bmatrix}.$$

Olkin and Press [12] considered three symmetries, namely circular, complete symmetrical and spherical and derived likelihood ratio tests and the asymptotic distributions under the null hypothesis and under the alternative. Olkin [11] generalizes the circular stationary model with a multivariate version in which each element was a vector and the covariance matrix can be written as a block circular matrix.

The covariance symmetries investigated by for example Wilks [18], Votaw [17] and Olkin and Press [12] are all special cases of *invariant normal models* considered by Anderson [2]. The invariant normal models include not only all models specified by symmetries of the covariance matrix, but also the linear models for the mean. The symmetry model defined by a group $\mathcal{G}$ is the family of covariance matrices given by

$$\mathcal{S}_{\mathcal{G}}^{+} = \left\{ \Sigma | \Sigma > 0, G\Sigma G' = \Sigma \text{ for all } G \in \mathcal{G} \right\},$$

i.e., this implies that if x is a random vector with $cov(x) = \Sigma$ such that $\Sigma \in \mathcal{S}_{\mathcal{G}}^{+}$ then there are symmetry restrictions on the covariance matrix, namely that $cov(x) = cov(Gx)$ for all $G \in \mathcal{G}$. Perlman [13] summarizes and discusses the group symmetry covariance models suggested by Anderson [2]. Furthermore, several examples of different symmetries, maximum likelihood estimators and likelihood ratio tests are given in [13].

In this paper we consider a special case of the block compound symmetry discussed by Szatrowski in [14] and [15]. We will say that the symmetric positive-definite matrix Σ of dimensions $mp \times mp$ has a double complete symmetric structure if we may write

$$\Sigma = (I_p \otimes R_0) + (J_p - I_p) \otimes R_1, \tag{2.1}$$

where

$$R_1 = cJ_m, \tag{2.2}$$

and where R_0 is itself a positive-definite complete symmetric matrix, of dimensions $m \times m$

$$R_0 = aI_m + b(J_m - I_m) = (a-b)I_m + bJ_m, \quad \text{with} \quad a > \max(0, b, (1-p)b). \tag{2.3}$$

For example for $m = 4$ and $p = 3$ the matrix Σ would have the following aspect

$$\Sigma = \begin{bmatrix} a & b & b & b & c & c & c & c & c & c & c & c \\ b & a & b & b & c & c & c & c & c & c & c & c \\ b & b & a & b & c & c & c & c & c & c & c & c \\ b & b & b & a & c & c & c & c & c & c & c & c \\ c & c & c & c & a & b & b & b & c & c & c & c \\ c & c & c & c & b & a & b & b & c & c & c & c \\ c & c & c & c & b & b & a & b & c & c & c & c \\ c & c & c & c & b & b & b & a & c & c & c & c \\ c & c & c & c & c & c & c & c & a & b & b & b \\ c & c & c & c & c & c & c & c & b & a & b & b \\ c & c & c & c & c & c & c & c & b & b & a & b \\ c & c & c & c & c & c & c & c & b & b & b & a \end{bmatrix}.$$

Restriction relations among a, b and c, in order to make Σ a positive-definite matrix, may be derived from the structure of R_0 and the result in Proposition 2.1 below, and these are stated in Appendix 1.

We are interested in testing the null hypothesis that Σ has the double complete symmetric structure given in (2.1), the alternative hypothesis being that Σ is just positive-definite, with no particular structure. In carrying out this test, the following Proposition is important.

Proposition 2.1 *To test*

$$H_0 : \Sigma = (I_p \otimes R_0) + (J_p - I_p) \otimes R_1 \tag{2.4}$$

for R_1 *in (2.2) and* R_0 *in (2.3), is equivalent to test*

$$H_0 : \Gamma \Sigma \Gamma' = diag\big(\lambda_1, \underbrace{\underbrace{\lambda_2, \dots, \lambda_2}_{m-1}, \lambda_3, \dots, \underbrace{\lambda_2, \dots, \lambda_2}_{m-1}}_{p-1 \text{ groups}}, \lambda_3, \underbrace{\lambda_2, \dots, \lambda_2}_{m-1}\big) \tag{2.5}$$

for

$$\Gamma = Q_p \otimes Q_m ,$$

where Q_p *and* Q_m *are Helmert matrices of order* p *and* m *respectively, and*

$$\lambda_1 = a - b + m(b - c) + mpc , \quad \lambda_2 = a - b , \quad \text{and} \quad \lambda_3 = a - b + m(b - c) .$$

Proof We may write, under H_0 in (2.4),

$$\begin{aligned} \Sigma &= I_p \otimes \Big((a - b)I_m + bJ_m\Big) + c(J_p - I_p) \otimes J_m \\ &= (a - b)(I_p \otimes I_m) + b(I_p \otimes J_m) + c(J_p \otimes J_m - I_p \otimes J_m) \\ &= (a - b)(I_p \otimes I_m) + (b - c)(I_p \otimes J_m) + c(J_p \otimes J_m) . \end{aligned}$$

Let us now take

$$\Gamma = Q_p \otimes Q_m ,$$

where Q_r is a Helmert matrix of order r, with

$$Q_r = \begin{bmatrix} r^{-1/2}\,\underline{1}_r' \\ \widetilde{Q}_r \end{bmatrix}$$

where $\underline{1}_r = [\,1, 1, \dots, 1\,]' : r \times 1$, and

$$Q_r' Q_r = Q_r Q_r' = I_r , \quad \widetilde{Q}_r \underline{1}_{r-1} = \underline{0}_{r-1} , \quad \widetilde{Q}_r \widetilde{Q}_r' = I_{r-1} .$$

Hence, we will have

$$
\begin{aligned}
Q_r J_r Q_r' = Q_r \underline{1}_r \underline{1}_r' Q_r' &= \begin{bmatrix} r^{-1/2}\,\underline{1}_r' \\ \widetilde{Q}_r \end{bmatrix} \underline{1}_r \underline{1}_r' \left[r^{-1/2}\,\underline{1}_r \middle| \widetilde{Q}_r' \right] \\
&= \begin{bmatrix} r^{1/2} \\ \underline{0}_{r-1} \end{bmatrix} \left[r^{1/2}\,\underline{0}_{r-1}' \right] = r\,\underline{r}_1 \underline{r}_1',
\end{aligned}
\tag{2.6}
$$

where

$$
\underset{(r\times 1)}{\underline{r}_1} = \begin{bmatrix} 1 \\ 0 \\ \vdots \\ 0 \end{bmatrix}, \qquad \underline{r}_1 \underline{r}_1' = diag(\underline{r}_1)\,, \tag{2.7}
$$

and we may write

$$
\begin{aligned}
\Gamma \Sigma \Gamma' &= (a-b)(I_p \otimes I_m) + (b-c)\big(I_p \otimes (Q_m J_m Q_m')\big) + c\big((Q_p J_p Q_p') \otimes (Q_m J_m Q_m')\big) \\
&= (a-b) I_{mp} + m(b-c)(I_p \otimes \underline{m}_1 \underline{m}_1') + mpc\big((\underline{p}_1 \underline{p}_1') \otimes (\underline{m}_1 \underline{m}_1')\big),
\end{aligned}
\tag{2.8}
$$

where $(\underline{p}_1 \underline{p}_1') \otimes (\underline{m}_1 \underline{m}_1') = \underline{mp}_1 (\underline{mp}_1)' = diag(\underline{mp}_1)$, which makes (2.8) the same as the expression in (2.5).

Since $\Gamma = Q_p \otimes Q_m$, the transpose of the matrix of eigenvalues of Σ, is not a function of the elements in Σ but only of the dimensions of its constituent blocks, that is, p and m, testing H_0 in (2.4) is equivalent to test H_0 in (2.5). □

2.2 Decomposition of the Null Hypothesis and Formulation of the Likelihood Ratio Test

The null hypothesis in (2.5) may be decomposed as

$$
H_0 \equiv H_{0b|a} \circ H_{0a}\,, \tag{2.9}
$$

where '∘' means 'after' and the hypotheses in the decomposition are

$$
H_{0a} : \Gamma \Sigma \Gamma' = diag(\lambda_1^*, \ldots, \lambda_{mp}^*) \tag{2.10}
$$

and

$$H_{0b|a}:\underbrace{\underbrace{\lambda_2^*=\cdots=\lambda_m^*}_{m-1}=\underbrace{\lambda_{m+2}^*=\cdots=\lambda_{2m}^*}_{m-1}=\cdots=\underbrace{\lambda_{(p-1)m+2}^*=\cdots=\lambda_{pm}^*}_{m-1}}_{p\text{ groups}}$$
$$\text{and}\quad\underbrace{\lambda_{m+1}^*=\lambda_{2m+1}^*=\cdots=\lambda_{(p-1)m+1}^*}_{p-1},$$
$$\text{assuming } H_{0a}\,. \tag{2.11}$$

See the note in Appendix 2 about the composition and decomposition of hypotheses.

In case the matrix $\Gamma\Sigma\Gamma'$ is taken as the variance-covariance matrix of a pm-multivariate normal random vector, the hypothesis H_{0a} in (2.10) is the hypothesis of independence of those pm random variables, and the likelihood ratio test (l.r.t.) statistic used to test H_{0a}, based on a sample of size $n > pm$, will be [1, Sec. 9.2]

$$\Lambda_a=\left(\frac{|A|}{\prod_{j=1}^{pm}|a_j|}\right)^{n/2},$$

where $A=\Gamma A^+\Gamma'$ is the maximum likelihood estimator (m.l.e.) of $\Gamma\Sigma\Gamma'$ (with A^+ being the m.l.e. of Σ), and a_j its j-th diagonal element.

Then $H_{0b|a}$ in (2.11) is the composition of two hypotheses, the first one of equality of $(p-1)(m-1)$ variances and the second one of $p-1$ variances, each one based on a sample of size n, and whose m.l.e.'s, once assumed H_{0a}, will be independent. As such, the l.r.t. statistic to test $H_{0b|a}$ in (2.11) will be [1, Sec. 10.2]

$$\Lambda_b=\Lambda_{b1}\Lambda_{b2}, \tag{2.12}$$

where

$$\Lambda_{b1}=\left((p(m-1))^{p(m-1)}\frac{\prod_{\ell=1}^{p}\prod_{k=1}^{m-1}a_{(\ell-1)m+1+k}}{(a_1^*)^{p(m-1)}}\right)^{n/2}, \tag{2.13}$$

and

$$\Lambda_{b2}=\left((p-1)^{p-1}\frac{\prod_{k=1}^{p-1}a_{km+1}}{(a_2^*)^{p-1}}\right)^{n/2}, \tag{2.14}$$

with

$$a_1^* = \sum_{\ell=1}^{p} \sum_{k=1}^{m-1} a_{(\ell-1)m+1+k}\,, \quad \text{and} \quad a_2^* = \sum_{k=1}^{p-1} a_{km+1}\,.$$

Then, through an extension of Lemma 10.3.1 in [1], the l.r.t. statistic to test H_0 in (2.2) will be

$$\Lambda = \Lambda_a \Lambda_b = \left((p(m-1))^{p(m-1)} (p-1)^{p-1} \frac{|A|}{a_1 \left(a_1^*\right)^{p(m-1)} \left(a_2^*\right)^{p-1}} \right)^{n/2}, \tag{2.15}$$

with

$$E\left(\Lambda^h\right) = E\left(\Lambda_a^h\right) E\left(\Lambda_{b1}^h\right) E\left(\Lambda_{b2}^h\right), \tag{2.16}$$

since on one hand, under H_{0a} Λ_a is independent of $\prod_{j=1}^{pm} a_j$ [6, 8], which makes Λ_a independent of Λ_{b1} and Λ_{b2}, while on the other hand, under H_{0a}, the a_j $(j = 1, \ldots, pm)$ are independent among themselves, which makes Λ_{b1} and Λ_{b2} independent because they are built on different a_j's.

2.3 On the Exact Distribution of the l.r.t. Statistic

Using the results in [4, 7, 9] we may write the h-th moment of Λ_a as

$$\begin{aligned} E\left(\Lambda_a^h\right) &= \prod_{k=1}^{pm-1} \frac{\Gamma\left(\frac{n-1}{2}\right) \Gamma\left(\frac{n-1-(pm-k)}{2} + \frac{n}{2}h\right)}{\Gamma\left(\frac{n-1-(pm-k)}{2}\right) \Gamma\left(\frac{n-1}{2} + \frac{n}{2}h\right)} \\ &= \underbrace{\left\{ \prod_{j=3}^{pm} \left(\frac{n-j}{n}\right)^{r_j} \left(\frac{n-j}{n} + h\right)^{-r_j} \right\}}_{=\Phi_{a,1}(h)} \underbrace{\left(\frac{\Gamma\left(\frac{n-1}{2}\right) \Gamma\left(\frac{n-2}{2} + \frac{n}{2}h\right)}{\Gamma\left(\frac{n-1}{2} + \frac{n}{2}h\right) \Gamma\left(\frac{n-2}{2}\right)} \right)^{k^*}}_{=\Phi_{a,2}(h)}, \end{aligned} \tag{2.17}$$

where

$$k^* = \left\lfloor \frac{pm}{2} \right\rfloor \tag{2.18}$$

and

$$r_j = \begin{cases} h_{j-2} + (-1)^j k^*, & \text{for } j = 3, 4, \\ r_{j-2} + h_{j-2}, & \text{for } j = 5, \ldots, pm, \end{cases} \tag{2.19}$$

with

$$h_j = \begin{cases} pm - 1, & \text{for } j = 1, \\ -1, & \text{for } j = 2, \ldots, pm - 2\,. \end{cases} \tag{2.20}$$

Using the results in [1, Sec. 10.2] we obtain the expression for the h-th moment of Λ_{b1} as

$$E\left(\Lambda_{b1}^h\right) = \underbrace{\prod_{k=2}^{p(m-1)} \frac{\Gamma\left(\frac{n-1}{2} + \frac{k-1}{p(m-1)}\right)\Gamma\left(\frac{n-1}{2} + \frac{n}{2}h\right)}{\Gamma\left(\frac{n-1}{2} + \frac{k-1}{p(m-1)} + \frac{n}{2}h\right)\Gamma\left(\frac{n-1}{2}\right)}}_{=\Phi_{b1}(h)} \tag{2.21}$$

while the h-th moment of Λ_{b2} may be obtained as

$$E\left(\Lambda_{b2}^h\right) = \underbrace{\prod_{k=2}^{p-1} \frac{\Gamma\left(\frac{n-1}{2} + \frac{k-1}{p-1}\right)\Gamma\left(\frac{n-1}{2} + \frac{n}{2}h\right)}{\Gamma\left(\frac{n-1}{2} + \frac{k-1}{p-1} + \frac{n}{2}h\right)\Gamma\left(\frac{n-1}{2}\right)}}_{=\Phi_{b2}(h)}\,. \tag{2.22}$$

Since the supports of Λ_a, Λ_{b1} and Λ_{b2} are delimited, their distributions are determined by their moments, and as such, from the first expression in (2.17) we may write

$$\Lambda_a \overset{st}{\sim} \prod_{k=1}^{pm-1} (X_k)^{n/2}\,, \qquad \text{with} \qquad X_k \sim Beta\left(\frac{n-1-(pm-k)}{2}, \frac{(pm-k)m}{2}\right), \tag{2.23}$$

where '$\overset{st}{\sim}$' means 'stochastically equivalent' and X_k $(k = 1, \ldots, pm-1)$ are independent random variables, while from (2.21) we may write

$$\Lambda_{b1} \overset{st}{\sim} \prod_{k=2}^{p(m-1)} \left(X_k^*\right)^{n/2}\,, \qquad \text{with} \qquad X_k^* \sim Beta\left(\frac{n-1}{2}, \frac{k-1}{p(m-1)}\right), \tag{2.24}$$

where X_k^* $(k = 1, \ldots, p(m-1))$ are independent, and from (2.22) we may write

$$\Lambda_{b2} \overset{st}{\sim} \prod_{k=2}^{p-1} \left(X_k^{**}\right)^{n/2}, \qquad \text{with} \qquad X_k^{**} \sim Beta\left(\frac{n-1}{2}, \frac{k-1}{p-1}\right), \tag{2.25}$$

where X_k^{**} $(k = 1, \ldots, p-1)$ are independent. Hence, we may write for the overall l.r.t. statistic, under H_0 in (2.9),

$$\Lambda \overset{st}{\sim} \left\{ \left(\prod_{k=1}^{pm-1} X_k\right)^{n/2} \times \left(\prod_{k=2}^{p(m-1)} X_k^*\right)^{n/2} \times \left(\prod_{k=2}^{p-1} X_k^{**}\right)^{n/2} \right\}, \tag{2.26}$$

where all random variables are independent.

On the other hand, based on the results in Appendix 3 and from the second expression in (2.17) we may write, for Λ_a,

$$\Lambda_a \overset{st}{\sim} \left(\prod_{j=3}^{pm} e^{-Z_j}\right) \times \left(\prod_{j=1}^{k^*} (W_j)^{n/2}\right), \tag{2.27}$$

where k^* is given by (2.18) and

$$Z_j \sim \Gamma\left(r_j, \frac{n-j}{n}\right) \quad \text{and} \quad W_j \sim Beta\left(\frac{n-2}{2}, \frac{1}{2}\right) \tag{2.28}$$

are all independent random variables.

Thus, we have the following Theorem.

Theorem 2.1 *The exact distribution of the overall l.r.t. statistic Λ in (2.15), used to test H_0 in (2.4) or (2.5) is, for general p and m, the same as that of*

$$\left(\prod_{j=3}^{pm} e^{-Z_j}\right) \times \left(\prod_{j=1}^{k^*} W_j\right)^{n/2} \times \left(\prod_{k=2}^{p(m-1)} X_k^*\right)^{n/2} \times \left(\prod_{k=2}^{p-1} X_k^{**}\right)^{n/2}, \tag{2.29}$$

where k^* is given by (2.18) and where the distributions of Z_j and W_j are given in (2.28) and those of X_k^* and X_k^{**} are respectively given in (2.24) and (2.25).

Proof The proof of the above theorem is rather trivial, from the previously established results. □

For either $p = 1$ or $m = 1$ the present test reduces to the common complete symmetry or equivariance-equicorrelation Wilks test [18]. The following Corollary gives the distribution of Λ for this case.

Corollary 2.1 *For $p = 1$ and $m > 1$, the statistic Λ_{b2} vanishes and the statistic Λ_{b1} assumes in this case the same distribution as that of the statistic Λ_{b2} for the case $p > 1$, while for $m = 1$ and $p > 1$ the statistic Λ_{b1} vanishes. As such, let us denote by q the value of the parameter, either p or m, that is greater than one. Then, for either $p = 1$ or $m = 1$, the exact distribution of the statistic Λ in (2.15) is the same as that of*

$$\left(\prod_{j=3}^{q} e^{-Z_j}\right) \times \left(\prod_{j=1}^{k^*} W_j\right)^{n/2} \times \left(\prod_{k=2}^{q-1} W_k^*\right)^{n/2}, \tag{2.30}$$

where Z_j and W_j still have the distributions in (2.28), $k^* = \lfloor q/2 \rfloor$, and

$$W_k^* \sim Beta\left(\frac{n-1}{2}, \frac{k-1}{q-1}\right).$$

The particular case of $p = 2$ should also be addressed since also in this case the statistic Λ_{b2} vanishes, and as such we have the following Corollary.

Corollary 2.2 *For $p = 2$, the statistic Λ_{b2} vanishes and as such, in this case the statistic Λ in (2.15) has the same distribution as that of*

$$\left(\prod_{j=3}^{2m} e^{-Z_j}\right) \times \left(\prod_{j=1}^{k^*} W_j\right)^{n/2} \times \left(\prod_{k=2}^{2(m-1)} X_k^*\right)^{n/2}, \tag{2.31}$$

where Z_j and W_j have the distributions in (2.28), $k^* = m$, and the distribution of X_k^* is given by (2.24).

2.4 The Characteristic Function of $W = -\log \Lambda$

The reason why in the previous section we actually obtain two equivalent representations for the exact distribution of Λ, which are the one in (2.26) and the one in Theorem 2.1 is because the first of these neither does it yield a manageable cumulative distribution function (c.d.f.) nor is it adequate to lead to a sharp approximation to the exact distribution of Λ.

In order to be able to obtain a very sharp and manageable approximation to the exact distribution of Λ, we will base our developments in the representation given by Theorem 2.1 and Corollaries 2.1 and 2.2.

From Theorem 2.1 and expressions (2.17)–(2.22) we may write the characteristic function (c.f.) of $W = -\log \Lambda$ as

$$\Phi_W(t) = E\left(e^{\mathrm{i}tW}\right) = E\left(\Lambda^{-\mathrm{i}t}\right)$$
$$= \underbrace{\left\{\prod_{j=3}^{pm}\left(\frac{n-j}{n}\right)^{r_j}\left(\frac{n-j}{n}-\mathrm{i}t\right)^{-r_j}\right\}}_{=\Phi_{W,1}(t)} \times \underbrace{\Phi_{a,2}(-\mathrm{i}t)\,\Phi_{b1}(-\mathrm{i}t)\,\Phi_{b2}(-\mathrm{i}t)}_{=\Phi_{W,2}(t)}, \tag{2.32}$$

where the r_j are given by (2.19) and (2.20), $\Phi_{a,2}(\cdot)$, $\Phi_{b1}(\cdot)$ and $\Phi_{b2}(\cdot)$ are defined in (2.17)–(2.22), and $\Phi_{W,1}(t)$ is equal to $\Phi_{a,1}(-\mathrm{i}t)$.

For $p = 1$ or $m = 1$, according to Corollary 2.1 in the previous section $\Phi_W(t)$ reduces to

$$\Phi_W(t) = \underbrace{\left\{\prod_{j=3}^{q}\left(\frac{n-j}{n}\right)^{r_j}\left(\frac{n-j}{n}-\mathrm{i}t\right)^{-r_j}\right\}}_{=\Phi_{W,1}(t)} \times \underbrace{\Phi_{a,2}(-\mathrm{i}t)\,\Phi_{b2}(-\mathrm{i}t)}_{=\Phi_{W,2}(t)}, \tag{2.33}$$

for r_j given by (2.19)–(2.20), $k^* = \lfloor q/2 \rfloor$ in $\Phi_{a,2}(\cdot)$, and where in $\Phi_{b2}(\cdot)$ we replace p by q, where q represents the value of the parameter, either m or p, which is not equal to 1.

For $p = 2$, according to Corollary 2.2 in the previous section $\Phi_W(t)$ reduces to

$$\Phi_W(t) = \underbrace{\left\{\prod_{j=3}^{2m}\left(\frac{n-j}{n}\right)^{r_j}\left(\frac{n-j}{n}-\mathrm{i}t\right)^{-r_j}\right\}}_{=\Phi_{W,1}(t)} \times \underbrace{\Phi_{a,2}(-\mathrm{i}t)\,\Phi_{b1}(-\mathrm{i}t)}_{=\Phi_{W,2}(t)}, \tag{2.34}$$

for r_j given by (2.19)–(2.20), $k^* = m$ in $\Phi_{a,2}(\cdot)$, and $p = 2$ in $\Phi_{b1}(\cdot)$.

Expressions (2.32)–(2.34), together with expressions (2.29) and (2.30), show that the exact distribution of $W = -\log \Lambda$ is the same as that of the sum of $pm - 2$ independent Gamma random variables with an independent sum of a number of independent Logbeta random variables.

As such, in building the near-exact distributions we will keep $\Phi_{W,1}(t)$ untouched and we will approximate $\Phi_{W,2}(t)$ asymptotically by the c.f. of a finite mixture of Gamma distributions.

2.5 Near-Exact Distributions

Indeed, based on the result in Section 5 of [16], we may, for increasing values of a, asymptotically replace the distribution of any $Logbeta(a, b)$ distributed random variable by an infinite mixture of $\Gamma(b+\ell, a)$ distributions ($\ell = 0, 1, \dots$). As such, we could replace $\Phi_{W,2}(t)$ in either (2.32), (2.33) or (2.34) by the c.f. of the sum of infinite mixtures of Gamma distributions, which would be the same as the c.f. of an infinite mixture of sums of Gamma distributions. Although it happens that these Gamma distributions would have different rate parameters, these parameters would anyway be of comparable magnitude. As such, in building our near-exact distributions for $W = -\log \Lambda$ and Λ, while we will leave $\Phi_{W,1}(t)$ unchanged, we will replace $\Phi_{W,2}(t)$ in either (2.32), (2.33) or (2.34), by

$$\Phi^*(t) = \sum_{\ell=0}^{m^*} \pi_\ell\, \lambda^{r+\ell}(\lambda - \mathrm{i}t)^{-(r+\ell)}, \tag{2.35}$$

which is the c.f. of a finite mixture of $\Gamma(r+\ell, \lambda)$ distributions, where, for the general case in (2.32), with $p > 2$ and $m > 1$, we will take

$$r = \frac{1}{2}\left\lfloor \frac{pm}{2} \right\rfloor + \sum_{k=2}^{p(m-1)} \frac{k-1}{p(m-1)} + \sum_{k=2}^{p-1} \frac{k-1}{p-1} = \frac{1}{2}\left(\left\lfloor \frac{pm}{2} \right\rfloor + pm - 3\right), \tag{2.36}$$

which is the sum of all the second parameters of the Logbeta distributions in $\Phi_{W,2}(t)$ in (2.32), while for the particular case in (2.33), for $q > 1$, we will take

$$r = \frac{1}{2}\left\lfloor \frac{q}{2} \right\rfloor + \sum_{k=2}^{q-1} \frac{k-1}{q-1} = \frac{1}{2}\left(\left\lfloor \frac{q}{2} \right\rfloor + q\right) - 1, \tag{2.37}$$

which is the sum of all the second parameters of the Logbeta distributions in $\Phi_{W,2}(t)$ in (2.33), and for the particular case in (2.34), for $m > 1$,

$$r = \frac{m}{2} + \sum_{k=2}^{2(m-1)} \frac{k-1}{2(m-1)} = \frac{3}{2}(m-1)\,. \tag{2.38}$$

The parameter λ in (2.35) is then taken as the rate parameter in

$$\Phi^{**}(t) = \theta\, \lambda^{s_1}(\lambda - \mathrm{i}t)^{-s_1} + (1-\theta)\lambda^{s_2}(\lambda - \mathrm{i}t)^{-s_2},$$

where θ, λ, s_1 and s_2 are determined in such a way that

$$\left.\frac{\partial \Phi_{W,2}(t)}{\partial t^h}\right|_{t=0} = \left.\frac{\partial \Phi^{**}(t)}{\partial t^h}\right|_{t=0} \quad \text{for} \quad h = 1, \ldots, 4,$$

while the weights π_ℓ ($\ell = 0, \ldots, m^* - 1$) in (2.35) will then be determined in such a way that

$$\left.\frac{\partial \Phi_{W,2}(t)}{\partial t^h}\right|_{t=0} = \left.\frac{\partial \Phi^{*}(t)}{\partial t^h}\right|_{t=0} \quad \text{for} \quad h = 1, \ldots, m^*,$$

with $\pi_{m^*} = 1 - \sum_{\ell=0}^{m^*-1} \pi_\ell$.

This procedure yields near-exact distributions for W which will match he first m^* exact moments of W and which have c.f.

$$\Phi_{W,1}(t)\Phi^*(t),$$

with $\Phi_{W,1}(t)$ given by (2.32), (2.33) or (2.34), according to the case, and $\Phi^*(t)$ by (2.35), where r, given by (2.36), (2.37) or (2.38) is always either an integer or a half-integer.

As such, the near-exact distributions developed yield, for W, distributions which, for non-integer r, are mixtures, with weights π_ℓ ($\ell = 0, \ldots, m^*$), of $m^* + 1$ Generalized Near-Integer Gamma (GNIG) distributions of depth $pm - 1$ with integer shape parameters r_j ($j = 3, \ldots, pm$) and real shape parameter r and corresponding rate parameters $(n - j)/n$ ($j = 3, \ldots, pm$) and λ, and which, for integer r, are similar mixtures but of Generalized Integer Gamma (GIG) distributions, with the same shape and rate parameters (see [3, 4] and Appendix 4 for further details on the GIG and GNIG distributions and their probability density and cumulative distribution functions).

Using the notation in Appendix 4 for the probability density and cumulative distribution functions of the GNIG distribution, the near-exact distributions obtained for $W = -\log \Lambda$, for the general case of $p > 2$ and $m > 1$ and for the case of non-integer r, will have probability density and cumulative distribution functions respectively of the form

$$f_W^*(w) = \sum_{\ell=0}^{m^*} \pi_\ell \, f^{GNIG}\left(w \,\middle|\, r_3, \ldots, r_{pm}; r + \ell; \frac{n-3}{n}, \ldots, \frac{n-pm}{n}; \lambda; pm - 1\right),$$

and

$$F_W^*(w) = \sum_{\ell=0}^{m^*} \pi_\ell \, F^{GNIG}\left(w \,\middle|\, r_3, \ldots, r_{pm}; r + \ell; \frac{n-3}{n}, \ldots, \frac{n-pm}{n}; \lambda; pm - 1\right),$$

for $w > 0$, while the near-exact probability density and cumulative distribution functions of Λ are respectively given by

$$f^*_\Lambda(z) = \sum_{\ell=0}^{m^*} \pi_\ell \, f^{\,GNIG}\left(-\log z \,\middle|\, r_3, \ldots, r_{pm}; r+\ell; \frac{n-3}{n}, \ldots, \frac{n-pm}{n}; \lambda; pm-1\right)\frac{1}{z},$$

and

$$F^*_\Lambda(z) = \sum_{\ell=0}^{m^*} \pi_\ell \left(1-F^{GNIG}\left(-\log z \,\middle|\, r_3, \ldots, r_{pm}; r+\ell; \frac{n-3}{n}, \ldots, \frac{n-pm}{n}; \lambda; pm-1\right)\right),$$

for $0 < z < 1$.

For the case $p = 1$ or $m = 1$ all we have to do is to use r given by (2.37), instead of r given by (2.36) and replace pm by q, while for the case of $p = 2$ we have to use r given by (2.38).

For integer r, all we have to do is to replace the GNIG probability density and cumulative distribution functions by their GIG counterparts (see Appendix 4), yielding near-exact distributions for W, for the general case, with probability density and cumulative distribution functions respectively of the form

$$f^*_W(w) = \sum_{\ell=0}^{m^*} \pi_\ell \, f^{\,GIG}\left(w;\, r_3, \ldots, r_{pm}, r+\ell;\, \frac{n-3}{n}, \ldots, \frac{n-pm}{n}, \lambda;\, pm-1\right),$$

and

$$F^*_W(w) = \sum_{\ell=0}^{m^*} \pi_\ell \, F^{GIG}\left(w;\, r_3, \ldots, r_{pm}, r+\ell;\, \frac{n-3}{n}, \ldots, \frac{n-pm}{n}, \lambda;\, pm-1\right),$$

for $w > 0$, while the near-exact probability density and cumulative distribution functions of Λ are respectively given by

$$f^*_\Lambda(z) = \sum_{\ell=0}^{m^*} \pi_\ell \, f^{\,GIG}\left(-\log z;\, r_3, \ldots, r_{pm}, r+\ell;\, \frac{n-3}{n}, \ldots, \frac{n-pm}{n}, \lambda;\, pm-1\right)\frac{1}{z},$$

and

$$F^*_\Lambda(z) = \sum_{\ell=0}^{m^*} \pi_\ell \left(1-F^{GIG}\left(-\log z;\, r_3, \ldots, r_{pm}, r+\ell;\, \frac{n-3}{n}, \ldots, \frac{n-pm}{n}, \lambda;\, pm-1\right)\right),$$

for $0 < z < 1$, and where, similar to what happens for the case of non-integer r, for the case $p = 1$ or $m = 1$ we have to use r given by (2.37), instead of r given by (2.36) and for the case $p = 2$ to use r given by (2.38).

2.6 Numerical Study

In order to assess the performance of the near-exact distributions developed we will use

$$\Delta = \frac{1}{2\pi} \int_{-\infty}^{+\infty} \left| \frac{\Phi_W(t) - \Phi_{W,1}(t)\Phi^*(t)}{t} \right| dt$$

with

$$\Delta \geq \max_{w} \left| F_W(w) - F_W^*(w) \right| ,$$

as a measure of proximity between the exact and the near-exact distributions, where $\Phi_W(t)$ is the exact c.f. of W in (2.32), (2.33) or (2.34) and $F_W(\cdot)$ and $F_W^*(\cdot)$ represent respectively the exact and near-exact cumulative distribution functions of W, corresponding respectively to $\Phi_W(t)$ and $\Phi_{W,1}(t)\Phi^*(t)$.

In Table 2.1 we may analyze values of Δ for different combinations of values of p and m and different sample sizes.

For each combination of values of p and m, at least three different sample sizes, that is, values of n, were used, exceeding pm by 2, 30 and 100. For larger combinations of values of p and m, some larger values of n were also used to illustrate the asymptotic behavior of the near-exact distributions in what concerns the sample size. For all near-exact distributions, values of m^* equal to 4, 6 and 10 were used, that is, we used for each case near-exact distributions matching the first 4, 6 and 10 exact moments of $W = -\log \Lambda$. Smaller values of Δ indicate a closer agreement with the exact distribution and as such, a better performance of the corresponding near-exact distribution.

We may see how the near-exact distributions developed show not only very sharp approximations to the exact distribution even for very small samples, that is, for sample sizes hardly exceeding the total number of variables involved, as well as they show clear asymptotic behaviors not only for increasing sample sizes, but also for increasing values of p and m, with the asymptotic behavior in terms of sample size being apparent for larger sample sizes as the values of p and m get larger.

Table 2.1 Values of the measure Δ for the near-exact distributions

n	m^*: 4	6	10	n	m^*: 4	6	10
	$p=1,\ m=2$				$p=1,\ m=5$		
4	7.93×10^{-3}	3.10×10^{-3}	3.22×10^{-5}	7	1.32×10^{-10}	5.31×10^{-12}	2.99×10^{-15}
32	2.52×10^{-7}	8.54×10^{-10}	3.36×10^{-14}	35	1.73×10^{-13}	6.75×10^{-17}	$6.56\times10^{\;23}$
102	5.71×10^{-10}	1.75×10^{-13}	9.18×10^{-20}	105	7.70×10^{-16}	2.46×10^{-20}	2.30×10^{-28}
	$p=2,\ m=2$				$p=1,\ m=10$		
6	4.11×10^{-7}	1.90×10^{-8}	2.85×10^{-10}	12	6.41×10^{-13}	2.03×10^{-15}	1.47×10^{-19}
34	3.33×10^{-11}	5.05×10^{-14}	4.14×10^{-19}	40	2.01×10^{-14}	9.00×10^{-18}	9.11×10^{-24}
104	1.28×10^{-13}	2.47×10^{-17}	5.64×10^{-24}	110	1.82×10^{-16}	8.64×10^{-21}	1.88×10^{-28}
	$p=2,\ m=5$				$p=2,\ m=10$		
12	2.33×10^{-14}	3.21×10^{-15}	3.43×10^{-19}	22	1.09×10^{-15}	7.93×10^{-20}	2.31×10^{-27}
40	4.86×10^{-15}	1.79×10^{-17}	1.90×10^{-23}	50	5.03×10^{-16}	1.37×10^{-20}	2.78×10^{-28}
110	7.37×10^{-17}	1.75×10^{-20}	4.27×10^{-28}	120	1.17×10^{-17}	5.63×10^{-23}	4.49×10^{-32}
	$p=5,\ m=2$				$p=5,\ m=5$		
12	2.38×10^{-13}	1.44×10^{-15}	1.81×10^{-19}	27	8.86×10^{-17}	1.09×10^{-21}	2.34×10^{-31}
40	3.90×10^{-15}	8.92×10^{-18}	9.32×10^{-24}	55	8.10×10^{-17}	5.71×10^{-22}	3.54×10^{-32}
110	1.48×10^{-17}	8.78×10^{-21}	2.17×10^{-28}	125	2.76×10^{-18}	4.42×10^{-24}	1.36×10^{-35}
	$p=5,\ m=10$				$p=10,\ m=2$		
52	9.99×10^{-20}	1.18×10^{-25}	4.47×10^{-37}	22	8.08×10^{-16}	4.52×10^{-20}	1.07×10^{-27}
80	4.46×10^{-19}	6.12×10^{-25}	3.02×10^{-36}	50	3.63×10^{-16}	7.82×10^{-21}	1.41×10^{-28}
150	5.89×10^{-20}	3.02×10^{-26}	2.00×10^{-38}	120	8.34×10^{-18}	3.24×10^{-23}	2.30×10^{-32}
	$p=10,\ m=5$				$p=10,\ m=10$		
52	9.62×10^{-20}	1.13×10^{-25}	4.39×10^{-37}	102	1.01×10^{-22}	1.14×10^{-29}	1.09×10^{-43}
80	4.31×10^{-19}	5.86×10^{-25}	2.97×10^{-36}	130	9.85×10^{-22}	1.90×10^{-28}	5.36×10^{-42}
150	5.69×10^{-20}	2.89×10^{-26}	1.96×10^{-38}	200	4.69×10^{-22}	5.55×10^{-29}	6.24×10^{-43}
				500	1.28×10^{-23}	3.05×10^{-31}	1.48×10^{-46}

2.7 Another View over the Double Compound Symmetry Test

The following Proposition shows another possible approach over the double complete symmetry test, which enables us to better place the test among other related tests.

Proposition 2.2 *To test the null hypothesis H_0 in (2.4), that is, to test*

$$H_0:\ \Sigma = (I_p \otimes R_0) + (J_p - I_p) \otimes R_1$$

for R_1 in (2.2) and R_0 in (2.3), is equivalent to test

$$H_0 : \Gamma^* \Sigma (\Gamma^*)' = block-diagonal(\Delta_1, \Delta_2, \ldots, \Delta_p), \tag{2.39}$$

where

$$\Delta_1 = R_0 + (p-1)R_1,$$
$$\Delta_2 = \cdots = \Delta_p = R_0 - R_1$$

and

$$\Gamma^* = Q_p \otimes I_m ,$$

where Q_p is a Helmert matrix of order p.

Proof The proof is very simple and it is only based on taking into account the relation in (2.6) and the notation used in (2.6) and (2.7), and writing

$$\begin{aligned}
\Gamma^* \Sigma (\Gamma^*)' &= (Q_p \otimes I_m)\Big((I_p \otimes R_0) + (J_p - I_p) \otimes R_1\Big)(Q_p \otimes I_m)' \\
&= (I_p \otimes R_0) + (Q_p J_p Q'_p - I_p) \otimes R_1 \\
&= (I_p \otimes R_0) + (p\, \underline{p}_1 \underline{p}'_1 - I_p) \otimes R_1 \\
&= (I_p \otimes R_0) - I_p \otimes R_1 + p\, diag(\underline{p}_1) \otimes R_1 \\
&= block\text{-}diagonal\Big(R_0 + (p-1)R_1, \underbrace{R_0 - R_1, \ldots, R_0 - R_1}_{p-1}\Big) .
\end{aligned}$$

Since the elements in $\Gamma^* = Q_p \otimes I_m$ are not a function of the elements in Σ but only of the dimensions of its constituent blocks, testing H_0 in (2.4) will be equivalent to test H_0 in (2.39). □

The result in Proposition 2.2 clearly shows that the double complete symmetry test may indeed be addressed in the framework of the eigenblock-eigenmatrix decomposition in [5], and as such, clearly related with the other tests mentioned in that same reference. In Fig. 2.1 we may see that the present test appears as test #7 in a chart that tries to encompass the double complete symmetry test into a set of other tests very much adequate to be addressed using the eigenblock-eigenmatrix decomposition in [5], where test # 1 is addressed.

However, it happens that in this case the matrices Λ_α in expression (2) in [5] do themselves have a compound or complete symmetric structure, being as such themselves suitable to be diagonalized. This is indeed what is done in Proposition 2.1, where Σ is actually submitted to a double diagonalization, say, the first of which is the one in Proposition 2.2 and a second one that diagonalizes the Δ matrices in Proposition 2.2.

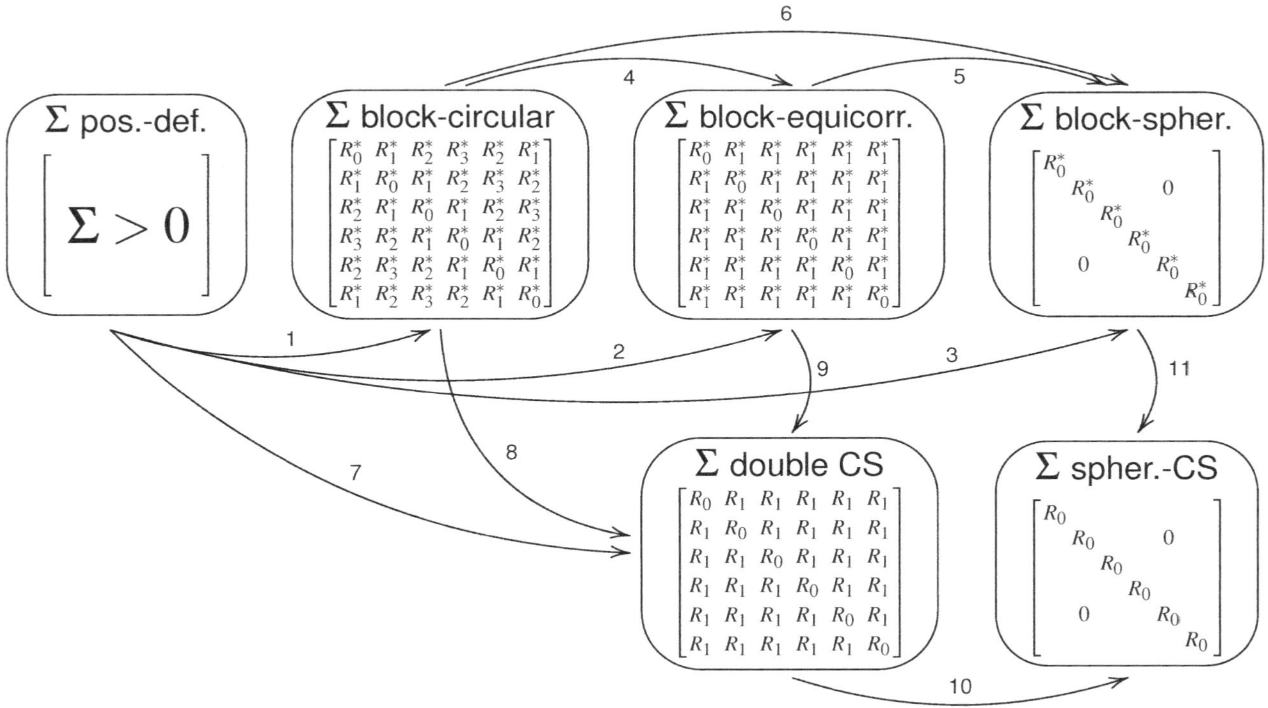

Fig. 2.1 Relations between the double complete symmetry structure and other block-structures: (a) $\Sigma > 0$ represents the absence of structure, with Σ being only positive-definite; (b) R_0 is a compound or complete symmetric matrix and $R_1 = cJ_m$; (c) R_0^* is an unstructured symmetric positive-definite matrix; (d) R_1^*, R_2^* and R_3^* are symmetric matrices with no particular structure. For each of the 11 possible tests of hypotheses depicted: the structure at the origin of the arrow represents H_1, while the structure at the tip of the arrow represents H_0

In order to keep the present manuscript focused, test 10, which, being the test to the hypothesis of $c = 0$, is also a test of interest, is intended to be addressed in future studies, together with tests 8 and 9 in Fig. 2.1.

2.8 Conclusions

We may see how the method used to handle the null hypothesis in (2.4), by first bringing it to the form in (2.5) and then using the decomposition in (2.9), enabled the development of very accurate near-exact distributions for the l.r.t. statistic. Actually the technique used to bring the null hypothesis in (2.4) to its form in (2.5) is a particular case of the eigenblock-eigenmatrix decomposition in [5].

From the results of the numerical studies carried out we see that the near-exact distributions developed show an interesting set of nice features. They not only have a good asymptotic behavior for increasing sample sizes but also an extraordinary performance for very small sample sizes, as for example for sample sizes exceeding only by 2 the overall number of variables. Furthermore, these near-exact distributions also display a marked asymptotic behavior for increasing values of p and m. All these features, together with their easy implementation, add up to make the near-exact approximations developed the best choice for practical applications of the test studied.

Moreover, given the results in Sections 9.11 and 10.11 of [1], the results presented concerning the exact distribution of the l.r.t. statistic as well the near-exact distributions developed may be made extensive to all cases where the underlying random vector has an elliptically contoured distribution.

For either $p = 1$ or $m = 1$, the present test reduces to the common test for compound symmetry or equivariance-equicorrelation developed by Wilks in [18].

Acknowledgments This research was partially supported by CMA/FCT/UNL, through projects UID/MAT/00297/2019 and UIDB/MAT/00297/2020.

The authors would also like to thank the reviewer of this paper for many valuable and helpful comments and suggestions.

Appendix 1: Restriction Relations Among *a*, *b* and *c* in Order to Keep *Σ* as a Positive-Definite Matrix

In order to keep Σ as a positive-definite matrix the following restrictions apply to the parameters a, b and c:

(I) for given $a > 0$ and $\frac{a}{1-m} < b < a$ we need to have $\frac{a+b(m-1)}{m(1-p)} < c < \frac{a+b(m-1)}{m}$

(II) for given $a > 0$ and $\frac{a}{1-p} < c < a$ we need to have $\max\left(\frac{cm-a}{m-1}, \frac{cm(1-p)-a}{m-1}\right) < b < a$

(III) for given $b < 0, c < 0$ we need to have $a > \max(0, b, c, c(1-p), (c-b)m + b - cmp, (c-b)m + b)$.

These restrictions may be easily derived from the facts that a is a variance, the matrix R_0 has to be a positive-definite matrix, and all three eigenvalues λ_1, λ_2 and λ_3 in Proposition 2.1 have to be positive.

Appendix 2: Brief Note on the Composition and Decomposition of Hypotheses

Let H_0 and H_1 denote respectively the null and the alternative hypothesis for a given test. Let then Ω_{H_0} and Ω_{H_1} denote respectively the parameter space for H_0 and the union of the parameter space for H_0 with the parameter space for H_1.

Then, to test H_0 versus H_1 will be the same as to test

$$H_{0b|a} \circ H_{0a} \quad \text{versus} \quad H_1 \,,$$

where '∘' means 'after', that is, to test the hypothesis H_0 will be equivalent to test the sequence of hypotheses $H_{0b|a} \circ H_{0a}$, if and only if

$$\Omega_{H_0} \equiv \Omega_{H_{0b|a}} \,, \qquad \Omega_{H_{1b|a}} \equiv \Omega_{H_{0a}} \,, \qquad \Omega_{H_{1a}} \equiv \Omega_{H_1} \,,$$

that is, if and only if

$$\Omega_{H_0} \equiv \Omega_{H_{0b|a}} \subset \Omega_{H_{1b|a}} \equiv \Omega_{H_{0a}} \subset \Omega_{H_{1a}} \equiv \Omega_{H_1} (\equiv \Omega),$$

where $\Omega_{H_{1b|a}} \equiv \Omega_{H_{0a}}$ is assured by the correct nesting of the hypotheses $H_{0b|a}$ and H_{0a} and where $H_{1b|a}$ denotes the alternative hypothesis when testing $H_{0b|a}$.

Appendix 3: Gamma Distribution and Some Related Results

We say that the random variable X follows a Gamma distribution with shape parameter $r > 0$ and rate parameter $\lambda > 0$, if the p.d.f. of X is

$$f_X(x) = \frac{\lambda^r}{\Gamma(r)} e^{-\lambda x} x^{r-1}, \quad (x > 0)$$

and we will denote this fact by $X \sim \Gamma(r, \lambda)$. Then we know that the moment generating function of X is

$$M_X(t) = \lambda^r (\lambda - t)^{-r} \,,$$

so that if we define $Z = e^{-X}$ we will have

$$E(Z^h) = E\left(e^{-hX}\right) = M_X(-h) = \lambda^r(\lambda+h)^{-r}\,.$$

Appendix 4: The GIG and GNIG Distributions

We will say that a r.v. Y has a GIG (Generalized Integer Gamma) distribution of depth p, with integer shape parameters r_j and rate parameters λ_j ($j = 1, \ldots, p$), if

$$Y = \sum_{j=1}^{p} Y_j,$$

where

$$Y_j \sim \Gamma(r_j, \lambda_j)\,, \quad r_j \in \mathbb{N},\ \lambda_j > 0,\ j = 1, \ldots, p$$

are p independent Gamma r.v.'s, with $\lambda_j \neq \lambda_{j'}$ for all $j \neq j'$, with $j, j' \in \{1, \ldots, p\}$.

The r.v. Y has p.d.f. and c.d.f. respectively given by (see [3]),

$$f^{GIG}(y; r_1, \ldots, r_p; \lambda_1, \ldots, \lambda_p; p) \;=\; K \sum_{j=1}^{p} P_j(y)\, e^{-\lambda_j y}\,, \quad (y > 0)$$

and

$$F^{GIG}(y; r_1, \ldots, r_p; \lambda_1, \ldots, \lambda_p; p) \;=\; 1 - K \sum_{j=1}^{p} P_j^*(y)\, e^{-\lambda_j y}\,, \quad (y > 0)$$

where $K = \prod_{j=1}^{p} \lambda_j^{r_j}$,

$$P_j(y) = \sum_{k=1}^{r_j} c_{j,k}\, y^{k-1}\,, \quad P_j^*(y) = \sum_{k=1}^{r_j} c_{j,k} \sum_{i=0}^{k-1} \frac{y^i\,(k-1)!}{i!\,\lambda_j^{k-i}}\,,$$

with

$$c_{j,r_j} = \frac{1}{(r_j - 1)!} \prod_{i=1, i\neq j}^{p} (\lambda_i - \lambda_j)^{-r_i}\,, \qquad j = 1, \ldots, p\,,$$

and, for $k = 1, \dots, r_j - 1;\ j = 1, \dots, p,$

$$c_{j,r_j-k} = \frac{1}{k} \sum_{i=1}^{k} \frac{(r_j - k + i - 1)!}{(r_j - k - 1)!} R(i, j, p)\, c_{j,r_j-(k-i)}\,,$$

where

$$R(i, j, p) = \sum_{k=1,k\neq j}^{p} r_k \left(\lambda_j - \lambda_k\right)^{-i} \quad (i = 1, \dots, r_j - 1)\,.$$

If Y_p has a Gamma distribution with a non-integer shape parameter r_p, then we will say that the r.v. Y has a GNIG (Generalized Near-Integer Gamma) distribution of depth p. The p.d.f. and c.d.f. of Y are, for $y > 0$, respectively given by (see [4])

$$f^{GNIG}(y \,|\, r_1, \dots, r_{p-1}; r_p;\ \lambda_1, \dots, \lambda_{p-1}; \lambda_p;\ p) =$$

$$K\lambda_p^{r_p} \sum_{j=1}^{p-1} e^{-\lambda_j y} \sum_{k=1}^{r_j} \left\{ c_{j,k} \frac{\Gamma(k)}{\Gamma(k+r)} y^{k+r_p-1}\, {}_1F_1(r_p, k+r_p, -(\lambda_p-\lambda_j)y) \right\},$$

and

$$F^{GNIG}(y \,|\, r_1, \dots, r_{p-1}; r_p;\ \lambda_1, \dots, \lambda_{p-1}; \lambda_p;\ p) = \frac{\lambda_p^{r_p} z^{r_p}}{\Gamma(r_p+1)}\, {}_1F_1(r_p, r_p+1, -\lambda_p z)$$

$$-K\lambda_p^{r_p} \sum_{j=1}^{p-1} e^{-\lambda_j y} \sum_{k=1}^{r_j} \frac{c_{j,k}\Gamma(k)}{\lambda_j^k} \sum_{i=0}^{k-1} \frac{z^{r_p+i}\lambda_j^i}{\Gamma(r_p+1+i)}\, {}_1F_1(r_p, r_p+1+i, -(\lambda_p - \lambda_j)y),$$

with $K = \prod_{j=1}^{p-1} \lambda_j^{r_j}$.

References

1. Anderson, T.W.: An Introduction to Multivariate Statistical Analysis, 3rd edn. Wiley, Hoboken (2003)
2. Andersson, S.A.: Invariant normal models. Ann. Stat. **3**, 132–154 (1975)
3. Coelho, C.A.: The generalized integer gamma distribution – a basis for distributions in Multivariate Statistics. J. Multivar. Anal. **64**, 86–102 (1998)
4. Coelho, C.A.: The generalized near-integer gamma distribution: a basis for 'near-exact' approximations to the distributions of statistics which are the product of an odd number of independent Beta random variables. J. Multivar. Anal. **89**, 191–218 (2004)
5. Coelho, C.A.: The eigenblock and eigenmatrix decomposition of a matrix: its usefulness in Statistics — application to the likelihood ratio test for block-circularity. Preprint #7, 2013. Departamento de Matemática - Faculdade de Ciências e Tecnologia da Universidade Nova de Lisboa (2013)

6. Coelho, C.A., Marques, F.J.: The multi-sample block-scalar sphericity test — exact and near-exact distributions for its likelihood ratio test statistic. Commun. Stat. Theory Methods **42**, 1153–1175 (2013)
7. Coelho, C.A., Arnold, B.C., Marques, F.J.: Near-exact distributions for certain likelihood ratio test statistics. J. Stat. Theory Pract. **4**, 711–725 (2010)
8. Marques, F.J., Coelho, C.A.: Near-exact distributions for the likelihood ratio test statistic of the multi-sample block-matrix sphericity test. Appl. Math. Comput. **219**, 2861–2874 (2012)
9. Marques, F.J., Coelho, C.A., Arnold, B.C.: A general near-exact distribution theory for the most common likelihood ratio test statistics used in Multivariate Analysis. Test **20**, 180–203 (2011)
10. Muirhead, R.J.: Aspects of Multivariate Statistical Theory. Wiley, Hoboken (2005)
11. Olkin, I.: Testing and estimation for structures which are circularly symmetric in blocks. In: Kabe, D.G., Gupta, R.P. (eds.) Multivariate Statistical Inference, pp. 183–195. North-Holland, Amsterdam (1973)
12. Olkin, I., Press, S.: Testing and estimation for a circular stationary model. Ann. Math. Stat. **40**, 1358–1373 (1969)
13. Perlman, M.D.: Group symmetry covariance models. Stat. Sci. **2**, 421–425 (1987)
14. Szatrowski, T.H.: Estimation and testing for block compound symmetry and other patterned covariance matrices with linear and nonlinear structure. Technical Report No. 107, Stanford University (1976)
15. Szatrowski, T.H.: Testing and estimation in the block compound symmetry problem. J. Edu. Stat. **7**, 3–18 (1982)
16. Tricomi, F.G., Erdélyi, A.: The asymptotic expansion of a ratio of Gamma functions. Pac. J. Math. **1**, 133–142 (1951)
17. Votaw, D. Jr.: Testing compound symmetry in a normal multivariate distribution. Ann. Math. Stat. **19**, 447–473 (1948)
18. Wilks, S.S.: Sample criteria for testing equality of means, equality of variances, and equality of covariances in a Normal multivariate distribution. Ann. Math. Stat. **17**, 257–281 (1946)

Chapter 3
Convexity of Sets Under Normal Distribution in the Structural Alloy Steel Standard

Kai-Tai Fang, Zhen Luo, and Yung Liang Tong

Abstract The paper is motivated by the structural alloy steel standard that has been used in China for a long period. This standard indicates the scope of several chemical elements in the steel and requests several mechanical properties for qualification. Fang and Wu (Acta Math Appl Sin 2:132–148, 1979) established the relationships between the percents of the controlled chemical elements and testing mechanical properties by a multivariate regression model, and proposed the algorithm for calculating qualification rate. Moreover, they proved the existence of the optimal chemical element combination. However, the uniqueness of the optimal solution for high dimensional case has been left. This open question is equivalent to showing the convexity of a type of probability sets under multivariate normal distribution. This paper proves that the open question is true.

3.1 Motivation

In 1973, the first author of this paper was invited to join an important project by the Ministry of Metallurgy of China to review the national standard for the structural alloy steel 20CrMnTi. In addition to controlling the carbon element, the structural alloy steel also needs chromium (Cr), manganese (Mn), nickel (Ni), molybdenum (Mo), silicon (Si), and titanium (Ti). The content of these elements must fall into the scope according to the national standards. Furthermore, the five mechanical properties such as strength, elasticity, etc., must exceed a certain threshold. Let

---

K.-T. Fang (✉)
Division of Science and Technology, BNU-HKBU United International College Zhuhai, Zhuhai, China
e-mail: ktfang@uic.edu.hk

Z. Luo
Pfizer China, Pudong, Shanghai, China
e-mail: zhen.luo@pfizer.com

Y. L. Tong
Department of Mathematics, Georgia Institute of Technology, Atlanta, GA, USA

© Springer Nature Switzerland AG 2020

T. Holgersson, M. Singull (eds.), *Recent Developments in Multivariate and Random Matrix Analysis*, https://doi.org/10.1007/978-3-030-56773-6_3

$X_1, \ldots, X_p$ denote the respective percents of the controlled chemical elements in the steel, and $Y_1, \ldots, Y_q$ the requested mechanical properties. The national standard indicates the range of the p elements $\mathscr{X} = \{a_i \le X_i \le b_i, i = 1, \ldots, p\}$. It is reasonable to consider that the q mechanical properties have a regression relationship with the p chemical elements, denoted as

$$\begin{cases} Y_1 = \beta_{10} + \beta_{11}X_1 + \beta_{12}X_2 + \cdots + \beta_{1p}X_p + \varepsilon_1, \\ Y_2 = \beta_{20} + \beta_{21}X_1 + \beta_{22}X_2 + \cdots + \beta_{2p}X_p + \varepsilon_2, \\ \quad \vdots \\ Y_q = \beta_{q0} + \beta_{q1}X_1 + \beta_{q2}X_2 + \cdots + \beta_{qp}X_p + \varepsilon_q, \end{cases}$$

where $\varepsilon_1, \ldots, \varepsilon_q$ are random errors. This can be expressed as

$$\mathbf{y} = \mathbf{B}\mathbf{x} + \boldsymbol{\varepsilon}, \tag{3.1}$$

where $\mathbf{x} = (1, X_1, \ldots, X_p)^T, \mathbf{y} = (Y_1, \ldots, Y_q)^T, \boldsymbol{\varepsilon} = (\varepsilon_1, \ldots, \varepsilon_q)^T$ and $\mathbf{B} = (\beta_{ij}, i = 1, \ldots, q, j = 0, 1, \ldots, p)$ is a $q \times (p+1)$ matrix. Assume that the random errors $\boldsymbol{\varepsilon}$ follows a multivariate normal distribution $N_q(\mathbf{0}, \Sigma)$, where Σ is positive definite covariance matrix. It is required by the national standard that a batch of steel material is qualified if

$$Y_i \ge T_i, i = 1, \ldots, q, \tag{3.2}$$

where T_i's are given by the national standard. If one or more $Y_i < T_i$, this batch of steel is unqualified.

Let $\widehat{\mathbf{B}}$ denote the least squares estimator of $\mathbf{B}$. Then $\mathbf{y} \sim N_q(\mathbf{B}\mathbf{x}, \Sigma)$ and $\widehat{\mathbf{y}} = \widehat{\mathbf{B}}\mathbf{x}$ is an estimate of $\mathbf{y}$. The domain of $\widehat{\mathbf{y}} = \widehat{\mathbf{B}}\mathbf{x}$, denoted by $\mathscr{Y}$, is a polyhedron as the domain of $\mathbf{x}$ is a rectangle $\mathscr{X} = [a_1, b_1] \times \cdots \times [a_p, b_p]$ in R^p. Therefore, the qualification rate (QR) for some given $\mathbf{x}_0 = (1, x_{10}, \ldots, x_{p0})^T$ can be calculated by

$$\begin{aligned} QR(\mathbf{x}_0) &= \int_{T_1}^{\infty} \cdots \int_{T_q}^{\infty} f_q(\mathbf{u}; \widehat{\mathbf{B}}\mathbf{x}_0, \Sigma) d\mathbf{u} \\ &= \int_{T_1}^{\infty} \cdots \int_{T_q}^{\infty} f_q(\mathbf{u}; \widehat{\mathbf{y}}_0, \Sigma) d\mathbf{u} \equiv QR(\widehat{\mathbf{y}}_0), \end{aligned} \tag{3.3}$$

where $\widehat{\mathbf{y}}_0 = \widehat{\mathbf{B}}\mathbf{x}_0$ and the multivariate normal distribution density

$$f_q(\mathbf{u}; \boldsymbol{\mu}, \Sigma) = (2\pi)^{-q/2}[det(\Sigma)]^{-1/2} \exp\{-\frac{1}{2}(\mathbf{u} - \boldsymbol{\mu})^T \Sigma^{-1}(\mathbf{u} - \boldsymbol{\mu})\}, \tag{3.4}$$

with $\boldsymbol{\mu} = \widehat{\mathbf{B}}\mathbf{x}_0 = \widehat{\mathbf{y}}_0$. If Σ is unknown, it can be estimated from the theory of regression analysis. For simplicity, we assume Σ is known in the paper. From formula (3.3) $QR(\mathbf{x})$ varies over $\mathscr{X}$. One wishes to find a maximum $\mathbf{x}^*$ such that $QR(\mathbf{x}^*) = \max_{\mathbf{x} \in \mathscr{X}} QR(\mathbf{x})$. Let $\widehat{\mathbf{y}}^* = \widehat{\mathbf{B}}\mathbf{x}^* \in \mathscr{Y}$, the latter is the domain of $\mathbf{y}$.

The national standard had been used by more than ten factories all over the country to produce the same kind of structure alloy steel 20CrMnTi. However, the results were inconsistent. While some steel mills produced a high proportion of qualified alloy steel, other steel mills had low proportions of qualified alloy steel even if percents of the controlled chemical elements were in full compliance with the national standard. For example, one steel mill can produced about 97% qualified steel while another mill only reach 38%. Thus, many steel mills entertained doubts about the national standard for the scope of chemical elements contents. This standard was introduced from the Soviet Union in 1950s. At that time, no one knew its principles, or the reasons for the inconsistent production qualified rate. Since refining a batch of low quality structural alloy steel is very expensive, the manufacturers not meeting the national standard would suffer heavy economic losses. Therefore, it was important and urgent to study whether the standard being used was reasonable or whether there was room for improvement.

In 1973, the Ministry of Metallurgy of China formed a team to review the national standard of structural alloy steel. The team involved senior engineers in metallurgy from different institutions and two statisticians from the Institute of Mathematics, Chinese Academy of Sciences including the last author of this paper.

Due to the lack of theory in metallurgy, the team decided to use statistics for checking whether the current standard is reasonable or not. They collected data from all the plants where the structural alloy steel had been produced. It was indeed a big dataset at the time of collection. After cleaning the data, a regression model (3.1) can be established that can fit all the data from different plants.

For calculation of QR in (3.3), denote $\Sigma = (\sigma_{ij})$ and the corresponding correlation matrix $\mathbf{R} = (\rho_{ij}, i, j = 1, \dots, q)$. Then $\sigma_{ij} = \sqrt{\sigma_{ii}\sigma_{ii}}\rho_{ij}, i, j = 1, \dots, q$ and $\widehat{\mathbf{y}} = (\widehat{y}_1, \dots, \widehat{y}_q)^T$. By the transformation $v_i = (u_i - \widehat{y}_1)/\sqrt{\sigma_{ii}}, i = 1, \dots, q$,

$$QR(\widehat{\mathbf{y}}) = \int_{w_1}^{\infty} \cdots \int_{w_q}^{\infty} f_q(\mathbf{v}; \mathbf{0}, \mathbf{R}) d\mathbf{v}, \tag{3.5}$$

where $w_i = (T_i - \widehat{y}_i)/\sqrt{\sigma_{ii}}, i = 1, \dots, q$. Fang and Wu [3] proved existence of maximum of QR in $\mathscr{Y}$. For uniqueness of maximum of QR in $\mathscr{Y}$, they raised the following conjecture:

Theorem 3.1 *For given $a > 0$ and positive definite matrix Σ of order q, $q \geq 2$, the following equation on a convex set C in $\mathscr{R}^q$*

$$\int_{x_1}^{\infty} \cdots \int_{x_q}^{\infty} \exp\left(-\frac{1}{2}\mathbf{t}^T \Sigma^{-1} \mathbf{t}\right) d\mathbf{t} = a \tag{3.6}$$

defines a function, $x_1 = h(x_2, \dots, x_q)$ say, and this function is concave in $\mathscr{R}^{q-1}$.

This conjecture was proven by Fang and Wu [3] for $q \leq 2$ but remained open for $q > 2$. However, their approach can't solve the above conjecture for $q > 2$. Therefore, Fang [4] proposed this open question in the journal, but without any valid solutions so far. In fact, this conjecture is equivalent to

Theorem 3.2 *For every $a > 0$ the set*

$$W_a = \left\{ \mathbf{x} \in \mathscr{R}^q : \int_{x_1}^{\infty} \cdots \int_{x_q}^{\infty} n_q(\mathbf{t}; \boldsymbol{\mu}, \Sigma) d\mathbf{t} \geq a \right\} \tag{3.7}$$

is convex in $\mathscr{R}^q$.

Theorem 3.2 holds true for any covariance matrix as long as it is positive definite because there is no other conditions imposed on it in the proof in Sect. 3.2.2. The main purpose of the paper is to give proofs for these two theorems and related discussion. First we introduce some concepts that are useful in the proofs.

3.2 Main Results

This section gives proofs of Theorems 3.1 and 3.2 separately.

Definition 3.1 A set $C \in \mathscr{R}^q$ is convex, if for any pair $\mathbf{x}, \mathbf{y} \in C$ and $\lambda \in (0, 1)$ we have

$$\lambda \mathbf{x} + (1 - \lambda)\mathbf{y} \in C.$$

Definition 3.2 A real function f on a convex set $C \in \mathscr{R}^q$ is said to be convex on C if for any pair $\mathbf{x}, \mathbf{y} \in C$ and $\lambda \in (0, 1)$ we have

$$f(\lambda \mathbf{x} + (1 - \lambda)\mathbf{y}) \leq \lambda f(\mathbf{x}) + (1 - \lambda) f(\mathbf{y}).$$

If $-f(\mathbf{x})$ is a convex function on C, the function $f(\mathbf{x})$ is called a concave function. For $q = 1$ a concave function $f(x)$ can be defined by

$$f(\lambda x + (1 - \lambda)y) \geq \lambda f(x) + (1 - \lambda) f(y),$$

where $x, y \in C, \lambda \in (0, 1)$.

3.2.1 Proof of Theorem 3.1

Lemma 3.1 *Let $p(\mathbf{x})$ be a continuous density function in $\mathscr{R}^q$ and let C is a convex set in $\mathscr{R}^q$. Denote*

$$\bar{F}(\mathbf{x}) = \bar{F}(x_1, \mathbf{x}_2) = \int_{x_1}^{\infty} \cdots \int_{x_q}^{\infty} p(\mathbf{t}) d\mathbf{t}, \tag{3.8}$$

where $\mathbf{x} = (x_1 \ldots, x_q) \equiv (x_1, \mathbf{x}_2), \mathbf{x}_2 = (x_2 \ldots, x_q), \mathbf{t} = (t_2, \ldots, t_q)$. For any $a > 0$ let $x_1 = h(\mathbf{x}_2)$ satisfy

$$\bar{F}(x_1, \mathbf{x}_2) = a. \tag{3.9}$$

Then we have the following equivalence relationships:

(a) If $A = \{\mathbf{x} | \bar{F}(\mathbf{x}) \geq a\}$ is a convex set in $\mathscr{R}^q$, then $x_1 = h(\mathbf{x}_2)$ is a concave function for any $a > 0$;
(b) Conversely, if $h(\mathbf{x}_2)$ is a concave function, then $A = \{\mathbf{x} | \bar{F}(\mathbf{x}) \geq a\}$ is a convex set.

Proof Let $\mathbf{x} = (x_1, \mathbf{x}_2)$ and $\mathbf{y} = (y_1, \mathbf{y}_2)$ satisfy

$$\bar{F}(x_1, \mathbf{x}_2) = \bar{F}(y_1, \mathbf{y}_2) = a.$$

For any fixed $\lambda \in (0, 1)$, let

$$\mathbf{z} = \lambda(x_1, \mathbf{x}_2) + (1 - \lambda)(y_1, \mathbf{y}_2) \equiv (z_1, \mathbf{z}_2).$$

Then $\mathbf{z} \in A$, so $\bar{F}(z_1, \mathbf{z}_2) \geq a$. Let $\mathbf{z}^* = (z_1^*, \mathbf{z}_2)$ be such that $\bar{F}(z_1^*, \mathbf{z}_2) = a$. Since for every fixed $\mathbf{z}_2$, $\bar{F}(u, \mathbf{z}_2)$ is a nonincreasing function for u, one must have $z_1^* \geq z_1$. That is

$$z_1^* = h(\mathbf{z}_2) = h(\lambda \mathbf{x}_2 + (1 - \lambda)\mathbf{y}_2) \geq z_1 = \lambda x_1 + (1 - \lambda) y_1 = \lambda h(\mathbf{x}_2) + (1 - \lambda) h(\mathbf{y}_2).$$

So h is a concave function. The proof of (b) is similar. □

Applying Lemma 3.1 to the density of multivariate normal distribution $N_q(\mathbf{0}, \Sigma)$, the assertion of Theorem 3.1 follows. By a similar approach, we have the following.

Corollary 3.1 *Let $p(\mathbf{x})$ be a continuous density function in $\mathscr{R}^q$ and let C be a convex set in R^q. Consider the function*

$$F(\mathbf{x}) = F(x_1, \mathbf{x}_2) = \int_{-\infty}^{x_1} \cdots \int_{-\infty}^{x_q} p(\mathbf{t}) d\mathbf{t} \tag{3.10}$$

on C. Let $x_1 = h(\mathbf{x}_2)$ *satisfy* $F(x_1, \mathbf{x}_2) = a > 0$. *Then h is a concave function if and only if* $A^c = \{\mathbf{x}|F(\mathbf{x}) \le a\}$ *for any* $a \in (0, 1)$ *is a convex set.*

3.2.2 Proof of Theorem 3.2

At first, we need more concepts given below. The reader can refer Eaton [2] for more details.

Definition 3.3 A function $f(\mathbf{x}), \mathbf{x} \in \mathscr{R}^q$, is called A-unimodal if for any real number a the set $\{\mathbf{x} \in \mathscr{R}^q : f(\mathbf{x}) \ge a\}$ is convex in $\mathscr{R}^q$.

Definition 3.4 A function $f(\mathbf{x})$ from $\mathscr{R}^q$ to $[0, \infty)$ is called log-concave if for any $\mathbf{x}, \mathbf{y} \in \mathscr{R}^q$ and $\lambda \in (0, 1)$ we have

$$f(\lambda \mathbf{x} + (1 - \lambda)\mathbf{y}) \ge f^{\lambda}(\mathbf{x}) f^{1-\lambda}(\mathbf{y}). \tag{3.11}$$

Lemma 3.2 *Assume* $h_1(\mathbf{x})$ *and* $h_2(\mathbf{x})$ *are two log-concave functions on* $\mathscr{R}^q$ *and for each* $\mathbf{x} \in \mathscr{R}^q$ *the function*

$$g(\mathbf{x}) \equiv \int h_1(\mathbf{x} - \mathbf{z}) h_2(\mathbf{z}) d\mathbf{z} \tag{3.12}$$

exists. Then the function g is a log-concave function on $\mathscr{R}^q$.

The proof can refer to Theorem 5.2 of Eaton [2]. The following lemma is easy to prove.

Lemma 3.3 *If* $f(\mathbf{x})$ *is a log-concave function from* $\mathscr{R}^q$ *to* $[0, \infty)$, *then the function f is A-unimodal.*

Proof of Theorem 3.2 We prove that the set W_a is convex for every $a \ge 0$. It is equivalent to show that the function

$$Q(\mathbf{x}) = \int_{x_1}^{\infty} \cdots \int_{x_q}^{\infty} n_q(\mathbf{t}; \boldsymbol{\mu}, \Sigma) d\mathbf{t} \ge a,$$

is A-unimodal from Definition 3.3. From Lemma 3.2, we have to prove that the function $Q(\mathbf{x})$ is log-concave. Rewrite $Q(\mathbf{x})$ as

$$Q(\mathbf{x}) = \int_{\mathscr{R}^q} I([x_1, \infty) \times \cdots \times [x_q, \infty))(\mathbf{t}) f_q(\mathbf{t}; \boldsymbol{\mu}, \Sigma) d\mathbf{t},$$

where $f_q(\mathbf{t}) = f_q(\mathbf{t}; \boldsymbol{\mu}, \Sigma)$ in (3.4), $I(A)(\mathbf{t})$ is the index function of the set A: $I(A)(\mathbf{t}) = 1$ if $\mathbf{t} \in A$, otherwise $I(A)(\mathbf{t}) = 0$. The function $Q(\mathbf{x})$ can be expressed as

$$Q(\mathbf{x}) = \int_{\mathscr{R}^q} I((-\infty, 0]^q)(\mathbf{t})(\mathbf{x} - \mathbf{t}) f_q(\mathbf{t}) d\mathbf{t}.$$

Note that

$$\log f_q(\mathbf{t}) = -\frac{1}{2}(\mathbf{t} - \boldsymbol{\mu})^T \Sigma^{-1}(\mathbf{t} - \boldsymbol{\mu}) + \text{ constant}$$

is a concave function in $\mathbf{t}$, and $(-\infty, 0]^q$ is a convex set in $\mathscr{R}^q$ and $I((-\infty, 0]^q)(\mathbf{t})$ is log-concave function of $\mathbf{t}$ on R^q. Both the functions $f_q(\mathbf{t})$ and $I((-\infty, 0]^q)(\mathbf{t})$ are log-concave. From Lemma 3.2, the function g is log-concave. Hence, the proof is complete. □

3.2.3 *Approach by Generalized Brunn-Minkowski Inequality*

This subsection gives some discussion on the problem in the paper by the generalization of Brunn-Minkowski-Lusternik inequality [1] as below.

Lemma 3.4 *Let f_0 and f_1 be two nonnegative measurable functions on $\mathscr{R}^q$ with non-empty supports S_0 and S_1, respectively. Assume that f_0 and f_1, are integrable with respect to the Lebesgue measure μ_n on R^q. Let $\theta (0 < \theta < 1)$ be a fixed number and f be a non-negative, measurable function on R^q such that*

$$f(\mathbf{x}) \geq M_\alpha[f_0(\mathbf{x}_0), f_1(\mathbf{x}_1); \theta],$$

whenever $\mathbf{x} = (1 - \theta)\mathbf{x}_0 + \theta\mathbf{x}_1$ with $\mathbf{x}_0 \in S_0, \mathbf{x}_1 \in S_1; -1/n \leq \alpha \leq \infty$. Then

$$\int_{(1-\theta)S_0+\theta S_1} f(\mathbf{x})d\mathbf{x} \geq M_{\alpha_q^*}\left[\int_{R^q} f_0(\mathbf{x})d\mathbf{x} \int_{R^q} f_1(\mathbf{x})d\mathbf{x}; \theta\right],$$

where

$$\alpha_q{}^* = \begin{cases} \alpha/(1+q\alpha), & \textit{for } -1/q < \alpha < \infty, \\ 1/q, & \textit{for } \alpha = +\infty, \\ -\infty, & \textit{for } \alpha = -1/q. \end{cases} \tag{3.13}$$

The generalized mean function M_α, is defined as

$$M_\alpha(a_0, a_1; \theta) = \begin{cases} [(1-\theta)a_0^\alpha + \theta a_1^\alpha]^{1/\alpha}, & \text{if } 0 < \alpha < \infty \text{ or if } -\infty < \alpha < 0 \\ & \text{and } a_0 a_1 \neq 0, \\ 0, & \text{if } -\infty < \alpha < 0 \text{ and } a_0 a_1 \neq 0, \\ a_0^{1-\theta} a_1^\theta, & \text{if } \alpha = 0, \\ \max(a_0, a_1), & \text{if } \alpha = \infty, \\ \min(a_0, a_1), & \text{if } \alpha = -\infty. \end{cases}$$

Definition 3.5 Let g be a non-negative function in R^q which satisfies

$$g((1-\theta)\mathbf{x}_0 + \theta\mathbf{x}_1) \geq M_\alpha[g(\mathbf{x}_0), g(\mathbf{x}_1); \theta], \forall \theta \in (0, 1), \mathbf{x}_0, \mathbf{x}_1 \in R^q. \quad (3.14)$$

Then g is called an α-unimodal function.

It is easy to see that a $(-\infty)$-unimodal function is a A-unimodal function and that 0-unimodal function is a log-concave function. For $\alpha_1 > \alpha_2$, if a function is an α_1-unimodal function then it is an α_2-unimodal function. The following Lemma is from Theorem 3 of Das Gupta [1].

Lemma 3.5 *Let g be a probability density function on $\mathscr{R}^q$ such that for $0 < \theta < 1$*

$$g(\mathbf{x}) \geq M_\alpha[g(\mathbf{x}_0), g(\mathbf{x}_1); \theta], \quad (3.15)$$

whenever $\mathbf{x} = (1-\theta)\mathbf{x}_0 + \theta\mathbf{x}_1$ and $\mathbf{x}_0, \mathbf{x}_1$ are in the support S of g and $-\frac{1}{q} \geq \alpha \geq \infty$. Then for any two non-empty measurable sets A_0 and A_1 in $\mathscr{R}^q$

$$\int_{(1-\theta)A_0+\theta A_1} g(\mathbf{x})d\mathbf{x} \geq M_{\alpha_q^*}\left[\int_{A_0} g(\mathbf{x})d\mathbf{x}, \int_{A_1} g(\mathbf{x})d\mathbf{x}; \theta\right], \quad (3.16)$$

where α_q^* is defined in (3.13), if $-\frac{1}{q} \leq \alpha \leq 0$, or $0 < \alpha \leq \infty$ and either both $A_0 \cap S$ and $A_1 \cap S$ are non-empty or both are empty.

Take

$$A_0 = [x_1^0, \infty) \times \cdots \times [x_q^0, \infty)$$
$$A_1 = [x_1^1, \infty) \times \cdots \times [x_q^1, \infty).$$

Then

$$(1-\theta)A_0 + \theta A_1 = [(1-\theta)x_1^0 + \theta x_1^1, \infty) \times \cdots \times [(1-\theta)x_q^0 + \theta x_q^1, \infty).$$

From Lemma 3.5 we have the following corollary.

Corollary 3.2 *Assume that $f(\mathbf{x})$ is an α-unimodal probability density function in $\mathscr{R}^q$, then*

$$g(\mathbf{x}) = \int_{[x_1,\infty)\times\cdots\times[x_q,\infty)} f(\mathbf{t})d\mathbf{t}$$

is an α_q^*-unimodal function.

This corollary can be used in the proof Theorem 3.2.

Acknowledgments The first author would like to thank Professor Dietrich von Rosen for a pleasant cooperation in the past decades and for his support. Happy 65th birthday to Dietrich!

References

1. Das Gupta, S.: Brunn-Minkowski inequality and its aftermath. J. Multivariate Anal. **10**, 296–318 (1980)
2. Eaton, M.L.: A review of selected topic in multivariate probability inequalities. Ann. Stat. **10**(1), 11–43 (1982)
3. Fang, K.T., Wu C.Y.: A probability extremum problem. Acta Math. Appl. Sin. **2**, 132–148 (1979). (in Chinese)
4. Fang, K.T.: An open question in applied mathematics. Math. Math. Pract. Theory **10**(3), 26 (1984)

Chapter 4
Comments on Maximum Likelihood Estimation and Projections Under Multivariate Statistical Models

Katarzyna Filipiak, Mateusz John, and Augustyn Markiewicz

Abstract Under the multivariate model with linearly structured covariance matrix with unknown variance components and known mean parameters (Szatrowski, Ann Stat 8:802–810, 1980) showed that the maximum likelihood estimators of variance components have explicit representation if and only if the space of covariance matrix is a quadratic subspace. The aim of this paper is to rewrite these results for models with unknown expectation and to give sufficient conditions for maximum likelihood estimator of covariance matrix to be a projection of the maximum likelihood estimator of unstructured covariance onto the space of structured matrices. The results will be illustrated by examples of structures suitable for multivariate models with general mean, growth curve models as well as doubly multivariate models.

4.1 Introduction and Preliminaries

Let $\mathbf{Y}$ be the $n \times p$ matrix of observations of a feature in p time points on n independent experimental subjects, with expectation $E(\mathbf{Y}) = \mathbf{1}_n\boldsymbol{\mu}'$, where $\boldsymbol{\mu}$ is a p-dimensional vector of means (the same for each subject), and with covariance matrix $D(\mathbf{Y}) = \boldsymbol{\Psi} \otimes \mathbf{I}_n$, where $\boldsymbol{\Psi}$ is a $p \times p$ symmetric positive-definite covariance matrix of time points (the same for each subject) and $\otimes$ denotes the Kronecker product. Such a model will be called standard multivariate model.

In the literature several specific structures of $\boldsymbol{\Psi}$ are considered, linear, such as e.g., compound symmetry (CS), (banded, circular) Toeplitz, or nonlinear, such as e.g., first-order autoregression (AR(1)). Identification of the covariance matrix structure as well as its estimation are commonly studied recently; cf. Cui et al. [1],

K. Filipiak (✉) · M. John
Institute of Mathematics, Poznań University of Technology, Poznań, Poland
e-mail: Katarzyna.Filipiak@put.poznan.pl; mateusz.john@put.poznan.pl

A. Markiewicz
Department of Mathematical and Statistical Methods, Poznań University of Life Sciences, Poznań, Poland
e-mail: amark@up.poznan.pl

© Springer Nature Switzerland AG 2020
T. Holgersson, M. Singull (eds.), *Recent Developments in Multivariate and Random Matrix Analysis*, https://doi.org/10.1007/978-3-030-56773-6_4

Filipiak, Klein [4], Filipiak et al. [9], Filipiak et al. [10], John, Mieldzioc [14], Lin et al. [18]. Note that the specification of the covariance structure very often allows to overcome the overparameterization problem which appears if $p \geq n$.

Besides a very general expectation structure, we may also assume e.g., a growth curve model, in which $E(\mathbf{Y}) = \mathbf{XBZ}$, where $\mathbf{X}$ and $\mathbf{Z}$ are some known design matrices of proper size and $\mathbf{B} : n_1 \times p_1$ is a matrix of unknown parameters. Estimation of parameters in the growth curve model with unstructured covariance matrix was considered e.g., by [15], while under assumption of CS covariance structure by e.g., Kopčová, Žežula [16] or under the assumption of CS and AR(1) covariance structure by e.g., Žežula [31].

Each model can be further extended for doubly-multivariate data, in which instead of one, q features are observed in p time/location points on n independent experimental subjects. Then, the observations can be presented in the form of three-indices $n \times q \times p$ matrix (tensor of order three), $\mathcal{Y}$. A $q \times p$ matrix of means of each subject is usually denoted by $\mathbf{M}$. A $qp \times qp$ covariance matrix of observations (the same for each subject) may have various structures, depending on the character of experiment. For example, if the covariance of features, say $\boldsymbol{\Sigma}$, does not change with time/location points, and the covariance of time/location points, $\boldsymbol{\Psi}$, does not change with features, the covariance $D(\mathcal{Y}) = \boldsymbol{\Psi} \otimes \boldsymbol{\Sigma} \otimes \mathbf{I}_n$ and such a structure is often called separable. Some estimators of a separable structure under the model with general mean were given e.g., by Lu and Zimmerman[19], by Filipiak et al. [7] in more compact form, or by Srivastava et al. [26] with the assumption of uniqueness of Kronecker product. Obviously, the matrices $\boldsymbol{\Psi}$ and $\boldsymbol{\Sigma}$ may remain unstructured, or can have some structure (for example one of the mentioned above). If one of the component of the Kronecker product, say $\boldsymbol{\Psi}$, is for example CS, then $\boldsymbol{\Psi}_{CS} \otimes \boldsymbol{\Sigma}$ is a natural extension of a classical CS covariance matrix of multivariate data for doubly-multivariate case.

Also some other separable covariance matrices with components structured in a specific way are considered in the literature; see e.g., Hao et al. [13], Szczepańska et al. [29].

Another covariance structure of doubly-multivariate data commonly studied in the literature is the block exchangeable structure (BCS), being a generalization of compound symmetry into the block case; cf. Olkin [22], Roy et al. [23], Roy et al. [24]. It means that any change of columns of the observation matrix of the ith subject does not change the covariance matrix. In contrast to separability, block exchangeable covariance structure remains linear. In the same way Toeplitz matrices can be extended into the doubly-multivariate case. The MLEs of block-circular Toeplitz were proposed e.g., by Olkin [22], Szatrowski [28, p. 808].

In this paper we are mainly interested in linear structures of the covariance matrix. Necessary and sufficient conditions for the existence of the maximum likelihood estimators (MLEs) of a linearly structured covariance matrix in explicit form, and for obtaining this explicit form in one iteration, under the model with known expectation, is given by Szatrowski [28]. In this paper we formulate sufficient conditions to ensure explicit representation of the MLE of linear covariance structure under the model with unknown expectation. Note that, in the case of separable structures, even if one component of separability is linear, the Kronecker product is not linear anymore, which often causes problems with estimation.

Similarly as in the standard multivariate model, in doubly-multivariate case some specific structure of expectation can be also considered. Estimation of unknown parameters under the generalized growth curve model with separable structure was considered by e.g., Filipiak, Klein [3], Srivastava et al. [27].

It should be pointed out that the number of unknown parameters to be estimated in unstructured covariance matrix increases very rapidly with the increase of either the number of features q, or the number of time/location points p. Moreover, when the sample size is not large enough, the sample covariance matrix or the MLE of true covariance matrix is singular or ill-conditioned. To avoid this problem, researchers usually rely on the structured covariance matrices which depend on a smaller number of unknown parameters. The problem, though, is to recognize the true structure using some regularization techniques: graphical, e.g., neural networks or mapping (Gilson et al. [12]) or graphical lasso algorithm (Devijver, Gallopin [2]) as well as algebraic (see e.g., Cui et al. [1], Lin et al. [18] for multivariate data, and Filipiak, Klein [4], Filipiak et al. [10] for doubly-multivariate data) can be used. On the other hand in many experiments the underlying structure follows from experimenter a-priori knowledge about the variables behavior which comes e.g., from the nature of the data. Nevertheless, in both cases the structure (regularized or a-priori identified) should be then tested. Filipiak et al. [5], Filipiak et al. [6] and Roy et al. [24] compared two procedures (based on likelihood ratio and Rao score) for testing one of the following structures: separability, separability with one component additionally structured as CS or AR(1), or BCS.

Since in the testing procedures very often the MLEs of unknown parameters are required, the normality of observations is assumed and the models are respectively denoted as

$$\mathbf{Y} \sim N_{n,p}\left(\mathbf{1}_n\boldsymbol{\mu}', \mathbf{I}_n, \boldsymbol{\Gamma}\right), \tag{4.1}$$

$$\mathbf{Y} \sim N_{n,p}\left(\mathbf{XBZ}, \mathbf{I}_n, \boldsymbol{\Gamma}\right), \tag{4.2}$$

$$\mathbf{Y} \sim N_{n,qp}\left(\mathbf{1}_n\text{vec}'\,\mathbf{M}, \mathbf{I}_n, \boldsymbol{\Gamma}\right) \tag{4.3}$$

with $\boldsymbol{\Gamma} \in \mathcal{G}$ is a specific structure of relevant order and $\mathbf{Y}$ in (4.3) being a transformation of a third order tensor $\mathcal{Y}$ into the $n \times pq$ matrix; cf. Filipiak, Klein [3] for more details. We call the models (4.1), (4.2) and (4.3) respectively *standard multivariate model*, *growth curve model* and *doubly multivariate model*.

Zehna [30] showed the invariance property of MLE. Briefly stating it was proven, that if $\widehat{\boldsymbol{\theta}}$ is the MLE of $\boldsymbol{\theta}$, then for arbitrary function u, not necessarily one-to-one, $u(\widehat{\boldsymbol{\theta}})$ is the MLE of $u(\boldsymbol{\theta})$. Thus, if by $\mathbf{S}$ we denote the MLE of unstructured covariance matrix, one can expect that the MLE of each structure should be expressed as a function of $\mathbf{S}$. The aim of this paper is to propose such a function (if exists in direct form) and to verify, under which conditions the MLE can be expressed as the projection of $\mathbf{S}$ onto the space of covariance structures.

Recall, that the projection of $\mathbf{S}$ onto the m-dimensional linear space with orthonormal basis $\mathscr{B}$ can be expressed as

$$\sum_{i=1}^{m} \langle \mathbf{S}, \mathbf{B}_i \rangle \mathbf{B}_i,$$

where $\mathbf{B}_i \in \mathscr{B}$, or as the minimum of Frobenius norm of the difference between $\mathbf{S}$ and the set of structures, $\mathscr{G}$, that is

$$\min_{\Gamma \in \mathscr{G}} ||\mathbf{S} - \mathbf{\Gamma}||_F^2$$

where $||\mathbf{A}||_F^2 = \text{tr}(\mathbf{A}\mathbf{A}')$.

Note that, in the literature in which the MLEs of covariance structures are considered, it usually does not matter whether the MLE has an explicit form or not. The question is whether the likelihood function attains the maximum in the interior of the parameter space. The aim of this paper is to formulate conditions under which the MLEs are projections of some matrix onto the covariance structures space, and thus possessing an explicit form which in turn allows for the MLE to be obtained in a single iteration.

We now give the notations used throughout the paper. The orthogonal projection onto the column space of a matrix $\mathbf{A}$ is denoted by $\mathbf{P}_A$, that is $\mathbf{P}_A = \mathbf{A}(\mathbf{A}'\mathbf{A})^{-}\mathbf{A}'$, while its orthocomplement is $\mathbf{Q}_A = \mathbf{I} - \mathbf{P}_A$, where $\mathbf{I}$ is the identity matrix of relevant order. In the very specific situation when $\mathbf{A} = \mathbf{1}_n$, we use the notations $\mathbf{P}_n$ and $\mathbf{Q}_n$. For a symmetric positive definite (p.d.) $\mathbf{B}$ we denote $\mathbf{P}_{A;B} = \mathbf{A}(\mathbf{A}'\mathbf{B}\mathbf{A})^{-}\mathbf{A}'\mathbf{B}$ and $\mathbf{Q}_{A;B} = \mathbf{I} - \mathbf{P}_{A;B}$. Vectorization of a matrix $\mathbf{A}$, that is rearrangement of $\mathbf{A}$ into the vector by writing one column under another in a sequence is denoted by vec $\mathbf{A}$.

The block trace and partial trace operators on $pq \times pq$ block matrix $\mathbf{A}$, are defined respectively as the sum of $q \times q$ diagonal blocks of $\mathbf{A}$, denoted $\text{BTr}_q(\mathbf{A})$, and the matrix with $q \times q$ blocks replaced by their traces, denoted $\text{PTr}_q(\mathbf{A})$; cf. Filipiak et al. [8]. Similarly, the notation $\text{BSum}_q(\mathbf{A})$ is used for the sum of all $q \times q$ blocks of $\mathbf{A}$.

This paper is organized as follows. Under the standard multivariate model (4.1) with a linearly structured covariance matrix, the sufficient conditions for MLE being the projection are presented as Theorem 4.1 in Sect. 4.2. The theorem is then illustrated by several examples. In Sect. 4.3 the MLE of a separable covariance structure is compared with projection under the growth curve model (4.2). Finally, in Sect. 4.4, the thesis of Theorem 4.1 is applied to the doubly-multivariate model (4.3). It is shown that, if in the standard multivariate model the linear structure coefficients (scalars) are replaced by matrices and matrix multiplications are replaced by Kronecker products, then all results can be extended to a doubly-multivariate model.

4.2 Parameterization, MLE and Projection Under the Standard Multivariate Model

In the mixed model theory, that is $\mathbf{y} = \mathbf{X}\boldsymbol{\beta} + \sum_{i=1}^{k} \mathbf{Z}_i\boldsymbol{\beta}_i + \mathbf{e}$, it is usually assumed that the covariance matrix has a linear structure, say $\mathbf{V} = \sum_{i=1}^{k} \sigma_i \mathbf{V}_i$, with unknown parameters σ_i and known symmetric p.d. matrices $\mathbf{V}_i$. Such a model is called *variance components model* and the problem under consideration is the estimation of σ_i. Note that, if in the above model every $\sigma_i > 0$ and $\mathbf{V}_i$ are pairwise orthogonal, then $\mathbf{V}$ is p.d.

In the multivariate model $\mathrm{E}(\mathbf{Y}) = \mathbf{XB}$, which can be transformed by vectorization to the univariate case, that is $\mathrm{E}(\mathbf{y}) = \mathbf{X}\boldsymbol{\beta}$ (where $\mathbf{y} = \operatorname{vec}\mathbf{Y}$ and $\boldsymbol{\beta} = \operatorname{vec}\mathbf{B}$) with $\mathrm{D}(\mathbf{y}) = \boldsymbol{\Sigma}$, the problem of estimating the covariance matrix can be expressed in similar way as in variance components model as long as $\boldsymbol{\Sigma}$ has linear structure depending on the vector of coefficients $\boldsymbol{\sigma}$.

It is known that if the projection onto the space of expectation commutes with the covariance matrix, then the likelihood equation for expectation does not depend on the covariance matrix and thus its solution is an ordinary least squares estimator. Indeed, in the model with covariance matrix $\mathbf{V}$ the likelihood equation for expectation has the form

$$\mathbf{X}'\mathbf{V}^{-1}(\mathbf{y} - \mathbf{X}\boldsymbol{\beta}) = \mathbf{0},$$

which can be equivalently expressed as

$$\mathbf{P}_X\mathbf{V}^{-1}(\mathbf{y} - \mathbf{X}\boldsymbol{\beta}) = \mathbf{0},$$

and if $\mathbf{P}_X\mathbf{V} = \mathbf{V}\mathbf{P}_X$ the MLE of $\boldsymbol{\beta}$ is obtained as

$$\mathbf{X}\widehat{\boldsymbol{\beta}} = \mathbf{P}_X\mathbf{y}.$$

Summing up, the Theorem 4 by Szatrowski [28] can be rewritten for unknown expectation satisfying commutativity assumption as follows.

Proposition 4.1 *Let* $\mathbf{y}$ *be a normally distributed random vector, with* $E(\mathbf{y}) = \mathbf{X}\boldsymbol{\beta}$ *and covariance matrix* $\mathbf{V} = \mathbf{V}(\boldsymbol{\sigma})$ *belonging to the linear space,* $\mathcal{V}$, *with unknown vector of linear combinations coefficients,* $\boldsymbol{\sigma}$. *If* $\mathbf{P}_X\mathbf{V} = \mathbf{V}\mathbf{P}_X$, *then the following two conditions are equivalent:*

(1) there exists an MLE of $\boldsymbol{\sigma}$ *in explicit form;*
(2) $\mathcal{V}$ *is a quadratic subspace, that is if* $\mathbf{V} \in \mathcal{V}$ *then* $\mathbf{V}^2 \in \mathcal{V}$.

Moreover, the following theorem holds.

Theorem 4.1 *Let* $\mathbf{y}$ *be a normally distributed random vector, with* $E(\mathbf{y}) = \mathbf{X}\boldsymbol{\beta}$ *and covariance matrix* $\mathbf{V} = \mathbf{V}(\boldsymbol{\sigma})$ *belonging to the linear space,* $\mathcal{V}$*, with unknown vector of linear combinations coefficients,* $\boldsymbol{\sigma}$*. Assume, that*

$$\mathbf{P}_X\mathbf{V} = \mathbf{V}\mathbf{P}_X \quad \textit{and} \quad \mathcal{V} \textit{ is a quadratic subspace.}$$

Then the MLE of $\mathbf{V}$ *is the projection of* $\mathbf{S}$ *onto* $\mathcal{V}$*, where* $\mathbf{S}$ *is the MLE of an unstructured covariance matrix, that is* $\mathbf{S} = \mathbf{Q}_X\mathbf{y}\mathbf{y}'\mathbf{Q}_X$.

Proof From the commutativity condition and $\mathcal{V}$ being a quadratic subspace we have from Proposition 4.1 that the MLE of $\boldsymbol{\sigma}$ has an explicit form. Now, since $\mathbf{V}^2 \in \mathcal{V}$, also $\mathbf{V}^{-1} \in \mathcal{V}$ (cf. Seely [25]), and hence $\mathbf{V}^{-1}$ is a linear combination of the basis vectors of $\mathcal{V}$, with the vector of combination coefficients, say $\boldsymbol{\alpha}$. If k is the dimension of $\mathcal{V}$ and if we denote by $\mathbf{V}_i$ the elements of orthonormal basis of $\mathcal{V}$, then the log-likelihood function can be expressed as

$$\ln L(\alpha) = -\frac{n}{2}\ln(2\pi) + \frac{1}{2}\ln\left|\sum_{i=1}^{k}\alpha_i\mathbf{V}_i\right| - \tfrac{1}{2}\left(\mathbf{y} - \mathbf{X}\widehat{\boldsymbol{\beta}}\right)\left(\sum_{i=1}^{k}\alpha_i\mathbf{V}_i\right)\left(\mathbf{y} - \mathbf{X}\widehat{\boldsymbol{\beta}}\right)',$$

and the derivatives with respect to each α_k gives the likelihood equations of the form

$$\mathrm{Tr}\left[\left(\sum\alpha_i\mathbf{V}_i\right)^{-1}\mathbf{V}_j\right] = \mathrm{Tr}\left(\mathbf{S}\mathbf{V}_j\right).$$

Since $\left(\sum\alpha_i\mathbf{V}_i\right)^{-1} = \sum\sigma_i\mathbf{V}_i$ and from orthonormality $\mathrm{Tr}\left(\mathbf{V}_i\mathbf{V}_j\right) = 0$ for $i \neq j$ and $\mathrm{Tr}\left(\mathbf{V}_i^2\right) = 1$, we obtain

$$\widehat{\sigma}_i = \mathrm{Tr}\left(\mathbf{V}_i\mathbf{S}\right).$$

On the other hand, projection of $\mathbf{S}$ onto the space $\mathcal{V}$ gives

$$\mathbf{V} = \sum\langle\mathbf{S}, \mathbf{V}_i\rangle\mathbf{V}_i = \sum\mathrm{Tr}\left(\mathbf{S}\mathbf{V}_i\right)\mathbf{V}_i$$

which ends the proof. □

Let us consider now model (4.1), whose vectorization gives

$$\mathrm{vec}\,\mathbf{Y} \sim N_{np}\left(\mathrm{vec}(\mathbf{1}_n\boldsymbol{\mu}'), \boldsymbol{\Psi}\otimes\mathbf{I}_n\right). \tag{4.4}$$

Since $\mathrm{vec}(\mathbf{1}_n\boldsymbol{\mu}') = (\mathbf{I}_p \otimes \mathbf{1}_n)\boldsymbol{\mu}$ and $\mathbf{P}_{I_p\otimes 1_n} = \mathbf{I}_p \otimes \mathbf{P}_n$, we have $\mathbf{P}_{I_p\otimes 1_n}(\boldsymbol{\Psi}\otimes\mathbf{I}_n) = \boldsymbol{\Psi}\otimes\mathbf{P}_n = (\boldsymbol{\Psi}\otimes\mathbf{I}_n)\mathbf{P}_{I_p\otimes 1_n}$, and by analogy to variance components models the MLE of $\boldsymbol{\mu}$ does not depend on $\boldsymbol{\Psi}$. Moreover, $(\mathbf{I}_p \otimes \mathbf{1}_n)\,\mathrm{vec}\,\widehat{\boldsymbol{\mu}}' = \mathbf{P}_{I_p\otimes 1_n}\,\mathrm{vec}\,\mathbf{Y}$ which implies $\mathbf{1}_n\widehat{\boldsymbol{\mu}}' = \mathbf{P}_n\mathbf{Y}$ and, after multiplication by $\mathbf{1}_n'$, $n\widehat{\boldsymbol{\mu}}' = \mathbf{1}_n'\mathbf{Y}$ giving finally $\widehat{\boldsymbol{\mu}} = \frac{1}{n}\mathbf{Y}'\mathbf{1}_n$, the sample mean.

Let us assume that $\boldsymbol{\Psi}$ has CS structure, that is

$$\boldsymbol{\Psi} = \boldsymbol{\Psi}_{\mathrm{CS}} = \sigma^2 \left[(1-\varrho)\mathbf{I}_p + \varrho \mathbf{1}_p \mathbf{1}_p' \right]. \tag{4.5}$$

Then the parameter ϱ is a correlation coefficient and should belong to the interval $(-1/(p-1), 1)$ (to assume positive definiteness of $\boldsymbol{\Psi}_{\mathrm{CS}}$). In such a case the maximum likelihood estimators have the form

$$\widehat{\sigma}^2 = \frac{1}{p}\mathrm{Tr}\mathbf{S}, \quad \widehat{\varrho} = \frac{1}{p-1}\left(\frac{\mathbf{1}_p'\mathbf{S}\mathbf{1}_p}{\mathrm{Tr}\mathbf{S}} - 1\right),$$

with $\mathbf{S} = \frac{1}{n}\mathbf{Y}\mathbf{Q}_n\mathbf{Y}'$, the MLE of unstructured $\boldsymbol{\Psi}$.

Note, that the equivalent form of $\boldsymbol{\Psi}_{\mathrm{CS}}$ is

$$\boldsymbol{\Psi}_{\mathrm{CS}} = \alpha \mathbf{I}_p + \beta(\mathbf{1}_p\mathbf{1}_p' - \mathbf{I}_p). \tag{4.6}$$

The MLEs of the parameters α and β can then be easily computed as

$$\widehat{\alpha} = \frac{1}{p}\mathrm{Tr}\mathbf{S}, \qquad \widehat{\beta} = \frac{1}{p(p-1)}\left(\mathbf{1}_p'\mathbf{S}\mathbf{1}_p - \mathrm{Tr}\mathbf{S}\right),$$

with $\mathbf{S}$ defined as above. Moreover, from this parameterization it can be easily seen that the space of CS matrices is linear, it is a quadratic subspace as $\boldsymbol{\Psi}_{CS}^2$ is also a CS matrix and the matrices $\frac{1}{\sqrt{p}}\mathbf{I}_p$ and $\frac{1}{\sqrt{p(p-1)}}\left(\mathbf{1}_p\mathbf{1}_p' - \mathbf{I}_p\right)$ forms the orthonormal basis. Thus, due to Theorem 4.1, the MLEs of α and β should be the projections of $\mathbf{S}$ onto the space generated by $\mathbf{I}_p$ and $\mathbf{1}_p\mathbf{1}_p' - \mathbf{I}_p$. Indeed, the projection has the form

$$\begin{aligned}
\widehat{\boldsymbol{\Psi}}_{\mathrm{CS}} &= \langle \mathbf{S}, \frac{1}{\sqrt{p}}\mathbf{I}_p\rangle \frac{1}{\sqrt{p}}\mathbf{I}_p + \langle \mathbf{S}, \frac{1}{\sqrt{p(p-1)}}\left(\mathbf{1}_p\mathbf{1}_p' - \mathbf{I}_p\right)\rangle \frac{1}{\sqrt{p(p-1)}}\left(\mathbf{1}_p\mathbf{1}_p' - \mathbf{I}_p\right) \\
&= \frac{1}{p}\,\mathrm{Tr}\,\mathbf{S}\cdot\mathbf{I}_p + \frac{1}{p(p-1)}\,\mathrm{Tr}\left[\mathbf{S}\left(\mathbf{1}_p\mathbf{1}_p' - \mathbf{I}_p\right)\right]\left(\mathbf{1}_p\mathbf{1}_p' - \mathbf{I}_p\right) \\
&= \frac{1}{p}\,\mathrm{Tr}\,\mathbf{S}\cdot\mathbf{I}_p + \frac{1}{p(p-1)}\left(\mathbf{1}_p'\mathbf{S}\mathbf{1}_p - \mathrm{Tr}\,\mathbf{S}\right)\left(\mathbf{1}_p\mathbf{1}_p' - \mathbf{I}_p\right).
\end{aligned} \tag{4.7}$$

Another parameterization is the following:

$$\boldsymbol{\Psi}_{\mathrm{CS}} = \gamma \mathbf{P}_p + \delta \mathbf{Q}_p. \tag{4.8}$$

In such a case the MLEs of γ and δ computed from log-likelihood function are of the form

$$\widehat{\gamma} = \frac{1}{p}\mathbf{1}_p'\mathbf{S}\mathbf{1}_p, \qquad \widehat{\delta} = \frac{1}{p(p-1)}\left(p\ \mathrm{Tr}\mathbf{S} - \mathbf{1}_p'\mathbf{S}\mathbf{1}_p\right)$$

with **S** defined as above. Similarly as in parameterization (4.6) it can be seen that the space of CS matrices is a quadratic subspace, and $\mathbf{P}_p$ and $\frac{1}{\sqrt{p-1}}\mathbf{Q}_p$ form an orthonormal basis. Thus, again from Theorem 4.1 it follows that

$$\begin{aligned}\widehat{\boldsymbol{\Psi}}_{\text{CS}} &= \langle \mathbf{S}, \mathbf{P}_p\rangle \mathbf{P}_p + \langle \mathbf{S}, \frac{1}{\sqrt{p-1}}\mathbf{Q}_p\rangle \frac{1}{\sqrt{p-1}}\mathbf{Q}_p \\ &= \text{Tr}(\mathbf{S}\mathbf{P}_p)\mathbf{P}_p + \frac{1}{p-1}\text{Tr}(\mathbf{S}\mathbf{Q}_p)\mathbf{Q}_p \\ &= \frac{1}{p}\mathbf{1}_p'\mathbf{S}\mathbf{1}_p\ \mathbf{P}_p + \frac{1}{p(p-1)}\left(p\ \text{Tr}\,\mathbf{S} - \mathbf{1}_p'\mathbf{S}\mathbf{1}_p\right)\mathbf{Q}_p.\end{aligned}$$

Obviously, all the different parameterizations give the same MLE of the covariance matrix, that is

$$p(p-1)\widehat{\boldsymbol{\Omega}}_{\text{CS}} = \begin{pmatrix} (p-1)\text{Tr}\mathbf{S} & \mathbf{1}_p'\mathbf{S}\mathbf{1}_p - \text{Tr}\mathbf{S} & \mathbf{1}_p'\mathbf{S}\mathbf{1}_p - \text{Tr}\mathbf{S} & \dots & \mathbf{1}_p'\mathbf{S}\mathbf{1}_p - \text{Tr}\mathbf{S} \\ \mathbf{1}_p'\mathbf{S}\mathbf{1}_p - \text{Tr}\mathbf{S} & (p-1)\text{Tr}\mathbf{S} & \mathbf{1}_p'\mathbf{S}\mathbf{1}_p - \text{Tr}\mathbf{S} & \dots & \mathbf{1}_p'\mathbf{S}\mathbf{1}_p - \text{Tr}\mathbf{S} \\ \vdots & \vdots & \ddots & & \vdots \\ \vdots & \vdots & & \ddots & \vdots \\ \mathbf{1}_p'\mathbf{S}\mathbf{1}_p - \text{Tr}\mathbf{S} & \mathbf{1}_p'\mathbf{S}\mathbf{1}_p - \text{Tr}\mathbf{S} & \dots & & (p-1)\text{Tr}\mathbf{S} \end{pmatrix},$$

which follows from invariance property of MLE shown by Zehna [30]. Thus, all the parameterizations can be expressed by e.g., $\widehat{\gamma}$ and $\widehat{\delta}$. Observe, however, that the complexity of procedures to obtain estimators from log-likelihood function differ with the parameterization. Note that (4.5) gives nonlinear likelihood equations while in the remaining parameterizations it is enough to solve linear equations. Moreover, in view of Theorem 4.1 it is not necessary to differentiate log-likelihood, but it is enough to project **S**, the MLE of unstructured $\boldsymbol{\Psi}$, onto the relevant space.

It should be also observed that the basis in parameterization (4.8) is orthogonal in the strong sense, that is $\mathbf{P}_p\mathbf{Q}_p = \mathbf{0}$, and both matrices are idempotent with $\mathbf{P}_p + \mathbf{Q}_p = \mathbf{I}_p$. By analogy to the variance components model, the estimators $\widehat{\gamma}$ and $\widehat{\delta}$ are nonnegative. Due to [25, Lemma 6] the necessary and sufficient condition for the existence of such an orthogonal basis is commutativity and quadracy of this subspace, that is for $\mathbf{V}_1, \mathbf{V}_2 \in \mathscr{V}$ we have $\mathbf{V}_1\mathbf{V}_2 = \mathbf{V}_2\mathbf{V}_1$ and $\mathbf{V}_i \in \mathscr{V}$, $i = 1, 2$. It can be easily seen that the space of CS matrices is a commutative quadratic subspace and hence the existence of a relevant basis follows.

A similar situation can be observed for circular Toeplitz covariance structure, that is

$$\boldsymbol{\Psi}_{\text{CT}} = \alpha_0\mathbf{I}_p + \sum_{i=1}^{\lfloor p/2\rfloor} \alpha_i(\mathbf{H}^i + \mathbf{H}'^i)$$

if p is odd and

$$\boldsymbol{\Psi}_{\mathrm{CT}} = \alpha_0 \mathbf{I}_p + \sum_{i=1}^{p/2-1} \alpha_i (\mathbf{H}^i + \mathbf{H}'^i) + \alpha_{p/2} \mathbf{H}^{p/2}$$

if p is even, where $\mathbf{H}$ is a binary matrix with 1s on the first supradiagonal and $h_{p1} = 1$. Then the MLE of α_i, $i = 0, 1, \ldots, \lfloor p/2 \rfloor$, is an average of elements from ith diagonal of $\mathbf{S}$, the MLE of unstructured $\boldsymbol{\Psi}$; cf. Szatrowski [28, p. 808]. Observe that we can also obtain these estimators from Theorem 4.1, as the space of CT matrices is a quadratic subspace and matrices $\frac{1}{\sqrt{p}} \mathbf{I}_p$ and $\frac{1}{\sqrt{2p}} (\mathbf{H}^i + \mathbf{H}'^i)$ (and additionally $\mathbf{H}^{p/2}$ for even p) form an orthonormal basis. Indeed

$$\begin{aligned}
\widehat{\boldsymbol{\Psi}}_{\mathrm{CT}} &= \langle \mathbf{S}, \frac{1}{\sqrt{p}} \mathbf{I}_p \rangle \frac{1}{\sqrt{p}} \mathbf{I}_p + \sum_{i=1}^{\lfloor p/2 \rfloor} \langle \mathbf{S}, \frac{1}{\sqrt{2p}} (\mathbf{H}^i + \mathbf{H}'^i) \rangle \frac{1}{\sqrt{2p}} (\mathbf{H}^i + \mathbf{H}'^i) \\
&= \frac{1}{p} \operatorname{Tr} \mathbf{S} \cdot \mathbf{I}_p + \frac{1}{2p} \sum_{i=1}^{\lfloor p/2 \rfloor} \operatorname{Tr} \left[\mathbf{S} (\mathbf{H}^i + \mathbf{H}'^i) \right] (\mathbf{H}^i + \mathbf{H}'^i)
\end{aligned}$$

if p is odd and

$$\begin{aligned}
\widehat{\Psi}_{\mathrm{CT}} &= \langle \mathbf{S}, \frac{1}{\sqrt{p}} \mathbf{I}_p \rangle \frac{1}{\sqrt{p}} \mathbf{I}_p \\
&\quad + \sum_{i=1}^{p/2-1} \langle \mathbf{S}, \tfrac{1}{\sqrt{2p}} (\mathbf{H}^i + \mathbf{H}'^i) \rangle \tfrac{1}{\sqrt{2p}} (\mathbf{H}^i + \mathbf{H}'^i) + \langle \mathbf{S}, \tfrac{1}{\sqrt{p}} \mathbf{H}^{p/2} \rangle \tfrac{1}{\sqrt{p}} \mathbf{H}^{p/2} \\
&= \frac{1}{p} \operatorname{Tr} \mathbf{S} \cdot \mathbf{I}_p + \frac{1}{2p} \sum_{i=1}^{p/2-1} \operatorname{Tr} \left[\mathbf{S} (\mathbf{H}^i + \mathbf{H}'^i) \right] (\mathbf{H}^i + \mathbf{H}'^i) + \tfrac{1}{p} \operatorname{Tr} \left(\mathbf{S} \mathbf{H}^{p/2} \right) \mathbf{H}^{p/2}
\end{aligned}$$

if p is even.

Note that the above basis is orthogonal in weak sense, that is the trace of the product of two different basic matrices is equal to zero. Nevertheless, the space of circular Toeplitz matrices can be also expressed in terms of unique idempotent and pairwise orthogonal matrices (cf. Marin, Dhorne [20]) which follows from the commutativity of this space (cf. Seely [25]). Observe that when estimating a structured $\boldsymbol{\Psi}$ in a multivariate model, the parameterization is not important (only orthogonality in weak sense is required), while in the case of variance components models nonnegative estimates of variance components are of interest and thus parameterization by unique idempotent basis should be more convenient. Summing up, commutativity of the space of covariance matrix is not a necessary condition for MLE to be a projection.

The last example which we present in this section is a block diagonal covariance structure,

$$\boldsymbol{\Psi}_d = \begin{pmatrix} \boldsymbol{\Psi}_1 & \mathbf{0} \\ \mathbf{0} & \boldsymbol{\Psi}_2 \end{pmatrix},$$

with $\boldsymbol{\Psi}_i, i = 1, 2$, being symmetric p.d. matrices. Obviously this space is a quadratic subspace, but not commutative. Due to Theorem 4.1 we have that the minimum of the Frobenius norm of $\mathbf{S} - \boldsymbol{\Psi}_d$ is obtained for $\widehat{\boldsymbol{\Psi}}_i$ equal to relevant diagonal blocks of $\mathbf{S}$.

The invariance property of MLE given by Zehna [30] shows that each function of MLE is also the MLE of this function. Thus, if the function transforming unstructured $\boldsymbol{\Psi}$ into the structured one is known (for example the projection), the MLE of the structure can be obtain by application of this function to $\mathbf{S}$, the MLE of unstructured $\boldsymbol{\Psi}$. On the other hand, due to Szatrowski [28, Theorem 4] or due to Proposition 4.1 of this paper, explicit solution exists only under special conditions, and thus projection (which gives explicit form of MLE) can be used if the space of covariance structures is a quadratic subspace. If the space is not a quadratic subspace, projection does not preserve definiteness (cf. Fuglede, Jensen [11]).

As an example, let us consider a banded version of a CT matrix, that is

$$\boldsymbol{\Psi}_{\mathrm{CT}} = \alpha_0 \mathbf{I}_p + \sum_{i=1}^{k} \alpha_i (\mathbf{H}^i + \mathbf{H}'^i),$$

with $k < \lfloor p/2 \rfloor$. Obviously, this space is not quadratic. The projection of $\mathbf{S}$ onto this space has the relevant form but does not need to be p.d. As the example let us assume that $\mathbf{S}$ is a 4×4 symmetric p.d. matrix with 1 on the diagonal and $-\frac{1}{3} < a < 1$ off-diagonal. Then the projection of $\mathbf{S}$ onto the space of tridiagonal circular Toeplitz matrices has the form

$$\begin{pmatrix} 1 & a & 0 & a \\ a & 1 & a & 0 \\ 0 & a & 1 & a \\ a & 0 & a & 1 \end{pmatrix},$$

which is indefinite for $a > 0.5$. A similar situation appears when we consider non-circular Toeplitz matrices (not necessarily banded); cf. Filipiak et al. [9]. One possible function transforming an unstructured covariance matrix into a structured one would be a projection onto the cone of symmetric p.d. structured matrices. However, such a transformation may not need to have explicit representation and, moreover, the cone of symmetric p.d. matrices is not a closed cone and thus the projection onto the closure of the cone may give a nonnegative definite (singular) matrix or a projection onto the open cone may result in an ill-conditioned matrix.

4.3 MLE and Projection Under the Growth Curve Model

Let us consider a growth curve model with

$$E(\mathbf{Y}) = \mathbf{XBZ},$$

with $\mathbf{Y} : n \times p$ being a matrix of observations, $\mathbf{X} : n_1 \times p_1$—matrix of unknown parameters, and $\mathbf{A} : n \times n_1$ and $\mathbf{B}$—known design matrices. Vectorizing of this model gives

$$E(\mathbf{y}) = (\mathbf{Z}' \otimes \mathbf{X})\boldsymbol{\beta},$$

where $\mathbf{y} = \operatorname{vec}\mathbf{X}$ and $\boldsymbol{\beta} = \operatorname{vec}\mathbf{B}$. To be consistent with the variance components model defined in the previous section, let us assume that there is no structure for the covariance matrix, that is $\mathrm{D}(\mathbf{y}) = \boldsymbol{\Sigma}$, with $\boldsymbol{\Sigma}$ being symmetric, positive definite. Then, the MLE of $\boldsymbol{\Sigma}$ is of the form $\mathbf{S} = \mathbf{Q}_{Z'\otimes X}\mathbf{y}\mathbf{y}'\mathbf{Q}_{Z'\otimes X}$.

Assume now that the covariance matrix has the form of a Kronecker product, $\boldsymbol{\Psi} \otimes \mathbf{I}_n$. Then, the MLE of $\boldsymbol{\Psi}$ is of the form

$$\mathbf{S}^* = \frac{1}{n}\left(\mathbf{Y}'\mathbf{Q}_X\mathbf{Y} + \mathbf{Q}_{Z';\tilde{S}^{-1}}\mathbf{Y}'\mathbf{P}_X\mathbf{Y}\mathbf{Q}'_{Z';\tilde{S}^{-1}}\right),$$

where $\tilde{S} = \mathbf{Y}'\mathbf{Q}_X\mathbf{Y}$; cf. Kollo, von Rosen [15], Kopčová, Žežula [16], Ohlson, von Rosen [21].

Let us now consider $\boldsymbol{\Psi} = \boldsymbol{\Psi}_{\mathrm{CS}}$ and assume that $\mathbf{1}_p$ belongs to the column space of $\mathbf{Z}'$. Then the MLEs of CS coefficients can be expressed through the reduced form of the above MLE, that is through

$$\mathbf{S}^{**} = \frac{1}{n}\left(\mathbf{Y}'\mathbf{Q}_X\mathbf{Y} + \mathbf{Q}_{Z'}\mathbf{Y}'\mathbf{P}_X\mathbf{Y}\mathbf{Q}_{Z'}\right),$$

cf. Žežula [31].

Let us now project $\mathbf{S}$, the MLE of completely unstructured $\boldsymbol{\Sigma}$, onto the space of Kronecker product matrices, $\boldsymbol{\Psi} \otimes \mathbf{I}_n$. Then we shall minimize the Frobenius norm of $\mathbf{S} - \boldsymbol{\Psi} \otimes \mathbf{I}_n$. Differentiating this function with respect to $\boldsymbol{\Psi}$ we obtain the equation

$$n \ \operatorname{vec}\boldsymbol{\Psi} = (\mathbf{I}_{p^2} \otimes \operatorname{vec}'\mathbf{I}_n)(\mathbf{I}_p \otimes \mathbf{K}_{p,n} \otimes \mathbf{I}_n)\operatorname{vec}\mathbf{S},$$

with $\mathbf{K}_{p,n}$ being a commutation matrix; cf. Kollo, von Rosen [15]. It can be seen from Filipiak et al. [8, Lemma 2.9] that the right hand side of the above can be expressed as a partial trace of $\mathbf{S}$. Thus we obtain

$$n\,\boldsymbol{\Psi} = \mathrm{PTr}_n\,\mathbf{S}.$$

If we now represent $\mathbf{S}$ in terms of the original observation matrix instead of its vectorized version, and if we use Lemma 2.8 of Filipiak et al. [8], we get

$$\begin{aligned}\mathrm{PTr}_n\,\mathbf{S} &= \mathbf{Y}'\mathbf{Y} - \mathbf{P}_{Z'}\mathbf{Y}'\mathbf{P}_X\mathbf{Y} - \mathbf{Y}'\mathbf{P}_X\mathbf{Y}\mathbf{P}_{Z'} + \mathbf{P}_{Z'}\mathbf{Y}'\mathbf{P}_X\mathbf{Y}\mathbf{P}_{Z'}\\ &= \left(\mathbf{Y}'\mathbf{Q}_X\mathbf{Y} + \mathbf{Q}_{Z'}\mathbf{Y}'\mathbf{P}_X\mathbf{Y}\mathbf{Q}_{Z'}\right) = n\,\mathbf{S}^{**}.\end{aligned}$$

Similarly as in previous section, if we project now $\mathbf{S}^{**}$ onto the space of CS matrices, we obtain the MLE of $\boldsymbol{\Psi}_{CS}$ given e.g., in Žežula [31].

Observe that the space of Kronecker products of symmetric p.d. matrix with identity, $\boldsymbol{\Psi} \otimes \mathbf{I}_n$, is a quadratic subspace. However, there is no commutativity between the space of expectation and $\boldsymbol{\Psi}$. Nevertheless, the MLE of $\boldsymbol{\Psi}$ has an explicit form; cf. Kollo, von Rosen [15]. However, an orthogonal projection of the MLE of unstructured $\boldsymbol{\Sigma}$ onto the space generated by $\boldsymbol{\Psi} \otimes \mathbf{I}_n$ is not an MLE of $\boldsymbol{\Psi} \otimes \mathbf{I}_n$. It shows that commutativity of the expectation space and covariance matrix might be a necessary condition for MLE to be a projection of $\mathbf{S}$ onto the space of covariance structures.

Let us assume now that $\boldsymbol{\Psi}$ is additionally structured as $\boldsymbol{\Psi}_{\mathrm{CS}}$ and that the condition $\mathbf{1}_p \in \mathscr{C}(\mathbf{Z}')$ holds. Then $\boldsymbol{\Psi}_{\mathrm{CS}}$ commutes with $\mathbf{P}_{Z'}$ (since $\mathbf{P}_{Z'}\mathbf{1}_p = \mathbf{1}_p$) and is obviously a quadratic subspace. Therefore, the conditions of Theorem 4.1 are satisfied and

$$\begin{aligned}\widehat{\boldsymbol{\Psi}}_{\mathrm{CS}} &= \langle \mathbf{S}^{**}, \frac{1}{\sqrt{p}}\mathbf{I}_p\rangle \frac{1}{\sqrt{p}}\mathbf{I}_p + \langle \mathbf{S}^{**}, \frac{1}{\sqrt{p(p-1)}}\left(\mathbf{1}_p\mathbf{1}_p' - \mathbf{I}_p\right)\rangle \frac{1}{\sqrt{p(p-1)}}\left(\mathbf{1}_p\mathbf{1}_p' - \mathbf{I}_p\right)\\ &= \frac{1}{p}\,\mathrm{Tr}\,\mathbf{S}^{**}\cdot\mathbf{I}_p + \frac{1}{p(p-1)}\,\mathrm{Tr}\left[\mathbf{S}^{**}\left(\mathbf{1}_p\mathbf{1}_p' - \mathbf{I}_p\right)\right]\left(\mathbf{1}_p\mathbf{1}_p' - \mathbf{I}_p\right)\\ &= \frac{1}{p}\,\mathrm{Tr}\,\mathbf{S}^{**}\cdot\mathbf{I}_p + \frac{1}{p(p-1)}\left(\mathbf{1}_p'\mathbf{S}^{**}\mathbf{1}_p - \mathrm{Tr}\,\mathbf{S}^{**}\right)\left(\mathbf{1}_p\mathbf{1}_p' - \mathbf{I}_p\right),\end{aligned}$$

which gives the same formula as (4.7) with $\mathbf{S}$ replaced by $\mathbf{S}^{**}$.

4.4 Doubly Multivariate Models

In this section we consider the doubly-multivariate model (4.3). For this model $E(\mathbf{Y}) = \mathbf{1}_n\,\mathrm{vec}'\,\mathbf{M}$, with $\mathbf{M} : q \times p$ being a mean matrix of each subject. Assume now that the covariance matrix of $\mathbf{y} = \mathrm{vec}\,\mathbf{Y}$ is $\boldsymbol{\Gamma} \otimes \mathbf{I}_n$. Then, similarly as in model (4.4), the space of expectation commutes with the space of covariance matrices, $\boldsymbol{\Gamma} \otimes \mathbf{I}_n$ generates a quadratic subspace. Hence, due to Theorem 4.1 the MLE of $\boldsymbol{\Gamma}$ can be expressed as the projection of

$$\mathbf{S} = \mathbf{Q}_{I_{qp}\otimes 1_n}\,\mathrm{vec}\,\mathbf{y}\,\mathrm{vec}'\,\mathbf{y}\,\mathbf{Q}_{I_{qp}\otimes 1_n} = \mathrm{vec}(\mathbf{Q}_n\mathbf{Y})\,\mathrm{vec}'(\mathbf{Q}_n\mathbf{Y}),$$

the MLE of unstructured covariance matrix, onto the space of Kronecker products, that is (by analogy to considerations from previous section, with the use of Lemmas 2.9 and 2.8 of Filipiak et al. [8])

$$\widehat{\boldsymbol{\Gamma}} = \frac{1}{n}\mathbf{Y}'\mathbf{Q}_n\mathbf{Y}.$$

Assume now that $\boldsymbol{\Gamma}$ is additionally structured as separability, that is $\boldsymbol{\Gamma} = \boldsymbol{\Psi} \otimes \boldsymbol{\Sigma}$. Note that this space is not linear anymore, and thus, even if $(\boldsymbol{\Psi} \otimes \boldsymbol{\Sigma})^2$ still belongs to the space of separable structures, it is not a quadratic subspace (it is just a quadratic set). In such a case Theorem 4.1 is not applicable. It should be pointed out, however, that the MLEs of $\boldsymbol{\Psi}$ and $\boldsymbol{\Sigma}$ do not exist in explicit form, but they can be expressed in terms of $\widehat{\boldsymbol{\Gamma}} = \mathbf{S}^{\#}$ as

$$q\,\widehat{\boldsymbol{\Psi}} = \mathrm{PTr}_q\left[\left(\mathbf{I}_p \otimes \widehat{\boldsymbol{\Sigma}}^{-1}\right)\mathbf{S}^{\#}\right], \qquad p\,\widehat{\boldsymbol{\Sigma}} = \mathrm{BTr}_q\left[\left(\widehat{\boldsymbol{\Psi}}^{-1} \otimes \mathbf{I}_q\right)\mathbf{S}^{\#}\right]$$

(cf. Filipiak et al. [7]). Obviously, if we assume additional structure for $\boldsymbol{\Psi}$ or $\boldsymbol{\Sigma}$, even the simplest (but different than proportional to identity and $p \neq 1$, $q \neq 1$), separable structure remains nonlinear. It should be noted that to determine the MLEs of unstructured separability components, iterative procedures are usually involved. However, for (without loss of generality) $\boldsymbol{\Psi} = \boldsymbol{\Psi}_{\mathrm{CS}}$, also non-iterative procedure can be proposed to find solutions of likelihood equations; for more details see Filipiak et al. [7].

Assume now that the covariance matrix $\boldsymbol{\Gamma}$ has a BCS structure, that is

$$\boldsymbol{\Gamma}_{\mathrm{BCS}} = \mathbf{I}_p \otimes \boldsymbol{\Gamma}_0 + \left(\mathbf{1}_p\mathbf{1}_p' - \mathbf{I}_p\right) \otimes \boldsymbol{\Gamma}_1,$$

where the $q \times q$ matrices $\boldsymbol{\Gamma}_0$ and $\boldsymbol{\Gamma}_1$ represent covariance matrices of the q response variables at any given time/location point, and between any two time/location points. To ensure positive definiteness of $\boldsymbol{\Gamma}$ it must be assumed that $\boldsymbol{\Gamma}_0$ is a symmetric p.d. matrix, while $\boldsymbol{\Gamma}_1$ only has to be symmetric. To verify positive definiteness as well as to determine the maximum likelihood estimators of $\boldsymbol{\Gamma}$ it is convenient to work with another parameterization, namely

$$\boldsymbol{\Gamma} = \mathbf{Q}_p \otimes \boldsymbol{\Delta}_1 + \mathbf{P}_p \otimes \boldsymbol{\Delta}_2,$$

where $\boldsymbol{\Delta}_1 = \boldsymbol{\Gamma}_0 - \boldsymbol{\Gamma}_1$ and $\boldsymbol{\Delta}_2 = \boldsymbol{\Gamma}_0 + (p-1)\boldsymbol{\Gamma}_1$. To ensure positive definiteness it is enough now to assume positive definiteness of $\boldsymbol{\Delta}_1$ and $\boldsymbol{\Delta}_2$.

Observe that in this case the expectation space commutes with BCS covariance structure, and BCS is a quadratic subspace. Thus, due to Theorem 4.1, the MLE of $\boldsymbol{\Gamma}_{\mathrm{BCS}}$ is an orthogonal projection of $\mathbf{S}^{\#}$ onto the space of BCS matrices. Indeed, if we differentiate the Frobenius norm of $\mathbf{S}^{\#} - \boldsymbol{\Gamma}_{\mathrm{BCS}}$ with respect to $\boldsymbol{\Delta}_1$ and $\boldsymbol{\Delta}_2$, we obtain the equations

$$\begin{cases} (p-1)\,\mathrm{vec}\,\boldsymbol{\Delta}_1 = \left(\mathrm{vec}'\,\mathbf{Q}_p \otimes \mathbf{I}_{q^2}\right)(\mathbf{I}_p \otimes \mathbf{K}_{p,q} \otimes \mathbf{I}_q)\,\mathrm{vec}\,\mathbf{S}^{\#} \\ \qquad\quad \mathrm{vec}\,\boldsymbol{\Delta}_2 = \left(\mathrm{vec}'\,\mathbf{P}_p \otimes \mathbf{I}_{q^2}\right)(\mathbf{I}_p \otimes \mathbf{K}_{p,q} \otimes \mathbf{I}_q)\,\mathrm{vec}\,\mathbf{S}^{\#}, \end{cases}$$

which from Corollary 2.10 and Lemma 2.7 of [8] is equivalent to

$$\begin{cases} (p-1)\widehat{\boldsymbol{\Delta}}_1 = \mathrm{BTr}_q\left[\left(\mathbf{Q}_p \otimes \mathbf{I}_q\right)\mathbf{S}^{\#}\right] \\ \qquad\quad \widehat{\boldsymbol{\Delta}}_2 = \mathrm{BTr}_q\left[\left(\mathbf{P}_p \otimes \mathbf{I}_q\right)\mathbf{S}^{\#}\right]. \end{cases}$$

These estimators are exactly the same as the MLEs obtained by differentiating log-likelihood function, presented e.g., by Roy et al. [24]. Note that the estimators also can be expressed in terms of $\widehat{\boldsymbol{\Gamma}}_0$ and $\widehat{\boldsymbol{\Gamma}}_1$ as

$$\widehat{\boldsymbol{\Gamma}}_0 = \frac{1}{p}\,\mathrm{BTr}_q\,\mathbf{S}^{\#} \quad \text{and} \quad \widehat{\boldsymbol{\Gamma}}_1 = \frac{1}{p(p-1)}\left(\mathrm{BSum}_q\mathbf{S}^{\#} - \mathrm{BTr}_q\,\mathbf{S}^{\#}\right).$$

Summing up, to obtain the MLEs of blocks of BCS covariance matrix it is enough to average all diagonal and all off-diagonal blocks of $\mathbf{S}^{\#}$, which is also confirmed by Szatrowski [28].

It is worth noting that the separable structure $\boldsymbol{\Psi}_{\mathrm{CS}} \otimes \boldsymbol{\Sigma}$ is a special case of BCS, with the additional parameter space restriction $\boldsymbol{\Gamma}_1 = \varrho\boldsymbol{\Gamma}_0$.

In a similar way we may determine the MLE of block circular Toeplitz (BCT) structure, in the which scalars α_i of CT structure are replaced by matrices $\mathbf{A}_i$ and the standard matrix products by Kronecker products (with the order of components as in BCS structure). Note that the space of BCT structures is a quadratic subspace and thus again Theorem 4.1 can be applied. Again, by projecting $\mathbf{S}^{\#}$ onto the space of BCT matrices we obtain that the MLE of $\mathbf{A}_i$ from BCT structure can be expressed as the average of blocks of $\mathbf{S}^{\#}$ from ith supra- and sub-diagonal.

Observe that the space of covariance structures of the form

$$\sum_{i=1}^{k} \mathbf{M}_i \otimes \boldsymbol{\Sigma}_i, \tag{4.9}$$

where $\mathbf{M}_i$ are symmetric idempotent and pairwise orthogonal matrices ($\mathbf{M}_i$ are orthogonal in strong sense) with $\sum_{i=1}^{k} \mathbf{M}_i = \mathbf{I}_p$ and $\boldsymbol{\Sigma}_i$ are symmetric matrices, is a quadratic subspace. Therefore, as long as commutativity of expectation space and the space of such covariance structures holds, Theorem 4.1 can be applied. Note that the mentioned structures, BCS and BCT are of the form (4.9). Moreover, when for example the three-level multivariate model with doubly BCS covariance structure is considered (see e.g., Kozioł et al. [17]), the MLE of covariance matrix can be obtained by a simple projection of a covariance matrix being an MLE of unstructured covariance matrix onto the space of doubly BCS structures.

Note that the structures of the form (4.9) are quadratic, but not commutative, subspaces, unless the order of symmetric $\mathbf{\Sigma}_i$ is equal to 1. It follows from the fact that two arbitrary symmetric matrices are not necessarily commuting. We may also see the lack of commutativity in a basis. The basis of (4.9) is a Kronecker product of every $\mathbf{M}_i$, $i = 1, \ldots, k$, and the basis of symmetric matrices, which is not orthogonal in the strong sense. From Lemma 6 of Seely [25] these two conditions are equivalent. Summing up, it can be seen that commutativity of the space of covariance structures is not a necessary condition for the existence of explicit representation of the MLE of a covariance matrix.

Acknowledgments The authors thank Stefan Banach International Mathematical Center, Institute of Mathematics, Polish Academy of Sciences, for supporting the Meetings of International Research Group on Multivariate and Mixed Linear Models during which this issue was studied.

This research is partially supported by Scientific Activities No. 04/43/SBAD/0115 (Katarzyna Filipiak, Mateusz John).

References

[1] Cui, X., Li, C., Zhao, J., Zeng, L., Zhang, D., Pan, J.: Covariance structure regularization via Frobenius norm discrepancy. Linear Algebra Appl. **510**, 124–145 (2016)
[2] Devijver, E., Gallopin, M.: Block-diagonal covariance selection for high-dimensional Gaussian graphical models. J. Am. Stat. Assoc. **113**, 306–314 (2018)
[3] Filipiak, K., Klein, D.: Estimation of parameters under a generalized growth curve model. J. Multivariate Anal. **158**, 73–86 (2017)
[4] Filipiak, K., Klein, D.: Approximation with a Kronecker product structure with one component as compound symmetry or autoregression. Linear Algebra Appl. **559**, 11–33 (2018)
[5] Filipiak, K., Klein, D., Roy. A.: Score test for a separable covariance structure with the first component as compound symmetric correlation matrix. J. Multivariate Anal. **150**, 105–124 (2016)
[6] Filipiak, K., Klein, D., Roy. A.: A comparison of likelihood ratio tests and Rao's score test for three separable covariance matrix structures. Biometr. J. **59**, 192–215 (2017)
[7] Filipiak, K., Klein, D., Mokrzycka, M.: Estimators comparison of separable covariance structure with one component as compound symmetry matrix. Electron. J. Linear Algebra **33**, 83–98 (2018)
[8] Filipiak, K., Klein, D., Vojtková, V.: The properties of partial trace and block trace operators of partitioned matrices. Electron. J. Linear Algebra **33**, 3–15 (2018)
[9] Filipiak, K., Markiewicz, A., Mieldzioc, A., Sawikowska, A.: On projection of a positive definite matrix on a cone of nonnegative definite Toeplitz matrices. Electron. J. Linear Algebra **33**, 74–82 (2018)
[10] Filipiak, K., Klein, D., Markiewicz, A., Mokrzycka, M.: Approximation with a Kronecker product structure via entropy loss function (2020, Submitted)
[11] Fuglede, B., Jensen, S.T.: Positive projections of symmetric matrices and Jordan algebras. Expo. Math. **31**, 295–303 (2013)
[12] Gilson, M., Dahmen, D., Moreno-Bote, R., Insabato, A., Helias, M.: The covariance perceptron: A new framework for classification and processing of time series in recurrent neural networks (2019). bioRxiv. https://doi.org/10.1101/562546

[13] Hao, C., Liang, Y., Mathew, T.: Testing variance parameters in models with a Kronecker product covariance structure. Stat. Probab. Lett. **118**, 182–189 (2016)
[14] John, M., Mieldzioc, A.: The comparison of the estimators of banded Toeplitz covariance structure under the high-dimensional multivariate model. Commun. Stat. Simul. Comput. (2019). https://doi.org/10.1080/03610918.2019.1614622
[15] Kollo, T., von Rosen, D.: Advanced Multivariate Statistics with Matrices. Springer, Dordrecht (2005)
[16] Kopčová, V., Žežula, I.: On intraclass structure estimation in the growth curve model. Stat. Papers (2017). https://doi.org/10.1007/s00362-017-0973-x
[17] Kozioł, A., Roy, A., Zmyślony, R., Leiva, R., Fonseca, M.: Best unbiased estimates for parameters of three-level multivariate data with doubly exchangeable covariance structure. Linear Algebra Appl. **535**, 87–104 (2017)
[18] Lin, L., Higham, N.J., Pan, J.: Covariance structure regularization via entropy loss function. Comput. Stat. Data Anal. **72**, 315–327 (2014)
[19] Lu, N., Zimmerman, D.: The likelihood ratio test for a separable covariance matrix. Stat. Probab. Lett. **73**, 449–457 (2005)
[20] Marin, J-M., Dhorne, T.: Linear Toeplitz covariance structure models with optimal estimators of variance components. Linear Algebra Appl. **354**, 195–212 (2002)
[21] Ohlson, M., von Rosen, D.: Explicit estimators of parameters in the growth curve model with linearly structured covariance matrices. J. Multivariate Anal. **101**, 1284–1295 (2010)
[22] Olkin, I.: Testing and estimation for structures which are circularly symmetric in blocks. In: Proceedings of the Symposium on Multivariate Analysis, 183–195, Dalhousie, Nova Scotia (1973)
[23] Roy, A., Zmyślony, R., Fonseca, M., Leiva, R.: Optimal estimation for doubly multivariate data in blocked compound symmetric covariance structure. J. Multivariate Anal. **144**, 81–90 (2016)
[24] Roy, A., Filipiak, K., Klein, D.: Testing a block exchangeable covariance matrix. Statistics **52**, 393–408 (2018)
[25] Seely, J.: Quadratic subspaces and completeness. Ann. Math. Stat. **42**, 710–721 (1971)
[26] Srivastava, M., von Rosen, T., von Rosen, D.: Models with a Kronecker product covariance structure: estimation and testing. Math. Methods Stat. **17**, 357–370 (2008)
[27] Srivastava, M., von Rosen, T., von Rosen, D.: Estimation and testing in general multivariate linear models with Kronecker product covariance structure. Sankhya **71-A**, 137–163 (2009)
[28] Szatrowski, T.H.: Necessary and sufficient conditions for explicit solutions in the multivariate normal estimation problem for patterned means and covariances. Ann. Stat. **8**, 802–810 (1980)
[29] Szczepańska, A., Hao, C., Liang, Y., von Rosen, D.: Estimation equations for multivariate linear models with Kronecker structured covariance matrices. Commun. Stat. Theory Methods **46**, 7902–7915 (2017)
[30] Zehna, P.W.: Invariance of maximum likelihood estimators. Ann. Math. Stat. **37**, 744 (1966)
[31] Žežula, I.: Special variance structures in the growth curve model. J. Multivariate Anal. **97**, 606–618 (2006)

Chapter 5
Growth Curve Model with Orthogonal Covariance Structure

Miguel Fonseca and Martin Singull

Abstract In this paper we study the Growth Curve model with orthogonal covariance structure and derive estimators for all parameters. The orthogonal covariance structure is a generalization of many known structures, e.g., compound symmetry covariance structure. Hence, we compare our estimators with earlier results found in the literature.

5.1 Introduction

Multivariate repeated measures data, which correspond to multiple measurements that are taken over time on each unit or subject, are common in various applications such as medicine, pharmacy, environmental research, engineering, business and finance etc. The Growth Curve model is a generalization of the multivariate analysis of variance model (MANOVA) and can be used to model repeated measures data with a mean which, for example, is polynomial in time.

The Growth Curve model introduced by Potthoff and Roy [15] has been extensively studied over many years. It is a generalized multivariate analysis of variance model (GMANOVA) which belongs to the curved exponential family. The mean structure for the Growth Curve model is bilinear in contrary to the ordinary MANOVA model where it is linear. For more details about the Growth Curve model see e.g., [7, 24].

For the Growth Curve model, when no assumption about the covariance matrix was made, Potthoff and Roy [15] originally derived a class of weighted estimators

M. Fonseca (✉)
Centro de Matemática e Aplicações (CMA), FCT, UNL, Almada, Portugal

Departamento de Matemática, FCT, UNL, Almada, Portugal
e-mail: fmig@fct.unl.pt

M. Singull
Department of Mathematics, Linköping University, Linköping, Sweden
e-mail: martin.singull@liu.se

© Springer Nature Switzerland AG 2020

T. Holgersson, M. Singull (eds.), *Recent Developments in Multivariate and Random Matrix Analysis*, https://doi.org/10.1007/978-3-030-56773-6_5

for the mean parameter matrix. Khatri [4] extended this result and showed that the maximum likelihood estimator is also a weighted estimator. Under a certain covariance structure, Rao [17] and Reinsel [19] have shown that the unweighted estimator is also the maximum likelihood estimator (MLE). Furthermore, Chinchilli and Walter Jr [2] have derived the likelihood ratio test for this type of covariance matrix.

In the MANOVA model, when dealing with measurements on k equivalent psychological tests, Wilks [26] was one of the first to consider patterned covariance matrices. A covariance matrix with equal diagonal elements and equal off-diagonal elements, i.e., a so-called compound symmetry (also known as uniform or intraclass) covariance structure, was studied. For the Growth Curve model, Khatri [5] derived the likelihood ratio test for the compound symmetry covariance structure and Lee [9] considered the model for prediction purposes.

More recently, Klein and Žežula [6], Žežula [31], and Ye and Wang [27] used the unbiased estimating equations to find estimators for the compound symmetry covariance structure, and Kopčová and Žežula [8] compared these with the MLEs for the structure under normality. Also, Srivastava and Singull [22] considered testing compound symmetry covariance structure under a Growth Curve model in high dimension.

Furthermore, Ohlson and von Rosen [14] gave a general estimation procedure for the class of linearly structured covariance matrices which can be applied to the compound symmetry structure. The idea in [14] was applied to the extended Growth Curve model by Nzabanita et al. [12, 13].

Closely connected to the intraclass covariance structure, with positive covariances, is the random effects covariance structure studied by Rao [16, 18], Reinsel [19, 20], Ware [25] and recently by Srivastava and Singull [21] for a parallel profile model. The random-effect covariance structure has also been considered for the mixed MANOVA-GMANOVA models and the Extended Growth Curve models, see e.g., [28–30].

In this paper we will study a model class with orthogonal covariance structure first proposed for univariate models in [10, 11] and further developed, for example, in [1, 3]. Several of the above discussed structures are special cases of the orthogonal covariance structure.

The paper is organized as follows. In Sect. 5.2 the Growth Curve model is given together with the maximum likelihood estimators without any extra structure on the design or covariance matrices. In Sect. 5.3 the Growth Curve model with orthogonal covariance structure is presented and in Sect. 5.4 the maximum likelihood estimators for this model is derived. In Sect. 5.5 an orthogonal structure on the design matrices is considered and in Sect. 5.6, the compound symmetry covariance structure, which is a special cases of the orthogonal covariance structure, is considered for the Growth Curve model and the result is compared with the one by Khatri [5] for the same model.

5.2 Growth Curve Model and Maximum Likelihood Estimators

As discussed in the introduction the Growth Curve (GC) model is a generalization of the multivariate analysis of variance (MANOVA) model and can be used to model repeated measures data. The GC model is given as

$$Y_{p\times n} = A_{p\times s}B_{s\times q}C_{q\times n} + E_{p\times n},$$

consisting of n multivariate observations with for p repeated measurements. The design matrices $A_{p\times s}$ and $C_{q\times n}$ are known, and $B_{s\times q}$ is unknown. The matrix $E_{p\times n}$ has a multivariate normal distribution

$$E_{p\times n} \sim \mathrm{MN}_{p,n}(0_{p\times n}, \Sigma_{p\times p}, \Psi_{n\times n}),$$

meaning that it has a null expected value matrix, the covariance matrix for elements in each column is $\Sigma_{p\times p}$ and the covariance matrix for elements in each row is $\Psi_{n\times n}$. Both of these matrices are symmetric positive definite and unknown. Hence, this model has an observations matrix with distribution

$$Y \sim \mathrm{MN}_{p,n}(ABC, \Sigma, \Psi).$$

For many situations we assume that we have independent observations, i.e., $\Psi = I$, where I is the identity matrix. Under this model the maximum likelihood estimator (MLE) for the mean parameter B in the GC model is given by many authors, e.g., see [4, 7, 24], and equals

$$\widehat{B} = \left(A'S^{-1}A\right)^{-} A'S^{-1}XC'\left(CC'\right)^{-} + (A')^{o}Z_1 + A'Z_2C^{o\prime}, \tag{5.1}$$

where Z_1 and Z_2 are arbitrary matrices and the matrix S is given by

$$S = X\left(I - C'\left(CC'\right)^{-}C\right)X' = XP_{C'}^{\perp}X', \tag{5.2}$$

where $P_{C'}^{\perp} = I - P_{C'}$ and $P_{C'} = C'\left(CC'\right)^{-}C$. We have used the notation A^{o} for any matrix of full rank which is spanning the orthogonal complement to $\mathscr{C}(A)$, i.e., $\mathscr{C}(A^{o}) = \mathscr{C}(A)^{\perp}$. If A and C are of full rank, i.e., $\mathrm{rank}(A) = s$ and $\mathrm{rank}(C) = q$, the estimator in (5.1) reduces to the unique estimator

$$\widehat{B} = \left(A'S^{-1}A\right)^{-1} A'S^{-1}XC'\left(CC'\right)^{-1}.$$

Furthermore, the maximum likelihood estimator of Σ is given by

$$n\widehat{\Sigma} = \left(X - A\widehat{B}C\right)\left(X - A\widehat{B}C\right)' = S + \left(I - P_{A,S}\right) X P_{C'} X' \left(I - P_{A,S}\right),$$

where $P_{A,S} = A\left(A'S^{-1}A\right)^{-} A'S^{-1}$ which always is unique since one can show that $A\widehat{B}C$ is unique for all choices of g-inverse.

5.3 GC Model with Orthogonal Covariance Structure

The above definition for the GC model is general, with no restrictions on the unknown parameters. Despite the fact that inference for the fixed effects B is established (see [7]), estimation for the covariance components Σ and Ψ can be difficult and unstructured covariance components may not adequately reflect the nature of data. In light of this problem, an often used model class is the orthogonal covariance structure (OCS) first proposed for univariate models in [10, 11] and further developed, for example, in [1, 3]. In this model class, a covariance matrix V is expressed as

$$V = \sum_j P_j \otimes \Sigma_j,$$

where $\otimes$ is the Kronecker product and P_j are orthogonal projection matrices such that

$$\sum_j P_j = I$$

and Σ_j are positive definite arbitrary matrices which are symmetric. Therefore, assume now that

$$\Sigma = \sum_{i=1}^{w} Q_i \otimes \Gamma_i, \tag{5.3}$$

where $\Gamma_i : p_0 \times p_0$ are unknown matrices, Q_i are idempotent and symmetric known matrices with $\mathrm{rank}(Q_i) = p_i$ such that

$$Q_i Q_{i'} = 0, \quad i \neq i';$$

$$\sum_{i=1}^{w} Q_i = I_{p_*},$$

and matrix H_i, of size $p_* \times p_i$, is such that $H_i H_i' = Q_i$ and $H_i' H_i = I_{p_i}$ with $p_* p_0 = p$.

Similarly, let

$$\Psi = \sum_{j=1}^{m} R_j \otimes \Delta_j,$$

with $\Delta_j : n_0 \times n_0$ are unknown matrices, R_j are idempotent and symmetric known matrices with $\text{rank}(R_j) = q_j$ such that

$$R_j R_{j'} = 0, \quad j \neq j';$$
$$\sum_{j=1}^{m} R_j = I_{n_*},$$

and matrix G_j, of size $n_* \times n_j$, is such that $G_j G_j' = R_j$ and $G_j' G_j = I_{n_j}$ with $n_* n_0 = n$. To guarantee estimability of the covariance structure, assume that

$$w\frac{p_0(p_0+1)}{2} + m\frac{n_0(n_0+1)}{2} + sq < \frac{pn(pn+1)}{2}$$

and that Ψ is a correlation matrix ($\psi_{ll} = 1, l = 1, \ldots, n$).

Consider now the matrices $Y_{ij} = (H_i' \otimes I_{p_0}) Y (G_j \otimes I_{n_0})$, $i = 1 \ldots, w$, $j = 1, \ldots, m$. Then,

$$Y_{ij} \sim \text{MN}\big((H_i' \otimes I_{p_0}) ABC(G_j \otimes I_{n_0}), I_{p_i} \otimes \Gamma_i, I_{n_j} \otimes \Lambda_j\big), \tag{5.4}$$

which are independent when $i \neq i'$ or $j \neq j'$.

Consider also the matrices

$$H_* = \begin{bmatrix} H_1 \otimes I_{p_0} & \cdots & H_w \otimes I_{p_0} \end{bmatrix},$$
$$G_* = \begin{bmatrix} G_1 \otimes I_{n_0} & \cdots & G_m \otimes I_{n_0} \end{bmatrix},$$

and the transformation

$$Y_* = H_*' Y G_* \sim \text{MN}(A_* B C_*, \Gamma_*, \Delta_*), \tag{5.5}$$

with the new design matrices

$$A_* = H_*' A, \tag{5.6}$$
$$C_* = C G_*, \tag{5.7}$$

and covariance matrices

$$\Gamma_* = H_*'\Sigma H_* = \mathrm{diag}(I_{p_1} \otimes \Gamma_1, \ldots, I_{p_w} \otimes \Gamma_w), \tag{5.8}$$

$$\Delta_* = G_*'\Psi G_* = \mathrm{diag}(I_{n_1} \otimes \Delta_1, \ldots, I_{n_m} \otimes \Delta_m). \tag{5.9}$$

In the next section will we find the MLEs for the parameters of model (5.5).

5.4 MLEs for GCM with OCS

In this paper we will assume a special form for matrices A and C. Matrices A and C will actuate on multivariate observations homogeneously.

Assume that A and C take the forms $A = A_0 \otimes I_{p_0}$ and $C = C_0 \otimes I_{n_0}$, where A_0 and C_0 are arbitrary matrices. Then the log-likelihood of Y_{ij} given in (5.4) is

$$\begin{aligned} L_{ij} &= n \log\left(|I_{p_i} \otimes \Gamma_i|\right) + p \log\left(|I_{n_j} \otimes \Delta_j|\right) \\ &\quad + \mathrm{tr}\big((I_{p_i} \otimes \Gamma_i^{-1})(Y_{ij} - A_i B C_j)(I_{n_j} \otimes \Delta_j^{-1})(Y_{ij} - A_i B C_j)'\big), \end{aligned}$$

with

$$\begin{aligned} A_i &= (H_i' A_0) \otimes I_{p_0}; \\ C_j &= (C_0 G_j) \otimes I_{n_0}. \end{aligned}$$

Furthermore, the log-likelihood of Y_* given in (5.5) follows as

$$\begin{aligned} L_* &= n \log\left(|\Gamma_*|\right) + p \log\left(|\Delta_*|\right) + \mathrm{tr}\big(\Gamma_*^{-1}(Y_* - A_* B C_*)\Delta_*^{-1}(Y_* - A_* B C_*)'\big) \\ &= \sum_{i=1}^{w} L_{i*} \\ &= \sum_{i=1}^{w} n_* n_0 \log\left(|I_{p_i} \otimes \Gamma_i|\right) + p_i p_0 \log\left(|\Delta_*|\right) \\ &\qquad + \mathrm{tr}\big((I_{p_i} \otimes \Gamma_i^{-1})(Y_{i*} - A_i B C_*)\Delta_*^{-1}(Y_{i*} - A_i B C_*)'\big), \end{aligned}$$

where

$$Y_{i*} = \begin{bmatrix} Y_{i1} & \cdots & Y_{im} \end{bmatrix}. \tag{5.10}$$

Note that

$$\begin{aligned} P_{A_i} &= P_{H_i' A_0} \otimes I_{p_0} = \left(H_i' A_0 (A_0' A_0)^{-1} A_0' H_i\right) \otimes I_{p_0}, \\ P_{C_j'} &= P_{C_0' G_i'} \otimes I_{p_0} = \left(G_j' C_0' (C_0 C_0')^{-1} C_0 G_j\right) \otimes I_{p_0}. \end{aligned}$$

Without loss of generality, take Ψ (and consequently Δ_*) to be known. Hence,

$$
\begin{aligned}
\sum_{i=1}^{w} L_{i*} &= \sum_{i=1}^{w} n_* n_0 \log\left(|I_{p_i} \otimes \Gamma_i|\right) + p_i p_0 \log\left(|\Delta_*|\right) \\
&\quad + \mathrm{tr}\left((I_{p_i} \otimes \Gamma_i^{-1})(Y_{i*} - A_i B C_*)\Delta_*^{-1}(Y_{i*} - A_i B C_*)'\right) \\
&= \sum_{i=1}^{w} n_* n_0 \log\left(|I_{p_i} \otimes \Gamma_i|\right) + p_i p_0 \log\left(|\Delta_*|\right) \\
&\quad + \mathrm{tr}\left((I_{p_i} \otimes \Gamma_i^{-1})(Y_{i*}\Delta_*^{-\frac{1}{2}} - A_i B C_* \Delta_*^{-\frac{1}{2}})()'\right) \\
&= \sum_{i=1}^{w} n_* n_0 \log\left(|I_{p_i} \otimes \Gamma_i|\right) + p_i p_0 \log\left(|\Delta_*|\right) \\
&\quad + \mathrm{tr}\left((I_{p_i} \otimes \Gamma_i^{-1})(Y_{i*}\Delta_*^{-\frac{1}{2}} P_{C'_*}^{\perp})()'\right) \\
&\quad + \mathrm{tr}\left((I_{p_i} \otimes \Gamma_i^{-1})(Y_{i*}\Delta_*^{-\frac{1}{2}} P_{C'_*} - A_i B C_* \Delta_*^{-\frac{1}{2}})()'\right) \\
&= \sum_{i=1}^{w} n_* n_0 \log\left(|I_{p_i} \otimes \Gamma_i|\right) + p_i p_0 \log\left(|\Delta_*|\right) \\
&\quad + \mathrm{tr}\Big((I_{p_i} \otimes \Gamma_i^{-1})\big((H_i \otimes I_{p_0})(Y_{i*}\Delta_*^{-\frac{1}{2}} P_{C_*}^{\perp})()' \\
&\quad + (P_{A_*}^{\perp} Y_{i*}\Delta_*^{-\frac{1}{2}} P_{C'_*})()'(H_i' \otimes I_{p_0})\big)\Big) \\
&\quad + \mathrm{tr}\left((I_{p_i} \otimes \Gamma_i^{-1})(P_{A_*} Y_{i*}\Delta_*^{-\frac{1}{2}} P_{C'_*} - A_i B C_* \Delta_*^{-\frac{1}{2}})()'\right),
\end{aligned}
$$

where $(A)(A)' = (A)()'$ for simplicity and $P_X^{\perp} = I - P_X$. Defining

$$
\begin{aligned}
S_i &= \sum_{l=1}^{p_i} \left(\delta'_{p^*}(\bar{p}(i-1)+l) \otimes I_{p_0}\right)\big((H_i \otimes I_{p_0})(Y_{i*}\Delta_*^{-\frac{1}{2}} P_{C_*}^{\perp})()' \\
&\quad + (P_{A_*}^{\perp} Y_{i*}\Delta_*^{-\frac{1}{2}} P_{C'_*})(P_{A_*}^{\perp} Y_{i*}\Delta_*^{-\frac{1}{2}} P_{C'_*})'(H_i' \otimes I_{p_0})\big)\left(\delta'_{p^*}(\bar{p}(i-1)+l) \otimes I_{p_0}\right) \\
&= \sum_{l=1}^{p_i} \left(\delta'_{p^*}(\bar{p}(i-1)+l) \otimes I_{p_0}\right)(Q_i \otimes I_{p_0})\Big(Y\left(\Psi^{-1} - \Psi^{-1}C'(C'\Psi^{-1}C)^{-1}C\Psi^{-1}\right)Y' \\
&\quad + P_A^{\perp} Y \Psi^{-1} C'(C'\Psi^{-1}C)^{-1} C\Psi^{-1} Y' P_A^{\perp}\Big)(Q_i \otimes I_{p_0})\left(\delta_{p^*}(\bar{p}(i-1)+l) \otimes I_{p_0}\right),
\end{aligned}
\tag{5.11}
$$

where $\boldsymbol{\delta}_{p^*}(l)$ is a vector of size p^* with the lth component equal to 1 and all others equal to 0, and

$$\bar{p}(i) = \sum_{k=1}^{i} p_i \quad \text{and} \quad \bar{p}(0) = 0,$$

which is invariant for the choice of eigenvectors of Q_i, one gets

$$\sum_{i=1}^{w} L_{i*} = \sum_{i=1}^{w} n_* n_0 \log\left(|I_{p_i} \otimes \Gamma_i|\right) + p_i p_0 \log\left(|\Delta_*|\right) + \mathrm{tr}(\Gamma_i^{-1} S_i)$$
$$+ \mathrm{tr}\left((I_{p_i} \otimes \Gamma_i^{-1})(P_{A_*} Y_{i*} \Delta_*^{-\frac{1}{2}} P_{C'_*} - A_i B C_* \Delta_*^{-\frac{1}{2}})()'\right).$$

For estimation of Γ_i minimization of L_{i*} gives the partial estimates

$$\widehat{B}_i = (A'_* A_*)^{-1} A'_* Y_{i*} \Delta_*^{-1} C'_* (C_* \Delta_*^{-1} C'_*)^{-1},$$
$$\widehat{\Gamma}_i = \frac{1}{p_i n} S_i,$$

where A_*, C_*, Δ_* and Y_{i*} are respectively given in (5.6), (5.7), (5.9) and (5.10), and S_i is given in (5.11). A full estimate for B can be obtained by the GLS as

$$\begin{aligned}\widehat{B} &= (A'_* \widehat{\Gamma}_*^{-1} A_*)^{-1} \widehat{\Gamma}_*^{-1} Y_* \Delta_*^{-1} C'_* (C_* \Delta_*^{-1} C'_*)^{-1} \\ &= (A' \widehat{\Sigma}^{-1} A)^{-1} \widehat{\Sigma}^{-1} Y \Psi^{-1} C' (C \Psi^{-1} C')^{-1}, \end{aligned} \tag{5.12}$$

with $\widehat{\Gamma}_*$ as the estimator of Γ_* in (5.8) as

$$\widehat{\Gamma}_* = \mathrm{diag}(I_{p_1} \otimes \widehat{\Gamma}_1, \ldots, I_{p_w} \otimes \widehat{\Gamma}_w).$$

Furthermore, the estimator of the covariance matrix Σ with OCS (5.3) is given by

$$\widehat{\Sigma} = \sum_{i=1}^{w} Q_i \otimes \widehat{\Gamma}_i. \tag{5.13}$$

To get an estimate for Δ_i, if unknown, one calculates the former estimates taking $\Delta_i = I_{n_0}$, repeats the procedure for the data matrix Y'_*, taking $\Gamma_* = \widehat{\Gamma}_*$, and then the initial step again with $\Delta_* = \widehat{\Delta}_*$. This cycle is continued using the estimates of the previous iteration until all the estimates converge. This procedure is known as the flip-flop algorithm and more details can be found in [23].

We summarize the above results in a theorem.

Theorem 5.1 *Consider the GC model $Y \sim MN_{p,n}(ABC, \Sigma, \Psi)$, with OCS for Σ, i.e., Σ is given as* (5.3) *and Ψ known. Assume that the design matrices A and C take the forms $A = A_0 \otimes I_{p_0}$ and $C = C_0 \otimes I_{n_0}$. Then the MLEs for B and Σ is given by* (5.12) *and* (5.13).

5.5 Orthogonal Model

In this section, an orthogonal model is considered, i.e., taking the orthogonal covariance structure (5.3) and the additional properties

$$P_A(Q_i \otimes \Gamma_i) = (Q_i \otimes \Gamma_i)P_A, \quad i = 1, \ldots, w, \tag{5.14}$$

$$P_{C'}(R_j \otimes \Delta_j) = (R_j \otimes \Delta_j)P_{C'}, \quad j = 1, \ldots, m. \tag{5.15}$$

One way to ensure this is restricting A and C to take the forms

$$A = A_0 \otimes I_{p_0},$$
$$C = C_0 \otimes I_{n_0},$$

so that

$$P_{A_0}Q_i = Q_i P_{A_0}, \quad i = 1, \ldots, w,$$
$$P_{C_0'}R_j = R_j P_{C_0'}, \quad j = 1, \ldots, m.$$

Then, the log-likelihood will be given as

$$L_* = n \log\left(|\Gamma_*|\right) + p \log\left(|\Delta_*|\right) + \operatorname{tr}\left(\Gamma_*^{-1}(Y_* - A_*BC_*)\Delta_*^{-1}(Y_* - A_*BC_*)'\right)$$

and we can see that

$$\begin{aligned}
&\operatorname{tr}\left(\Gamma_*^{-1}(Y_* - A_*BC_*)\Delta_*^{-1}(Y_* - A_*BC_*)'\right)\\
&= \operatorname{tr}\left(\Gamma_*^{-1}(Y_* - A_*BC_*)P_{C_*'}\Delta_*^{-1}P_{C_*'}(Y_* - A_*BC_*)'\right)\\
&\quad + \operatorname{tr}\left(\Gamma_*^{-1}(Y_* - A_*BC_*)(I_n - P_{C_*'})\Delta_*^{-1}(I_n - P_{C_*'})(Y_* - A_*BC_*)'\right)\\
&= \operatorname{tr}\left(\Gamma_*^{-1}(P_{A_*}Y_*P_{C_*'} - A_*BC_*)\Delta_*^{-1}(P_{A_*}Y_*P_{C_*'} - A_*BC_*)'\right)\\
&\quad + \operatorname{tr}\left(\Gamma_*^{-1}Y_*(I_n - P_{C_*'})\Delta_*^{-1}(I_n - P_{C_*'})Y_*'\right)\\
&\quad + \operatorname{tr}\left(\Gamma_*^{-1}(I_p - P_{A_*})Y_*P_{C_*'}\Delta_*^{-1}P_{C_*'}Y_*'(I_p - P_{A_*})\right)
\end{aligned}$$

$$= \operatorname{tr}\big(\Gamma_*^{-1}(P_{A_*}Y_*P_{C_*'} - A_*BC_*)\Delta_*^{-1}(P_{A_*}Y_*P_{C_*'} - A_*BC_*)'\big)$$

$$+ \sum_{i=1}^{w}\sum_{l=1}^{p_i} \operatorname{tr}\Big(\Gamma_i^{-1}\Big(\delta_{p^*}\big(\bar{\mathrm{p}}(i-1)+l\big)\otimes I_{p_0}\Big)\big(Y_*(I_n - P_{C_*'})\Delta_*^{-1}(I_n - P_{C_*'})Y_*'$$

$$+ (I_p - P_{A_*})Y_*P_{C_*'}\Delta_*^{-1}P_{C_*'}Y_*'(I_p - P_{A_*})\big)\Big(\delta_{p^*}'\big(\bar{\mathrm{p}}(i-1)+l\big)\otimes I_{p_0}\Big)\Big),$$

where

$$\bar{\mathrm{p}}(i) = \sum_{k=1}^{i} p_i \quad \text{and} \quad \bar{\mathrm{p}}(0) = 0.$$

The estimator for Γ_i is

$$\widehat{\Gamma}_i = \frac{1}{np_i}\sum_{l=1}^{p_i}\Big(\delta_{p^*}\big(\bar{\mathrm{p}}(i-1)+l\big)\otimes I_{p_0}\Big)\big(Y_*(I_n - P_{C_*'})\Delta_*^{-1}(I_n - P_{C_*'})Y_*'$$

$$+ (I_p - P_{A_*})Y_*P_{C_*'}\Delta_*^{-1}P_{C_*'}Y_*'(I_p - P_{A_*})\big)\Big(\delta_{p^*}'\big(\bar{\mathrm{p}}(i-1)+l\big)\otimes I_{p_0}\Big) \tag{5.16}$$

and the estimator for B is

$$\widehat{B} = (A_*'A_*)^{-1}A_*'Y_*C_*'(C_*C_*')^{-1}, \tag{5.17}$$

where Y_*, A_* and C_* are given in (5.5), (5.6), and (5.7), respectively. Again, we summarize the results in a theorem.

Theorem 5.2 *Given the orthogonal GC model $Y \sim MN_{p,n}(ABC, \Sigma, \Psi)$, with known Ψ, OCS for Σ, i.e., $\Sigma = \sum_{i=1}^{w} Q_i \otimes \Gamma_i$, as in* (5.3)*, and* (5.14) *and* (5.15) *are fulfilled, then the MLEs for B and Γ_i are given by* (5.17) *and* (5.16).

5.6 GC Model with Compound Symmetry Covariance Structure

5.6.1 General Model

Consider the particular case of the Growth Curve model, the compound symmetry model, i.e.,

$$Y \sim \mathrm{MN}_{t,r}\big(ABC, \Sigma_{t\times t}, I_r\big), \tag{5.18}$$

consisting of r multivariate observations with t components, and where

$$\Sigma = (\omega_0 - \omega_1)I_t + \omega_1 t J_t = \gamma_1 K_t + \gamma_2 J_t = \gamma_1 I_t + (\gamma_2 - \gamma_1)J_t, \tag{5.19}$$

with $J_s = s^{-1}\mathbf{1}_s\mathbf{1}_s'$. Putting $K_s = I_s - J_s$, one gets

$$Q_1 = K_t; \quad \Gamma_1 = \gamma_1 = \omega_0 - \omega_1; \quad H_1 = M_t;$$
$$Q_2 = J_t; \quad \Gamma_2 = \gamma_2 = \omega_0 - (t-1)\omega_1; \quad H_2 = L_t;$$
$$A_* = \begin{bmatrix} M_t' \\ L_t' \end{bmatrix} A;$$
$$C_* = C;$$
$$I - P_{A_*} = \begin{bmatrix} M_t' \\ L_t' \end{bmatrix} A(A'A)^{-1}A' \begin{bmatrix} M_t & L_t \end{bmatrix}$$

denoting $L_t = (\sqrt{t})^{-1}\mathbf{1}_t$ and M_t as a $t \times (t-1)$ matrix such that $M_t M_t' = K_t$ and $M_t' M_t = I_{t-1}$ with

$$p_0 = 1; \quad p_1 = t-1; \quad p_2 = 1;$$
$$n_0 = 1; \quad n_1 = r.$$

Using Theorem 5.1 we have the following corollary.

Corollary 5.1 *Given the GC model* $Y \sim \mathrm{MN}_{t,r}(ABC, \Sigma_{t\times t}, I_r)$, *with compound symmetry covariance structure* $\Sigma = \gamma_1 K_t + \gamma_2 J_t$, *the MLEs for the variance components* γ_1 *and* γ_2 *are given by*

$$\widehat{\gamma_1} = \frac{1}{(t-1)n}\mathrm{tr}\Big(K_t\big(S + (I - P_A)Y P_{C'} Y'(I - P_A)\big)K_t\Big),$$
$$\widehat{\gamma_2} = \frac{1}{tn}\mathbf{1}_t'\big(S + (I - P_A)Y P_{C'} Y'(I - P_A)\big)\mathbf{1}_t,$$

where $S = XP_C^{\perp}X'$ *as in* (5.2).

One can easily conclude that the estimators $\widehat{\gamma_1}$ and $\widehat{\gamma_2}$ in Corollary 5.1 coincide with the estimators derived by Khatri [5].

5.6.2 Orthogonal Model

Consider the particular case of the Growth Curve model, the compound symmetry model with orthogonal mean structure

$$Y \sim \mathrm{MN}_{t,r}\big((\mathbf{1}_t \otimes \mathbf{1}_r')\beta,\ \Sigma_{t\times t},\ I_r\big), \tag{5.20}$$

which are r multivariate observations with t components, and where

$$\Sigma = (\omega_0 - \omega_1)I_t + \omega_1 t J_t = \gamma_1 K_t + \gamma_2 J_t, \tag{5.21}$$

with $K_t = I_t - J_t$ and $J_t = t^{-1}\mathbf{1}_t\mathbf{1}_t'$. Translating the above model in the notation of the previous sections, we have

$$A = \mathbf{1}_t; \quad B = \beta; \quad C = \mathbf{1}_r'.$$

Following the notation above, one gets

$$\begin{aligned}
Q_1 &= K_t; \quad \Gamma_1 = \gamma_1 = \omega_0 - \omega_1; \quad H_1 = M_t;\\
Q_2 &= J_t; \quad \Gamma_2 = \gamma_2 = \omega_0 - (t-1)\omega_1; \quad H_2 = L_t;\\
A_* &= (\mathbf{0}_{p-1}' \ \sqrt{p})';\\
C_* &= \mathbf{1}_n';\\
I - P_{A_*} &= diag(1, \ldots, 1, 0)
\end{aligned}$$

denoting $L_t = (\sqrt{t})^{-1}\mathbf{1}_t$ and M_t as a $t \times (t-1)$ matrix such that $M_t M_t' = K_t$ and $M_t' M_t = I_{t-1}$ with

$$\begin{aligned}
p_0 &= 1; \quad p_1 = t-1; \quad p_2 = 1;\\
n_0 &= 1; \quad n_1 = r.
\end{aligned}$$

Through some computation,

$$Y_*(I_n - P_{C_*'})\Delta_*^{-1}(I_n - P_{C_*'})Y_*' = Y_* K_r Y_*' = \begin{bmatrix} M_t' \\ L_t' \end{bmatrix} Y K_r Y' \begin{bmatrix} M_t & L_t \end{bmatrix}$$

and

$$\begin{aligned}(I_t - P_{A_*})Y_* P_{C_*'} \Delta_*^{-1} P_{C_*'} Y_*'(I_t - P_{A_*}) &= (I_t - P_{A_*})Y_* J_r Y_*'(I_t - P_{A_*}) \\ &= (I_t - P_{A_*}) \begin{bmatrix} M_t' \\ L_t' \end{bmatrix} Y J_r Y' \begin{bmatrix} M_t & L_t \end{bmatrix} (I_t - P_{A_*}) \\ &= \begin{bmatrix} M_t' \\ 0_t' \end{bmatrix} Y J_r Y' \begin{bmatrix} M_t & 0_t \end{bmatrix}.\end{aligned}$$

Hence, following Theorem 5.2 we have

$$\begin{aligned}\widehat{\gamma_1} &= \frac{1}{r(t-1)} \sum_{l=1}^{t-1} \delta_t'(l) \left(\begin{bmatrix} M_t' \\ L_t' \end{bmatrix} Y K_r Y' \begin{bmatrix} M_t & L_t \end{bmatrix} + \begin{bmatrix} M_t' \\ 0_t' \end{bmatrix} Y J_r Y' \begin{bmatrix} M_t & 0_t \end{bmatrix} \right) \delta_t(l) \\ &= \frac{1}{r(t-1)} \mathrm{tr}(M_t' Y K_r Y' M_t + M_t' Y_* J_r Y_*' M_t) = \frac{1}{r(t-1)} \mathrm{tr}(M_t' Y Y' M_t) \\ &= \frac{1}{r(t-1)} \mathrm{tr}(Y' K_t Y), \\ \widehat{\gamma_2} &= \frac{1}{r} \delta_t'(t) \left(\begin{bmatrix} M_t' \\ L_t' \end{bmatrix} Y K_r Y' \begin{bmatrix} M_t & L_t \end{bmatrix} + \begin{bmatrix} M_t' \\ 0_t' \end{bmatrix} Y J_r Y' \begin{bmatrix} M_t & 0_t \end{bmatrix} \right) \delta_t(t) \\ &= \frac{1}{r} \mathrm{tr}\left(L_t' Y K_r Y' L_t\right) = \frac{1}{rt} \mathrm{tr}\left(\mathbf{1}_t' Y K_r Y' \mathbf{1}_t\right)\end{aligned}$$

and

$$\widehat{\beta} = \frac{1}{rt} \mathbf{1}_t' Y \mathbf{1}_r.$$

We can conclude with a final corollary.

Corollary 5.2 *Given the compound symmetry model with orthogonal mean structure $Y \sim \mathrm{MN}_{t,r}\left((\mathbf{1}_t \otimes \mathbf{1}_r')\beta, \Sigma, I_r\right)$, with $\Sigma = \gamma_1 K_t + \gamma_2 J_t$, the estimators for the variance components γ_1 and γ_2 are given by*

$$\widehat{\gamma_1} = \frac{1}{r(t-1)} tr(Y' K_t Y),$$

$$\widehat{\gamma_2} = \frac{1}{rt} tr\left(\mathbf{1}_t' Y K_r Y' \mathbf{1}_t\right)$$

and for the mean parameter $\widehat{\beta} = \dfrac{1}{rt} \mathbf{1}_t' Y \mathbf{1}_r$.

One can easily conclude that the estimators given in Corollary 5.2 just are the OLS estimators.

Acknowledgments For Miguel Fonseca, this work was partially supported by the Fundação para a CiŘncia e a Tecnologia (Portuguese Foundation for Science and Technology) through the project UIDB/00297/2020 (Centro de Matemática e Aplicações).

The authors would also like to thank the reviewer of this paper for several valuable and helpful suggestions and comments.

References

1. Andersson, S.A.: The lattice structure of orthogonal linear models and orthogonal variance component models. Scand. J. Stat. **17**, 287–319 (1990)
2. Chinchilli, V.M., Walter Jr., H.C.: A likelihood ratio test for a patterned covariance matrix in a multivariate Growth Curve model. Biometrics **40**, 151–156 (1984)
3. Fonseca, M., Mexia, J.T., Zmyzlony, R.: Binary operations on Jordan algebras and orthogonal normal models. Linear Algebra Appl. **417**, 75–86 (2006)
4. Khatri, C.G.: A note on a MANOVA model applied to problems in Growth Curve. Ann. Inst. Stat. Math. **18**, 75–86 (1966)
5. Khatri, C.G.: Testing some covariance structures under a Growth Curve model. J. Multivariate Anal. **3**(1), 102–116 (1973)
6. Klein, D., Žežula, I.: Orthogonal decompositions in Growth Curve models. Acta Comment. Univ. Tartu. Math. **14**, 35–44 (2010)
7. Kollo, T., von Rosen, D.: *Advanced Multivariate Statistics with Matrices*. Mathematics and Its Applications, vol. 579. Springer, Dordrecht (2005)
8. Kopčová, V., Žežula, I.: On intraclass structure estimation in the Growth Curve model. Stat. Papers **61**, 1085–1106 (2020)
9. Lee, J.C.: Prediction and estimation of growth curves with special covariance structures. J. Am. Stat. Assoc. **83**, 432–440 (1988)
10. Nelder, J.A.: The analysis of randomized experiments with orthogonal block structure. I. Block structure and the null analysis of variance. Proc. R. Soc. Lond. A Math. Phys. Sci. **283**, 147–162 (1965)
11. Nelder, J.A.: The analysis of randomized experiments with orthogonal block structure. II. Treatment structure and the general analysis of variance. Proc. R. Soc. Lond. A Math. Phys. Sci. **283**, 163–178 (1965)
12. Nzabanita, J., Singull, M., von Rosen, D.: Estimation of parameters in the extended Growth Curve model with a linearly structured covariance matrix. Acta Comment. Univ. Tartu. Math. **16**(1), 13–32 (2012)
13. Nzabanita, J., von Rosen, D., Singull, M.: Extended GMANOVA model with a linearly structured covariance matrix. Math. Methods Stat. **24**(4), 280–291 (2015)
14. Ohlson, M., von Rosen, D.: Explicit estimators of parameters in the Growth Curve model with linearly structured covariance matrices. J. Multivariate Anal. **101**(5), 1284–1295 (2010)
15. Potthoff, R.F., Roy, S.: A generalized multivariate analysis of variance model useful especially for growth curve problems. Biometrika **51**(3–4), 313–326 (1964)
16. Rao, C.R.: The theory of least squares when the parameters are stochastic and its application to the analysis of growth curves. Biometrika **53**(3/4), 447–458 (1965)
17. Rao, C.R.: Least squares theory using an estimated dispersion matrix and its application to measurement of signals. In: LeCam, L.M., Neyman, J. (eds.) Proceedings of the Fifth Berkeley Symposium on Mathematial Statistics and Probability, vol. I. University of California Press, Berkeley, pp. 355–372 (1967)
18. Rao, C.R.: Simultaneous estimation of parameters in different linear models and applications to biometric problems. Biometrics **31**(2), 545–554 (1975)
19. Reinsel, G.: Multivariate repeated measurement or Growth Curve models with multivariate random-effects covariance structure. J. Am. Stat. Assoc. **77**(377), 190–195 (1982)

20. Reinsel, G.: Estimation and prediction in a multivariate random effects generalized linear model. J. Am. Stat. Assoc. **79**, 406–414 (1984)
21. Srivastava, M.S., Singull, M.: Profile analysis with random-effects covariance structure. J. Japan Stat. Soc. **42**(2), 145–164 (2012)
22. Srivastava, M.S., Singull, M.: Testing sphericity and intraclass covariance structures under a Growth Curve model in high dimension. Commun. Stat. Simul. Comput. **46**(7), 5740–5751 (2017)
23. Srivastava, M.S., von Rosen, T., von Rosen, D.: Models with a Kronecker product covariance structure: estimation and testing. Math. Methods Stat. **17**(4), 357–370 (2009)
24. von Rosen, D.: Bilinear Regression Analysis. Lecture Notes in Statistics, vol. 220. Springer, New York (2018)
25. Ware, J.H.: Linear models for the analysis of longitudinal studies linear models for the analysis of longitudinal studies. Am. Stat. **39**(2), 95–101 (1985)
26. Wilks, S.S.: Sample criteria for testing equality of means, equality of variances, and equality of covariances in a normal multivariate distribution. Ann. Math. Stat. **17**, 257–281 (1946)
27. Ye, R.-D., Wang, S.-G.: Estimating parameters in extended Growth Curve models with special covariance structures. J. Stat. Plan. Inference **139**(8), 2746–2756 (2009)
28. Yokoyama, T.: Statistical inference on some mixed MANOVA-GMANOVA models with random effects. Hiroshima Math. J. **25**(3), 441–474 (1995)
29. Yokoyama, T.: Extended Growth Curve models with random-effects covariance structures. Commun. Stat. Theory Methods **25**(3), 571–584 (1996)
30. Yokoyama, T.: Tests for a family of random-effects covariance structures in a multivariate Growth Curve model. J. Stat. Plan. Inference **65**(2), 281–292 (1997)
31. Žežula, I.: Special variance structures in the Growth Curve model. J. Multivariate Anal. **97**(3), 606–618 (2006)

Chapter 6
Holonomic Gradient Method for the Cumulative Distribution Function of the Largest Eigenvalue of a Complex Wishart Matrix with Noncentrality Matrix of Rank One

Yuta Fukasawa and Akimichi Takemura

Abstract We apply the holonomic gradient method for evaluation of the cumulative distribution function of the largest eigenvalue of a complex Wishart matrix with the noncentrality matrix of rank one. This problem appears in the context of Rician fading of multiple-input/multiple-output (MIMO) wireless communications systems. We also give a brief survey of the use of multivariate analysis in wireless communication and the holonomic gradient method for statistical problems of performance evaluation in wireless communication.

6.1 Introduction

In this paper we present an application of the holonomic gradient method for a distributional problem in the study of MIMO wireless communication systems. In statistical evaluation of wireless communication systems the complex normal distribution and the complex Wishart distribution are used, because the signal and the noise are modeled by complex normal distribution. Some notable results on complex normal distribution have been discovered in wireless communication area, because statisticians usually study the real normal distribution and the real Wishart distribution. In this paper we look at the result of Kang and Alouini [7] on the distribution of the largest eigenvalue of a complex noncentral Wishart distribution and discuss how to compute the distribution function by the holonomic gradient method.

Y. Fukasawa
The Risk Data Bank of Japan, Limited, Tokyo, Japan
e-mail: yuta_fukasawa@forev.co.jp

A. Takemura (✉)
Faculty of Data Science, Shiga University, Shiga, Japan
e-mail: a-takemura@biwako.shiga-u.ac.jp

© Springer Nature Switzerland AG 2020
T. Holgersson, M. Singull (eds.), *Recent Developments in Multivariate and Random Matrix Analysis*, https://doi.org/10.1007/978-3-030-56773-6_6

The holonomic gradient method (HGM) was proposed in [9] and since then it was applied to many distributional problems in statistics. In particular we were successful to apply HGM to the evaluation of the hypergeometric function ${}_1F_1$ of a matrix argument [2]. We also applied the holonomic gradient method to wireless communication in [13] and [14]. This paper gives another application of HGM to a problem in MIMO wireless communication.

In Sect. 6.2 we give a brief summary of the statistical setup of MIMO problems and in Sect. 6.3 we give a short introduction to HGM. In Sect. 6.4 we apply HGM to the result of Kang and Alouini [7]. Our main result is the Pfaffian equations in Sect. 6.4.6. Finally in Sect. 6.5 we show results of some numerical experiments.

6.2 Multivariate Distribution Theory Used in the Study of MIMO Wireless Communication Systems

Nowadays multiple antennas are commonly used both in the transmitting and receiving stations because of practical advantages. Suppose that there are m transmitting antennas sending an m-dimensional signal x and there are n antennas receiving an n-dimensional signal y. Then the relations between x and y are modeled as $y = Ax + v$, where A is an $n \times m$ "channel matrix" and v is a n-dimensional noise vector following a complex normal distribution. In a particular communication between a transmitting station and a receiving station, the channel matrix A is considered to be fixed. However when the communications are repeated with different conditions, A is considered a random matrix and usually the elements of A are also assumed to follow complex normal distribution. Then the performance of the MIMO system is evaluated in terms of the distribution of $A^{\mathrm{H}}A$, where A^{H} denotes the conjugate transpose of A. Under standard assumptions $A^{\mathrm{H}}A$ follows a complex noncentral Wishart distribution. Hence the study of complex Wishart distribution is important for MIMO communication. In [11] and [12] we studied Schur complement of a complex Wishart matrix, which is important for MIMO "Zero-Forcing detection" analysis. In this paper we study the largest eigenvalue of a complex Wishart matrix, which is important for MIMO "Maximal-Ratio Combining" analysis.

Note that the distributions of real noncentral Wishart distribution is difficult, often involving zonal polynomials and hypergeometric function of a matrix argument (e.g., [6, 8, 15]). Results are often easier in the complex case than the real case and this may be one reason that complex noncentral Wishart distribution is studied in wireless communication literature. For example, the Schur polynomial for the complex case has an explicit determinantal expression, whereas the zonal polynomial for the real case does not have such an expression.

For general introduction to MIMO systems see [3] and [5].

6.3 The Holonomic Gradient Method in the Study of Multivariate Distribution Theory

The holonomic gradient method can be used to distributional problems when the density function is "holonomic". A holonomic function satisfies a system of partial differential equations with rational function coefficients. For example, the bivariate normal density

$$f(x_1, x_2) = c \times \exp(-\frac{x_1^2}{2} - \frac{x_2^2}{2})$$

is holonomic because it satisfies the following differential equations

$$\partial_{x_1} f(x_1, x_2) = -x_1 f(x_1, f_2), \qquad \partial_{x_2} f(x_1, x_2) = -x_2 f(x_1, f_2).$$

In this paper we write these two equations as

$$(\partial_{x_1} + x_1) \bullet f = 0, \quad (\partial_{x_2} + x_2) \bullet f = 0,$$

where $\bullet$ means the operation of the differential operator $\partial_{x_i} + x_i$, $i = 1, 2$.

More precisely a smooth function $f(x_1, \ldots, x_n) = f(\boldsymbol{x})$ is holonomic if there are differential operators $l_1, \ldots, l_N$ $(N \geq n)$ such that $\{l_1, \ldots, l_N\}$ generates "a zero-dimensional left ideal" in the ring of differential operators with rational function coefficients. When f is holonomic, then by Gröbner basis computation of $l_1, \ldots, l_N$, we can find a finite set (called "standard monomials") of lower order partial derivatives of f

$$\mathbf{f} = (f, \partial_{x_1} f, \ldots, \partial_{x_n} f, \partial_{x_1}^2 f, \ldots, \partial_{x_1}^{\alpha_1} \cdots \partial_{x_n}^{\alpha_n} f)', \tag{6.1}$$

such that

$$\partial_{x_i} \mathbf{f} = Q_i \mathbf{f} + \mathbf{u}_i, \quad i = 1, \ldots, n, \tag{6.2}$$

where Q_i is a square matrix with rational function elements and $\mathbf{u}_i$ is a vector of smooth functions. (6.2) are called Pfaffian equations. When $\mathbf{u}_i = 0$ for all i the equations are homogeneous and when $\mathbf{u}_i \neq 0$ for some i they are inhomogeneous. In this paper we need inhomogeneous equations. Our main result of this paper in Sect. 6.4.6 is the Pfaffian equations for a function appearing in the cumulative distribution function of the largest eigenvalue of a complex Wishart matrix with noncentrality matrix of rank one.

The Gröbner basis computation is very useful, because the set of standard monomials in (6.1) can be algorithmically obtained from the set of differential operators $\{l_1, \ldots, l_N\}$ by the Gröbner basis computation. Often this derivation is too difficult for hand computation and we need to use a symbolic computer algebra system. In the problem studied in this paper, we can compute the set of standard monomials without resorting to a symbolic computer algebra system, although our derivation is somewhat lengthy. Once the set of standard monomials is obtained, the Pfaffian system (6.2) can be mechanically obtained by applying the differential operators $\{l_1, \ldots, l_N\}$ to the set of standard monomials, although this process may also be cumbersome. Given the Pfaffian equations, we can numerically integrate (6.2) to obtain the value of $\mathbf{f}$ at any point $\boldsymbol{x}$ starting from a convenient initial point $\boldsymbol{x}_0$.

See Chapter 6 of [4] for a comprehensive treatment of the Gröbner basis theory of the ring of differential operators with rational function coefficients. Section 2 of [10] gives a readable introduction to HGM for statisticians.

6.4 Distribution of the Largest Eigenvalue of Non-central Wishart Matrix

In this section we apply the holonomic gradient method to the result of Kang and Alouini [7] for the case of the complex Wishart matrix with noncentrality matrix of rank one. First in Sect. 6.4.1 we setup some notation needed for stating our results. In Sect. 6.4.2 we present the result of Kang and Alouini [7]. In Sect. 6.4.3 we transform their result so that the noncentrality is separated into a function $F(x, \lambda; n, m, h)$ defined in Sect. 6.4.1. In Sects. 6.4.4 and 6.4.5 we obtain the set of differential operators for $F(x, \lambda; n, m, h)$ and its integral $g(x, \lambda; n, m, h)$ with respect to x. Finally in Sect. 6.4.6 we obtain the Pfaffian equations for $g(x, \lambda; n, m, h)$, which is the main result of this paper.

6.4.1 Some Notation

First we setup some notation used in this section. The incomplete gamma function is denoted by

$$\gamma(k, x) = \int_0^x y^{k-1} e^{-y} dy.$$

We define $F(x, \lambda; n, m, h)$ by

$$F(x, \lambda; n, m, h) = \sum_{k=0}^{\infty} \frac{\lambda^k e^{-\lambda}}{(k+n-1)!(k+m-1)!} \gamma(n+k+m-1-h, x) \tag{6.3}$$

and its integral by

$$g(x, \lambda; n, m, h) = \int_0^x F(y, \lambda; n, m, h)dy. \tag{6.4}$$

We also define the following $m \times m$ determinant.

$$H_m(n_1, \dots, n_m; x) = \begin{vmatrix} \gamma(n_1 + m - 1, x) & \gamma(n_2 + m - 1, x) & \dots & \gamma(n_m + m - 1, x) \\ \gamma(n_1 + m - 2, x) & \gamma(n_2 + m - 2, x) & \dots & \gamma(n_m + m - 2, x) \\ \vdots & \vdots & \dots & \vdots \\ \gamma(n_1, x) & \gamma(n_2, x) & \dots & \gamma(n_m, x) \end{vmatrix}. \tag{6.5}$$

6.4.2 The Result of Kang and Alouini

Here we present the result of Kang and Alouini [7]. Let A denote an $n \times m$ random matrix with independent rows having the m-dimensional complex normal distribution $\mathscr{C}\mathscr{N}_m(\mu_i, \Sigma)$, $i = 1, \dots, n$, with common covariance matrix Σ. Let $M := (\mu_1, \dots, \mu_n)'$ denote the mean matrix. In the following we assume $m \leq n$. The case $m > n$ can be treated by interchanging the roles of m and n. In this paper we consider the distribution of the largest eigenvalue of $S = \Sigma^{-1} A^{\mathrm{H}} A$. Note that by the left-multiplication of $A^{\mathrm{H}} A$ by Σ^{-1}, we are effectively considering the special case of $\Sigma = I$.

Denote the nonzero eigenvalues of S by $0 < \phi_1 < \phi_2 < \cdots < \phi_m$. Kang and Alouini [7] derived the cumulative distribution function $\Pr(\phi_m \leq x)$ of the largest eigenvalue ϕ_m when $\Sigma^{-1} M^{\mathrm{H}} M$ have m distinct nonzero eigenvalues $0 < \lambda_1 < \lambda_2 < \cdots < \lambda_m$. As a corollary Kang and Alouini [7] derived the cumulative distribution function of ϕ_m for the case of $1 = \operatorname{rank} \Sigma^{-1} M^{\mathrm{H}} M = \operatorname{rank} M$, by a limiting argument. The rank one case is of some practical importance as discussed in [14].

We studied the holonomic gradient method for the case of m distinct eigenvalues in [16] for $m = 2$ and $m = 3$. Unfortunately obtaining results for general dimension m is very difficult. Hence in this paper we study the case of rank one M and derive results applicable for general m. For the rank one case we denote $0 = \lambda_1 = \cdots = \lambda_{m-1}$, $\lambda_m = \lambda > 0$ and the cumulative distribution function as $F(x, \lambda) = \Pr(\phi_m \leq x)$. Then the result by Kang and Alouini [7] is given as follows.

Proposition 6.1 ([7]) *Suppose that $\Sigma^{-1} M^{\mathrm{H}} M$ has only one nonzero eigenvalue λ. Then the cumulative distribution function $F(x, \lambda)$ of the largest eigenvalue ϕ_m of S is given by*

$$F(\lambda, x) = \frac{1}{\Gamma(n - m + 1) \prod_{l=1}^{m-1} \Gamma(n - l)\Gamma(m - l)} \times \frac{e^{-\lambda}}{\lambda^{m-1}} |\Psi(x, \lambda)|, \tag{6.6}$$

where $\Psi(x, \lambda)$ *is an* $m \times m$ *matrix with the elements*

$$\Psi(x, \lambda)_{i1} = \int_0^x y^{n-i} e^{-y} {}_0F_1(n-m+1; \lambda y) dy, \qquad i = 1, \ldots, m,$$

$$\Psi(x, \lambda)_{ij} = \gamma(n+m-i-j+1, x), \qquad i = 2, \ldots, m, \quad j = 1, \ldots, m.$$

6.4.3 Transformation of the Cumulative Distribution Function

Here we transform (6.6) into the form where the noncentrality is separated into $F(x, \lambda; n, m, h)$ of (6.3). Then we can apply HGM to $F(x, \lambda; n, m, h)$, independent of the dimension m.

In (6.6), the elements of the first column $\Psi(x, \lambda)$ involves the hypergeometric function ${}_0F_1(\cdot; \cdot)$ and it is difficult to directly derive the differential equation satisfied by $F(x, \lambda)$. We first rewrite $F(x, \lambda)$ as an infinite series. By the definition of ${}_0F_1(\cdot; \cdot)$,

$$\begin{aligned}
\Psi(x, \lambda)_{i1} &= \int_0^x y^{n-i} e^{-y} \sum_{k=0}^{\infty} \frac{1}{(n-m+1)_k} \frac{\lambda^k}{k!} y^k dy \\
&= \sum_{k=0}^{\infty} \frac{1}{(n-m+1)_k} \frac{\lambda^k}{k!} \int_0^x y^{n-i+k} e^{-y} dy \\
&= \sum_{k=0}^{\infty} \frac{1}{(n-m+1)_k} \frac{\lambda^k}{k!} \gamma(n-i+1+k, x).
\end{aligned}$$

Hence by linearity of the determinant we have

$$\begin{aligned}
|\Psi(x, \lambda)| &= \begin{vmatrix}
\sum_{k=0}^{\infty} \frac{1}{(n-m+1)_k} \frac{\lambda^k}{k!} \gamma(n+k, x) & \gamma(n+m-2, x) \ldots & \gamma(n, x) \\
\sum_{k=0}^{\infty} \frac{1}{(n-m+1)_k} \frac{\lambda^k}{k!} \gamma(n-1+k, x) & \gamma(n+m-3, x) \ldots & \gamma(n-1, x) \\
\vdots & \vdots \quad \ldots & \vdots \\
\sum_{k=0}^{\infty} \frac{1}{(n-m+1)_k} \frac{\lambda^k}{k!} \gamma(n-m+1+k, x) & \gamma(n-1, x) \ \ldots & \gamma(n-m+1, x)
\end{vmatrix} \\
&= \sum_{k=0}^{\infty} \frac{1}{(n-m+1)_k} \frac{\lambda^k}{k!} \\
&\qquad \times \begin{vmatrix}
\gamma(n+k, x) & \gamma(n+m-2, x) \ldots & \gamma(n, x) \\
\gamma(n-1+k, x) & \gamma(n+m-3, x) \ldots & \gamma(n-1, x) \\
\vdots & \vdots \quad \ldots & \vdots \\
\gamma(n-m+1+k, x) & \gamma(n-1, x) \ \ldots & \gamma(n-m+1, x)
\end{vmatrix}.
\end{aligned}$$

Note that the determinant for $k = 0, \dots, m-2$ vanishes, because in these matrices the first column is equal to the $(m-k)$-th column. Therefore by replacing the running index k by $k-m-1$, $|\Psi(x,\lambda)|$ is written as

$$|\Psi(x,\lambda)| = \sum_{k=m-1}^{\infty} \frac{1}{(n-m+1)_k} \frac{\lambda^k}{k!} H_m(n+k, n-1, \dots, n-m+1; x)$$

and $F(x,\lambda)$ is expressed as

$$F(x,\lambda) = C \sum_{k=0}^{\infty} \frac{\lambda^k e^{-\lambda}}{(k+n-1)!(k+m-1)!} H_m(n+k, n-1, \dots, n-m+1; x), \tag{6.7}$$

where $C = 1/(\prod_{l=1}^{m-1} \Gamma(n-l)\Gamma(m-l))$.

We now want to separate $n+k$ in H_m and replace H_m by H_{m-1} We express the incomplete gamma function in the integral form and again use the linearity of the determinant on the right-hand side of (6.7). Then we obtain

$$\begin{aligned}
&H_m(n+k, n-1, \dots, n-m+1; x) \\
&= \begin{vmatrix} \int_0^x y_1^{n+m-2+k} e^{-y_1} dy_1 & \gamma(n+m-2,x) & \dots & \gamma(n,x) \\ \int_0^x y_1^{n+m-3+k} e^{-y_1} dy_1 & \gamma(n+m-3,x) & \dots & \gamma(n-1,x) \\ \vdots & \vdots & \dots & \vdots \\ \int_0^x y_1^{n-1+k} e^{-y_1} dy_1 & \gamma(n-1,x) & \dots & \gamma(n-m+1,x) \end{vmatrix} \\
&= \int_0^x y_1^{n+k-1} e^{-y_1} \begin{vmatrix} y_1^{m-1} & \gamma(n+m-2,x) & \dots & \gamma(n,x) \\ y_1^{m-2} & \gamma(n+m-3,x) & \dots & \gamma(n-1,x) \\ \vdots & \vdots & \dots & \vdots \\ 1 & \gamma(n-1,x) & \dots & \gamma(n-m+1,x) \end{vmatrix} dy_1 \\
&= \int_0^x \cdots \int_0^x y_1^{n+k-1} y_2^{n-2} \dots y_m^{n-m} e^{-y_1-\dots-y_m} \begin{vmatrix} y_1^{m-1} & y_2^{m-1} & \dots & y_m^{m-1} \\ y_1^{m-2} & y_2^{m-2} & \dots & y_m^{m-2} \\ \vdots & \vdots & \dots & \vdots \\ 1 & 1 & \dots & 1 \end{vmatrix} dy_1 \dots dy_m \\
&= \int_0^x \cdots \int_0^x y_1^{n+k-1} y_2^{n-2} \dots y_m^{n-m} e^{-y_1-\dots-y_m} \prod_{i<j} (y_i - y_j) dy_1 \dots dy_m.
\end{aligned}$$

We expand the Vandermonde determinant with respect to y_1 as

$$\begin{aligned}\prod_{1\le i<j\le m}(y_i-y_j) &= \prod_{j=2}^{m}(y_1-y_j)\prod_{2\le i<j\le m}(y_i-y_j)\\ &= \left(y_1^{m-1}-y_1^{m-2}e_1'+y_1^{m-3}e_2'+\cdots+(-1)^{m-1}e_{m-1}'\right)\\ &\times\prod_{2\le i<j\le m}(y_i-y_j),\end{aligned}$$

where e_h' is the h-th elementary symmetric function of $y_2,\ldots,y_n$

$$e_h' = \sum_{2\le i_1<\cdots<i_h\le n} y_{i_1}\cdots y_{i_h}.$$

We now separate y_1 and evaluate

$$\int_0^x\cdots\int_0^x e_h' y_2^{n-2}\cdots y_m^{n-m}e^{-y_2-\cdots-y_m}\prod_{2\le i<j\le m}(y_i-y_j)dy_2\ldots dy_m. \tag{6.8}$$

By anti-symmetry of $\prod_{2\le i<j\le m}(y_i-y_j)$ it is easily seen that in e_h' only the leftmost term $y_2\ldots y_{2+h-1}$ remains and other terms vanish by integration. Hence (6.8) is equal to

$$H_{m-1}(n,\ldots,n-h+1,n-h-1,\ldots,n-m+1;x)$$

and we have the following result.

Proposition 6.2 *The cumulative distribution function $F(x,\lambda)$ is written as*

$$F(x,\lambda)=\sum_{h=0}^{m-1}(-1)^h F(x,\lambda;n,m,h)H_{m-1}(n,\ldots,n-h+1,n-h-1,\ldots,n-m+1;x), \tag{6.9}$$

where $F(x,\lambda;n,m,h)$ is defined in (6.3) *and H_m is defined in* (6.5).

Note that this expression involves $2m$ functions $F(x,\lambda;n,m,h)$, $H_{m-1}(n,\ldots,n-h+1,n-h-1,\ldots,n-m+1;x)$. The difficulty with large λ is now separated to $F(x,\lambda;n,m,h)$ in (6.3). We found that computing the determinant H_{m-1} is relatively straightforward for moderate m. Hence we apply the holonomic gradient method for $F(x,\lambda;n,m,h)$. Our approach is a hybrid approach, where HGM is applied only to $F(x,\lambda;n,m,h)$.

6.4.4 Partial Differential Equation for $F(x, \lambda; n, m, h)$

Here we derive partial differential equations satisfied by $F(x, \lambda; n, m, h)$.

By differentiating (6.10) by x

$$f(x, \lambda; n, m, h) = \partial_x F(x, \lambda; n, m, h) = e^{-\lambda - x} \sum_{k=0}^{\infty} \frac{\lambda^k x^{n+k+m-h-2}}{(k+n-1)!(k+m-1)!}. \tag{6.10}$$

Define an infinite series in x and λ by

$$q(x, \lambda; n, m, h) = \sum_{k=0}^{\infty} \frac{\lambda^k x^{n+k+m-h-2}}{(k+n-1)!(k+m-1)!}. \tag{6.11}$$

For an infinite series we can use the Euler's method [1] for deriving the partial differential equations satisfied by the series. Let $\theta_\lambda = \lambda \partial_\lambda, \theta_x = x \partial_x$ denote the Euler operators. From

$$\theta_\lambda \bullet q = \sum_{k=0}^{\infty} k \times \frac{\lambda^k x^{n+k+m-h-2}}{(k+n-1)!(k+m-1)!},$$

$$\theta_x \bullet q = \sum_{k=0}^{\infty} (n+k+m-h-2) \times \frac{\lambda^k x^{n+k+m-h-2}}{(k+n-1)!(k+m-1)!},$$

we have

$$(\theta_x - \theta_\lambda) \bullet q = (n+m-h-2)q. \tag{6.12}$$

Furthermore

$$(\theta_\lambda + m - 1) \bullet q = \sum_{k=0}^{\infty} (k+m-1) \times \frac{\lambda^k x^{n+k+m-h-2}}{(k+n-1)!(k+m-1)!},$$

$$(\theta_x - (m-1-h)) \bullet q = \sum_{k=0}^{\infty} (n+k-1) \times \frac{\lambda^k x^{n+k+m-h-2}}{(k+n-1)!(k+m-1)!},$$

implies

$$(\theta_\lambda + m - 1)(\theta_x - (m-1-h)) \bullet q(x, \lambda; n, m, h)$$

$$= \sum_{k=0}^{\infty} (k+n-1)(k+m-1) \times \frac{\lambda^k x^{n+k+m-h-2}}{(k+n-1)!(k+m-1)!}$$

$$
\begin{aligned}
&= x\lambda \sum_{k=1}^{\infty} \frac{\lambda^{k-1} x^{n+k+m-h-3}}{(k+n-2)!(k+m-2)!} + \frac{x^{n+m-h-2}}{(n-2)!(m-2)!} \\
&= x\lambda q(x, \lambda; n, m, h) + \frac{x^{n+m-h-2}}{(n-2)!(m-2)!}.
\end{aligned}
\tag{6.13}
$$

By the relation

$$
f(x, \lambda; n, m, h) = e^{-\lambda - x} q(x, \lambda; n, m, h)
$$

we have

$$
\begin{aligned}
\partial_x q &= e^{\lambda+x}(1+\partial_x) f, \qquad & \theta_x \bullet q &= e^{\lambda+x}(x + x\partial_x) f, \\
\partial_\lambda q &= e^{\lambda+x}(1+\partial_\lambda) f, \qquad & \theta_\lambda \bullet q &= e^{\lambda+x}(\lambda + \lambda\partial_\lambda) f.
\end{aligned}
$$

By substituting this into (6.12), (6.13) and by dividing by $e^{\lambda+x}$ we obtain the partial differential equations satisfied by f.

$$
(x\partial_x - \lambda\partial_\lambda + x - \lambda - (n+m-h-2)) \bullet f = 0,
$$

$$
\begin{aligned}
&(x\lambda\partial_x\partial_\lambda + x(\lambda+m-1)\partial_x + \lambda(x-(m-h-1))\partial_\lambda \\
&\quad + x(m-1) - \lambda(m-h-1) - (m-1)(m-h-1)) \bullet f = \frac{x^{n+m-h-2} e^{-x-\lambda}}{(n-2)!(m-2)!}.
\end{aligned}
$$

Hence if we define differential operators l_1, l_2 and function u_1, u_2 by

$$
\begin{aligned}
l_1 &:= x\partial_x - \lambda\partial_\lambda + x - \lambda - (n+m-h-2), \\
l_2 &:= x\lambda\partial_x\partial_\lambda + x(\lambda+m-1)\partial_x + \lambda(x-(m-h-1))\partial_\lambda, \\
&\qquad + x(m-1) - \lambda(m-h-1) - (m-1)(m-h-1), \\
u_1 &:= 0, \\
u_2 &:= \frac{x^{n+m-h-2} e^{-x-\lambda}}{(n-2)!(m-2)!},
\end{aligned}
$$

then we can concisely write the above differential equations as

$$
l_1 \bullet f = u_1, \quad l_2 \bullet f = u_2. \tag{6.14}
$$

Let $R_2 = \mathbb{C}(x, \lambda)\langle \partial_x, \partial_\lambda \rangle$ denote the ring of differential operators with rational function coefficients. We adopt the the graded reverse lexicographic order with $x \succ \lambda$.

We now compute the Gröbner bases of the left ideal generated by l_1, l_2 in R_2. By the Buchberger's algorithm, if we add

$$l_3 := \lambda\partial_\lambda^2 + (2\lambda + (m+n-1))\partial_\lambda - \frac{x\lambda - \lambda^2 - (m+n-1)\lambda - (n-1)(m-1)}{\lambda},$$

then $\{l_1, l_2, l_3\}$ is a Gröbner basis. Also we can check that $\{l_1, l_3\}$ is a reduced Gröbner basis. Hence the system of differential equations (6.14) is equivalent to

$$l_1 \bullet f = u_1, \quad l_3 \bullet f = u_3,$$

where

$$\begin{aligned} u_3 =& l_3 \bullet f = -\left(\partial_\lambda + \frac{\lambda + (m-1)}{\lambda}\right) l_1 \bullet f + \frac{1}{\lambda} l_2 \bullet f \\ =& \frac{1}{(n-2)!(m-2)!} \frac{x^{n+m-h-2}}{\lambda} e^{-x-\lambda}. \end{aligned}$$

6.4.5 *Partial Differential Equations for* $g(x, \lambda; n, m, h)$

Here we derive partial differential equations satisfied by $g(x, \lambda; n, m, h)$ of (6.4).

We integrate $l_1 \bullet f = u_1, l_3 \bullet f = u_3$, which are written as

$$(x\partial_x - \lambda\partial_\lambda + x - \lambda - (n+m-h-2)) \bullet f = 0, \tag{6.15}$$

$$\begin{aligned} &\left(\lambda\partial_\lambda^2 + (2\lambda + (m+n-1))\partial_\lambda - \frac{x\lambda - \lambda^2 - (m+n-1)\lambda - (n-1)(m-1)}{\lambda}\right) \\ &\qquad \bullet f = \frac{1}{(n-2)!(m-2)!} \frac{x^{n+m-h-2}}{\lambda} e^{-x-\lambda}, \end{aligned} \tag{6.16}$$

with respect to x in the interval $[0, x]$. By integration by parts we have

$$\int_0^x y\partial_y F(y)dy = xF(x) - \int_0^x F(y)dy,$$

and $f = \partial_x F$, integrating (6.15) and (6.16), we obtain

$$xf - F - \lambda\partial_\lambda F + xF - \int_0^x F dx - \lambda F - (n+m-h-2)F = 0, \tag{6.17}$$

$$\lambda\partial_\lambda^2 F + (2\lambda + (n+m-1))\partial_\lambda F - xF + \int_0^x F dx \tag{6.18}$$

$$+ \lambda F + (m+n-1)F + \frac{(n-1)(m-1)}{\lambda} F = \frac{\gamma(n+m-h-1, x)}{(n-2)!(m-2)!} \frac{e^{-\lambda}}{\lambda}.$$

Let $g(x, \lambda; n, m, h)$ be defined as in (6.4). Then by $F = \partial_x g, = \partial_x^2 g$, we can express (6.17), (6.18) as the differential equations:

$$(x\partial_x^2 - \lambda\partial_x\partial_\lambda + (x - \lambda - (n+m-h-1)\partial_x - 1) \bullet g = 0, \tag{6.19}$$

$$\Bigg(\lambda\partial_x\partial_\lambda^2 + (2\lambda + (n+m-1))\partial_x\partial_\lambda$$

$$- \left(x - \lambda - (m+n-1) - \frac{(n-1)(m-1)}{\lambda}\right)\partial_x + 1\Bigg) \tag{6.20}$$

$$\bullet g = \frac{\gamma(n+m-h-1, x)}{(n-2)!(m-2)!} \frac{e^{-\lambda}}{\lambda}. \tag{6.21}$$

Defining

$$L_1 := x\partial_x^2 - \lambda\partial_x\partial_\lambda + (x - \lambda - (n+m-h-1)\partial_x - 1,$$

$$L_3 := \lambda\partial_x\partial_\lambda^2 + (2\lambda + (n+m-1))\partial_x\partial_\lambda$$

$$- \left(x - \lambda - (m+n-1) - \frac{(n-1)(m-1)}{\lambda}\right)\partial_x + 1,$$

$$U_1 := 0,$$

$$U_3 := \frac{\gamma(n+m-h-1, x)}{(n-2)!(m-2)!} \frac{e^{-\lambda}}{\lambda},$$

we can concisely express (6.19), (6.21) as

$$L_1 \bullet g = U_1, \qquad L_3 \bullet g = U_3.$$

We compute the Gröbner bases of the left ideal generated by $L_1, L_3 \in R_2$. By Buchberger's algorithm, we add

$$L_4 := \partial_\lambda^2 + \frac{x}{\lambda}\partial_x + \frac{\lambda + (m+n-1)}{\lambda}\partial_\lambda + \frac{(h-1)\lambda + (n-1)(m-1)}{\lambda^2}.$$

Then $\{L_1, L_3, L_4\}$ is a Gröbner basis. We can check that $\{L_1, L_4\}$ is the reduced Gröbner basis. Also we can show

$$
\begin{aligned}
L_4 \bullet g &= -\left(\partial_\lambda^2 + \frac{2\lambda + (m+n-1)}{\lambda}\partial_\lambda - \frac{x\lambda - \lambda^2 - (m+n-1)\lambda - (n-1)(m-1)}{\lambda^2}\right) U_1 \\
&\quad + \left(\frac{x}{\lambda}\partial_x - \partial_\lambda + \frac{x - \lambda - (m+n-h-1)}{\lambda}\right) U_3 \\
&= \frac{1}{(n-2)!(m-2)!}\left(x^{n+m-h-1}e^{-x} + \left(x - (m+n-h-1)\right)\gamma(n+m-h-1, x)\right)\frac{e^{-\lambda}}{\lambda^2} \\
&=: U_4.
\end{aligned}
$$

6.4.6 *Pfaffian Equations*

Now we are going to derive the Pfaffian equations for $g(x, \lambda; n, m, h)$ of (6.4), which is the main result of this paper.

We compute the normal form of

$$
\partial_x, \partial_x^2, \partial_x\partial_\lambda, \partial_x^2\partial_\lambda, \partial_\lambda, \partial_\lambda^2, \partial_x\partial_\lambda^2,
$$

when they are divided by the Gröbner basis $\{L_1, L_3, L_4\}$. The normal forms are expressed as linear combinations of $1, \partial_x, \partial_\lambda, \partial_x\partial_\lambda$ with rational function coefficients.

First, $\partial_x, \partial_x\partial_\lambda, \partial_\lambda$ are themselves in the normal form. Furthermore by the definition of L_1, L_2, L_4 we easily obtain

$$
\begin{aligned}
\partial_x^2 &= \frac{1}{x}L_1 + \frac{\lambda}{x}\partial_x\partial_\lambda - \frac{x - \lambda - n - m + h + 1}{x}\partial_x + \frac{1}{x}, \\
\partial_\lambda^2 &= L_4 - \frac{x}{\lambda}\partial_x - \frac{\lambda + (m+n-1)}{\lambda}\partial_\lambda - \frac{(h-1)\lambda + (n-1)(m-1)}{\lambda^2}, \\
\partial_x\partial_\lambda^2 &= \frac{1}{\lambda}L_3 - \frac{(2\lambda + (m+n-1))}{\lambda}\partial_x\partial_\lambda \\
&\quad + \frac{x\lambda - \lambda^2 - (m+n-1)\lambda - (n-1)(m-1)}{\lambda^2}\partial_x - \frac{1}{\lambda}.
\end{aligned}
$$

The normal form of $\partial_x^2\partial_\lambda$ needs actual division as

$$
\begin{aligned}
\partial_x^2\partial_\lambda &= \frac{1}{x}\partial_x L_1 + \frac{1}{x}L_3 \\
&\quad - \frac{x + \lambda + h - 1}{x}\partial_x\partial_\lambda
\end{aligned}
$$

$$+\frac{x\lambda-\lambda^2-(m+n-2)\lambda-(n-1)(m-1)}{x\lambda}\partial_x+\frac{1}{x}\partial_\lambda-\frac{1}{x}.$$

Summarizing above computations, we have the following theorem.

Theorem 6.1 *Pfaffian equations for g are given as*

$$\partial_x\begin{bmatrix} g \\ \partial_x g \\ \partial_\lambda g \\ \partial_x\partial_\lambda g \end{bmatrix}=\begin{bmatrix} 0 & 1 & 0 & 0 \\ \frac{1}{x} & -\frac{x-\lambda-n-m+h+1}{x} & 0 & \frac{\lambda}{x} \\ 0 & 0 & 0 & 1 \\ -\frac{1}{x} & \frac{x\lambda-\lambda^2-(m+n-2)\lambda-(n-1)(m-1)}{x\lambda} & \frac{1}{x} & -\frac{x+\lambda+h-1}{x} \end{bmatrix}\begin{bmatrix} g \\ \partial_x g \\ \partial_\lambda g \\ \partial_x\partial_\lambda g \end{bmatrix}$$
$$+\begin{bmatrix} 0 \\ 0 \\ 0 \\ \frac{\gamma(n+m-h-1,x)e^{-\lambda}}{(n-2)!(m-2)!x\lambda} \end{bmatrix}, \tag{6.22}$$

$$\partial_\lambda\begin{bmatrix} g \\ \partial_x g \\ \partial_\lambda g \\ \partial_x\partial_\lambda g \end{bmatrix}$$
$$=\begin{bmatrix} 0 & 0 & 1 & 0 \\ 0 & 0 & 0 & 1 \\ -\frac{(h-1)\lambda+(n-1)(m-1)}{\lambda^2} & -\frac{x}{\lambda} & -\frac{\lambda+(m+n-1)}{\lambda} & 0 \\ -\frac{1}{\lambda} & \frac{x\lambda-\lambda^2-(m+n-1)\lambda-(n-1)(m-1)}{\lambda^2} & 0 & -\frac{2\lambda+(m+n-1)}{\lambda} \end{bmatrix}\begin{bmatrix} g \\ \partial_x g \\ \partial_\lambda g \\ \partial_x\partial_\lambda g \end{bmatrix}$$
$$+\begin{bmatrix} 0 \\ 0 \\ \frac{1}{(n-2)!(m-2)!}\left(x^{n+m-h-1}e^{-x}+\left(x-(m+n-h-1)\right)\gamma(n+m-h-1,x)\right)\frac{e^{-\lambda}}{\lambda^2} \\ \frac{\gamma(n+m-h-1,x)e^{-\lambda}}{(n-2)!(m-2)!\lambda^2} \end{bmatrix}. \tag{6.23}$$

Note that $F=\partial_x g$ is obtained as the second element when we numerically solve (6.22) and (6.23).

6.5 Numerical Results

In this section we present some numerical experiments.

6.5.1 Series Expansion and Initial Values

By the Pfaffian equations obtained in the last section, once we have the initial values $(g, \partial_x g, \partial_\lambda g, \partial_x \partial_\lambda g)|_{x=x_0, \lambda=\lambda_0}$ at some point (x_0, λ_0), we can use the standard ODE solver, such as the Runge-Kutta method, to evaluate $(g, \partial_x g, \partial_\lambda g, \partial_x \partial_\lambda g)$ at any point (x, λ). Since

$$g(x, \lambda; n, m, h) = xF(x, \lambda; n, m, h) - F(x, \lambda; n, m, h-1),$$

$g, \partial_x g, \partial_\lambda g, \partial_x \partial_\lambda g$ are written as the following infinite series.

$$\begin{aligned} g(x, \lambda; n, m, h) &= \sum_{k=0}^{\infty} \frac{\lambda^k e^{-\lambda}}{(k+n-1)!(k+m-1)!} \\ &\quad \times \Big(x\gamma(n+k+m-1-h, x) - \gamma(n+k+m-h, x) \Big), \\ \partial_x g(x, \lambda; n, m, h) &= \sum_{k=0}^{\infty} \frac{\lambda^k e^{-\lambda}}{(k+n-1)!(k+m-1)!} \gamma(n+k+m-1-h, x), \\ \partial_\lambda g(x, \lambda; n, m, h) &= \sum_{k=0}^{\infty} \frac{(k\lambda^{k-1} - \lambda^k)e^{-\lambda}}{(k+n-1)!(k+m-1)!} \\ &\quad \times \Big(x\gamma(n+k+m-1-h, x) - \gamma(n+k+m-h, x) \Big), \\ \partial_x \partial_\lambda g(x, \lambda; n, m, h) &= \sum_{k=0}^{\infty} \frac{(k\lambda^{k-1} - \lambda^k)e^{-\lambda}}{(k+n-1)!(k+m-1)!} \gamma(n+k+m-1-h, x). \end{aligned}$$

For the initial values at (x_0, λ_0), we choose a sufficiently large positive integer K and truncate the above infinite series by K terms.

By the truncation we have numerical errors in the initial values, in particular when λ_0 is larger than 1. In Fig. 6.1 we compute the successive truncations of the infinite series expression of

$$F(x, \lambda) = C \sum_{k=0}^{\infty} \frac{\lambda^k e^{-\lambda}}{(k+n-1)!(k+m-1)!} H_m(n+k, n-1, \ldots, n-m+1; x), \tag{6.24}$$

$$C = \frac{1}{\prod_{l=1}^{m-1} \Gamma(n-l)\Gamma(m-l)}$$

for $k = 0, 1, \ldots, K$. From this figure we see that for $\lambda \gg 1$ we need a large K.

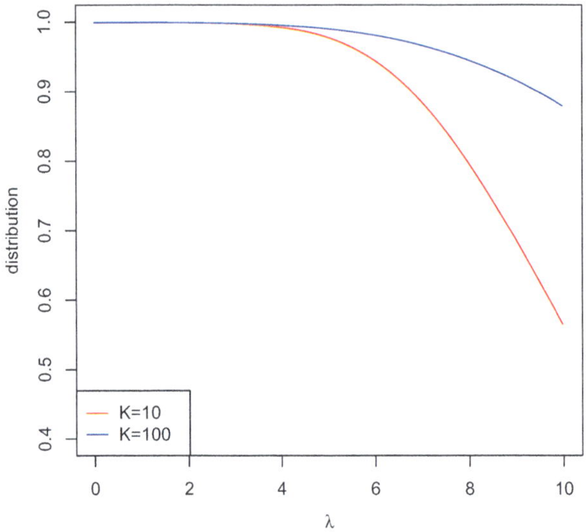

Fig. 6.1 CDF for $x = 20, n = 3, m = 2$. The horizontal axis is λ with the red line for $K = 10$ and the blue line for $K = 100$. We see that we need large K for $\lambda \gg 1$

6.5.2 *Numerical Integration by HGM*

6.5.2.1 Integration with Respect to *x*

We use the statistical software *R* with the package "deSolve" with $\Delta x = 0.01$ in (6.22). The numerical solution of (6.22) is presented in Fig. 6.2. Furthermore by $G_2(x, \lambda; n, m, h) = \partial_x g(x, \lambda; n, m, h) = F(x, \lambda; n, m, h)$ we can compute the cumulative distribution function by G and

$$F(x, \lambda) = \sum_{h=0}^{m-1} (-1)^h F(x, \lambda; n, m, h) H_{m-1}(n, \ldots, n-h+1, n-h-1, \ldots, n-m+1; x) \tag{6.25}$$

as in Fig. 6.3. Figure 6.3 shows that the cumulative distribution function converges to 1 as $x \to \infty$.

6.5.2.2 Integration with Respect to λ

For checking the accuracy of HGM, we compare the values of the cumulative distribution function by HGM in Fig. 6.4 and a truncation of the infinite series

$$F(x, \lambda; n, m, h) = \sum_{k=0}^{\infty} \frac{\lambda^k e^{-\lambda}}{(k+n-1)!(k+m-1)!} \gamma(n+k+m-1-h, x) \tag{6.26}$$

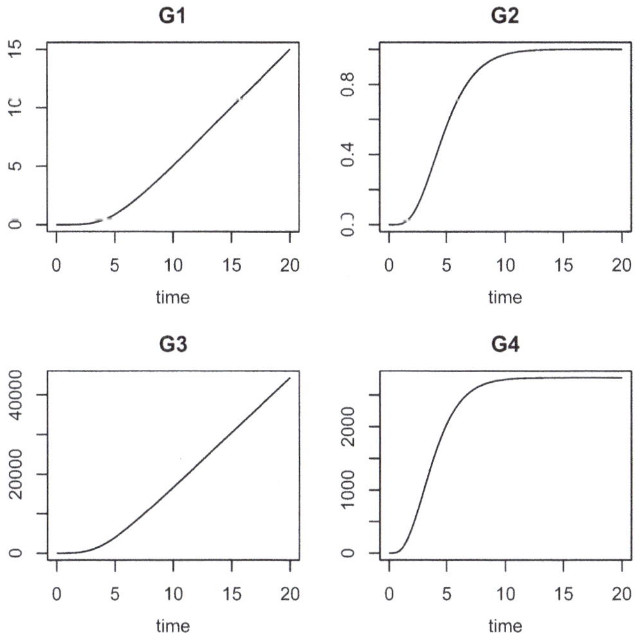

Fig. 6.2 Computation of G by integration of x, $n=5, m=2, h=1, \lambda=10^{-7}, x_0=10^{-7}, K=100, \Delta x=0.01$. Recall $G1=g, G2=\partial_x g, G3=\partial_\lambda g, G4=\partial_x\partial_\lambda g$

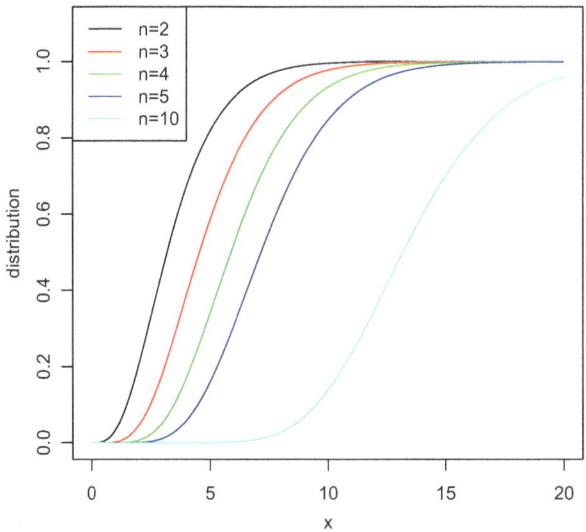

Fig. 6.3 CDF for the case $m=2, \lambda=10^{-7}$ by HGM. CDF converges to 1 as $x \to \infty$

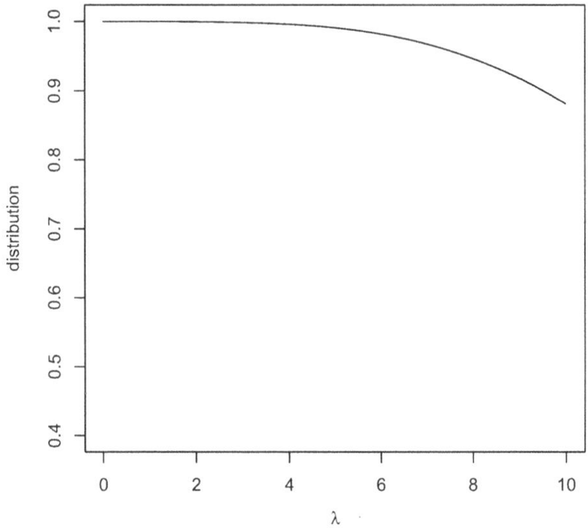

Fig. 6.4 CDF by HGM. We used $K = 10$ terms at $x_0 = 20$, $\lambda_0 = 10^{-7}$

Table 6.1 Values of CDF

$x = 20$	HGM	Series ($K = 10$)	Series ($K = 100$)
$\lambda = 10^{-7}$	0.9999921	0.9999921	0.9999921
$\lambda = 10$	0.8788772	0.5658010	0.8793867

in Fig. 6.1.

As discussed above, the series (6.26) converges slowly for large noncentrality λ. We take $K = 10$ terms and compare the values of the cumulative distribution function at $10^{-7} < \lambda \leq 10$. For parameter values $m = 2, n = 3, x = 20$, the results of HGM is depicted in Fig. 6.4 and the truncation at $K = 10$ terms of (6.26) is depicted in Fig. 6.1. In Fig. 6.1 we overlay the result of truncation at $K = 100$.

Table 6.1 shows the numerical values. For $\lambda = 10^{-7}$,the difference of truncations of the infinite series at $K = 10$ and at $K = 100$ is smaller than 10^{-7}. However for $\lambda = 10$ these truncations are totally different. On the other hand, HGM with the initial value at $\lambda = 10^{-7}$ computed with the truncation $K = 10$ and $\Delta\lambda = 0.01$, coincides with the series truncated at $K = 100$ at $\lambda = 10$ with two digits.

From these results we see that HGM needs only small number of terms for the initial value at small λ and gives an accurate values for large λ, whereas the infinite series needs large K for large λ.

Acknowledgments We would like to thank Constantin Siriteanu for useful comments and suggestions. We are grateful to the referee for constructive comments. This work was supported by JSPS KAKENHI Grant Number 18H04092.

References

1. Ascher, U.M., Petzold, L.R.: Computer Methods for Ordinary Differential Equations and Differential-Algebraic Equations. Society for Industrial and Applied Mathematics (SIAM), Philadelphia (1998)
2. Hashiguchi, H., Numata, Y., Takayama, N., Takemura, A.: The holonomic gradient method for the distribution function of the largest root of a Wishart matrix. J. Multivariate Anal. **117**, 296–312 (2013)
3. Heath, R., Lozano, A.: Foundations of MIMO Communication. Cambridge University Press, Cambridge (2019)
4. Hibi, T. (ed.): Grobner Bases. Statistics and Software Systems. Springer, Tokyo (2013)
5. Hong, Y.P., Huang, W., Kuo, C.J.: Cooperative Communications and Networking: Technologies and System Design. Springer Science+Business Media, New York (2010)
6. James, A.T.: Distributions of matrix variates and latent roots derived from normal samples. Ann. Math. Stat. **35**, 475–501 (1964)
7. Kang, M., Alouini, M.S.: Largest eigenvalue of complex Wishart matrices and performance analysis of MIMO MRC systems. IEEE J. Sel. Areas Commun. **21**(3), 418–426 (2003)
8. Muirhead, R.I.: Aspects of Multivariate Statistical Theory. Wiley, Hoboken (2005)
9. Nakayama, H., Nishiyama, K., Noro, M., Ohara, K., Sei, T., Takayama, N., Takemura, A.: Holonomic gradient descent and its application to the Fisher–Bingham integral. Adv. Appl. Math. **47**(3), 639–658 (2011)
10. Sei, T., Kume, A.: Calculating the normalising constant of the Bingham distribution on the sphere using the holonomic gradient method. Stat. Comput. **25**, 321–332 (2015)
11. Siriteanu, C., Blostein, S.D., Takemura, A., Shin, H., Yousefi, S., Kuriki, S.: Exact MIMO zero-forcing detection analysis for transmit-correlated Rician fading. IEEE Trans. Wirel. Commun. **13**(3), 1514–1527 (2014)
12. Siriteanu, C., Takemura, A., Kuriki, S., Richards, D., Shin, H.: Schur complement based analysis of MIMO zero-forcing for Rician fading. IEEE Trans. Wirel. Commun. **14**(4), 1757–1771 (2015)
13. Siriteanu, C., Takemura, A., Kuriki, S., Shin, H., Koutschan, C.: MIMO zero-forcing performance evaluation using the holonomic gradient method. IEEE Trans. Wirel. Commun. **14**(4), 2322–2335 (2015)
14. Siriteanu, C., Takemura, A., Koutschan, C., Kuriki, S., Richards, D., Shin, H.: Exact ZF analysis and computer-algebra-aided evaluation in rank-1 LoS Rician fading. IEEE Trans. Wirel. Commun. **15**(8), 5245–5259 (2016)
15. Takemura, A.: Zonal Polynomials. IMS Lecture Notes. IMS, Hayward (1984)
16. Vidunas, R., Takemura, A.: Differential relations for the largest root distribution of complex non-central Wishart matrices. In: Hibi, T. (ed.) The 50th Anniversary of Gröbner Basis. Advanced Studies in Pure Mathematics, vol. 77, pp. 411–436 (2018)

Chapter 7
Some Tests for the Extended Growth Curve Model and Applications in the Analysis of Clustered Longitudinal Data

Jemila S. Hamid and Sayantee Jana

Abstract The Growth Curve Model (GCM) is a Generalized Multivariate Analysis of Variance (GMANOVA) model especially useful in the analysis of longitudinal data, growth curves as well as other response curves. The model is a natural extension of the classical Multivariate Analysis of Variance (MANOVA) model and among other assumptions, relies on the assumption that the mean for each group can be represented as a polynomial of degree q. The assumption that the mean over time for all groups follows a polynomial of the same degree is not always satisfied, since individuals across the different groups may respond differently. An excellent scenario is when we have clustered longitudinal data, where the response over time can be represented by polynomials of different degrees. In such situations, the natural extension is to use the Extended Growth Curve Model (EGCM), where one can assume different shapes to represent different groups or clusters. In this paper, we formulate hypotheses motivated by real life scenarios involving clustered longitudinal data, and propose tests that are motivated by residuals in the EGCM. We then mathematically derive the tests and evaluate performances using simulations. We provide real data examples as illustrations.

7.1 Introduction

The Extended Growth Curve Model (EGCM) is an extension of the Growth Curve Model (GCM), and arises in situations where there are linear restrictions on the mean parameter of the model [22, 26, 27, 30, 31]. Both the GCM and EGCM are also referred to as Generalized Multivariate Analysis of Variance (GMANOVA) models, simply because they are indeed generalizations of the classical Multivariate

J. S. Hamid (✉)
University of Ottawa, Ottawa, ON, Canada
e-mail: jhamid@uottawa.ca

S. Jana
Indian Institute of Management Nagpur, Nagpur, India
e-mail: sayantee@iimnagpur.ac.in

© Springer Nature Switzerland AG 2020

T. Holgersson, M. Singull (eds.), *Recent Developments in Multivariate and Random Matrix Analysis*, https://doi.org/10.1007/978-3-030-56773-6_7

Analysis of Variance (MANOVA) models [1, 9–12, 26]. The term bilinear regression will be used frequently in describing the GCM and its extensions, to indicate the presence of two design matrices: the within and between design matrices, and to describe the bilinear nature of the projections corresponding to these two design matrices. The linear restrictions on MANOVA models lead to a bilinear model, where the model involves projections with respect to two design matrices, as already mentioned [3, 4, 16, 30].

The GCM has been demonstrated to be useful in the analysis of longitudinal data, especially for short to moderate time series, and the analysis is performed assuming that the mean for each of the groups follows a polynomial of degree q [11, 21, 23]. There is extensive literature in the area and various distributions and covariance structures have been considered [6, 13, 14, 18–20, 28]. High-dimensional extensions have also been considered in recent years [5, 8, 24, 25]. In practical applications, however, the mean for different groups might be represented by polynomials of different degrees. This, for instance, happens when we have clustered longitudinal data, where the response over time follows different shapes for different clusters or groups. In such situations, the EGCM is useful since it allows different degrees of polynomials to be fitted within one modeling framework [3, 7, 15, 17, 26, 31].

Consider a GCM with m groups, where measurements are taken from each of the n individuals at p different time points. Suppose also, that the mean (across time) for the ith ($i = 1, 2, \ldots, k$) group can be represented by a polynomial function of degree $q - 1$ and can be described as

$$b_{0,i} + b_{1,i}t + +b_{2,i}t^2 \cdots + b_{q,i}t^{q-1}, \quad t = t_1, t_2, \ldots, t_p.$$

The GCM in matrix format, is given as

$$\mathbf{Y} = \mathbf{ZBX} + \mathbf{E},$$

where $\mathbf{Y} : p \times n$ represents the observation (outcome) matrix, $\mathbf{Z} : p \times q$ and $\mathbf{X} : m \times n$ are the between and within individual design matrices, and $\mathbf{B} : q \times m$ represents the parameter matrix with the coefficients of the polynomials. The columns of the error matrix $\mathbf{E} : p \times n$ are assumed to be distributed as a p-variate normal distribution with mean zero and positive definite covariance matrix $\boldsymbol{\Sigma}$. Description of the various matrices in the model above are given by

$$\mathbf{Z}' = \begin{pmatrix} 1 & 1 & 1 & \cdots & 1 \\ t_1 & t_2 & t_3 & \cdots & t_p \\ t_1^2 & t_2^2 & t_3^2 & \cdots & t_p^2 \\ \vdots & \vdots & \vdots & \ddots & \vdots \\ t_1^{q-1} & t_2^{q-1} & t_3^{q-1} & \cdots & t_p^{q-1} \end{pmatrix}, \quad \mathbf{X} = \begin{pmatrix} \mathbf{1}_{n_1} & \mathbf{0}_{n_2} & \mathbf{0}_{n_3} & \cdots & \mathbf{0}_{n_m} \\ \mathbf{0}_{n_1} & \mathbf{1}_{n_2} & \mathbf{0}_{n_3} & \cdots & \mathbf{0}_{n_m} \\ \mathbf{0}_{n_1} & \mathbf{0}_{n_2} & \mathbf{1}_{n_3} & \cdots & \mathbf{0}_{n_m} \\ \vdots & \vdots & \vdots & \ddots & \vdots \\ \mathbf{0}_{n_1} & \mathbf{0}_{n_2} & \mathbf{0}_{n_3} & \cdots & \mathbf{1}_{n_m} \end{pmatrix},$$

$$\mathbf{Y} = \begin{pmatrix} y_{11} & y_{12} & y_{13} & \cdots & y_{1n} \\ y_{21} & y_{22} & y_{23} & \cdots & y_{2n} \\ y_{31} & y_{32} & y_{33} & \cdots & y_{3n} \\ \vdots & \vdots & \vdots & \ddots & \vdots \\ y_{p1} & y_{p2} & y_{p3} & \cdots & y_{pn} \end{pmatrix}, \quad \mathbf{B} = \begin{pmatrix} b_{01} & b_{02} & b_{03} & \cdots & b_{0m} \\ b_{11} & b_{12} & b_{13} & \cdots & b_{1m} \\ b_{21} & b_{22} & b_{23} & \cdots & b_{2m} \\ \vdots & \vdots & \vdots & \ddots & \vdots \\ b_{(q-1)1} & b_{(q-1)2} & b_{(q-1)3} & \cdots & b_{(q-1)m} \end{pmatrix}$$

and it is assumed that $q \leq p$ and $rank(\mathbf{X}) + p \leq n$ and $n = n_1 + n_2 + \ldots + n_m$.

Inferences for the GCM have been considered by many, and likelihood estimators for both the mean parameter and covariance matrices are already available [4, 6, 11]. The likelihood ratio test for testing the general linear hypotheses was also derived by Khatri [11], under some full rank restrictions. Residuals in the GCM were considered and mathematical decompositions were performed to provide some insights on the characteristics of the various components [29, 30]. This work was further extended to the EGCM, where better understanding of the design space as well as the residual space were achieved [3, 30].

Consider now the decomposed residuals in the GCM [29] and the simple hypothesis $\mathbf{B} = \mathbf{0}$ (mean is zero), an extension of the well-known Lawley-Hotelling trace test was previously derived for testing this hypothesis [4]. The test turned out to be functions of the decomposed residuals, which in turn facilitated appropriate interpretations of the decomposed residuals as well as better understanding of the distribution of the test statistic. In their paper, Hamid and colleagues showed that the distributions of the test statistics for the simple hypothesis of $\mathbf{B} = \mathbf{0}$ as well as the general linear hypothesis $\mathbf{GBF} = \mathbf{0}$ are free of the unknown covariance matrix $\boldsymbol{\Sigma}$ [4]. This means that the distributions can be generated empirically and the critical value for the tests can be calculated through simulations or parametric bootstrapping. Furthermore, the authors also showed that the distribution of the test can be represented as weighted sums of chi-square random variables, which allowed the authors to provide appropriate approximations.

In this paper, we consider the EGCM and provide some tests for the special case of the model, where we assume the mean growth curves to be clustered in two categories. The mean (over time) for the two clusters follow polynomials of different degrees, e.g., one cluster consisting of groups with linear growth curves and the other cluster consisting of groups with quadratic curves. We first present and discuss the decomposed residuals in the EGCM [3, 30], and formulate two hypotheses based on their practical relevance in analysis of clustered longitudinal data, requiring fitting of the EGCM. We propose potential statistical tests motivated by the residuals, and using a formulation similar to the trace test in the GCM, which as mentioned above is an extension of the well-known Lawley-Hotelling trace test in the MANOVA model. We then provide formal mathematical derivation using restricted maximum likelihood approach and provide some distributional properties of the test statistics. We evaluate performance using extensive simulations and provide real data illustration.

7.2 Residuals in the Extended Growth Curve Model

Suppose we have m groups and a total of n individuals from whom measurements are taken at p time points. Without loss of generality, consider the EGCM, where groups can be categorized into two clusters. Suppose, again without loss of generality, that the mean for the first cluster, consisting of m_1 groups, follows a polynomial of degree $q_1 - 1$ and the mean for the second cluster (consisting of m_2 groups) follows a polynomial of degree $q_1 + q_2 - 1$. For presentation purposes, let $m = m_1 + m_2$ and $q = q_1 + q_2$. The EGCM, in matrix formulations, can be represented as

$$\mathbf{Y} = \mathbf{Z}_1\mathbf{B}_1\mathbf{X}_1 + \mathbf{Z}_2\mathbf{B}_2\mathbf{X}_2 + \mathbf{E}, \tag{7.1}$$

where $q_1, q_2 \leq p$, $rank(\mathbf{X}_1) + p \leq n$ and $\mathscr{C}(\mathbf{X}_2') \subseteq \mathscr{C}(\mathbf{X}_1')$, where $\mathscr{C}(.)$ represents the column space of a matrix. The observation matrix $\mathbf{Y}$ has the same dimension and representation as the one described above for the GCM, and the n columns of the error matrix $\mathbf{E}$ are assumed to be distributed as a p-variate normal random variable with mean 0 and covariance $\boldsymbol{\Sigma}$. The within individual design matrix $\mathbf{Z}_1$, now of dimension $p \times q_1$, also has the same representation as $\mathbf{Z}$ in GCM, but with q_1 replacing q; the parameter matrix $\mathbf{B}_1$, of dimension $q_1 \times m$ represents the coefficients of the q_1 polynomials that are common for both clusters consisting of all m groups. The between individual design matrix $\mathbf{X}_1$, of dimension $m \times n$, has the same representation as $\mathbf{X}$ in GCM. The descriptions of the remaining matrices involved in the EGCM are given as

$$\mathbf{Z}_2' : q_2 \times p = \begin{pmatrix} t_1^{q_1} & t_2^{q_1} & t_3^{q_1} & \cdots & t_p^{q_1} \\ t_1^{q_1+1} & t_2^{q_1+1} & t_3^{q_1+1} & \cdots & t_p^{q_1+1} \\ t_1^{q_1+2} & t_2^{q_1+2} & t_3^{q_1+2} & \cdots & t_p^{q_1+2} \\ \vdots & \vdots & \vdots & \ddots & \vdots \\ t_1^{q_1+q_2-1} & t_2^{q_1+q_2-1} & t_3^{q_1+q_2-1} & \cdots & t_p^{q_1+q_2-1} \end{pmatrix},$$

$$\mathbf{X}_2 : m_2 \times n = \begin{pmatrix} \mathbf{0}_{n_1} & \mathbf{0}_{n_2} & \cdots & \mathbf{1}_{n_{m_1+1}} & \mathbf{0}_{n_{m_1+2}} & \cdots & \mathbf{0}_{n_m} \\ \mathbf{0}_{n_1} & \mathbf{0}_{n_2} & \cdots & \mathbf{0}_{n_{m_1+1}} & \mathbf{1}_{n_{m_1+2}} & \cdots & \mathbf{0}_{n_m} \\ \vdots & \vdots & \vdots & \vdots & \vdots & \ddots & \vdots \\ \mathbf{0}_{n_1} & \mathbf{0}_{n_2} & \cdots & \mathbf{0}_{n_{m_1+1}} & \mathbf{0}_{n_{m_1+2}} & \cdots & \mathbf{1}_{n_m} \end{pmatrix},$$

$$\mathbf{B}_2 : q_2 \times m_2 = \begin{pmatrix} b_{q_1 1} & b_{q_1 2} & b_{q_1 3} & \cdots & b_{q_1 m_2} \\ b_{(q_1+1)1} & b_{(q_1+1)2} & b_{(q_1+1)3} & \cdots & b_{(q_1+1)m_2} \\ b_{(q_1+2)1} & b_{(q_1+2)2} & b_{(q_1+2)3} & \cdots & b_{(q_1+2)m_2} \\ \vdots & \vdots & \vdots & \ddots & \vdots \\ b_{(q_1+q_2-1)1} & b_{(q_1+q_2-1)2} & b_{(q_1+q_2-1)3} & \cdots & b_{(q_1+q_2-1)m_2} \end{pmatrix}.$$

Considerable literature is available on GMANOVA models in general, and EGCM in particular [3, 7, 12, 15, 26, 30, 31]. Solutions to the likelihood functions are also provided, and explicit formulae for the estimators are provided [12, 26]. Nevertheless, there is limited work related to hypothesis testing in the context of the mean parameters of EGCM, and hence limited applications of the model exist despite the fact that longitudinal data in practice follow different functions over time across the different groups. The focus of this paper is, therefore, to contribute towards hypothesis testing for the mean parameters of the EGCM. In doing so, we are particularly interested in the decomposed residuals, and we consider a special case of the model, without loss of generality. We use the residuals as motivations to formulate hypotheses relevant to real world applications. We then mathematically derive corresponding test statistics. As such, let us first consider the estimated model, derived using the maximum likelihood (ML) approach, which is always unique [3, 26, 27]. The mathematical expression for the estimated model is given by

$$\begin{aligned}\hat{\mathbf{Y}} &= \mathbf{Z}_1\widehat{\mathbf{B}}_1\mathbf{X}_1 + \mathbf{Z}_2\widehat{\mathbf{B}}_2\mathbf{X}_2\\ &= (\mathbf{I}-\mathbf{T}_1)\mathbf{Y}\mathbf{X}_1'(\mathbf{X}_1\mathbf{X}_1')^-\mathbf{X}_1 + (\mathbf{I}-\mathbf{T}_2)\mathbf{Y}\mathbf{X}_2'(\mathbf{X}_2\mathbf{X}_2')^-\mathbf{X}_2, \end{aligned} \tag{7.2}$$

where $\mathbf{A}^-$ represents a generalized inverse of any matrix $\mathbf{A}$ and

$$\begin{aligned}\mathbf{T}_1 &= \mathbf{I} - \mathbf{Z}_1(\mathbf{Z}_1'\mathbf{S}_1^{-1}\mathbf{Z}_1)^-\mathbf{Z}_1'\mathbf{S}_1^{-1},\\ \mathbf{T}_2 &= \mathbf{I} - \mathbf{T}_1\mathbf{Z}_2(\mathbf{Z}_2'\mathbf{T}_1'\mathbf{S}_2^{-1}\mathbf{T}_1\mathbf{Z}_2)^-\mathbf{Z}_2'\mathbf{T}_1'\mathbf{S}_2^{-1};\\ \mathbf{S}_1 &= \mathbf{Y}(\mathbf{I} - \mathbf{X}_1'(\mathbf{X}_1\mathbf{X}_1')^-\mathbf{X}_1\mathbf{Y}';\\ \mathbf{S}_2 &= \mathbf{S}_1 + \mathbf{T}_1\mathbf{Y}\mathbf{X}_1'(\mathbf{X}_1\mathbf{X}_1')^-\mathbf{X}_1(\mathbf{I} - \mathbf{X}_2'(\mathbf{X}_2\mathbf{X}_2')^-\mathbf{X}_2)\\ &\quad \times\mathbf{X}_1'(\mathbf{X}_1\mathbf{X}_1')^-\mathbf{X}_1\mathbf{Y}'\mathbf{T}_1'.\end{aligned}$$

In vectorized form, this can be re-written as

$$Vec\hat{\mathbf{Y}} = [\mathbf{P}_{\mathbf{X}_1} \otimes \mathbf{P}_{\mathbf{Z}_1}]Vec\mathbf{Y} + [\mathbf{P}_{\mathbf{X}_2} \otimes \mathbf{P}_{\mathbf{Z}_2}]Vec\mathbf{Y} = \mathbf{P}Vec\mathbf{Y},$$

where Vec represents a vectorized form a matrix, $\otimes$ represents the Kronecker product and

$$\begin{aligned}\mathbf{P}_{\mathbf{X}_1} &= \mathbf{X}_1'(\mathbf{X}_1\mathbf{X}_1')^-\mathbf{X}_1,\\ \mathbf{P}_{\mathbf{X}_2} &= \mathbf{X}_2'(\mathbf{X}_2\mathbf{X}_2')^-\mathbf{X}_2,\\ \mathbf{P}_{\mathbf{Z}_1} &= \mathbf{Z}_1(\mathbf{Z}_1'\mathbf{S}_1^{-1}\mathbf{Z}_1)^-\mathbf{Z}_1'\mathbf{S}_1^{-1},\\ \mathbf{P}_{\mathbf{Z}_2} &= \mathbf{T}_1\mathbf{Z}_2(\mathbf{Z}_2'\mathbf{T}_1'\mathbf{S}_2^{-1}\mathbf{T}_1\mathbf{Z}_2)^-\mathbf{Z}_2'\mathbf{T}_1'\mathbf{S}_2^{-1}.\end{aligned}$$

It is evident that the estimated model is a bilinear projection with respect to the four design matrices, which is equivalent to the column space of the matrix $\mathbf{P}$ [3, 30]. This column space can be represented as sum of two tensor product spaces

$$\mathscr{C}(\mathbf{P}) = \mathscr{C}(\mathbf{X}_1') \otimes \mathscr{C}_{\mathbf{S}_1}(\mathbf{Z}_1) + \mathscr{C}(\mathbf{X}_2') \otimes \mathscr{C}_{\mathbf{S}_2}(\mathbf{T}_1\mathbf{Z}_2),$$

where $\otimes$ represents tensor product between two spaces. The relationship between Kronecker product of two matrices and tensor product space are presented in more details in Kollo and von Rosen [12], and its applications in decomposition of residuals in GMANOVA models is provided in Hamid and von Rosen [3].

Now consider the residuals in the EGCM, which are defined on the orthogonal complement of the design space, that is a projection onto $(\mathscr{C}(\mathbf{X}_1') \otimes \mathscr{C}_{\mathbf{S}_1}(\mathbf{Z}_1) + \mathscr{C}(\mathbf{X}_2') \otimes \mathscr{C}_{\mathbf{S}_2}(\mathbf{T}_1\mathbf{Z}_2))^{\perp}$ [3, 30]. This space was mathematically decomposed and four residuals are provided in a previous work. Graphical elucidation of the design and residual spaces as well as the corresponding formulas for the residuals are provided below (Fig. 7.1). More mathematical details about the residuals in the EGCM can be found in Hamid and von Rosen [3].

Note that, some of the spaces in Fig. 7.1 can be decomposed further (e.g., note the broken lines across $\mathbf{R}_2$ and $\mathbf{R}_3$). Although further decompositions can also be interpreted in terms of the model characteristics with respect to what the model was not able to explain (i.e., residuals), we focused on these four residuals for now, mainly because of their practical relevance in applications of GMANOVA models:

$$\begin{aligned}
\mathbf{R}_1 &= (\mathbf{I} - \mathbf{T}_1)\mathbf{Y}(\mathbf{I} - \mathbf{X}_1'(\mathbf{X}_1\mathbf{X}_1')^-\mathbf{X}_1),\\
\mathbf{R}_2 &= \mathbf{T}_1\mathbf{Y}(\mathbf{I} - \mathbf{X}_1'(\mathbf{X}_1\mathbf{X}_1')^-\mathbf{X}_1),\\
\mathbf{R}_3 &= \mathbf{T}_1\mathbf{Y}(\mathbf{X}_1'(\mathbf{X}_1\mathbf{X}_1')^-\mathbf{X}_1 - \mathbf{X}_2'(\mathbf{X}_2\mathbf{X}_2')^-\mathbf{X}_2),\\
\mathbf{R}_4 &= (\mathbf{T}_1 + \mathbf{T}_2 - \mathbf{I})\mathbf{Y}\mathbf{X}_2'(\mathbf{X}_2\mathbf{X}_2')^-\mathbf{X}_2.
\end{aligned}$$

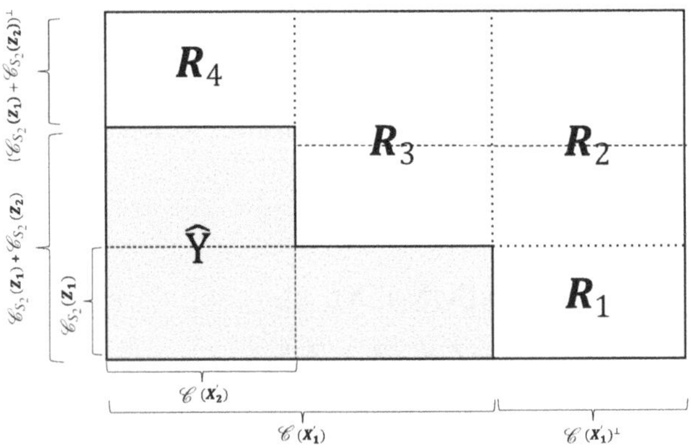

Fig. 7.1 The spaces representing the fitted model (design space) and the residuals in the EGCM

As we can see from the figure and mathematical characteristics of the residuals, $\mathbf{R}_{12} = \mathbf{R}_1 + \mathbf{R}_2$ is equivalent to the residual in MANOVA models and appears in the GCM as a sum of two of the decomposed residuals [29, 30]. This residual represents the difference between the observations and the group mean $(\mathbf{Y} - \mathbf{Y}\mathbf{X}_1'(\mathbf{X}_1\mathbf{X}_1')^{-}\mathbf{X}_1) = \mathbf{Y}(\mathbf{I} - \mathbf{X}_1'(\mathbf{X}_1\mathbf{X}_1')^{-}\mathbf{X}_1)$ and is distributed as a multivariate normal random variable, and hence useful in assessing between individual assumptions as well as distributional assumptions. Note further, that $\mathbf{R}_{12}\mathbf{R}_{12}' = \mathbf{S}_1$, and that the sample covariance matrix is $\frac{1}{n-1}\mathbf{S}_1$. On the other hand, $\mathbf{R}_{34} = \mathbf{R}_3 + \mathbf{R}_4$ provides information about overall model fit, while $\mathbf{R}_3$ and $\mathbf{R}_4$ individually provide information on different components of the polynomial fit.

For instance, consider two clusters, where the mean for one of the clusters can be represented by a linear function over time and the mean of the second cluster follows a quadratic curve. For this scenario, $\mathbf{R}_3$ provides information on the linear component of the model (for both clusters) and $\mathbf{R}_4$ provides information on the quadratic component of the second cluster. More on this will come while deriving and evaluating the proposed tests.

7.3 Some Tests for the Extended Growth Curve Model

Consider the EGCM defined in (7.1). Without loss of generality, consider only two groups, where each group is considered as a cluster. Suppose measurements from each group are taken at different time points. Suppose the mean for the first group can be represented by a linear function over time and the mean for the second group follows quadratic curve. Note that, this is just to simplify presentation in this paper, nevertheless the tests are derived under general assumptions, and are not restricted to two groups or linear/quadratic polynomials.

Consider now the simple hypothesis that the mean is zero (testing overall significance of the model), which can be formulated as

$$\begin{aligned} H_o &: \mathbf{B}_1 = \mathbf{0}, \mathbf{B}_2 = \mathbf{0}, \\ H_1 &: \mathbf{B}_1 \neq \mathbf{0}, \mathbf{B}_2 \neq \mathbf{0}. \end{aligned} \tag{7.3}$$

Recall the Lawley-Hotelling trace test for MANOVA and GCM [4, 5]. Both tests are weighted functions of the observed mean and the corresponding residuals (the part of the observed mean that is left unexplained by the fitted model), where the weight is the between individual variation represented by the sample variance-covariance matrix in GCM, which is a function of $\mathbf{S}$. Using analogous arguments, one can suggest that a test statistic for the simple hypothesis presented above, in the EGCM framework, will be a weighted function of $\mathbf{Y}\mathbf{X}_1'(\mathbf{X}_1\mathbf{X}_1')^{-}\mathbf{X}_1\mathbf{Y}'$ and $\mathbf{R}_{34}\mathbf{R}_{34}'$, and will have the format

$$\frac{f(\mathbf{Y}\mathbf{X}_1'(\mathbf{X}_1\mathbf{X}_1')^{-}\mathbf{X}_1\mathbf{Y}')}{f(\mathbf{R}_{34}\mathbf{R}_{34}')}.$$

Below, we will provide mathematical derivation for the test. In doing so, we write the likelihood function as a product of three independent terms. We maximize a certain part of the likelihood to get an estimator for the unknown covariance matrix, which then replaces the covariance matrix in the likelihood function to give the estimated likelihood.

Consider first the EGCM in (7.1), we can rewrite it as

$$\begin{aligned}\mathbf{Y} &= \mathbf{Z}_1(\mathbf{B}_{11} : \mathbf{B}_{12})\begin{pmatrix}\mathbf{X}_{11}\\ \mathbf{X}_{12}\end{pmatrix} + \mathbf{Z}_2\mathbf{B}_2\mathbf{X}_2 + \mathbf{E}\\ &= \mathbf{Z}_1\mathbf{B}_{11}\mathbf{X}_{11} + \mathbf{Z}_1\mathbf{B}_{12}\mathbf{X}_{12} + \mathbf{Z}_2\mathbf{B}_2\mathbf{X}_2 + \mathbf{E},\end{aligned} \tag{7.4}$$

where $\mathbf{B}_1 = (\mathbf{B}_{11} : \mathbf{B}_{12})$ and $\mathbf{X}_1' = (\mathbf{X}_{11}' : \mathbf{X}_{12}')$. Considering the two clusters (in our case, two groups as well) separately, the model reduces to

$$\mathbf{Y}_1 = \mathbf{Z}_1\mathbf{B}_{11}\mathbf{X}_{11}^1 + \mathbf{E}_1,$$

and

$$\mathbf{Y}_2 = \mathbf{Z}_1\mathbf{B}_{12}\mathbf{X}_{12}^2 + \mathbf{Z}_2\mathbf{B}_2\mathbf{X}_{22} + \mathbf{E}_2,$$

for Group I and Group II, respectively. Here $\mathbf{Y}_1$, $\mathbf{X}_{11}^1$ and $\mathbf{E}_1$ are matrices consisting of the first n_1 columns of $\mathbf{Y}$, $\mathbf{X}_{11}$ and $\mathbf{E}$, respectively. The matrices $\mathbf{Y}_2$, $\mathbf{X}_{12}^2$, $\mathbf{X}_{22}$ and $\mathbf{E}_2$ consist of the last n_2 columns of $\mathbf{Y}$, $\mathbf{X}_{12}$, $\mathbf{X}_2$ and $\mathbf{E}$, respectively. Observe that $\mathbf{X}_{12} = \mathbf{X}_2$. Moreover, it is possible to show that

$$\mathbf{X}_{11}'(\mathbf{X}_{11}\mathbf{X}_{11}')^-\mathbf{X}_{11} = \mathbf{X}_1'(\mathbf{X}_1\mathbf{X}_1')^-\mathbf{X}_1 - \mathbf{X}_2'(\mathbf{X}_2\mathbf{X}_2')^-\mathbf{X}_2.$$

Now consider the likelihood function for the EGCM

$$L = \gamma|\boldsymbol{\Sigma}|^{-\frac{n}{2}}e^{-\frac{1}{2}tr\{\boldsymbol{\Sigma}^{-1}(\mathbf{Y}-(\mathbf{Z}_1\mathbf{B}_1\mathbf{X}_1+\mathbf{Z}_2\mathbf{B}_2\mathbf{X}_2))(\mathbf{Y}-(\mathbf{Z}_1\mathbf{B}_1\mathbf{X}_1+\mathbf{Z}_2\mathbf{B}_2\mathbf{X}_2))'\}}, \tag{7.5}$$

where $\gamma = (2\pi)^{-\frac{1}{2}np}$. It can be rewritten as a product of three terms,

$$L = L_1 \times L_2 \times L_3,$$

where

$$\begin{aligned}L_1 &= \gamma\exp\{-\frac{1}{2}tr\{\boldsymbol{\Sigma}^{-1}(\mathbf{Y}\mathbf{X}_2'(\mathbf{X}_2\mathbf{X}_2')^-\mathbf{X}_2 - (\mathbf{Z}_1\mathbf{B}_{12}\mathbf{X}_{12} + \mathbf{Z}_2\mathbf{B}_2\mathbf{X}_2))\\ &\quad\times(\mathbf{Y}\mathbf{X}_2'(\mathbf{X}_2\mathbf{X}_2')^-\mathbf{X}_2 - (\mathbf{Z}_1\mathbf{B}_{12}\mathbf{X}_{12} + \mathbf{Z}_2\mathbf{B}_2\mathbf{X}_2))'\}\},\\ L_2 &= \exp\{-\frac{1}{2}tr\{\boldsymbol{\Sigma}^{-1}(\mathbf{Y}(\mathbf{X}_1'(\mathbf{X}_1\mathbf{X}_1')^-\mathbf{X}_1 - \mathbf{X}_2'(\mathbf{X}_2\mathbf{X}_2')^-\mathbf{X}_2) - (\mathbf{Z}_1\mathbf{B}_{11}\mathbf{X}_{11}))\\ &\quad\times(\mathbf{Y}(\mathbf{X}_1'(\mathbf{X}_1\mathbf{X}_1')^-\mathbf{X}_1 - \mathbf{X}_2'(\mathbf{X}_2\mathbf{X}_2')^-\mathbf{X}_2) - (\mathbf{Z}_1\mathbf{B}_{11}\mathbf{X}_{11}))'\}\},\\ L_3 &= |\boldsymbol{\Sigma}|^{-\frac{n}{2}}\exp\{-\frac{1}{2}tr\{\boldsymbol{\Sigma}^{-1}\mathbf{Y}(\mathbf{I} - \mathbf{X}_1'(\mathbf{X}_1\mathbf{X}_1')^-\mathbf{X}_1)\mathbf{Y}'\}\}.\end{aligned}$$

Let us now consider L_3, which is free of the parameters specified in the hypothesis and maximize the expression to get an estimator for the unknown covariance matrix $\boldsymbol{\Sigma}$. It is possible to show that the estimator that maximizes L_3 is $(n-1)\hat{\boldsymbol{\Sigma}} = \mathbf{S}_1$. We now replace $\boldsymbol{\Sigma}$ in (7.5) by its estimator, to get the estimated likelihood, denoted by EL, and then maximize EL under H_o and $H_o \cup H_1$, to get the desired test. The maximum of the EL under H_o and $H_o \cup H_1$ are respectively given by

$$\gamma_1 |\mathbf{S}_1|^{-\frac{n}{2}} e^{-\frac{1}{2} tr\{nS_1^{-1}\mathbf{Y}\mathbf{X}_1'(\mathbf{X}_1\mathbf{X}_1')^{-}\mathbf{X}_1\mathbf{Y}'\}} \tag{7.6}$$

and

$$\gamma_1 |\mathbf{S}_1|^{-\frac{n}{2}} \exp\{-\frac{1}{2} ntr\{\mathbf{S}_1^{-1}(\mathbf{Y}\mathbf{X}_1'(\mathbf{X}_1\mathbf{X}_1')^{-}\mathbf{X}_1 - (\mathbf{Z}_1\hat{\mathbf{B}}_1\mathbf{X}_1 + \mathbf{Z}_2\hat{\mathbf{B}}_2\mathbf{X}_2)) \\ \times (\mathbf{Y}\mathbf{X}_1'(\mathbf{X}_1\mathbf{X}_1')^{-}\mathbf{X}_1 - (\mathbf{Z}_1\hat{\mathbf{B}}_1\mathbf{X}_1 + \mathbf{Z}_2\hat{\mathbf{B}}_2\mathbf{X}_2))'\}\}, \tag{7.7}$$

where $\gamma_1 = n^{\frac{n}{2}}(2\pi)^{-\frac{1}{2}np} e^{-\frac{1}{2}np}$. Note that we can rewrite $\mathbf{R}_3$ and $\mathbf{R}_4$ as follows:

$$\mathbf{R}_3 = \mathbf{S}_1\mathbf{Z}_1^o(\mathbf{Z}_1^{o\prime}\mathbf{S}_1\mathbf{Z}_1^o)^{-}\mathbf{Z}_1^{o\prime}\mathbf{Y}(\mathbf{X}_1'(\mathbf{X}_1\mathbf{X}_1')^{-}\mathbf{X}_1 - \mathbf{X}_2'(\mathbf{X}_2\mathbf{X}_2')^{-}\mathbf{X}_2), \tag{7.8}$$

$$\mathbf{R}_4 = \mathbf{S}_2\mathbf{Z}^o(\mathbf{Z}^{o\prime}\mathbf{S}_2\mathbf{Z}^o)^{-}\mathbf{Z}^{o\prime}\mathbf{Y}\mathbf{X}_2'(\mathbf{X}_2\mathbf{X}_2')^{-}\mathbf{X}_2,$$

where $\mathbf{Z}_1^o$ and $\mathbf{Z}^o$ are matrices of full rank spanning the orthogonal complements of the column spaces of the matrices $\mathbf{Z}_1$ and $\mathbf{Z} = (\mathbf{Z}_1 : \mathbf{T}_1\mathbf{Z}_2)$, respectively. It is possible to show that $\mathbf{R}_{34}$, which denotes the sum of the residuals $\mathbf{R}_3$ and $\mathbf{R}_4$, can be written as a difference between the observed and estimated means, i.e.,

$$\mathbf{R}_{34} = \mathbf{Y}\mathbf{X}_1'(\mathbf{X}_1\mathbf{X}_1')^{-}\mathbf{X}_1 - (\mathbf{Z}_1\hat{\mathbf{B}}_1\mathbf{X}_1 + \mathbf{Z}_2\hat{\mathbf{B}}_2\mathbf{X}_2).$$

A test statistic is defined by taking the ratio between (7.6) and (7.7), which can be simplified as

$$\frac{e^{-\frac{1}{2}ntr\{\mathbf{S}_1^{-1}\mathbf{Y}\mathbf{X}_1'(\mathbf{X}_1\mathbf{X}_1')^{-}\mathbf{X}_1\mathbf{Y}'\}}}{e^{-\frac{1}{2}ntr\{\mathbf{S}_1^{-1}\mathbf{R}_{34}\mathbf{R}_{34}'\}}}, \tag{7.9}$$

where the hypothesis is rejected when the value of the ratio is small, i.e., close to zero. Note that the ratio has values between zero and one. One can also define an equivalent test by taking the logarithm of the test statistic, which can be re-written as

$$tr\{\mathbf{S}_1^{-1}\mathbf{Y}\mathbf{X}_1'(\mathbf{X}_1\mathbf{X}_1')^{-}\mathbf{X}_1\mathbf{Y}'\} - tr\{\mathbf{S}_1^{-1}\mathbf{R}_{34}\mathbf{R}_{34}'\}, \tag{7.10}$$

and the hypothesis will be rejected for large values of (7.10). This formulation allows a relatively easier understanding of some distributional characteristics of the proposed test. As such, consider the first term in (7.10) and write it as a sum of two terms as follows:

$$tr\{\mathbf{S}_1^{-1}\mathbf{Y}\mathbf{X}_1'(\mathbf{X}_1\mathbf{X}_1')^{-}\mathbf{X}_1\mathbf{Y}'\} = tr\{\mathbf{S}_1^{-1}\mathbf{Y}\mathbf{X}_{11}'(\mathbf{X}_{11}\mathbf{X}_{11}')^{-}\mathbf{X}_{11}\mathbf{Y}'\} + tr\{\mathbf{S}_1^{-1}\mathbf{Y}\mathbf{X}_2'(\mathbf{X}_2\mathbf{X}_2')^{-}\mathbf{X}_2\mathbf{X}_2'\}, \tag{7.11}$$

where $\mathbf{X}_{11}$ is as in (7.4). Similarly, use the expressions for $\mathbf{R}_3$ and $\mathbf{R}_4$ and write the second term in (7.10) as

$$tr\{\mathbf{S}_1^{-1}\mathbf{R}_{34}\mathbf{R}_{34}'\} = tr\{\mathbf{Y}\mathbf{X}_{11}'(\mathbf{X}_{11}\mathbf{X}_{11}')^{-}\mathbf{X}_{11}\mathbf{Y}'\mathbf{S}_1^{-1}\mathbf{Z}_1(\mathbf{Z}_1'\mathbf{S}_1^{-1}\mathbf{Z}_1)^{-}\mathbf{Z}_1'\mathbf{S}_1^{-1}\} + tr\{\mathbf{Y}\mathbf{X}_2'(\mathbf{X}_2\mathbf{X}_2')^{-}\mathbf{X}_2\mathbf{Y}'\mathbf{S}_2^{-1}\mathbf{Z}(\mathbf{Z}'\mathbf{S}_1^{-1}\mathbf{Z})^{-}\mathbf{Z}'\mathbf{S}_1^{-1}\}. \tag{7.12}$$

By subtracting (7.12) from (7.11), the test statistic reduces to

$$\phi_1(\mathbf{Y}) = tr\{\mathbf{Y}\mathbf{X}_{11}'(\mathbf{X}_{11}\mathbf{X}_{11}')^{-}\mathbf{X}_{11}\mathbf{Y}'\mathbf{S}_1^{-1}\mathbf{Z}_1(\mathbf{Z}_1'\mathbf{S}_1^{-1}\mathbf{Z}_1)^{-}\mathbf{Z}_1'\mathbf{S}_1^{-1}\} + tr\{\mathbf{Y}\mathbf{X}_2'(\mathbf{X}_2\mathbf{X}_2')^{-}\mathbf{X}_2\mathbf{Y}'\mathbf{S}_2^{-1}\mathbf{Z}(\mathbf{Z}'\mathbf{S}_1^{-1}\mathbf{Z})^{-}\mathbf{Z}'\mathbf{S}_1^{-1}\}, \tag{7.13}$$

where $\mathbf{Z} = (\mathbf{Z}_1 : \mathbf{T}_1\mathbf{Z}_2)$. The hypothesis is rejected when the value of $\phi_1(\mathbf{Y})$ is large. Here it is important to note that the column spaces of $(\mathbf{Z}_1 : \mathbf{Z}_2)$ and $(\mathbf{Z}_1 : \mathbf{T}_1\mathbf{Z}_2)$ are identical [2, 3].

The test statistic given in (7.13) is always greater or equal to zero. Moreover, it is possible to see from the expression in (7.9) that the numerator is a function of $\mathbf{Y}\mathbf{X}_1'(\mathbf{X}_1\mathbf{X}_1')^{-}\mathbf{X}_1$, which is the observed mean. On the other hand, in the denominator we have a function of $\mathbf{R}_{34}$, which is the residual obtained by subtracting the estimated mean from the observed mean. This shows that the test compares the observed and estimated means, in some weighted fashion, and rejects the hypothesis when the difference between them is "small", which is quite intuitive and in agreement with the formulation of the Lawley-Hotelling trace tests in the MANOVA and GMANOVA models.

Below, we show that the distribution of $\phi_1(\mathbf{Y})$ under the null hypothesis is independent of the unknown covariance matrix, $\boldsymbol{\Sigma}$. This is extremely important in applications, since the empirical distribution can be generated without the knowledge of the covariance matrix, and hence significance testing can be performed and p-values can be provided. Performance evaluation through simulations can also be performed by assuming, without loss of generality, that $\boldsymbol{\Sigma} = \mathbf{I}$. On the other hand, it is possible to show that the distribution under the alternative depends on $\boldsymbol{\Sigma}$, and hence the power of the test depends on the variance-covariance matrix. We will empirically show this through the simulations.

Now, to show that the distribution of the test statistic in (7.13) under the null is independent of $\boldsymbol{\Sigma}$, consider the first part of the expression and let $\mathbf{Z}_1^o$ be a matrix of full rank spanning the orthogonal complement to the space generated by the columns of $\mathbf{Z}_1$. We can write first expression in (7.13) as

$$\begin{aligned} &tr\{\mathbf{Y}\mathbf{X}_{11}'(\mathbf{X}_{11}\mathbf{X}_{11}')^{-}\mathbf{X}_{11}\mathbf{Y}'\mathbf{S}_1^{-1}\} \\ &\quad - tr\{\mathbf{Y}\mathbf{X}_{11}'(\mathbf{X}_{11}\mathbf{X}_{11}')^{-}\mathbf{X}_{11}\mathbf{Y}'\mathbf{Z}_1^o(\mathbf{Z}_1^{o\prime}\mathbf{S}_1\mathbf{Z}_1^o)^{-1}\mathbf{Z}_1^{o\prime}\}. \end{aligned} \tag{7.14}$$

The first term in (7.14) is invariant under the transformation $\boldsymbol{\Sigma}^{-\frac{1}{2}}\mathbf{Y}$. It is, therefore, possible to replace $\mathbf{Y}$ by $\boldsymbol{\Sigma}^{-\frac{1}{2}}\mathbf{Y}$, which shows that the distribution of this term in is independent of $\boldsymbol{\Sigma}$. For the second term, by using the property of the trace function, we can rewrite it as

$$tr\{\mathbf{X}_{11}'(\mathbf{X}_{11}\mathbf{X}_{11}')^{-}\mathbf{X}_{11}\mathbf{Y}'\mathbf{Z}_1^o(\mathbf{Z}_1^{o\prime}\mathbf{S}_1\mathbf{Z}_1^o)^{-1}\mathbf{Z}_1^{o\prime}\mathbf{Y}\}.$$

Now, write $\mathbf{Z}_1^{o\prime}\mathbf{Y}$ as

$$(\mathbf{Z}_1^{o\prime}\boldsymbol{\Sigma}\mathbf{Z}_1^o)^{\frac{1}{2}}(\mathbf{Z}_1^{o\prime}\boldsymbol{\Sigma}\mathbf{Z}_1^o)^{-\frac{1}{2}}\mathbf{Z}_1^{o\prime}\mathbf{Y}$$

and observe that we can rewrite $(\mathbf{Z}_1^{o\prime}\boldsymbol{\Sigma}\mathbf{Z}_1^o)^{\frac{1}{2}}(\mathbf{Z}_1^{o\prime}\mathbf{S}_1\mathbf{Z}_1^o)^{-1}(\mathbf{Z}_1^{o\prime}\boldsymbol{\Sigma}\mathbf{Z}_1^o)^{\frac{1}{2}}$ as

$$((\mathbf{Z}_1^{o\prime}\boldsymbol{\Sigma}\mathbf{Z}_1^o)^{-\frac{1}{2}}\mathbf{Z}_1^{o\prime}\mathbf{Y}(\mathbf{I}-\mathbf{X}_{11}{}'(\mathbf{X}_{11}\mathbf{X}_{11}{}')^{-}\mathbf{X}_{11})\mathbf{Y}'\mathbf{Z}_1^o(\mathbf{Z}_1^{o\prime}\boldsymbol{\Sigma}\mathbf{Z}_1^o)^{-\frac{1}{2}})^{-1}.$$

Consequently, it remains to show that the distribution of $(\mathbf{Z}_1{}^{o\prime}\boldsymbol{\Sigma}\mathbf{Z}_1{}^o)^{-\frac{1}{2}}\mathbf{Z}_1{}^{o\prime}\mathbf{Y}$ is independent of $\boldsymbol{\Sigma}$. Note that the expression is a linear function of a multivariate normal random variable. As a result, it is enough to show that the mean and dispersion matrices are independent of $\boldsymbol{\Sigma}$, which are shown below.

Under the null hypothesis $E[\mathbf{Y}] = \mathbf{Z}_1\mathbf{B}_1\mathbf{X}_1 + \mathbf{Z}_2\mathbf{B}_2\mathbf{X}_2 = \mathbf{0}$ which implies

$$E[(\mathbf{Z}_1{}^{o\prime}\boldsymbol{\Sigma}\mathbf{Z}_1{}^o)^{-\frac{1}{2}}\mathbf{Z}_1{}^{o\prime}\mathbf{Y}] = \mathbf{0}.$$

The dispersion matrix, $\mathbf{D}$, is given by

$$D[(\mathbf{Z}_1{}^{o\prime}\boldsymbol{\Sigma}\mathbf{Z}_1{}^o)^{-\frac{1}{2}}\mathbf{Z}_1{}^{o\prime}\mathbf{Y}] = (\mathbf{Z}_1{}^{o\prime}\boldsymbol{\Sigma}\mathbf{Z}_1{}^o)^{-\frac{1}{2}}\mathbf{Z}_1{}^{o\prime}\boldsymbol{\Sigma}\mathbf{Z}_1{}^o(\mathbf{Z}_1{}^{o\prime}\boldsymbol{\Sigma}\mathbf{Z}_1{}^o)^{-\frac{1}{2}} = \mathbf{I}.$$

Suppose now, that we want to check if the quadratic term in the growth curves of the individuals in the second group (cluster) is significantly different from zero. We can formulate the hypotheses (in terms of the mean parameters) as

$$\begin{aligned} H_o &: \mathbf{B}_2 = \mathbf{0} \\ H_1 &: \mathbf{B}_2 \neq \mathbf{0}. \end{aligned} \tag{7.15}$$

Note that this hypothesis is associated with the residual, $\mathbf{R}_4$. Using a similar motivation as in the Lawley-Hotelling trace test in MANOVA, the trace test in the

GCM and the test provided in (7.13), one can argue that the statistic for testing the hypothesis in (7.15) should have the format

$$\frac{f(\mathbf{YX}_2'(\mathbf{X}_2\mathbf{X}_2')^{-}\mathbf{X}_2\mathbf{Y}')}{f(\mathbf{R}_4\mathbf{R}_4')}.$$

In order to provide a formal derivation, recall the likelihood function of the EGCM given in (7.5) and maximize the product, $L_2 \times L_3$, which is independent of $\mathbf{R}_4$. This gives us the estimator $\mathbf{S}_2$ for $n\boldsymbol{\Sigma}$. Update the likelihood by using the estimator and proceed by taking the ratio of the maximum of the estimated likelihood under H_o and $H_o \cup H_1$. The test statistic provided in (7.17) can be obtained by taking the logarithm of the ratio and doing some algebraic manipulations similar to $\phi_1(\mathbf{Y})$, including using the property of the trace function (e.g., for any two matrices $tr(AB) = tr(BA)$) as well as the fact that

$$\begin{aligned}\mathbf{I} - \mathbf{Z}_1(\mathbf{Z}_1'\mathbf{S}_2{}^{-1}\mathbf{Z}_1)^{-}\mathbf{Z}_1'\mathbf{S}_2^{-1} &= \mathbf{T}_1\mathbf{Z}_2(\mathbf{Z}_2{}'\mathbf{T}_1{}'\mathbf{S}_2{}^{-1}\mathbf{T}_1\mathbf{Z}_2)^{-}\mathbf{Z}_2{}'\mathbf{T}_1{}'\mathbf{S}_2{}^{-1} \\ &\quad + \mathbf{S}_2\mathbf{Z}^o(\mathbf{Z}^{o\prime}\mathbf{S}_2\mathbf{Z}^o)\mathbf{Z}^{o\prime},\end{aligned} \tag{7.16}$$

where $\mathbf{T}_1$ and $\mathbf{Z}_1{}^o$ are as presented in (7.2) and (7.8), respectively. Note that the two terms on the right hand side of (7.16) are orthogonal to each other. The test statistic for testing the hypothesis in (7.15) can be formally written as

$$\phi_2(\mathbf{Y}) = tr\{\mathbf{YX}_2'(\mathbf{X}_2\mathbf{X}_2')^{-}\mathbf{X}_2\mathbf{Y}'\mathbf{S}_2^{-1}\mathbf{T}_1\mathbf{Z}_2(\mathbf{Z}_2'\mathbf{T}_1'\mathbf{S}_2^{-1}\mathbf{T}_1\mathbf{Z}_2)^{-}\mathbf{Z}_2'\mathbf{T}_1'\mathbf{S}_2^{-1}\}, \tag{7.17}$$

and the hypothesis is rejected when the value of $\phi_2(\mathbf{Y})$ is large. The above test statistic is always greater or equal to zero. Similar to $\phi_1(\mathbf{Y})$, the distribution of $\phi_2(\mathbf{Y})$ is under the null hypothesis independent of the unknown covariance matrix, $\boldsymbol{\Sigma}$. To show this, note that the test statistic in (7.17) can be written as the difference between two terms as

$$\begin{aligned}&tr\{\mathbf{YX}_2'(\mathbf{X}_2\mathbf{X}_2')^{-}\mathbf{X}_2\mathbf{Y}'\mathbf{G}_1(\mathbf{G}_1'\mathbf{W}_2\mathbf{G}_1)^{-1}\mathbf{G}_1'\} \\ &\quad - tr\{\mathbf{YX}_2'(\mathbf{X}_2\mathbf{X}_2')^{-}\mathbf{X}_2\mathbf{Y}'\mathbf{G}_2(\mathbf{G}_2'\mathbf{W}_2\mathbf{G}_2)^{-1}\mathbf{G}_2'\},\end{aligned} \tag{7.18}$$

where

$$\begin{aligned}\mathbf{G}_{r+1} &= \mathbf{G}_r(\mathbf{G}_r'\mathbf{Z}_{r+1})^o, \quad \mathbf{G}_0 = \mathbf{I}, \\ \mathbf{W}_{r+1} &= \mathbf{Y}(\mathbf{I} - \mathbf{X}_r'(\mathbf{X}_r\mathbf{X}_r')^{-}\mathbf{X}_r)\mathbf{Y}', \ r = 0, 1, 2, \ldots m-1.\end{aligned}$$

Such approaches have been discussed in a more general form in von Rosen [27]. For special cases, we refer to Hamid [2] and Hamid and von Rosen [3]. The two terms in (7.18) can be, respectively, rewritten as

$$tr\{\mathbf{X}_2'(\mathbf{X}_2\mathbf{X}_2')^{-}\mathbf{X}_2\mathbf{Y}'\mathbf{G}_1(\mathbf{G}_1'\mathbf{W}_2\mathbf{G}_1)^{-1}\mathbf{G}_1'\mathbf{Y}\}$$

and

$$tr\{\mathbf{X}_2'(\mathbf{X}_2\mathbf{X}_2')^{-}\mathbf{X}_2\mathbf{Y}'\mathbf{G}_2(\mathbf{G}_2'\mathbf{W}_2\mathbf{G}_2)^{-1}\mathbf{G}_2'\mathbf{Y}\}.$$

In order to show that the distribution of $\phi_2(\mathbf{Y})$ under the null hypothesis is independent of $\boldsymbol{\Sigma}$, we want to show that the distributions of the above two expressions under the null hypothesis are independent of $\boldsymbol{\Sigma}$, which is equivalent to showing that the distributions of $\mathbf{G}_1'\mathbf{Y}$ and $\mathbf{G}_2'\mathbf{Y}$ under the null hypothesis are independent of $\boldsymbol{\Sigma}$. As such, we write $\mathbf{G}_1'\mathbf{Y}$ as

$$(\mathbf{G}_1\boldsymbol{\Sigma}\mathbf{G}_1')^{\frac{1}{2}}(\mathbf{G}_1\boldsymbol{\Sigma}\mathbf{G}_1')^{-\frac{1}{2}}\mathbf{G}_1'\mathbf{Y}',$$

and it remains to show that the distribution of $(\mathbf{G}_1\boldsymbol{\Sigma}\mathbf{G}_1')^{-\frac{1}{2}}\mathbf{G}_1'\mathbf{Y}'$, which is a linear function of a multivariate normal random variable, is independent of $\boldsymbol{\Sigma}$. Once again, because of normality, it is enough to show that the mean and dispersion matrices are independent of $\boldsymbol{\Sigma}$.

Under the null hypothesis, $\mathbf{B}_2 = \mathbf{0}$, we have that $\mathbf{G}_1'\mathbf{Z}_1 = \mathbf{Z}_1^{o\prime}\mathbf{Z}_1 = \mathbf{0}$. Consequently,

$$E[(\mathbf{G}_1\boldsymbol{\Sigma}\mathbf{G}_1')^{-\frac{1}{2}}\mathbf{G}_1'\mathbf{Y}] = (\mathbf{G}_1\boldsymbol{\Sigma}\mathbf{G}_1')^{-\frac{1}{2}}\mathbf{G}_1'(\mathbf{Z}_1\mathbf{B}_1\mathbf{X}_1 + \mathbf{Z}_2\mathbf{B}_2\mathbf{X}_2) = \mathbf{0}.$$

Furthermore,

$$D[\mathbf{G}_1\boldsymbol{\Sigma}\mathbf{G}_1')^{-\frac{1}{2}}\mathbf{G}_1'\mathbf{Y}] = (\mathbf{G}_1\boldsymbol{\Sigma}\mathbf{G}_1')^{-\frac{1}{2}}\mathbf{G}_1'\boldsymbol{\Sigma}\mathbf{G}_1(\mathbf{G}_1\boldsymbol{\Sigma}\mathbf{G}_1')^{-\frac{1}{2}} = \mathbf{I}.$$

Similar calculations can show that the distribution of $\mathbf{G}_2'\mathbf{Y}$ is independent of $\boldsymbol{\Sigma}$.

7.4 Simulations

We performed simulations to evaluate the performance of the tests. We first generated the empirical distributions of both tests under the corresponding null hypotheses, and calculated the critical values for the tests based on 50,000 simulations. The empirical level and power of the tests were then calculated from second sets of 10,000 simulations. Several scenarios were considered in terms of sample size (n), departure from the null hypotheses and number of time points, p. We have also considered several scenarios in terms of the covariance matrix $\boldsymbol{\Sigma}$.

For the null hypothesis in (7.3), the distribution of the test statistic, $\phi_1(\mathbf{Y})$ under the null hypothesis is skewed to the left (Fig. 7.2), which is consistent across different sample sizes, dimensions of p, degrees of polynomial q and magnitudes of $\boldsymbol{\Sigma}$.

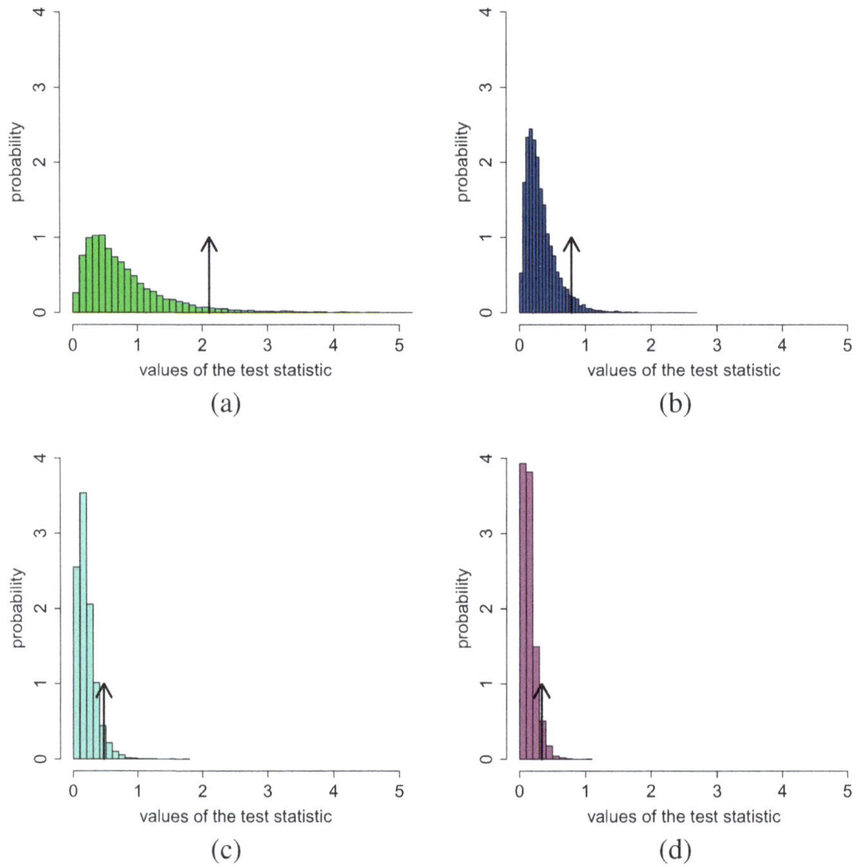

Fig. 7.2 The null distribution of $\phi_1(Y)$, where $p = 8$, and (**a**) n=20, (**b**) n=30, (**c**) n=40 and (**d**) n=50. The arrows show the critical values corresponding to the tests

The results of the simulation show that $\phi_1(Y)$ performs well and possesses all the desirable properties of a test statistic. Similar to the trace test in the GCM, our simulation results show that the test maintains the nominal level $\alpha = 0.05$ (Fig. 7.3), for all the scenarios considered.

The results also show that the test is unbiased, symmetric, monotone with respect to both n and departures from the null hypothesis (Fig. 7.4). Similar to our previous studies on the GCM and EGCM, departure from the null hypotheses is measured using the Euclidean norm of the parameter matrix [5–8]. The results also show that

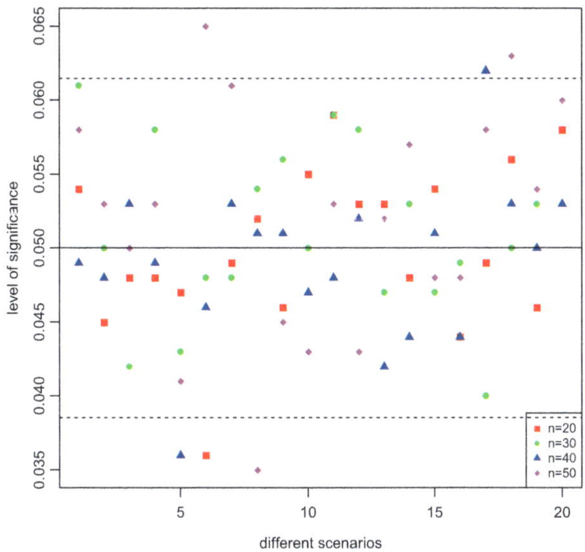

Fig. 7.3 Empirical level of $\phi_1(Y)$ and the corresponding 95% confidence bands

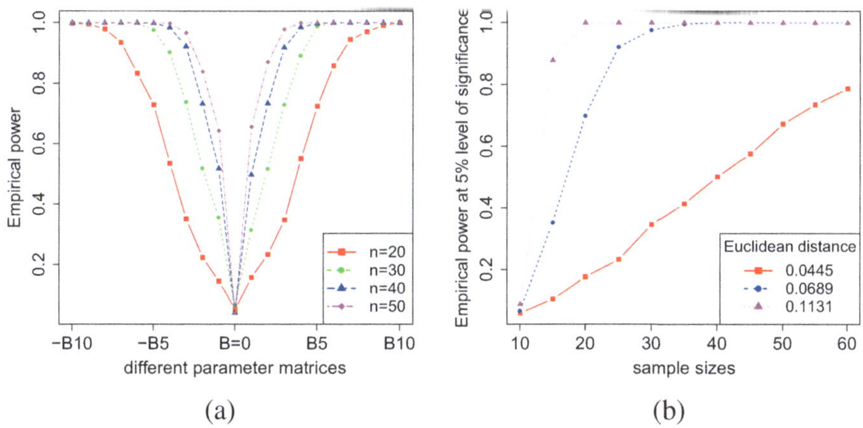

Fig. 7.4 Empirical power of $\phi_1(Y)$ with x-axis depicting (**a**) positive and negative scenarios to show symmetry, (**b**) Euclidean norm of the parameter matrix, used to measure departures from the null

the test has a reasonably good statistical power, where the test detects very small departures from the null hypothesis, even with relatively small sample sizes (Fig. 7.4 and Table 7.1). Please note, the results are consistent across all the scenarios we considered.

Table 7.1 Empirical power of $\phi_1(\mathbf{Y})$ for multiple n, p and $\mathbf{B}$

	$p = 8$				$p = 4$			
Euclidean distance	$n = 20$	$n = 30$	$n = 40$	$n = 50$	$n = 20$	$n = 30$	$n = 40$	$n = 50$
0	0.052	0.064	0.04	0.042	0.045	0.06	0.051	0.042
0.054	0.349	0.729	0.918	0.979	0.112	0.167	0.221	0.32
0.061	0.551	0.891	0.985	1.000	0.157	0.234	0.333	0.415
0.069	0.726	0.988	0.998	1.000	0.188	0.304	0.417	0.556
0.077	0.859	0.999	1.000	1.000	0.239	0.395	0.521	0.684
0.086	0.945	1.000	1.000	1.000	0.289	0.501	0.669	0.795
0.095	0.971	1.000	1.000	1.000	0.371	0.593	0.757	0.884
0.104	0.992	1.000	1.000	1.000	0.42	0.723	0.859	0.944
0.113	1.000	1.000	1.000	1.000	0.521	0.800	0.935	0.981

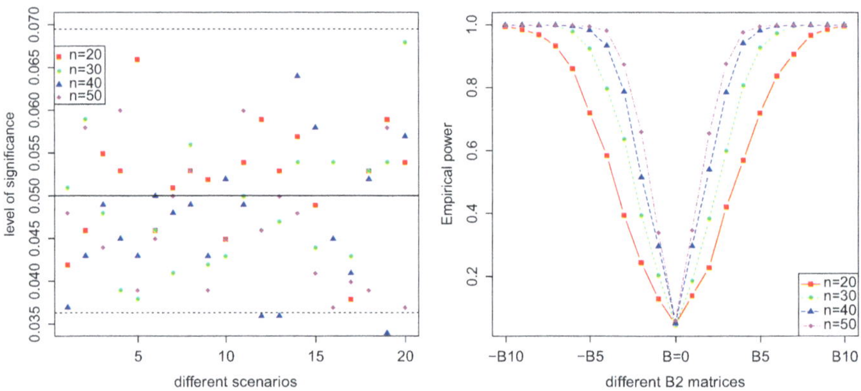

(a) (b)

Fig. 7.5 (**a**) Empirical level and (**b**) empirical power, of $\phi_2(Y)$

Similar results were also obtained for $\phi_2(\mathbf{Y})$, where the simulation results show that the distribution of the test statistic is skewed to the left. The test statistic maintains the nominal level and has all the other desirable properties such as unbiasedness, symmetry and monotonicity (Figs. 7.5 and 7.6). Furthermore, our simulations demonstrate that the test has a very good performance, as it detects extremely small departures from the null hypothesis, with a reasonably small sample size (Fig. 7.6).

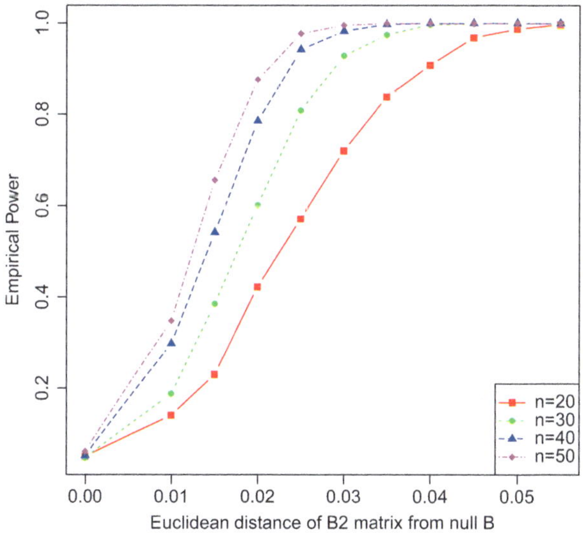

Fig. 7.6 Empirical power of $\phi_2(Y)$ with respect to Euclidean norm of the parameter matrix across four sample sizes

7.5 Empirical Example

As an empirical illustration, we consider the glucose data [32]. Data consists of measurements taken at 8 time points from 13 controls and 20 obese patients. The mean profile plots for the two groups are provided in Fig. 7.7 below. For illustration purposes, we assume two clusters, where the mean of one of the clusters (consisting of individuals in the obese group) follows a quadratic growth curve and the mean for the second cluster (consisting of individuals in the control group) follows a cubic growth curve.

Results show that the null hypothesis of zero mean for both groups (i.e., $\mathbf{B}_1 = \mathbf{0}$, $\mathbf{B}_2 = \mathbf{0}$) is rejected, with observed test statistic = 62.76, critical value = 0.80, and p-value $r < 0.0001$. For the hypothesis that the coefficient of the cubic term is zero, the observed test statistic value is 0.142 and the critical value is 0.173 (Fig. 7.8), indicating that there is no evidence to reject the null hypothesis (p-value = 0.07328) at 5% level of significance.

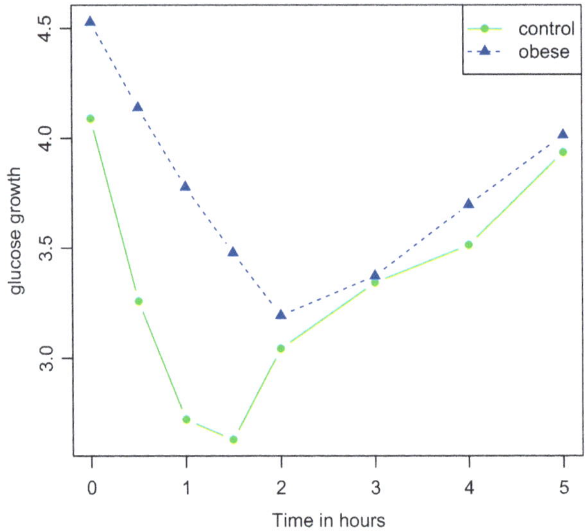

Fig. 7.7 Mean profile plots for the glucose data

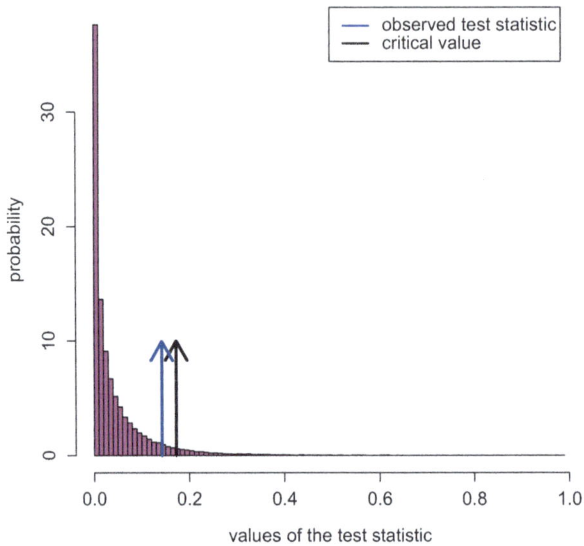

Fig. 7.8 The null distribution of $\phi_2(\mathbf{Y})$ for the glucose data

7.6 Summary

We considered the Extended Growth Curve Model, which can be applied in situations where longitudinal measurements from groups of individuals can be clustered into several groups. Using decomposed residuals as a motivation, we derived test statistics for two hypotheses related to the mean parameters of the model. The two corresponding tests can be shown to be extensions of the trace test in the Growth Curve Model, which in turn is the Lawley-Hotelling trace test.

Simulation results showed that the two tests possess all the desirable properties such as unbiasedness, symmetry and monotonicity with respect to both sample size and departures from the null hypotheses, where departure is measured using Euclidean norm. Results also demonstrate that the tests have good performances, where the tests were able to detect very small departures from the null hypotheses. The results were consistent under all scenarios considered (e.g., different covariance matrices, different values of p and q).

Although the tests are derived under special scenarios, where we assumed two clusters, the approach used in our manuscript allows extensions to several clusters (the general EGCM) to be made. Moreover, the formulation we provided in this study can also be extended to allow a more general linear hypothesis to be tested.

Acknowledgments Parts of this work was done under the supervision of Professor Dietrich von Rosen and we would like to acknowledge his contributions. This work is supported by funds from the Natural Sciences and Engineering Council of Canada (NSERC).

References

1. Fujikoshi, Y: Asymptotic expansions of the non-null distributions of three statistics in GMANOVA. Ann. Inst. Stat. Math. **26**, 289–297 (1974)
2. Hamid, J.S.: New residuals in multivariate bilinear models. PhD dissertation, Swedish University of Agricultural Sciences (2005). ISBN 91-576-6982-1
3. Hamid, J.S., von Rosen, D.: Residuals in the extended growth curve model. Scand. J. Stat. **33**(1), 121–138 (2006)
4. Hamid, J.S., Beyene, J., von Rosen, D.: A novel trace test for the mean parameters in a multivariate growth curve model. J. Multivar. Anal. **102**(2), 238–251 (2011)
5. Jana, S., Balakrishnan, N., Hamid, J.S.: High dimensional extension of the growth curve model and its application in genetics. Stat. Methods Appl. **6**(2), 273–292 (2017)
6. Jana, S., Balakrishnan, N., Hamid, J.S.: Inference in the growth curve model under multivariate skew normal distribution. Sankhya B. (2018). https://doi.org/10.1007/s13571-018-0174-1
7. Jana, S., Balakrishnan, N., Hamid, J.S.: Estimation of the parameters of the extended growth curve model under multivariate skew normal distribution. J. Multivar. Anal. **166**(C), 111–128 (2018)
8. Jana, S., Balakrishnan, N., Hamid, J.S.: Bayesian growth curve model useful for high-dimensional longitudinal data. J. Appl. Stat. **46**(5), 814–834 (2019)
9. Kariya, T.: The general MANOVA problem. Ann. Stat. **6**(1), 200–214 (1978)
10. Kariya, T.: Testing in the Multivariate General Linear Model. Kinokuniya Co., Tokyo (1985)

11. Khatri, C.G.: A note on a MANOVA model applied to the problems in growth curve. Ann. Inst. Stat. Math. **18**, 75–86 (1966)
12. Kollo, T., von Rosen, D.: Advanced Multivariate Statistics with Matrices. Springer, New York (2005)
13. Kshirsagar, A.M., Smith, W.B.: Growth Curves. Marcel Dekker, New York (1995)
14. Liski, E.P.: Detecting influential measurements in a growth curve model. Biometrics **47**, 659–668 (1991)
15. Nzabanita, J., Singull, M., von Rosen, D.: Estimation of parameters in the extended growth curve model with a linearly structured covariance matrix. Acta Comment. Univ. Tartu. Math. **16**, 13–32 (2012)
16. Nzabanita, J., von Rosen, D., Singull, M.: Bilinear regression model with Kronecker and linear structures for the covariance matrix. Afr. Stat. **10**, 827–837 (2015)
17. Nzabanita, J., von Rosen, D., Singull, M.: Extended GMANOVA model with a linearly structured covariance matrix. Math. Methods Stat. **24**, 280–291 (2015)
18. Ohlson, M., von Rosen, D.: Explicit estimators of parameters in the growth curve model with linearly structured covariance matrices. J. Multivar. Anal. **101**(5), 1284–1295 (2010)
19. Pan, J.X., Fang, K.T.: Multiple outlier detection in growth curve model with unstructured covariance matrix. Ann. Inst. Stat. Math. **47**, 137–153 (1995)
20. Pan, J.X., Fang, K.T.: Influential observations in the growth curve model with unstructured covariance matrix. Comput. Stat. Data Anal. **22**, 71–87 (1996)
21. Pan, J.X., Fang, K.T.: Growth Curve Model and Statistical Diagnostics. Springer, New York (2002)
22. Potthoff, R.F., Roy, S.N.: A generalized multivariate analysis of variance model useful especially for growth curves. Biometrika **51**, 313–326 (1964)
23. Rao, C.R.: The theory of least squares when the parameters are stochastic and its application to the analysis of growth curves. Biometrika **52**, 447–458 (1965)
24. Srivastava, M.S., Singull, M.: Test for the mean matrix in a growth curve model for high dimensions. Commun. Stat. Theory Methods. **46**(13), 6668–6683 (2017)
25. Srivastava, M.S., Singull, M.: Testing sphericity and intraclass covariance structures under a growth curve model in high dimension. Commun. Stat. Simul. Comput. **46**(7), 5740–5751 (2017)
26. von Rosen, D.: Maximum likelihood estimators in multivariate linear normal models. J. Multivar. Anal. **31**, 187–200 (1989)
27. von Rosen, D.: Moments of a multivariate linear model with an application to the growth curve model. J. Multivar. Anal. **35**, 243–259 (1990)
28. von Rosen, D.: The growth curve model: a review. Commun. Stat. Theory Methods. **20**, 2791–1822 (1991)
29. von Rosen, D.: Residuals in the growth curve model. Ann. Inst. Stat. Math. **47**, 129–136 (1995)
30. von Rosen, D.: Bilinear Regression Analysis: An Introduction. Springer International Publishing, Part of Springer Nature, Cham (2018)
31. Verbyla, A.P., Venables, W.N.: An extension of the growth curve model. Biometrika **75**(1), 129–138 (1988)
32. Zerbe, G.O.: Randomization analysis of the completely randomized design extended to growth and response curves. J. Am. Stat. Assoc. **74**, 215–221 (1979)

Chapter 8
Properties of BLUEs and BLUPs in Full vs. Small Linear Models with New Observations

Stephen J. Haslett, Augustyn Markiewicz, and Simo Puntanen

Abstract In this article we consider the partitioned linear model $\mathscr{M}_{12} = \{\mathbf{y}, \mathbf{X}_1\boldsymbol{\beta}_1 + \mathbf{X}_2\boldsymbol{\beta}_2, \mathbf{V}\}$, where $\boldsymbol{\mu} = \mathbf{X}_1\boldsymbol{\beta}_1 + \mathbf{X}_2\boldsymbol{\beta}_2$, and the corresponding small model $\mathscr{M}_1 = \{\mathbf{y}, \mathbf{X}_1\boldsymbol{\beta}_1, \mathbf{V}\}$, where $\boldsymbol{\mu}_1 = \mathbf{X}_1\boldsymbol{\beta}_1$. These models are supplemented with the new unobservable random vector $\mathbf{y}_*$, coming from $\mathbf{y}_* = \mathbf{K}\boldsymbol{\beta}_1 + \boldsymbol{\varepsilon}_*$, where the covariance matrix of $\mathbf{y}_*$ is known as well as the cross-covariance matrix between $\mathbf{y}_*$ and $\mathbf{y}$. We focus on comparing the BLUEs of $\boldsymbol{\mu}_1$ and $\boldsymbol{\mu}$, and BLUPs of $\mathbf{y}_*$ and $\boldsymbol{\varepsilon}_*$ under $\mathscr{M}_{12}$ and $\mathscr{M}_1$.

8.1 Introduction

In this paper we consider the partitioned linear model $\mathbf{y} = \mathbf{X}_1\boldsymbol{\beta}_1 + \mathbf{X}_2\boldsymbol{\beta}_2 + \boldsymbol{\varepsilon}$ and so-called small model (submodel) $\mathbf{y} = \mathbf{X}_1\boldsymbol{\beta}_1 + \boldsymbol{\varepsilon}$, or shortly

$$\mathscr{A}_{12} = \{\mathbf{y}, \mathbf{X}\boldsymbol{\beta}, \mathbf{V}\} = \{\mathbf{y}, \mathbf{X}_1\boldsymbol{\beta}_1 + \mathbf{X}_2\boldsymbol{\beta}_2, \mathbf{V}\}, \quad \mathscr{A}_1 = \{\mathbf{y}, \mathbf{X}_1\boldsymbol{\beta}_1, \mathbf{V}\}. \tag{8.1}$$

Here $\mathbf{y}$ is an n-dimensional observable response variable, and $\boldsymbol{\varepsilon}$ is an unobservable random error with a known covariance matrix $\operatorname{cov}(\boldsymbol{\varepsilon}) = \mathbf{V} = \operatorname{cov}(\mathbf{y})$ and expectation $\mathrm{E}(\boldsymbol{\varepsilon}) = \mathbf{0}$. The matrix $\mathbf{X}$ is a known $n \times p$ matrix, i.e., $\mathbf{X} \in \mathbb{R}^{n\times p}$, partitioned columnwise as $\mathbf{X} = (\mathbf{X}_1 : \mathbf{X}_2)$, $\mathbf{X}_i \in \mathbb{R}^{n\times p_i}$, $i = 1, 2$. Vector

S. J. Haslett
School of Fundamental Sciences, Massey University, Palmerston North, New Zealand
e-mail: s.j.haslett@massey.ac.nz

A. Markiewicz
Department of Mathematical and Statistical Methods, Poznań University of Life Sciences, Poznań, Poland
e-mail: amark@up.poznan.pl

S. Puntanen (✉)
Faculty of Information Technology and Communication Sciences, Tampere University, Tampere, Finland
e-mail: simo.puntanen@tuni.fi

© Springer Nature Switzerland AG 2020

T. Holgersson, M. Singull (eds.), *Recent Developments in Multivariate and Random Matrix Analysis*, https://doi.org/10.1007/978-3-030-56773-6_8

$\boldsymbol{\beta} = (\boldsymbol{\beta}_1', \boldsymbol{\beta}_2')' \in \mathbb{R}^p$ is a vector of fixed (but unknown) parameters; here symbol $'$ stands for the transpose.

Let the new unknown q-dimensional future response $\mathbf{y}_*$ be

$$\mathbf{y}_* = \mathbf{X}_*\boldsymbol{\beta} + \boldsymbol{\varepsilon}_* = \mathbf{K}\boldsymbol{\beta}_1 + \boldsymbol{\varepsilon}_*\,, \quad \mathbf{X}_* = (\mathbf{K} : \mathbf{0})\,, \quad \mathbf{K} \in \mathbb{R}^{q \times p_1}, \tag{8.2}$$

and

$$\operatorname{cov}\begin{pmatrix}\mathbf{y}\\ \mathbf{y}_*\end{pmatrix} = \begin{pmatrix}\mathbf{V} & \mathbf{V}_{12}\\ \mathbf{V}_{21} & \mathbf{V}_{22}\end{pmatrix}. \tag{8.3}$$

Of course, the word "new" need not be taken here literally. Putting $\mathscr{A}_{12}$, $\mathscr{A}_1$ and (8.2) together, we can denote the models shortly as

$$\mathscr{M}_1 = \left\{\begin{pmatrix}\mathbf{y}\\ \mathbf{y}_*\end{pmatrix}, \begin{pmatrix}\mathbf{X}_1\\ \mathbf{K}\end{pmatrix}\boldsymbol{\beta}_1, \begin{pmatrix}\mathbf{V} & \mathbf{V}_{12}\\ \mathbf{V}_{21} & \mathbf{V}_{22}\end{pmatrix}\right\}, \tag{8.4a}$$

$$\begin{aligned}\mathscr{M}_{12} &= \left\{\begin{pmatrix}\mathbf{y}\\ \mathbf{y}_*\end{pmatrix}, \begin{pmatrix}\mathbf{X}\\ \mathbf{X}_*\end{pmatrix}\boldsymbol{\beta}, \begin{pmatrix}\mathbf{V} & \mathbf{V}_{12}\\ \mathbf{V}_{21} & \mathbf{V}_{22}\end{pmatrix}\right\}\\ &= \left\{\begin{pmatrix}\mathbf{y}\\ \mathbf{y}_*\end{pmatrix}, \begin{pmatrix}\mathbf{X}_1 & \mathbf{X}_2\\ \mathbf{K} & \mathbf{0}\end{pmatrix}\boldsymbol{\beta}, \begin{pmatrix}\mathbf{V} & \mathbf{V}_{12}\\ \mathbf{V}_{21} & \mathbf{V}_{22}\end{pmatrix}\right\}.\end{aligned} \tag{8.4b}$$

Thus: $\mathscr{M}_{12}$ is the full model with new observations and $\mathscr{M}_1$ is the small model with new observations. We may drop off the subscripts from $\mathscr{M}_{12}$ if the partitioning is not essential in the context. We are interested in estimating $\boldsymbol{\mu}_* = \mathbf{X}_*\boldsymbol{\beta}$ and predicting $\mathbf{y}_*$ and $\boldsymbol{\varepsilon}_*$ on the basis of $\mathbf{y}$.

As for notations, the symbols $\mathrm{r}(\mathbf{A})$, $\mathbf{A}^-$, $\mathbf{A}^+$, $\mathscr{C}(\mathbf{A})$, and $\mathscr{C}(\mathbf{A})^\perp$, denote, respectively, the rank, a generalized inverse, the (unique) Moore–Penrose inverse, the column space, and the orthogonal complement of the column space of the matrix $\mathbf{A}$. By $\mathbf{A}^\perp$ we denote any matrix satisfying $\mathscr{C}(\mathbf{A}^\perp) = \mathscr{C}(\mathbf{A})^\perp$. Furthermore, we will write $\mathbf{P}_\mathbf{A} = \mathbf{P}_{\mathscr{C}(\mathbf{A})} = \mathbf{A}\mathbf{A}^+ = \mathbf{A}(\mathbf{A}'\mathbf{A})^-\mathbf{A}'$ to denote the orthogonal projector (with respect to the standard inner product) onto $\mathscr{C}(\mathbf{A})$. The orthogonal projector onto $\mathscr{C}(\mathbf{A})^\perp$ is denoted as $\mathbf{Q}_\mathbf{A} = \mathbf{I}_a - \mathbf{P}_\mathbf{A}$, where $\mathbf{I}_a$ refers to the $a \times a$ identity matrix and a is the number of rows of $\mathbf{A}$. It appears convenient to use the short notations

$$\mathbf{M} = \mathbf{I}_n - \mathbf{P}_\mathbf{X}\,, \quad \mathbf{M}_i = \mathbf{I}_n - \mathbf{P}_{\mathbf{X}_i}\,, \; i = 1, 2. \tag{8.5}$$

One obvious choice for $\mathbf{X}^\perp$ is $\mathbf{M}$.

When using generalized inverses it is important to know whether the expressions are independent of the choice of the generalized inverses involved. Lemma 8.1 below gives some invariance conditions; cf. Rao and Mitra [21, Lemma 2.2.4].

Lemma 8.1 *For nonnull matrices* $\mathbf{A}$ *and* $\mathbf{C}$ *the following holds:*

(a) $\mathbf{AB^-C} = \mathbf{AB^+C}$ *for all* $\mathbf{B}^-$ $\iff$ $\mathscr{C}(\mathbf{C}) \subset \mathscr{C}(\mathbf{B})$ & $\mathscr{C}(\mathbf{A}') \subset \mathscr{C}(\mathbf{B}')$.
(b) $\mathbf{AA^-C} = \mathbf{C}$ *for some (and hence for all)* $\mathbf{A}^-$ $\iff$ $\mathscr{C}(\mathbf{C}) \subset \mathscr{C}(\mathbf{A})$.

Let the set $\mathscr{W}$ of nonnegative definite matrices be defined as

$$\mathscr{W} = \left\{\mathbf{W} \in \mathbb{R}^{n\times n} : \mathbf{W} = \mathbf{V} + \mathbf{XUU'X'},\ \mathscr{C}(\mathbf{W}) = \mathscr{C}(\mathbf{X} : \mathbf{V})\right\}. \tag{8.6}$$

In (8.6), $\mathbf{U}$ can be any matrix comprising p rows as long as $\mathscr{C}(\mathbf{W}) = \mathscr{C}(\mathbf{X} : \mathbf{V})$ is satisfied. Lemma 8.2 collects together some important properties of the class $\mathscr{W}$; see, e.g., Baksalary et al. [1, Th. 2] and Puntanen et al. [18, Sect. 12.3].

Lemma 8.2 *Let* $\mathbf{V}$ *be an* $n \times n$ *nonnegative definite matrix, let* $\mathbf{X}$ *be an* $n \times p$ *matrix, and define* $\mathbf{W}$ *as* $\mathbf{W} = \mathbf{V} + \mathbf{XUU'X'}$, *where* $\mathbf{U}$ *is a* $p \times p$ *matrix, i.e.,* $\mathbf{W} \in \mathscr{W}$. *Then the following statements are equivalent:*

(a) $\mathscr{C}(\mathbf{X} : \mathbf{V}) = \mathscr{C}(\mathbf{W})$,
(b) $\mathscr{C}(\mathbf{X}) \subset \mathscr{C}(\mathbf{W})$,
(c) $\mathscr{C}(\mathbf{X'W^-X}) = \mathscr{C}(\mathbf{X}')$ *for any choice of* $\mathbf{W}^-$,
(d) $\mathbf{X(X'W^-X)^-X'W^-X} = \mathbf{X}$ *for any choices of* $\mathbf{W}^-$ *and* $(\mathbf{X'W^-X})^-$.

For the partitioned linear model $\mathscr{M}_{12}$ we will say that $\mathbf{W} \in \mathscr{W}$ if the following properties hold:

$$\mathbf{W} = \mathbf{V} + \mathbf{XUU'X'} = \mathbf{V} + \mathbf{X}_1\mathbf{U}_1\mathbf{U}_1'\mathbf{X}_1' + \mathbf{X}_2\mathbf{U}_2\mathbf{U}_2'\mathbf{X}_2', \tag{8.7a}$$

$$\mathbf{W}_i = \mathbf{V} + \mathbf{X}_i\mathbf{U}_i\mathbf{U}_i'\mathbf{X}_i', \quad i = 1, 2, \tag{8.7b}$$

$$\mathscr{C}(\mathbf{W}) = \mathscr{C}(\mathbf{X} : \mathbf{V}), \quad \mathscr{C}(\mathbf{W}_i) = \mathscr{C}(\mathbf{X}_i : \mathbf{V}), \quad i = 1, 2. \tag{8.7c}$$

The particular choice of $\mathbf{U} = (\mathbf{X}_1 : \mathbf{U}_2)$ does not matter in our considerations and for simplicity we have put $\mathbf{U}_1'\mathbf{U}_2 = \mathbf{0}$.

By the consistency of the model $\mathscr{M}$ it is meant that $\mathbf{y}$ lies in $\mathscr{C}(\mathbf{X} : \mathbf{V})$ with probability 1; see, e.g., Baksalary et al. [2]. Hence we assume that under the consistent model $\mathscr{M}$ the observed numerical value of $\mathbf{y}$ satisfies

$$\mathbf{y} \in \mathscr{C}(\mathbf{X} : \mathbf{V}) = \mathscr{C}(\mathbf{X} : \mathbf{VX}^{\perp}) = \mathscr{C}(\mathbf{X} : \mathbf{VM}) = \mathscr{C}(\mathbf{X}) \oplus \mathscr{C}(\mathbf{VM}), \tag{8.8}$$

where "$\oplus$" refers to the direct sum, implying that $\mathscr{C}(\mathbf{X}) \cap \mathscr{C}(\mathbf{VX}^{\perp}) = \{\mathbf{0}\}$. For the equality $\mathscr{C}(\mathbf{X} : \mathbf{V}) = \mathscr{C}(\mathbf{X} : \mathbf{VM})$, we refer to Rao [20, Lemma 2.1]. There is a related decomposition, see, e.g., Puntanen et al. [18, Th. 8]: for any conformable matrices $\mathbf{A}$ and $\mathbf{B}$ we have

$$\mathscr{C}(\mathbf{A} : \mathbf{B}) = \mathscr{C}(\mathbf{A} : \mathbf{Q_A B}), \text{ and thereby } \mathbf{P}_{(\mathbf{A}:\mathbf{B})} = \mathbf{P}_{\mathbf{A}} + \mathbf{P}_{\mathbf{Q_A B}}. \tag{8.9}$$

Thus we can obtain part (a) of Lemma 8.3 below.

Lemma 8.3 *Consider* $\mathbf{X} = (\mathbf{X}_1 : \mathbf{X}_2)$ *and let* $\mathbf{M}_2 = \mathbf{I}_n - \mathbf{P}_{\mathbf{X}_2}$. *Then*

(a) $\mathbf{M} = \mathbf{I}_n - \mathbf{P}_{(\mathbf{X}_1:\mathbf{X}_2)} = \mathbf{I}_n - (\mathbf{P}_{\mathbf{X}_2} + \mathbf{P}_{\mathbf{M}_2\mathbf{X}_1}) = \mathbf{M}_2\mathbf{Q}_{\mathbf{M}_2\mathbf{X}_1} = \mathbf{Q}_{\mathbf{M}_2\mathbf{X}_1}\mathbf{M}_2$,
(b) $\mathscr{C}(\mathbf{X}_2) \subset \mathscr{C}(\mathbf{X}_1 : \mathbf{V}) \iff \mathscr{C}(\mathbf{M}_1\mathbf{X}_2) \subset \mathscr{C}(\mathbf{M}_1\mathbf{V})$.

For the following lemma, see, e.g., Isotalo et al. [12], Puntanen et al. [18, Prop. 15.2] and Markiewicz and Puntanen [16, Sec. 4].

Lemma 8.4 *Consider the partitioned linear model* $\{\mathbf{y}, \mathbf{X}_1\boldsymbol{\beta}_1 + \mathbf{X}_2\boldsymbol{\beta}_2, \mathbf{V}\}$, *let* $\mathbf{W} = \mathbf{V} + \mathbf{XUU}'\mathbf{X}' \in \mathscr{W}$ *and denote* $\mathbf{M}_1 = \mathbf{I}_n - \mathbf{P}_{\mathbf{X}_1}$ *and*

$$\dot{\mathbf{M}} = \mathbf{M}(\mathbf{MVM})^-\mathbf{M}, \quad \dot{\mathbf{M}}_1 = \mathbf{M}_1(\mathbf{M}_1\mathbf{VM}_1)^-\mathbf{M}_1. \tag{8.10a}$$

Then the following equalities hold:

(a) $$\begin{aligned} \mathbf{X}(\mathbf{X}'\mathbf{W}^-\mathbf{X})^-\mathbf{X}'\mathbf{W}^+ &= \mathbf{P}_\mathbf{W} - \mathbf{VM}(\mathbf{MVM})^-\mathbf{MP}_\mathbf{W} \\ &= \mathbf{P}_\mathbf{W} - \mathbf{V}\dot{\mathbf{M}}\mathbf{P}_\mathbf{W} \\ &= \mathbf{P}_\mathbf{W} - \mathbf{VM}(\mathbf{MVM})^+ \\ &= \mathbf{P}_\mathbf{W} - \mathbf{VM}(\mathbf{MVM})^+\mathbf{M}. \end{aligned}$$
(b) *Replacing* $\mathbf{X}, \mathbf{W}, \mathbf{M}$ *and* $\dot{\mathbf{M}}$ *with* $\mathbf{X}_1, \mathbf{W}_1, \mathbf{M}_1$ *and* $\dot{\mathbf{M}}_1$ *in (a), the corresponding expressions for* $\mathbf{X}_1(\mathbf{X}_1'\mathbf{W}_1^-\mathbf{X}_1)^-\mathbf{X}_1'\mathbf{W}_1^+$ *can be obtained.*

A couple of clarifying words about Lemma 8.4 may be in place. We observe that

$$\begin{aligned} \mathbf{VM}(\mathbf{MVM})^-\mathbf{MP}_\mathbf{W} &= \mathbf{VM}(\mathbf{MVM})^-\mathbf{P}_{\mathbf{MV}} \\ &= \mathbf{VM}(\mathbf{MVM})^+\mathbf{P}_{\mathbf{MV}} \\ &= \mathbf{VM}(\mathbf{MVM})^+, \end{aligned} \tag{8.11}$$

where we have used Lemmas 8.1 and 8.3, which gives

$$\mathbf{MP}_\mathbf{W} = \mathbf{M}(\mathbf{P}_\mathbf{X} + \mathbf{P}_{\mathbf{MV}}) = \mathbf{MP}_{\mathbf{MV}} = \mathbf{P}_{\mathbf{MV}}. \tag{8.12}$$

In addition, it is noteworthy that the matrix $\dot{\mathbf{M}} = \mathbf{M}(\mathbf{MVM})^-\mathbf{M}$ is unique with respect to the choice of $(\mathbf{MVM})^-$ if and only if $\mathbb{R}^n = \mathscr{C}(\mathbf{X} : \mathbf{V})$, see Isotalo et al. [12, p. 1439]. For the Moore–Penrose inverse the following holds:

$$\mathbf{M}(\mathbf{MVM})^+\mathbf{M} = (\mathbf{MVM})^+\mathbf{M} = \mathbf{M}(\mathbf{MVM})^+ = (\mathbf{MVM})^+. \tag{8.13}$$

Let $\mathbf{A}$ and $\mathbf{B}$ be arbitrary $m \times n$ matrices. Then, in the consistent linear model $\mathscr{M}$, the estimators $\mathbf{Ay}$ and $\mathbf{By}$ are said to be equal with probability 1 if

$$\mathbf{Ay} = \mathbf{By} \quad \text{for all } \mathbf{y} \in \mathscr{C}(\mathbf{X} : \mathbf{V}) = \mathscr{C}(\mathbf{W}), \tag{8.14}$$

where $\mathbf{W} \in \mathscr{W}$. Thus, if $\mathbf{A}$ and $\mathbf{B}$ satisfy (8.14), then $\mathbf{A} - \mathbf{B} = \mathbf{CQ}_\mathbf{W}$ for some matrix $\mathbf{C}$. When talking about the equality of estimators like $\mathbf{Ay} = \mathbf{By}$, we often drop off the phrase "with probability 1".

The properties of the BLUE deserve particular attention when $\mathscr{C}(\mathbf{X} : \mathbf{V}) = \mathbb{R}^n$ does not hold: then there is an infinite number of multipliers $\mathbf{B}$ such that $\mathbf{By}$ is BLUE but for all such multipliers the vector $\mathbf{By}$ itself is unique (with probability 1, which is the phrase in this context). In the case of two linear models, $\mathscr{B}_i = \{\mathbf{y}, \mathbf{X}\boldsymbol{\beta}, \mathbf{V}_i\}$, $i = 1, 2$, Mitra and Moore [17] divide the problems into three questions:

(a) When is a specific linear representation of the BLUE of $\boldsymbol{\mu} = \mathbf{X}\boldsymbol{\beta}$ under $\mathscr{B}_1$ also a BLUE under $\mathscr{B}_2$?
(b) When does $\boldsymbol{\mu} = \mathbf{X}\boldsymbol{\beta}$ have a common BLUE under $\mathscr{B}_1$ and $\mathscr{B}_2$?
(c) When is the BLUE of $\boldsymbol{\mu} = \mathbf{X}\boldsymbol{\beta}$ under $\mathscr{B}_1$ irrespective of the linear representation used in its expression, also a BLUE under $\mathscr{B}_2$?

The purpose of this paper is to do considerations in the spirit of Mitra and Moore [17] regarding the models $\mathscr{M}_{12}$ and $\mathscr{M}_1$. We pick up particular fixed representations for the BLUEs and BLUPs under these two models, study the conditions under which they are equal for all values of $\mathbf{y} \in \mathscr{C}(\mathbf{X}_1 : \mathbf{X}_2 : \mathbf{V})$ or $\mathbf{y} \in \mathscr{C}(\mathbf{X}_1 : \mathbf{V})$. Moreover, we review the conditions under which *all* representations of the BLUEs and BLUPs in one model continue to be valid in the other model. Corresponding relations between the covariance matrices of the BLUEs, BLUPs and prediction errors are characterized. The well-known (or pretty well-known) results are given as Lemmas, while the new (or at least not so well-known) results are represented as Propositions. As this paper is more like a review-type, though providing some new characterizations, we provide a reasonable background and matrix tools, to make the article more self-contained, i.e., easier to read. Most of this background material is in the first two sections.

8.2 Fundamental BLUE and BLUP Equations

A linear statistic $\mathbf{By}$ is said to be linear unbiased estimator (LUE) for $\boldsymbol{\mu}_* = \mathbf{X}_*\boldsymbol{\beta}$ in $\mathscr{M}_{12}$ if its expectation is equal to $\boldsymbol{\mu}_*$, which happens if and only if $\mathbf{X}_*' = \mathbf{X}'\mathbf{B}'$. For our purposes, the parametric function $\boldsymbol{\mu}_* = \mathbf{K}\boldsymbol{\beta}_1$ must be estimable in $\mathscr{M}_{12}$ and $\mathscr{M}_1$ as well. Now, see, e.g., Groß and Puntanen [4, Lemma 1],

$$\boldsymbol{\mu}_* = \mathbf{X}_*\boldsymbol{\beta} = (\mathbf{K} : \mathbf{0})\boldsymbol{\beta} = \mathbf{K}\boldsymbol{\beta}_1 \quad \text{is estimable under } \mathscr{M}_{12} \tag{8.15}$$

if and only if $\mathscr{C}(\mathbf{K}') \subset \mathscr{C}(\mathbf{X}_1'\mathbf{M}_2)$, i.e., $\mathbf{K} = \mathbf{J}\mathbf{M}_2\mathbf{X}_1$. Thus

$$\mathbf{X}_* = (\mathbf{K} : \mathbf{0}) = (\mathbf{J}\mathbf{M}_2\mathbf{X}_1 : \mathbf{0}) = \mathbf{J}\mathbf{M}_2\mathbf{X} = \mathbf{L}\mathbf{X}\,, \text{ where } \mathbf{L} = \mathbf{J}\mathbf{M}_2\,. \tag{8.16}$$

This means that for our purpose it is essential to consider the best LUE, i.e., the BLUE of $\mathbf{M}_2\mathbf{X}_1\boldsymbol{\beta}_1$. Obviously (8.16) means that $\mathbf{K}\boldsymbol{\beta}_1$ is estimable also under $\mathscr{M}_1$.

The LUE $\mathbf{By}$ is the best linear unbiased estimator, BLUE, of estimable $\mathbf{X}_*\boldsymbol{\beta}$ if $\mathbf{By}$ has the smallest covariance matrix in the Löwner sense among all LUEs of $\mathbf{X}_*\boldsymbol{\beta}$:

$$\operatorname{cov}(\mathbf{By}) \leq_{\mathrm{L}} \operatorname{cov}(\mathbf{B}_{\#}\mathbf{y}) \quad \text{for all } \mathbf{B}_{\#} : \mathbf{B}_{\#}\mathbf{X} = \mathbf{X}_* \,. \tag{8.17}$$

Correspondingly, the linear predictor $\mathbf{Ay}$ is said to be unbiased for $\mathbf{y}_*$ if the expected prediction error is zero, i.e., $\mathrm{E}(\mathbf{y}_* - \mathbf{Ay}) = \mathbf{0}$ for all $\boldsymbol{\beta} \in \mathbb{R}^p$, which happens if and only if $\mathbf{X}_*' = \mathbf{X}'\mathbf{A}'$. When $\mathscr{C}(\mathbf{X}_*') \subset \mathscr{C}(\mathbf{X}')$ holds, we will say that $\mathbf{y}_*$ is predictable under $\mathscr{M}$. Now a linear unbiased predictor $\mathbf{Ay}$ is the best linear unbiased predictor, BLUP, for $\mathbf{y}_*$, if we have the Löwner ordering

$$\operatorname{cov}(\mathbf{y}_* - \mathbf{Ay}) \leq_{\mathrm{L}} \operatorname{cov}(\mathbf{y}_* - \mathbf{A}_{\#}\mathbf{y}) \quad \text{for all } \mathbf{A}_{\#} : \mathbf{A}_{\#}\mathbf{X} = \mathbf{X}_* \,. \tag{8.18}$$

Consider then the BLUP of $\boldsymbol{\varepsilon}_*$. Obviously $\mathbf{Dy}$ is an unbiased predictor for $\boldsymbol{\varepsilon}_*$ if and only if $\mathbf{DX} = \mathbf{0}$, i.e., $\mathbf{D} = \mathbf{FM}$ for some $\mathbf{L}$. Thus the unbiased $\mathbf{Dy}$ is the BLUP for $\boldsymbol{\varepsilon}_*$ if and only if

$$\operatorname{cov}(\boldsymbol{\varepsilon}_* - \mathbf{Dy}) \leq_{\mathrm{L}} \operatorname{cov}(\boldsymbol{\varepsilon}_* - \mathbf{FMy}) \quad \text{for all } \mathbf{F} \in \mathbb{R}^{q\times n}. \tag{8.19}$$

For Lemma 8.5, characterizing the BLUE, see, e.g., Rao [19, p. 282], and the BLUP, see, e.g., Christensen [3, p. 294], and Isotalo and Puntanen [11, p. 1015]. For part (d), see Isotalo et al. [10, Th. 3.1]. For the general reviews of the BLUP-properties, see, e.g., Tian [23, 24], [9], and Markiewicz and Puntanen [15].

Lemma 8.5 *Consider the linear model with new observations defined as* $\mathscr{M}_{12}$ *where* $\mathscr{C}(\mathbf{X}_*') \subset \mathscr{C}(\mathbf{X}')$, *i.e.,* $\mathbf{y}_*$ *is predictable. Then the following statements hold:*

(a) $\mathbf{Ay} = \mathrm{BLUP}(\mathbf{y}_*) \iff \mathbf{A}(\mathbf{X} : \mathbf{VX}^{\perp}) = (\mathbf{X}_* : \mathbf{V}_{21}\mathbf{X}^{\perp})\,,$ *i.e.,* $\mathbf{A} \in \{\mathbf{P}_{\mathbf{y}_*|\mathscr{M}_{12}}\}$.
(b) $\mathbf{By} = \mathrm{BLUE}(\mathbf{X}_*\boldsymbol{\beta}) \iff \mathbf{B}(\mathbf{X} : \mathbf{VX}^{\perp}) = (\mathbf{X}_* : \mathbf{0})\,,$ *i.e.,* $\mathbf{B} \in \{\mathbf{P}_{\mathbf{X}_*|\mathscr{M}_{12}}\}$.
(c) $\mathbf{Cy} = \mathrm{BLUE}(\mathbf{X}\boldsymbol{\beta}) \iff \mathbf{C}(\mathbf{X} : \mathbf{VX}^{\perp}) = (\mathbf{X} : \mathbf{0})\,,$ *i.e.,* $\mathbf{C} \in \{\mathbf{P}_{\mathbf{X}|\mathscr{M}_{12}}\}$.
(d) $\mathbf{Dy} = \mathrm{BLUP}(\boldsymbol{\varepsilon}_*) \iff \mathbf{D}(\mathbf{X} : \mathbf{VX}^{\perp}) = (\mathbf{0} : \mathbf{V}_{21}\mathbf{X}^{\perp})\,,$ *i.e.,* $\mathbf{D} \in \{\mathbf{P}_{\boldsymbol{\varepsilon}_*|\mathscr{M}_{12}}\}$.

The sets $\{\mathbf{P}_{\mathbf{y}_*|\mathscr{M}_1}\}$, $\{\mathbf{P}_{\mathbf{X}_1|\mathscr{M}_1}\}$ and $\{\mathbf{P}_{\boldsymbol{\varepsilon}_*|\mathscr{M}_1}\}$ are defined in the corresponding way. Putting (b) and (d) of Lemma 8.5 together yields

$$\begin{pmatrix}\mathbf{B}\\ \mathbf{D}\end{pmatrix}(\mathbf{X} : \mathbf{VX}^{\perp}) = \begin{pmatrix}\mathbf{X}_* & \mathbf{0}\\ \mathbf{0} & \mathbf{V}_{21}\mathbf{X}^{\perp}\end{pmatrix}, \tag{8.20}$$

which implies that

$$\mathbf{A}(\mathbf{X} : \mathbf{VX}^{\perp}) = (\mathbf{B} + \mathbf{D})(\mathbf{X} : \mathbf{VX}^{\perp}) = (\mathbf{X}_* : \mathbf{V}_{21}\mathbf{X}^{\perp})\,, \tag{8.21}$$

and thereby $(\mathbf{B} + \mathbf{D})\mathbf{y}$ is the BLUP for $\mathbf{y}_*$ and we have

$$\mathrm{BLUP}(\mathbf{y}_*) = \mathrm{BLUE}(\mathbf{X}_*\boldsymbol{\beta}) + \mathrm{BLUP}(\boldsymbol{\varepsilon}_*)\,, \quad \text{i.e., } \tilde{\mathbf{y}}_* = \tilde{\boldsymbol{\mu}}_* + \tilde{\boldsymbol{\varepsilon}}_* \,. \tag{8.22}$$

Using Lemma 8.2 we can obtain, for example, the following well-known solutions to $\mathbf{B}$ and $\mathbf{C}$ in Lemma 8.5:

$$\mathbf{X}_*(\mathbf{X}'\mathbf{W}^-\mathbf{X})^-\mathbf{X}'\mathbf{W}^- \in \{\mathbf{P}_{\mathbf{X}_*|\mathscr{M}}\}, \quad \mathbf{X}(\mathbf{X}'\mathbf{W}^-\mathbf{X})^-\mathbf{X}'\mathbf{W}^- \in \{\mathbf{P}_{\mathbf{X}|\mathscr{M}}\}, \tag{8.23}$$

where $\mathbf{W} \in \mathscr{W}$ and we can freely choose the generalized inverses involved. Expression $\mathbf{X}(\mathbf{X}'\mathbf{W}^-\mathbf{X}')^-\mathbf{X}'\mathbf{W}^-$ is not necessarily unique with respect to the choice of $\mathbf{W}^-$ but

$$\mathbf{G} = \mathbf{X}(\mathbf{X}'\mathbf{W}^-\mathbf{X}')^-\mathbf{X}'\mathbf{W}^+ \in \{\mathbf{P}_{\mathbf{X}|\mathscr{M}}\} \tag{8.24}$$

is unique whatever choice of $\mathbf{W}^-$ we have. The *general* solution for $\mathbf{C}$ in Lemma 8.5, can be expressed, for example, as

$$\mathbf{P}_{\mathbf{X}|\mathscr{M}} = \mathbf{G} + \mathbf{N}\mathbf{Q}_\mathbf{W}, \quad \text{where } \mathbf{N} \in \mathbb{R}^{n\times n} \text{ is free to vary,} \tag{8.25}$$

and $\mathbf{Q}_\mathbf{W} = \mathbf{I}_n - \mathbf{P}_\mathbf{W}$. Thus the solution for $\mathbf{C}$ is unique if and only if $\mathscr{C}(\mathbf{X} : \mathbf{V}) = \mathbb{R}^n$.

In particular, in view of Lemma 8.4, we have the following:

$$\begin{aligned}\mathbf{G} &= \mathbf{X}(\mathbf{X}'\mathbf{W}^-\mathbf{X})^-\mathbf{X}'\mathbf{W}^+ = \mathbf{P}_\mathbf{W} - \mathbf{VM}(\mathbf{MVM})^-\mathbf{MP}_\mathbf{W} \\ &= \mathbf{P}_\mathbf{W} - \mathbf{VM}(\mathbf{MVM})^+ = \mathbf{P}_\mathbf{W} - \mathbf{VM}(\mathbf{MVM})^+\mathbf{M},\end{aligned} \tag{8.26}$$

and thus

$$\mathbf{I}_n - \mathbf{G} = \mathbf{Q}_\mathbf{W} + \mathbf{VM}(\mathbf{MVM})^-\mathbf{MP}_\mathbf{W} = \mathbf{Q}_\mathbf{W} + \mathbf{VM}(\mathbf{MVM})^+\mathbf{M}. \tag{8.27}$$

Corresponding expressions for $\mathbf{G}_1$ can be obtained by replacing $\mathbf{X}$, $\mathbf{M}$ and $\mathbf{W}$ with $\mathbf{X}_1$, $\mathbf{M}_1$ and $\mathbf{W}_1$, respectively, in (8.26). Premultiplying (8.26) by $\mathbf{P}_\mathbf{X}$ gives

$$\mathbf{G} = \mathbf{P}_\mathbf{X} - \mathbf{P}_\mathbf{X}\mathbf{VM}(\mathbf{MVM})^+\mathbf{M}, \tag{8.28}$$

and correspondingly,

$$\mathbf{G}_1 = \mathbf{P}_{\mathbf{X}_1} - \mathbf{P}_{\mathbf{X}_1}\mathbf{VM}_1(\mathbf{M}_1\mathbf{VM}_1)^+\mathbf{M}_1. \tag{8.29}$$

Notice that by Lemma 8.1,

- $\mathbf{G} = \mathbf{X}(\mathbf{X}'\mathbf{W}^-\mathbf{X})^-\mathbf{X}'\mathbf{W}^+$ is unique for any choice of $\mathbf{W}^-$ and $(\mathbf{X}'\mathbf{W}^-\mathbf{X})^-$,
- $\mathbf{VM}(\mathbf{MVM})^-\mathbf{MP}_\mathbf{W}$ is unique for any choice of $(\mathbf{MVM})^-$,
- $\mathbf{VM}(\mathbf{MVM})^-\mathbf{M}$ is unique for any choice of $(\mathbf{MVM})^-$ if and only if $\mathrm{r}(\mathbf{MV}) = \mathrm{r}(\mathbf{M})$, i.e., $\mathrm{r}(\mathbf{X} : \mathbf{V}) = n$.

In order to consider the BLUP($\boldsymbol{\varepsilon}_*$) under $\mathscr{M}_{12}$, we observe that

$$\mathbf{V}_{21}\mathbf{V}^+(\mathbf{I}_n - \mathbf{G})(\mathbf{X} : \mathbf{VM}) = (\mathbf{0} : \mathbf{V}_{21}\mathbf{M}), \tag{8.30}$$

and thus $\mathbf{V}_{21}\mathbf{V}^{+}(\mathbf{I}_n - \mathbf{G}) \in \{\mathbf{P}_{\boldsymbol{\varepsilon}_*|\cdot\mathscr{M}_{12}}\}$. On the other hand, in view of (8.27),

$$\mathbf{V}_{21}\mathbf{V}^{+}(\mathbf{I}_n - \mathbf{G}) = \mathbf{V}_{21}\mathbf{M}(\mathbf{MVM})^{-}\mathbf{MP}_{\mathbf{W}} = \mathbf{V}_{21}\mathbf{M}(\mathbf{MVM})^{+}\mathbf{M}\,, \tag{8.31}$$

and so

$$\begin{aligned}\tilde{\boldsymbol{\varepsilon}}_* &= \mathrm{BLUP}(\boldsymbol{\varepsilon}_* \mid \mathscr{M}_{12}) = \mathbf{V}_{21}\mathbf{V}^{+}(\mathbf{I}_n - \mathbf{G})\mathbf{y} \\ &= \mathbf{V}_{21}\mathbf{M}(\mathbf{MVM})^{-}\mathbf{MP}_{\mathbf{W}}\mathbf{y} \\ &= \mathbf{V}_{21}\mathbf{M}(\mathbf{MVM})^{+}\mathbf{My} \\ &= \mathbf{Ey},\end{aligned} \tag{8.32}$$

where we have denoted

$$\mathbf{E} = \mathbf{V}_{21}\mathbf{V}^{+}(\mathbf{I}_n - \mathbf{G}) = \mathbf{V}_{21}\mathbf{M}(\mathbf{MVM})^{+}\mathbf{M} \in \{\mathbf{P}_{\boldsymbol{\varepsilon}_*|\cdot\mathscr{M}_{12}}\}\,. \tag{8.33}$$

Equation (8.32) holds for *any* $\mathbf{y} \in \mathbb{R}^n$. In particular, if $\mathbf{y} \in \mathscr{C}(\mathbf{W})$, then we can replace $(\mathbf{MVM})^{+}$ with any $(\mathbf{MVM})^{-}$. In the case of the small model we denote

$$\mathbf{E}_1 = \mathbf{V}_{21}\mathbf{V}^{+}(\mathbf{I}_n - \mathbf{G}_1) = \mathbf{V}_{21}\mathbf{M}_1(\mathbf{M}_1\mathbf{VM}_1)^{+}\mathbf{M}_1 \in \{\mathbf{P}_{\boldsymbol{\varepsilon}_*|\cdot\mathscr{M}_1}\}\,. \tag{8.34}$$

Moreover, it can be observed that

$$\mathbf{E} = \mathbf{V}_{21}\mathbf{W}^{+}(\mathbf{I}_n - \mathbf{G})\,, \quad \mathbf{E}_1 = \mathbf{V}_{21}\mathbf{W}_1^{+}(\mathbf{I}_n - \mathbf{G}_1)\,. \tag{8.35}$$

Let us denote $\mathbf{L} = \mathbf{JM}_2$ and $\mathbf{S} = \mathbf{L} - \mathbf{V}_{21}\mathbf{V}^{+}$. Then we can write

$$\begin{aligned}\tilde{\mathbf{y}}_* &= \mathrm{BLUP}(\tilde{\mathbf{y}}_* \mid \mathscr{M}_{12}) \\ &= \mathbf{LGy} + \mathbf{V}_{21}\mathbf{V}^{+}(\mathbf{y} - \mathbf{Gy}) = \mathbf{LGy} + \mathbf{Ey} \\ &= (\mathbf{L} - \mathbf{V}_{21}\mathbf{V}^{+})\mathbf{Gy} + \mathbf{V}_{21}\mathbf{V}^{+}\mathbf{y} = \mathbf{SGy} + \mathbf{V}_{21}\mathbf{V}^{+}\mathbf{y} \\ &= \mathbf{Ty},\end{aligned} \tag{8.36}$$

where

$$\begin{aligned}\mathbf{T} &= \mathbf{LG} + \mathbf{V}_{21}\mathbf{M}(\mathbf{MVM})^{-}\mathbf{MP}_{\mathbf{W}} = \mathbf{LG} + \mathbf{E} \\ &= \mathbf{SG} + \mathbf{V}_{21}\mathbf{V}^{+} \in \{\mathbf{P}_{\mathbf{y}_*|\cdot\mathscr{M}_{12}}\}\,.\end{aligned} \tag{8.37}$$

Let us put our results together:

Lemma 8.6 *Let* $\mathbf{y}_*$ *be predictable under* $\mathscr{M}_{12}$*, so that*

$$\mathbf{X}_* = (\mathbf{K} : \mathbf{0}) = \mathbf{JM}_2\mathbf{X} = \mathbf{LX} = (\mathbf{JM}_2\mathbf{X}_1 : \mathbf{0}) = (\mathbf{LX}_1 : \mathbf{0}) \tag{8.38}$$

for some $\mathbf{J} \in \mathbb{R}^{q\times n}$, $\mathbf{L} = \mathbf{J}\mathbf{M}_2$, $\mathbf{S} = \mathbf{L} - \mathbf{V}_{21}\mathbf{V}^+$ *and*

$$\mathbf{G} = \mathbf{X}(\mathbf{X}'\mathbf{W}^-\mathbf{X}')^-\mathbf{X}'\mathbf{W}^+, \qquad \mathbf{G}_1 = \mathbf{X}_1(\mathbf{X}_1'\mathbf{W}^-\mathbf{X}_1)^-\mathbf{X}_1'\mathbf{W}^+, \tag{8.39a}$$

$$\mathbf{E} = \mathbf{V}_{21}\mathbf{M}(\mathbf{M}\mathbf{V}\mathbf{M})^+\mathbf{M}, \qquad \mathbf{E}_1 = \mathbf{V}_{21}\mathbf{M}_1(\mathbf{M}_1\mathbf{V}\mathbf{M}_1)^-\mathbf{M}_1. \tag{8.39b}$$

Then the BLUP($\mathbf{y}_*$) *under* $\mathscr{M}_{12}$ *can be written as*

$$\begin{aligned} \mathrm{BLUP}(\mathbf{y}_* \mid \mathscr{M}_{12}) &= \tilde{\mathbf{y}}_* \\ &= \mathbf{L}\mathbf{G}\mathbf{y} + \mathbf{V}_{21}\mathbf{V}^+(\mathbf{I}_n - \mathbf{G})\mathbf{y} \\ &= \mathbf{L}\mathbf{G}\mathbf{y} + \mathbf{V}_{21}\mathbf{M}(\mathbf{M}\mathbf{V}\mathbf{M})^-\mathbf{M}\mathbf{P}_{\mathbf{W}}\mathbf{y} \\ &= \mathbf{L}\mathbf{G}\mathbf{y} + \mathbf{V}_{21}\mathbf{M}(\mathbf{M}\mathbf{V}\mathbf{M})^+\mathbf{M}\mathbf{y} \\ &= \mathbf{L}\mathbf{G}\mathbf{y} + \mathbf{E}\mathbf{y} \\ &= (\mathbf{L} - \mathbf{V}_{21}\mathbf{V}^+)\mathbf{G}\mathbf{y} + \mathbf{V}_{21}\mathbf{V}^+\mathbf{y} \\ &= \mathbf{S}\mathbf{G}\mathbf{y} + \mathbf{V}_{21}\mathbf{V}^+\mathbf{y} \\ &= \mathbf{T}\mathbf{y}, \end{aligned} \tag{8.40}$$

or shortly,

$$\tilde{\mathbf{y}}_* = \tilde{\boldsymbol{\mu}}_* + \tilde{\boldsymbol{\varepsilon}}_*. \tag{8.41}$$

Corresponding expressions for the BLUP($\mathbf{y}_*$) *under* $\mathscr{M}_1$, *i.e., for* $\tilde{\mathbf{y}}_{*1} = \mathbf{T}_1\mathbf{y}$ *can be obtained by replacing* $\mathbf{G}$, $\mathbf{X}$, $\mathbf{M}$ *and* $\mathbf{W}$ *with* $\mathbf{G}_1$, $\mathbf{X}_1$, $\mathbf{M}_1$ *and* $\mathbf{W}_1$, *respectively, in* (8.40)*; shortly,*

$$\tilde{\mathbf{y}}_{*1} = \tilde{\boldsymbol{\mu}}_{*1} + \tilde{\boldsymbol{\varepsilon}}_{*1}. \tag{8.42}$$

For the covariance matrices we get

$$\mathrm{cov}(\tilde{\boldsymbol{\mu}}) = \mathrm{cov}(\mathbf{G}\mathbf{y}) = \mathbf{G}\mathbf{V}\mathbf{G}' = \mathbf{V} - \mathbf{V}\mathbf{M}(\mathbf{M}\mathbf{V}\mathbf{M})^-\mathbf{M}\mathbf{V}, \tag{8.43a}$$

$$\mathrm{cov}(\tilde{\boldsymbol{\mu}}_*) = \mathrm{cov}(\mathbf{L}\mathbf{G}\mathbf{y}) = \mathbf{L}\mathbf{G}\mathbf{V}\mathbf{G}'\mathbf{L}', \tag{8.43b}$$

$$\mathrm{cov}(\tilde{\boldsymbol{\varepsilon}}_*) = \mathrm{cov}(\mathbf{E}\mathbf{y}) = \mathbf{E}\mathbf{V}\mathbf{E}' = \mathbf{V}_{21}\mathbf{M}(\mathbf{M}\mathbf{V}\mathbf{M})^-\mathbf{M}\mathbf{V}_{12}. \tag{8.43c}$$

For an extensive review of the BLUE's covariance matrix, see Isotalo et al. [13]. The random vectors $\tilde{\boldsymbol{\mu}}_*$ and $\tilde{\boldsymbol{\varepsilon}}_*$ are uncorrelated,

$$\mathrm{cov}(\tilde{\boldsymbol{\mu}}_*, \tilde{\boldsymbol{\varepsilon}}_*) = \mathrm{cov}(\mathbf{L}\mathbf{G}\mathbf{y}, \mathbf{E}\mathbf{y}) = \mathbf{L}\mathbf{G}\mathbf{V}\mathbf{M}(\mathbf{M}\mathbf{V}\mathbf{M})^+\mathbf{M} = \mathbf{0}, \tag{8.44}$$

where we have used the fact that $\mathbf{GVM} = \mathbf{0}$ and thereby

$$\operatorname{cov}(\tilde{\mathbf{y}}_*) = \operatorname{cov}(\tilde{\boldsymbol{\mu}}_*) + \operatorname{cov}(\tilde{\boldsymbol{\varepsilon}}_*) = \mathbf{LGVG'L'} + \mathbf{V}_{21}\mathbf{M}(\mathbf{MVM})^{-}\mathbf{MV}_{12}\,. \tag{8.45}$$

8.3 Equalities of the BLUEs Under the Full and Small Models

Let us start by considering the equality between $\operatorname{cov}(\mathbf{Gy})$ and $\operatorname{cov}(\mathbf{G}_1\mathbf{y})$. In light of $\mathbf{GG}_1 = \mathbf{G}_1$, we have

$$\operatorname{cov}(\tilde{\boldsymbol{\mu}}_1 \mid \mathscr{M}_1) = \mathbf{G}_1\mathbf{VG}_1' = \mathbf{GG}_1\mathbf{VG}_1'\mathbf{G}', \tag{8.46}$$

and thus

$$\begin{aligned} \operatorname{cov}(\tilde{\boldsymbol{\mu}} \mid \mathscr{M}_{12}) - \operatorname{cov}(\tilde{\boldsymbol{\mu}}_1 \mid \mathscr{M}_1) &= \mathbf{GVG}' - \mathbf{GG}_1\mathbf{VG}_1'\mathbf{G}' \\ &= \mathbf{G}(\mathbf{V} - \mathbf{G}_1\mathbf{VG}_1')\mathbf{G}' \\ &= \mathbf{GVM}_1(\mathbf{M}_1\mathbf{VM}_1)^{-}\mathbf{M}_1\mathbf{VG}' \\ &= \mathbf{GV\dot{M}}_1\mathbf{VG}', \end{aligned} \tag{8.47}$$

where we have used

$$\operatorname{cov}(\mathbf{G}_1\mathbf{y}) = \mathbf{G}_1\mathbf{VG}_1' = \mathbf{V} - \mathbf{VM}_1(\mathbf{M}_1\mathbf{VM}_1)^{-}\mathbf{M}_1\mathbf{V}. \tag{8.48}$$

Clearly $\mathbf{GV\dot{M}}_1\mathbf{VG}'$ is nonnegative definite and thereby

$$\operatorname{cov}(\tilde{\boldsymbol{\mu}}_1 \mid \mathscr{M}_1) \leq_{\mathrm{L}} \operatorname{cov}(\tilde{\boldsymbol{\mu}} \mid \mathscr{M}_{12})\,. \tag{8.49}$$

It is obvious that the equality $\operatorname{cov}(\tilde{\boldsymbol{\mu}}_1 \mid \mathscr{M}_1) = \operatorname{cov}(\tilde{\boldsymbol{\mu}} \mid \mathscr{M}_{12})$ holds if and only if

$$\mathbf{GVM}_1 = \mathbf{0}\,. \tag{8.50}$$

Actually, the above equality (8.50) is a necessary and sufficient condition for $\mathbf{Gy}$ being the BLUE for $\boldsymbol{\mu}_1$ under the small model $\mathscr{M}_1$. Recall that the fundamental BLUE equation in this case is

$$\mathbf{G}(\mathbf{X}_1 : \mathbf{VM}_1) = (\mathbf{X}_1 : \mathbf{0})\,, \tag{8.51}$$

where the left-hand part $\mathbf{GX}_1 = \mathbf{X}_1$ trivially holds. Thus (8.50) is equivalent to $\mathbf{G} \in \{\mathbf{P}_{\mathbf{X}_1|\mathscr{M}_1}\}$. The general expression for a member of the class $\{\mathbf{P}_{\mathbf{X}|\mathscr{M}}\}$ is

$$\mathbf{P}_{\mathbf{X}|\mathscr{M}} = \mathbf{G} + \mathbf{NQ_W}\,, \quad \text{where } \mathbf{N} \text{ is free to vary.} \tag{8.52}$$

It is easy to confirm that $\{\mathbf{P}_{\mathbf{X}|\mathscr{M}}\} \subset \{\mathbf{P}_{\mathbf{X}_1|\mathscr{M}_1}\}$ if and only if $\mathbf{GVM}_1 = \mathbf{0}$. In other words, every representation of the BLUE of $\boldsymbol{\mu}$ under $\mathscr{M}_{12}$ is BLUE also under $\mathscr{M}_1$, for which we can use notation

$$\{\mathrm{BLUE}(\boldsymbol{\mu} \mid \mathscr{M}_{12})\} \subset \{\mathrm{BLUE}(\boldsymbol{\mu}_1 \mid \mathscr{M}_1)\}, \quad \text{i.e.,} \quad \{\mathbf{P}_{\mathbf{X}|\mathscr{M}_{12}}\} \subset \{\mathbf{P}_{\mathbf{X}_1|\mathscr{M}_1}\}. \tag{8.53}$$

It may be mentioned that writing up the condition $\mathbf{GVM}_1 = \mathbf{0}$ we obtain

$$\mathbf{X}(\mathbf{X}'\mathbf{W}^-\mathbf{X}')^-\mathbf{X}'\mathbf{W}^+\mathbf{VM}_1 = \mathbf{0}, \tag{8.54}$$

which is equivalent to

$$\mathbf{X}'\mathbf{W}^+\mathbf{VM}_1 = \mathbf{0}. \tag{8.55}$$

Consider then the covariance matrix of $\mathbf{Gy} - \mathbf{G}_1\mathbf{y}$:

$$\mathrm{cov}(\mathbf{Gy} - \mathbf{G}_1\mathbf{y}) = \mathbf{GVG}' + \mathbf{G}_1\mathbf{VG}_1' - \mathbf{GVG}_1' - \mathbf{G}_1\mathbf{VG}'. \tag{8.56}$$

In view of $\mathbf{G}_1 = \mathbf{P}_{\mathbf{W}_1} - \mathbf{VM}_1(\mathbf{M}_1\mathbf{VM}_1)^+\mathbf{M}_1$, we have

$$\begin{aligned} \mathbf{G}_1\mathbf{VG}' &= [\mathbf{P}_{\mathbf{W}_1} - \mathbf{VM}_1(\mathbf{M}_1\mathbf{VM}_1)^+\mathbf{M}_1]\mathbf{VG}' \\ &= [\mathbf{V} - \mathbf{VM}_1(\mathbf{M}_1\mathbf{VM}_1)^+\mathbf{M}_1\mathbf{V}]\mathbf{G}' \\ &= \mathbf{G}_1\mathbf{VG}_1'\mathbf{G}' = \mathbf{G}_1\mathbf{VG}_1', \end{aligned} \tag{8.57}$$

where we have used $\mathbf{GG}_1 = \mathbf{G}_1$ and $\mathbf{G}_1\mathbf{VG}_1' = \mathbf{V} - \mathbf{V}\dot{\mathbf{M}}_1\mathbf{V}$. Thus

$$\begin{aligned} \mathrm{cov}(\mathbf{Gy} - \mathbf{G}_1\mathbf{y}) &= \mathrm{cov}(\mathbf{Gy}) - \mathrm{cov}(\mathbf{G}_1\mathbf{y}) \\ &= \mathbf{GVM}_1(\mathbf{M}_1\mathbf{VM}_1)^-\mathbf{M}_1\mathbf{VG}' \\ &= \mathrm{cov}\,[\mathbf{GVM}_1(\mathbf{M}_1\mathbf{VM}_1)^-\mathbf{M}_1\mathbf{y}]. \end{aligned} \tag{8.58}$$

Notice that the matrix $\mathbf{GV}\dot{\mathbf{M}}_1$ may not be unique but $\mathrm{cov}(\mathbf{GV}\dot{\mathbf{M}}_1\mathbf{y})$ is unique with respect to the choice of $(\mathbf{M}_1\mathbf{VM}_1)^-$ in $\dot{\mathbf{M}}_1 = \mathbf{M}_1(\mathbf{M}_1\mathbf{VM}_1)^-\mathbf{M}_1$.

We can now put our findings together:

Proposition 8.1 *The following statements are equivalent:*

(a) $\mathbf{Gy} = \mathbf{G}_1\mathbf{y}$ *for all* $\mathbf{y} \in \mathscr{C}(\mathbf{W}_1) = \mathscr{C}(\mathbf{X}_1 : \mathbf{VM}_1)$,
(b) $\mathrm{cov}(\tilde{\boldsymbol{\mu}}_1 \mid \mathscr{M}_1) = \mathrm{cov}(\tilde{\boldsymbol{\mu}} \mid \mathscr{M}_{12})$,
(c) $\mathbf{GVM}_1 = \mathbf{0}$,
(d) $\mathbf{X}'\mathbf{W}^+\mathbf{VM}_1 = \mathbf{0}$,
(e) $\mathbf{G} \in \{\mathbf{P}_{\mathbf{X}_1|\mathscr{M}_1}\}$,
(f) $\{\mathbf{P}_{\mathbf{X}|\mathscr{M}_{12}}\} \subset \{\mathbf{P}_{\mathbf{X}_1|\mathscr{M}_1}\}$, *i.e.,* $\{\mathrm{BLUE}(\boldsymbol{\mu} \mid \mathscr{M}_{12})\} \subset \{\mathrm{BLUE}(\boldsymbol{\mu}_1 \mid \mathscr{M}_1)\}$.

Moreover, the following properties hold:

(g) $\mathbf{G}_1\mathbf{V}\mathbf{G}' = \mathbf{G}_1\mathbf{V}\mathbf{G}_1'$,
(h) $\mathrm{cov}(\mathbf{G}\mathbf{y} - \mathbf{G}_1\mathbf{y}) = \mathrm{cov}(\mathbf{G}\mathbf{y}) - \mathrm{cov}(\mathbf{G}_1\mathbf{y}) = \mathrm{cov}(\mathbf{G}\mathbf{V}\dot{\mathbf{M}}_1\mathbf{y})$,
(i) $\mathrm{cov}(\mathbf{G}_1\mathbf{y}) \leq_{\mathrm{L}} \mathrm{cov}(\mathbf{G}\mathbf{y})$.

What about the equality

$$\mathbf{G}\mathbf{y} = \mathbf{G}_1\mathbf{y} \quad \text{for all } \mathbf{y} \in \mathscr{C}(\mathbf{W}) = \mathscr{C}(\mathbf{X}_1 : \mathbf{X}_2 : \mathbf{V}\mathbf{M}) , \tag{8.59}$$

i.e.,

$$\mathbf{G}_1(\mathbf{X}_1 : \mathbf{X}_2 : \mathbf{V}\mathbf{M}) = (\mathbf{X}_1 : \mathbf{X}_2 : \mathbf{0}) . \tag{8.60}$$

Noting that $\mathbf{G}_1\mathbf{X}_1 = \mathbf{X}_1$ and $\mathbf{G}_1\mathbf{V}\mathbf{M} = \mathbf{G}_1\mathbf{V}\mathbf{M}_1\mathbf{Q}_{\mathbf{M}_1\mathbf{X}_2} = \mathbf{0}$, we conclude that (8.60) holds, i.e., $\mathbf{G}_1 \in \{\mathbf{P}_{\mathbf{X}|\mathscr{M}_{12}}\}$, if and only if

$$\mathbf{G}_1\mathbf{X}_2 = \mathbf{X}_2 , \quad \text{i.e.,} \quad \mathbf{X}_1(\mathbf{X}_1'\mathbf{W}_1^-\mathbf{X}_1')^-\mathbf{X}_1'\mathbf{W}_1^+\mathbf{X}_2 = \mathbf{X}_2 . \tag{8.61}$$

It is clear that (8.61) implies

$$\mathscr{C}(\mathbf{X}_2) \subset \mathscr{C}(\mathbf{X}_1) . \tag{8.62}$$

On the other hand, if (8.62) holds then $\mathbf{X}_2 = \mathbf{X}_1\mathbf{A}$ for some $\mathbf{A}$ which further implies (8.61). Assuming that (8.62) holds, i.e., $\mathscr{C}(\mathbf{X}) = \mathscr{C}(\mathbf{X}_1)$, we can use (8.28) and (8.29) and conclude that $\mathbf{G} = \mathbf{G}_1$ holds if and only if (8.62) holds.

Let us consider the general expression for the member of the class $\{\mathbf{P}_{\mathbf{X}_1|\mathscr{M}_1}\}$:

$$\mathbf{P}_{\mathbf{X}_1|\mathscr{M}_1} = \mathbf{G}_1 + \mathbf{N}\mathbf{Q}_{\mathbf{W}_1} \quad \text{for some } \mathbf{N} . \tag{8.63}$$

The equality

$$\mathbf{G}\mathbf{y} = (\mathbf{G}_1 + \mathbf{N}\mathbf{Q}_{\mathbf{W}_1})\mathbf{y} \quad \text{for all } \mathbf{y} \in \mathscr{C}(\mathbf{X}_1 : \mathbf{X}_2 : \mathbf{V}\mathbf{M}) , \tag{8.64}$$

i.e., $(\mathbf{G}_1 + \mathbf{N}\mathbf{Q}_{\mathbf{W}_1})(\mathbf{X}_1 : \mathbf{X}_2 : \mathbf{V}\mathbf{M}) = (\mathbf{X}_1 : \mathbf{X}_2 : \mathbf{0})$, simplifies to

$$\mathbf{G}_1\mathbf{X}_2 + \mathbf{N}\mathbf{Q}_{\mathbf{W}_1}\mathbf{X}_2 = \mathbf{X}_2 . \tag{8.65}$$

Requesting (8.65) to hold for *any* $\mathbf{N}$ yields $\mathscr{C}(\mathbf{X}_2) \subset \mathscr{C}(\mathbf{W}_1)$, and consequently, $\mathbf{G}_1\mathbf{X}_2 = \mathbf{X}_2$. Thus we conclude the following:

$$\{\mathbf{P}_{\mathbf{X}_1|\mathscr{M}_1}\} \subset \{\mathbf{P}_{\mathbf{X}|\mathscr{M}_{12}}\} \iff \mathscr{C}(\mathbf{X}_2) \subset \mathscr{C}(\mathbf{X}_1) . \tag{8.66}$$

Moreover, the inclusion $\mathscr{C}(\mathbf{X}_2) \subset \mathscr{C}(\mathbf{X}_1)$ implies that $\mathbf{GVM}_1 = \mathbf{0}$, which further, by Proposition 8.1, implies that $\{\mathbf{P}_{\mathbf{X}|\mathscr{M}_{12}}\} \subset \{\mathbf{P}_{\mathbf{X}_1|\mathscr{M}_1}\}$.

We can also pose a question under which the set $\{\mathbf{P}_{\mathbf{X}|\mathscr{M}_{12}}\} \cap \{\mathbf{P}_{\mathbf{X}_1|\mathscr{M}_1}\}$ is not empty. This happens whenever the equation

$$\mathbf{A}(\mathbf{X}_1 : \mathbf{X}_2 : \mathbf{VM} : \mathbf{X}_1 : \mathbf{VM}_1) = (\mathbf{X}_1 : \mathbf{X}_2 : \mathbf{0} : \mathbf{X}_1 : \mathbf{0})\,, \tag{8.67}$$

i.e., $\mathbf{A}(\mathbf{X}_1 : \mathbf{X}_2 : \mathbf{VM}_1) = (\mathbf{X}_1 : \mathbf{X}_2 : \mathbf{0})$ has a solution for $\mathbf{A}$. This happens if and only if

$$\mathscr{C}\begin{pmatrix}\mathbf{X}' \\ \mathbf{0}\end{pmatrix} \subset \mathscr{C}\begin{pmatrix}\mathbf{X}' \\ \mathbf{M}_1\mathbf{V}\end{pmatrix}, \tag{8.68}$$

which can be expressed equivalently as $\mathscr{C}(\mathbf{X}) \cap \mathscr{C}(\mathbf{VM}_1) = \{\mathbf{0}\}$; see Puntanen et al. [18, Ch. 16].

Thus we have proved the following:

Proposition 8.2 *The following statements are equivalent:*

(a) $\mathbf{Gy} = \mathbf{G}_1\mathbf{y}$ *for all* $\mathbf{y} \in \mathscr{C}(\mathbf{W}) = \mathscr{C}(\mathbf{X}_1 : \mathbf{X}_2 : \mathbf{VM})$,
(b) $\mathbf{Gy} = \mathbf{G}_1\mathbf{y}$ *for all* $\mathbf{y} \in \mathbb{R}^n$,
(c) $\mathbf{G}_1\mathbf{X}_2 = \mathbf{X}_2$,
(d) $\mathbf{G}_1 \in \{\mathbf{P}_{\mathbf{X}|\mathscr{M}_{12}}\}$,
(e) $\mathscr{C}(\mathbf{X}_2) \subset \mathscr{C}(\mathbf{X}_1)$,
(f) $\{\mathbf{P}_{\mathbf{X}_1|\mathscr{M}_1}\} \subset \{\mathbf{P}_{\mathbf{X}|\mathscr{M}_{12}}\}$, *i.e.*, $\{\mathrm{BLUE}(\boldsymbol{\mu}_1 \mid \mathscr{M}_1)\} \subset \{\mathrm{BLUE}(\boldsymbol{\mu} \mid \mathscr{M}_{12})\}$,
(g) $\{\mathbf{P}_{\mathbf{X}_1|\mathscr{M}_1}\} = \{\mathbf{P}_{\mathbf{X}|\mathscr{M}_{12}}\}$, *i.e.*, $\{\mathrm{BLUE}(\boldsymbol{\mu}_1 \mid \mathscr{M}_1)\} = \{\mathrm{BLUE}(\boldsymbol{\mu} \mid \mathscr{M}_{12})\}$.

Moreover,

(h) $\{\mathbf{P}_{\mathbf{X}|\mathscr{M}_{12}}\} \cap \{\mathbf{P}_{\mathbf{X}_1|\mathscr{M}_1}\} \neq \{\emptyset\} \iff \mathscr{C}(\mathbf{X}) \cap \mathscr{C}(\mathbf{VM}_1) = \{\mathbf{0}\}$.

In this context it is convenient to refer to the possible equality of $\tilde{\boldsymbol{\mu}}_1(\mathscr{M}_1)$ and $\tilde{\boldsymbol{\mu}}_1(\mathscr{M}_{12})$. Notice that in this section we have put our attention on the equality between $\tilde{\boldsymbol{\mu}}_1(\mathscr{M}_1)$ and $\tilde{\boldsymbol{\mu}}(\mathscr{M}_{12})$. Haslett and Puntanen Haslett and Puntanen [7] showed that if $\boldsymbol{\mu}_1 = \mathbf{X}_1\boldsymbol{\beta}_1$ is estimable under $\mathscr{M}_{12}$ and $\mathscr{C}(\mathbf{X}_2) \subset \mathscr{C}(\mathbf{W}_1)$, then

$$\tilde{\boldsymbol{\mu}}_1(\mathscr{M}_{12}) = \tilde{\boldsymbol{\mu}}_1(\mathscr{M}_1) - \mathbf{X}_1(\mathbf{X}_1'\mathbf{W}_1^-\mathbf{X}_1)^-\mathbf{X}_1'\mathbf{W}_1^+\,\tilde{\boldsymbol{\mu}}_2(\mathscr{M}_{12})\,. \tag{8.69}$$

From (8.69) it can be concluded that $\tilde{\boldsymbol{\mu}}_1(\mathscr{M}_{12}) = \tilde{\boldsymbol{\mu}}_1(\mathscr{M}_1)$ holds if and only if $\mathbf{X}_1'\mathbf{W}_1^+\mathbf{X}_2 = \mathbf{0}$; see, e.g., Markiewicz and Puntanen [16, Sec. 4]. Some related considerations, using different approach, appear in Lu et al. [14] and Tian and Zhang [25].

8.4 Equalities of the BLUPs of the Error Term

The covariance matrices for $\tilde{\boldsymbol{\varepsilon}}_1$ and $\tilde{\boldsymbol{\varepsilon}}_{*1}$ are

$$\operatorname{cov}(\tilde{\boldsymbol{\varepsilon}}_*) = \mathbf{EVE}' = \mathbf{V}_{21}\mathbf{M}(\mathbf{MW}_1\mathbf{M})^-\mathbf{MV}_{12}\,, \tag{8.70a}$$

$$\operatorname{cov}(\tilde{\boldsymbol{\varepsilon}}_{*1}) = \mathbf{E}_1\mathbf{VE}_1' = \mathbf{V}_{21}\mathbf{M}_1(\mathbf{M}_1\mathbf{W}_1\mathbf{M}_1)^-\mathbf{M}_1\mathbf{V}_{12}\,. \tag{8.70b}$$

Thus the the difference $\operatorname{cov}(\tilde{\boldsymbol{\varepsilon}}_{*1}) - \operatorname{cov}(\tilde{\boldsymbol{\varepsilon}}_*)$ can be expressed as

$$\begin{aligned}\operatorname{cov}(\tilde{\boldsymbol{\varepsilon}}_{*1}) - \operatorname{cov}(\tilde{\boldsymbol{\varepsilon}}_*) &= \mathbf{V}_{21}\mathbf{W}_1^{+1/2}(\mathbf{P}_{\mathbf{W}_1^{1/2}\mathbf{M}_1} - \mathbf{P}_{\mathbf{W}_1^{1/2}\mathbf{M}})\mathbf{W}_1^{+1/2}\mathbf{V}_{12} \\ &= \mathbf{V}_{21}\mathbf{W}_1^{+1/2}\mathbf{P}_{\mathbf{A}}\mathbf{W}_1^{+1/2}\mathbf{V}_{12}\,,\end{aligned} \tag{8.71}$$

where

$$\mathscr{C}(\mathbf{A}) = \mathscr{C}(\mathbf{W}_1^{1/2}\mathbf{M}_1) \cap \mathscr{C}(\mathbf{W}_1^{1/2}\mathbf{M})^{\perp}. \tag{8.72}$$

In (8.71) $\mathbf{W}_1^{1/2}$ refers to the nonnegative definite square root of $\mathbf{W}_1$ and $\mathbf{W}_1^{+1/2}$ is the Moore–Penrose inverse of $\mathbf{W}_1^{1/2}$, and so $\mathbf{W}_1^{1/2}\mathbf{W}_1^{+1/2} = \mathbf{P}_{\mathbf{W}_1}$. For the difference of two orthogonal projectors, see, e.g., Puntanen et al. [18, Ch. 7].

In light of (8.71), we observe that $\operatorname{cov}(\tilde{\boldsymbol{\varepsilon}}_*) \leq_{\mathrm{L}} \operatorname{cov}(\tilde{\boldsymbol{\varepsilon}}_{*1})$, while $\operatorname{cov}(\tilde{\boldsymbol{\varepsilon}}_*) = \operatorname{cov}(\tilde{\boldsymbol{\varepsilon}}_{*1})$ holds if and only if

$$\mathscr{C}(\mathbf{W}_1^{+1/2}\mathbf{V}_{12}) \subset \mathscr{C}(\mathbf{A})^{\perp} = \mathscr{C}[(\mathbf{W}_1^{1/2}\mathbf{M}_1)^{\perp} : \mathbf{W}_1^{1/2}\mathbf{M}]\,. \tag{8.73}$$

In view of Lemma 4 of Markiewicz and Puntanen [16],

$$\mathscr{C}(\mathbf{W}_1^{1/2}\mathbf{M}_1)^{\perp} = \mathscr{C}(\mathbf{W}_1^{+1/2}\mathbf{X}_1 : \mathbf{Q}_{\mathbf{W}_1})\,. \tag{8.74}$$

Thus the equality $\operatorname{cov}(\tilde{\boldsymbol{\varepsilon}}_*) = \operatorname{cov}(\tilde{\boldsymbol{\varepsilon}}_{*1})$ holds if and only if

$$\begin{aligned}\mathscr{C}(\mathbf{W}_1^{+1/2}\mathbf{V}_{12}) &\subset \mathscr{C}[(\mathbf{W}_1^{1/2}\mathbf{M}_1)^{\perp} : \mathbf{W}_1^{1/2}\mathbf{M}] \\ &= \mathscr{C}(\mathbf{W}_1^{+1/2}\mathbf{X}_1 : \mathbf{Q}_{\mathbf{W}_1} : \mathbf{W}_1^{1/2}\mathbf{M})\,.\end{aligned} \tag{8.75}$$

Premultiplying (8.75) by $\mathbf{W}_1^{1/2}$ yields

$$\mathscr{C}(\mathbf{V}_{12}) \subset \mathscr{C}(\mathbf{X}_1 : \mathbf{W}_1\mathbf{M}) = \mathscr{C}(\mathbf{X}_1 : \mathbf{VM})\,, \tag{8.76}$$

which is a necessary and sufficient condition for the equality $\operatorname{cov}(\tilde{\boldsymbol{\varepsilon}}_*) = \operatorname{cov}(\tilde{\boldsymbol{\varepsilon}}_{*1})$.

We can further show the following:

$$\begin{aligned}\operatorname{cov}(\tilde{\boldsymbol{\varepsilon}}_{*1}, \tilde{\boldsymbol{\varepsilon}}_*) &= \mathbf{E}_1\mathbf{V}\mathbf{E}' \\ &= \mathbf{V}_{21}\mathbf{M}_1(\mathbf{M}_1\mathbf{V}\mathbf{M}_1)^+\mathbf{M}_1\mathbf{V}\mathbf{M}(\mathbf{M}\mathbf{V}\mathbf{M})^+\mathbf{M}\mathbf{V}_{12} \\ &= \mathbf{V}_{21}\mathbf{M}_1(\mathbf{M}_1\mathbf{V}\mathbf{M}_1)^+\mathbf{M}_1\mathbf{V}\mathbf{M}_1\mathbf{M}(\mathbf{M}\mathbf{V}\mathbf{M})^+\mathbf{M}\mathbf{V}_{12} \\ &= \mathbf{V}_{21}\mathbf{M}_1\mathbf{M}(\mathbf{M}\mathbf{V}\mathbf{M})^+\mathbf{M}\mathbf{V}_{12} \\ &= \mathbf{E}\mathbf{V}\mathbf{E}' = \operatorname{cov}(\tilde{\boldsymbol{\varepsilon}}_*)\,, \end{aligned} \tag{8.77}$$

where we have used $\mathbf{M} = \mathbf{M}_1\mathbf{M}$ and $\mathbf{V}_{21}\mathbf{M}_1(\mathbf{M}_1\mathbf{V}\mathbf{M}_1)^+\mathbf{M}_1\mathbf{V}\mathbf{M}_1 = \mathbf{V}_{21}\mathbf{M}_1$. Thus we have proved the following:

Proposition 8.3 *Denote*

$$\mathbf{E} = \mathbf{V}_{21}\mathbf{M}(\mathbf{M}\mathbf{V}\mathbf{M})^+\mathbf{M}\,, \quad \mathbf{E}_1 = \mathbf{V}_{21}\mathbf{M}_1(\mathbf{M}_1\mathbf{V}\mathbf{M}_1)^+\mathbf{M}_1\,. \tag{8.78}$$

The following statements are equivalent:

(a) $\operatorname{cov}(\tilde{\boldsymbol{\varepsilon}}_*) = \operatorname{cov}(\tilde{\boldsymbol{\varepsilon}}_{*1})$, *i.e.*, $\mathbf{E}\mathbf{V}\mathbf{E}' = \mathbf{E}_1\mathbf{V}\mathbf{E}_1'$,
(b) $\mathscr{C}(\mathbf{V}_{12}) \subset \mathscr{C}(\mathbf{X}_1 : \mathbf{V}\mathbf{M})$.

Moreover, the following statements hold:

(c) $\mathbf{E}_1\mathbf{V}\mathbf{E}' = \mathbf{E}\mathbf{V}\mathbf{E}'$,
(d) $\operatorname{cov}(\tilde{\boldsymbol{\varepsilon}}_{*1} - \tilde{\boldsymbol{\varepsilon}}_*) = \operatorname{cov}(\tilde{\boldsymbol{\varepsilon}}_{*1}) - \operatorname{cov}(\tilde{\boldsymbol{\varepsilon}}_*)$,
(e) $\operatorname{cov}(\tilde{\boldsymbol{\varepsilon}}_*) \leq_{\mathrm{L}} \operatorname{cov}(\tilde{\boldsymbol{\varepsilon}}_{*1})$.

Here is an extended version of Proposition 8.3.

Proposition 8.4 *The following statements are equivalent:*

(a) $\mathbf{E}\mathbf{y} = \mathbf{E}_1\mathbf{y}$ *for all* $\mathbf{y} \in \mathscr{C}(\mathbf{W}_1) = \mathscr{C}(\mathbf{X}_1 : \mathbf{V}\mathbf{M}_1)$,
(b) $\mathbf{E}\mathbf{V}\mathbf{M}_1 = \mathbf{V}_{21}\mathbf{M}_1$,
(c) $\operatorname{cov}(\tilde{\boldsymbol{\varepsilon}}_*) = \operatorname{cov}(\tilde{\boldsymbol{\varepsilon}}_{*1})$, *i.e.*, $\mathbf{E}\mathbf{V}\mathbf{E}' = \mathbf{E}_1\mathbf{V}\mathbf{E}_1'$,
(d) $\mathscr{C}(\mathbf{V}_{12}) \subset \mathscr{C}(\mathbf{X}_1 : \mathbf{V}\mathbf{M})$,
(e) $\mathbf{E} \in \{\mathbf{P}_{\boldsymbol{\varepsilon}_*|\mathscr{M}_1}\}$,
(f) $\{\mathbf{P}_{\boldsymbol{\varepsilon}_*|\mathscr{M}_{12}}\} \subset \{\mathbf{P}_{\boldsymbol{\varepsilon}_*|\mathscr{M}_1}\}$, *i.e.*, $\{\mathrm{BLUP}(\boldsymbol{\varepsilon}_* \mid \mathscr{M}_{12})\} \subset \{\mathrm{BLUP}(\boldsymbol{\varepsilon}_* \mid \mathscr{M}_1)\}$,
(g) $\mathscr{C}\begin{pmatrix}\mathbf{V}\mathbf{M}_1 \\ \mathbf{V}_{21}\mathbf{M}_1\end{pmatrix} \subset \mathscr{C}\begin{pmatrix}\mathbf{X}_1 & \mathbf{X}_2 & \mathbf{V}\mathbf{M} \\ \mathbf{0} & \mathbf{0} & \mathbf{V}_{21}\mathbf{M}\end{pmatrix}$.

Proof Consider the equality $\mathbf{E}\mathbf{y} = \mathbf{E}_1\mathbf{y}$ for all $\mathbf{y} \in \mathscr{C}(\mathbf{X}_1 : \mathbf{V}\mathbf{M}_1)$. Now

$$\mathbf{E}(\mathbf{X}_1 : \mathbf{V}\mathbf{M}_1) = \mathbf{E}_1(\mathbf{X}_1 : \mathbf{V}\mathbf{M}_1) = (\mathbf{0} : \mathbf{V}_{21}\mathbf{M}_1) \tag{8.79}$$

holds if and only if

$$\mathbf{E}\mathbf{V}\mathbf{M}_1 = \mathbf{V}_{21}\mathbf{M}(\mathbf{M}\mathbf{V}\mathbf{M})^+\mathbf{M}\mathbf{V}\mathbf{M}_1 = \mathbf{V}_{21}\mathbf{M}_1\,. \tag{8.80}$$

Postmultiplying (8.80) by $(\mathbf{M}_1\mathbf{V}\mathbf{M}_1)^-\mathbf{M}_1\mathbf{V}_{12}$ yields

$$\mathbf{E}\mathbf{V}\mathbf{E}_1' = \mathbf{E}\mathbf{V}\mathbf{E}' = \mathbf{E}_1\mathbf{V}\mathbf{E}_1' \,, \tag{8.81}$$

i.e., $\operatorname{cov}(\tilde{\boldsymbol{\varepsilon}}_*) = \operatorname{cov}(\tilde{\boldsymbol{\varepsilon}}_{*1})$. On the other hand, suppose that the equality $\operatorname{cov}(\tilde{\boldsymbol{\varepsilon}}_*) = \operatorname{cov}(\tilde{\boldsymbol{\varepsilon}}_{*1})$ holds. Then by part (b) of Proposition 8.3,

$$\mathbf{V}_{12} = \mathbf{X}_1\mathbf{A} + \mathbf{V}\mathbf{M}\mathbf{B} \quad \text{for some } \mathbf{A} \text{ and } \mathbf{B} \,. \tag{8.82}$$

Straightforward calculation shows that (8.82) implies (8.80). Thus we have shown the equivalence of (a), ..., (e).

An arbitrary member of $\{\mathbf{P}_{\boldsymbol{\varepsilon}_*|\cdot\mathscr{M}_{12}}\}$ can be expressed as $\mathbf{P}_{\boldsymbol{\varepsilon}_*|\cdot\mathscr{M}_{12}} = \mathbf{E} + \mathbf{N}\mathbf{Q}_\mathbf{W}$, where $\mathbf{N}$ is free to vary. Clearly

$$(\mathbf{E} + \mathbf{N}\mathbf{Q}_\mathbf{W})(\mathbf{X}_1 : \mathbf{V}\mathbf{M}_1) = (\mathbf{0} : \mathbf{V}_{21}\mathbf{M}_1) \tag{8.83}$$

for any $\mathbf{N}$ if and only if $\mathbf{E}\mathbf{V}\mathbf{M}_1 = \mathbf{V}_{21}\mathbf{M}_1$. This proves the equivalence between (b) and (f).

Obviously $\mathscr{C}(\mathbf{V}\mathbf{M}_1) \subset \mathscr{C}(\mathbf{W})$ and so

$$\mathbf{V}\mathbf{M}_1 = (\mathbf{X}_1 : \mathbf{X}_2 : \mathbf{V}\mathbf{M})\mathbf{D} \,, \tag{8.84}$$

for some $\mathbf{D} = (\mathbf{A}' : \mathbf{B}' : \mathbf{C}')'$. Thus $\mathbf{E}\mathbf{V}\mathbf{M}_1 = \mathbf{V}_{21}\mathbf{M}_1$ gets the form

$$\begin{aligned} \mathbf{E}\mathbf{V}\mathbf{M}_1 &= \mathbf{E}(\mathbf{X}_1 : \mathbf{X}_2 : \mathbf{V}\mathbf{M})\mathbf{D} = (\mathbf{0} : \mathbf{0} : \mathbf{V}_{21}\mathbf{M})\mathbf{D} \\ &= \mathbf{V}_{21}\mathbf{M}\mathbf{C} = \mathbf{V}_{21}\mathbf{M}_1 \,, \end{aligned} \tag{8.85}$$

and thereby (g) is a necessary condition for (b). Its sufficiency follows by postmultiplying

$$\mathbf{E}(\mathbf{X}_1 : \mathbf{X}_2 : \mathbf{V}\mathbf{M}) = (\mathbf{0} : \mathbf{0} : \mathbf{V}_{21}\mathbf{M}) \tag{8.86}$$

by

$$\begin{pmatrix} \mathbf{I}_{p_1} & \mathbf{A} \\ \mathbf{0} & \mathbf{B} \\ \mathbf{0} & \mathbf{C} \end{pmatrix} \tag{8.87}$$

which yields $\mathbf{E}(\mathbf{X}_1 : \mathbf{V}\mathbf{M}_1) = (\mathbf{0} : \mathbf{V}_{21}\mathbf{M}_1)$. □

Consider then the equality

$$\mathbf{E}\mathbf{y} = \mathbf{E}_1\mathbf{y} \quad \text{for all } \mathbf{y} \in \mathscr{C}(\mathbf{W}) = \mathscr{C}(\mathbf{X}_1 : \mathbf{X}_2 : \mathbf{V}\mathbf{M}) \,, \tag{8.88}$$

i.e.,

$$\mathbf{E}_1(\mathbf{X}_1 : \mathbf{X}_2 : \mathbf{VM}) = (\mathbf{0} : \mathbf{0} : \mathbf{V}_{21}\mathbf{M})\,. \tag{8.89}$$

In view of $\mathbf{M} = \mathbf{M}_1\mathbf{Q}_{\mathbf{M}_1\mathbf{X}_2}$, we have

$$\mathbf{E}_1\mathbf{VM} = \mathbf{E}_1\mathbf{VM}_1\mathbf{Q}_{\mathbf{M}_1\mathbf{X}_2} = \mathbf{V}_{21}\mathbf{M}_1\mathbf{Q}_{\mathbf{M}_1\mathbf{X}_2} = \mathbf{V}_{21}\mathbf{M}\,, \tag{8.90}$$

and thus (8.89) becomes

$$\mathbf{E}_1\mathbf{X}_2 = \mathbf{V}_{21}\mathbf{M}_1(\mathbf{M}_1\mathbf{VM}_1)^+\mathbf{M}_1\mathbf{X}_2 = \mathbf{0}\,. \tag{8.91}$$

In this context we may mention that requesting

$$\mathbf{V}_{21}\mathbf{M}_1(\mathbf{M}_1\mathbf{VM}_1)^-\mathbf{M}_1\mathbf{X}_2 = \mathbf{0} \quad \text{for any } (\mathbf{M}_1\mathbf{VM}_1)^-, \tag{8.92}$$

yields $\mathscr{C}(\mathbf{M}_1\mathbf{X}_2) \subset \mathscr{C}(\mathbf{M}_1\mathbf{V})$, i.e., $\mathscr{C}(\mathbf{X}_2) \subset \mathscr{C}(\mathbf{X}_1 : \mathbf{V})$, where we have used Lemma 8.3.

An arbitrary member of the class $\{\mathbf{P}_{\varepsilon_*|\mathscr{M}_1}\}$ can be expressed as

$$\mathbf{P}_{\varepsilon_*|\mathscr{M}_1} = \mathbf{E}_1 + \mathbf{NQ}_{\mathbf{W}_1}\,, \tag{8.93}$$

where $\mathbf{N}$ is free to vary. Thereby, if the equality

$$(\mathbf{E}_1 + \mathbf{NQ}_{\mathbf{W}_1})(\mathbf{X}_1 : \mathbf{X}_2 : \mathbf{VM}) = (\mathbf{0} : \mathbf{0} : \mathbf{V}_{21}\mathbf{M}) \tag{8.94}$$

holds for any matrix $\mathbf{N}$, then necessarily

$$\mathscr{C}(\mathbf{X}_2) \subset \mathscr{C}(\mathbf{W}_1)\,, \tag{8.95}$$

in which case (8.94) simplifies to

$$\mathbf{E}_1\mathbf{X}_2 = \mathbf{V}_{21}\mathbf{M}_1(\mathbf{M}_1\mathbf{VM}_1)^+\mathbf{M}_1\mathbf{X}_2 = \mathbf{0}\,. \tag{8.96}$$

In view of (8.95), we have

$$\mathbf{X}_2 = \mathbf{X}_1\mathbf{A} + \mathbf{VM}_1\mathbf{B} \tag{8.97}$$

for some $\mathbf{A}$ and $\mathbf{B}$. Substituting (8.97) into (8.96) yields

$$\begin{aligned} \mathbf{E}_1\mathbf{X}_2 &= \mathbf{V}_{21}\mathbf{M}_1(\mathbf{M}_1\mathbf{VM}_1)^+\mathbf{M}_1\mathbf{VM}_1\mathbf{B} \\ &= \mathbf{V}_{21}\mathbf{M}_1\mathbf{B} \\ &= \mathbf{0}\,. \end{aligned} \tag{8.98}$$

Putting (8.97) and (8.98) together, gives

$$\mathscr{C}\begin{pmatrix}\mathbf{X}_2\\ \mathbf{0}\end{pmatrix} \subset \mathscr{C}\begin{pmatrix}\mathbf{X}_1 & \mathbf{VM}_1\\ \mathbf{0} & \mathbf{V}_{21}\mathbf{M}_1\end{pmatrix}. \tag{8.99}$$

Thus (8.99) is a necessary condition for $\{\mathbf{P}_{\boldsymbol{\varepsilon}_*|\cdot\mathscr{M}_1}\} \subset \{\mathbf{P}_{\boldsymbol{\varepsilon}_*|\cdot\mathscr{M}_{12}}\}$. Its sufficiency is straightforward to show. Thus we have proved the following:

Proposition 8.5 *The following statements are equivalent:*

(a) $\mathbf{Ey} = \mathbf{E}_1\mathbf{y}$ *for all* $\mathbf{y} \in \mathscr{C}(\mathbf{W}) = \mathscr{C}(\mathbf{X}_1 : \mathbf{X}_2 : \mathbf{VM})$,
(b) $\mathbf{E}_1\mathbf{X}_2 = \mathbf{0}$.

Moreover, the following statements are equivalent:

(c) $\mathbf{Ey} = (\mathbf{E}_1 + \mathbf{NQ}_{\mathbf{W}_1})\mathbf{y}$ *for all* $\mathbf{y} \in \mathscr{C}(\mathbf{W})$ *and for all* $\mathbf{N}$,
(d) $\mathbf{E}_1\mathbf{X}_2 = \mathbf{0}$ *and* $\mathscr{C}(\mathbf{X}_2) \subset \mathscr{C}(\mathbf{W}_1)$,
(e) $\mathbf{E}_1\mathbf{X}_2 = \mathbf{V}_{21}\mathbf{M}_1(\mathbf{M}_1\mathbf{VM}_1)^-\mathbf{M}_1\mathbf{X}_2 = \mathbf{0}$ *for all* $(\mathbf{M}_1\mathbf{VM}_1)^-$,
(f) $\{\mathbf{P}_{\boldsymbol{\varepsilon}_*|\cdot\mathscr{M}_1}\} \subset \{\mathbf{P}_{\boldsymbol{\varepsilon}_*|\cdot\mathscr{M}_{12}}\}$,
(g) $\mathscr{C}\begin{pmatrix}\mathbf{X}_2\\ \mathbf{0}\end{pmatrix} \subset \mathscr{C}\begin{pmatrix}\mathbf{X}_1 & \mathbf{VM}_1\\ \mathbf{0} & \mathbf{V}_{21}\mathbf{M}_1\end{pmatrix}$.

The following result can be straightforwardly confirmed.

Proposition 8.6 *The following statements are equivalent:*

(a) $\{\mathbf{P}_{\boldsymbol{\varepsilon}_*|\cdot\mathscr{M}_1}\} = \{\mathbf{P}_{\boldsymbol{\varepsilon}_*|\cdot\mathscr{M}_{12}}\}$,
(b) $\mathscr{C}\begin{pmatrix}\mathbf{X}_1 & \mathbf{X}_2 & \mathbf{VM}\\ \mathbf{0} & \mathbf{0} & \mathbf{V}_{21}\mathbf{M}\end{pmatrix} = \mathscr{C}\begin{pmatrix}\mathbf{X}_1 & \mathbf{VM}_1\\ \mathbf{0} & \mathbf{V}_{21}\mathbf{M}_1\end{pmatrix}$.

8.5 Properties of the BLUPs of the Future Response

Let us recall the notations $\tilde{\mathbf{y}}_* = \mathbf{Ty}$ and $\tilde{\mathbf{y}}_{*1} = \mathbf{T}_1\mathbf{y}$, where

$$\mathbf{T} = \mathbf{LG} + \mathbf{E} = \mathbf{SG} + \mathbf{V}_{21}\mathbf{V}^+, \tag{8.100a}$$

$$\mathbf{T}_1 = \mathbf{LG}_1 + \mathbf{E}_1 = \mathbf{SG}_1 + \mathbf{V}_{21}\mathbf{V}^+, \tag{8.100b}$$

and $\mathbf{S} = \mathbf{L} - \mathbf{V}_{21}\mathbf{V}^+$. Thus $\tilde{\mathbf{y}}_* = \mathbf{Ty}$ is one representation for the BLUP($\mathbf{y}_* \mid \mathscr{M}_{12}$) and $\mathbf{y}_* - \tilde{\mathbf{y}}_*$ is the corresponding prediction error. In this section we pay particular attention on the covariance matrices of $\tilde{\mathbf{y}}_*$ and $\tilde{\mathbf{y}}_{*1}$ and of the corresponding prediction errors.

In view of (8.45), we have

$$\operatorname{cov}(\tilde{\mathbf{y}}_*) = \operatorname{cov}(\tilde{\boldsymbol{\mu}}_*) + \operatorname{cov}(\tilde{\boldsymbol{\varepsilon}}_*) = \mathbf{LGVG'L'} + \mathbf{EVE'}, \tag{8.101a}$$

$$\operatorname{cov}(\tilde{\mathbf{y}}_{*1}) = \operatorname{cov}(\tilde{\boldsymbol{\mu}}_{*1}) + \operatorname{cov}(\tilde{\boldsymbol{\varepsilon}}_{*1}) = \mathbf{L}\mathbf{G}_1\mathbf{V}\mathbf{G}_1'\mathbf{L}' + \mathbf{E}_1\mathbf{V}\mathbf{E}_1' \,. \tag{8.101b}$$

Moreover,

$$\begin{aligned}\operatorname{cov}(\tilde{\mathbf{y}}_*, \tilde{\mathbf{y}}_{*1}) &= \operatorname{cov}(\mathbf{L}\mathbf{G}\mathbf{y} + \mathbf{E}\mathbf{y},\ \mathbf{L}\mathbf{G}_1\mathbf{y} + \mathbf{E}_1\mathbf{y}) \\ &= \mathbf{L}\mathbf{G}\mathbf{V}\mathbf{G}_1\mathbf{L}' + \mathbf{L}\mathbf{G}\mathbf{V}\mathbf{E}_1' + \mathbf{E}\mathbf{V}\mathbf{G}_1'\mathbf{L}' + \mathbf{E}\mathbf{V}\mathbf{E}_1' \\ &= \mathbf{L}\mathbf{G}_1\mathbf{V}\mathbf{G}_1\mathbf{L}' + \mathbf{E}\mathbf{V}\mathbf{E}' + \mathbf{L}\mathbf{G}\mathbf{V}\mathbf{E}_1' \,, \end{aligned} \tag{8.102}$$

where we have used $\mathbf{G}\mathbf{V}\mathbf{G}_1' = \mathbf{G}_1\mathbf{V}\mathbf{G}_1'$, $\mathbf{E}\mathbf{V}\mathbf{E}_1' = \mathbf{E}\mathbf{V}\mathbf{E}'$, and

$$\begin{aligned}\mathbf{E}\mathbf{V}\mathbf{G}_1' &= \mathbf{V}_{21}\mathbf{M}(\mathbf{M}\mathbf{V}\mathbf{M})^+\mathbf{M}\mathbf{V}\mathbf{W}_1^+\mathbf{X}_1(\mathbf{X}_1'\mathbf{W}_1^+\mathbf{X}_1)^-\mathbf{X}_1' \\ &= \mathbf{V}_{21}\mathbf{M}(\mathbf{M}\mathbf{V}\mathbf{M})^+\mathbf{M}\mathbf{W}_1\mathbf{W}_1^+\mathbf{X}_1(\mathbf{X}_1'\mathbf{W}_1^+\mathbf{X}_1)^-\mathbf{X}_1' \\ &= \mathbf{0} \,. \end{aligned} \tag{8.103}$$

Thus the equality

$$\begin{aligned}\operatorname{cov}(\tilde{\mathbf{y}}_* - \tilde{\mathbf{y}}_{*1}) &= \operatorname{cov}(\tilde{\mathbf{y}}_*) + \operatorname{cov}(\tilde{\mathbf{y}}_{*1}) - \operatorname{cov}(\tilde{\mathbf{y}}_*, \tilde{\mathbf{y}}_{*1}) - \operatorname{cov}(\tilde{\mathbf{y}}_{*1}, \tilde{\mathbf{y}}_1) \\ &= \operatorname{cov}(\tilde{\mathbf{y}}_*) - \operatorname{cov}(\tilde{\mathbf{y}}_{*1}) + [2\operatorname{cov}(\mathbf{y}_{*1}) - \operatorname{cov}(\tilde{\mathbf{y}}_*, \tilde{\mathbf{y}}_{*1}) - \operatorname{cov}(\tilde{\mathbf{y}}_{*1}, \tilde{\mathbf{y}}_1)] \\ &= \operatorname{cov}(\tilde{\mathbf{y}}_*) - \operatorname{cov}(\tilde{\mathbf{y}}_{*1}) \,, \end{aligned} \tag{8.104}$$

holds if and only if $2\operatorname{cov}(\tilde{\mathbf{y}}_{*1}) = \operatorname{cov}(\tilde{\mathbf{y}}_*, \tilde{\mathbf{y}}_{*1}) + \operatorname{cov}(\tilde{\mathbf{y}}_{*1}, \tilde{\mathbf{y}}_1)$, i.e.,

$$2(\mathbf{L}\mathbf{G}_1\mathbf{V}\mathbf{G}_1'\mathbf{L}' + \mathbf{E}_1\mathbf{V}\mathbf{E}_1') = 2(\mathbf{L}\mathbf{G}_1\mathbf{V}\mathbf{G}_1\mathbf{L}' + \mathbf{E}\mathbf{V}\mathbf{E}') + \mathbf{L}\mathbf{G}\mathbf{V}\mathbf{E}_1' + \mathbf{E}_1\mathbf{V}\mathbf{G}'\mathbf{L}', \tag{8.105}$$

which further can be written as

$$2(\mathbf{E}_1\mathbf{V}\mathbf{E}_1' - \mathbf{E}\mathbf{V}\mathbf{E}') = \mathbf{L}\mathbf{G}\mathbf{V}\mathbf{E}_1' + \mathbf{E}_1\mathbf{V}\mathbf{G}'\mathbf{L}'. \tag{8.106}$$

In passing we may mention that it be shown that

$$\mathbf{G}\mathbf{V}\mathbf{E}_1' = \mathbf{V}(\mathbf{E}_1 - \mathbf{E})'. \tag{8.107}$$

Let us calculate $\operatorname{cov}(\tilde{\mathbf{y}}_* - \tilde{\mathbf{y}}_{*1})$ in another way. In view of

$$\tilde{\mathbf{y}}_* - \tilde{\mathbf{y}}_{*1} = \mathbf{T}\mathbf{y} - \mathbf{T}_1\mathbf{y} = \mathbf{S}(\mathbf{G} - \mathbf{G}_1)\mathbf{y}, \tag{8.108}$$

and (8.58), we have

$$\operatorname{cov}(\tilde{\mathbf{y}}_* - \tilde{\mathbf{y}}_{*1}) = \mathbf{S}\mathbf{G}\mathbf{V}\mathbf{M}_1(\mathbf{M}_1\mathbf{V}\mathbf{M}_1)^-\mathbf{M}_1\mathbf{V}\mathbf{G}'\mathbf{S}'. \tag{8.109}$$

We observe from (8.109) that $\mathrm{cov}(\tilde{\mathbf{y}}_* - \tilde{\mathbf{y}}_{*1}) = \mathbf{0}$ if and only if $\mathbf{SGVM}_1 = \mathbf{0}$. Moreover, the equality $\tilde{\mathbf{y}}_* = \tilde{\mathbf{y}}_{*1}$ holds for all $\mathbf{y} \in \mathscr{C}(\mathbf{W}_1)$ if and only if

$$\mathbf{SG}(\mathbf{X}_1 : \mathbf{VM}_1) = \mathbf{SG}_1(\mathbf{X}_1 : \mathbf{VM}_1) = \mathbf{S}(\mathbf{X}_1 : \mathbf{0}), \tag{8.110}$$

i.e., $\mathbf{SGVM}_1 = \mathbf{0}$. Correspondingly, the equality $\tilde{\mathbf{y}}_* = \tilde{\mathbf{y}}_{*1}$ for all $\mathbf{y} \in \mathscr{C}(\mathbf{W})$ yields the requirement $\mathbf{SG}_1\mathbf{X}_2 = \mathbf{SX}_2$.

Following Sengupta and Jammalamadaka [22, p. 292] and Haslett et al. [5, p. 553], we can write the prediction errors as

$$\mathbf{y}_* - \tilde{\mathbf{y}}_* = \mathbf{y}_* - \mathbf{Ty} = (\mathbf{y}_* - \mathbf{V}_{21}\mathbf{V}^+\mathbf{y}) - \mathbf{SGy}, \tag{8.111a}$$

$$\mathbf{y}_* - \tilde{\mathbf{y}}_{*1} = \mathbf{y}_* - \mathbf{T}_1\mathbf{y} = (\mathbf{y}_* - \mathbf{V}_{21}\mathbf{V}^+\mathbf{y}) - \mathbf{SG}_1\mathbf{y}. \tag{8.111b}$$

The random vectors $\mathbf{y}_* - \mathbf{V}_{21}\mathbf{V}^+\mathbf{y}$ and $\mathbf{SGy}$ are uncorrelated,

$$\mathrm{cov}(\mathbf{SGy}, \mathbf{y}_* - \mathbf{V}_{21}\mathbf{V}^+\mathbf{y}) = \mathbf{SGV}_{12} - \mathbf{SGVV}^+\mathbf{V}_{12} = \mathbf{0}, \tag{8.112}$$

and hence

$$\begin{aligned}\mathrm{cov}(\mathbf{y}_* - \tilde{\mathbf{y}}_*) &= \mathrm{cov}(\mathbf{y}_* - \mathbf{V}_{21}\mathbf{V}^+\mathbf{y}) + \mathrm{cov}(\mathbf{SGy}) \\ &= \mathbf{V}_{22} - \mathbf{V}_{21}\mathbf{V}^+\mathbf{V}_{12} + \mathbf{SGVG}'\mathbf{S}' \\ &= \boldsymbol{\Sigma}_{22\cdot 1} + \mathbf{SGVG}'\mathbf{S}'. \end{aligned} \tag{8.113}$$

The first term $\boldsymbol{\Sigma}_{22\cdot 1} = \mathbf{V}_{22} - \mathbf{V}_{21}\mathbf{V}^+\mathbf{V}_{12}$ in (8.113) is the Schur complement of $\mathbf{V}$ in

$$\boldsymbol{\Sigma} = \begin{pmatrix} \mathbf{V} & \mathbf{V}_{12} \\ \mathbf{V}_{21} & \mathbf{V}_{22} \end{pmatrix}, \tag{8.114}$$

and as Sengupta and Jammalamadaka [22, p. 293] point out, it is the covariance matrix of the prediction error associated with the best linear predictor (supposing that $\mathbf{X}\boldsymbol{\beta}$ were known) while the second term represents the increase in the covariance matrix of the prediction error due to estimation of $\mathbf{X}\boldsymbol{\beta}$.

Remark 5.1 Suppose that $\boldsymbol{\mu} = \mathbf{X}\boldsymbol{\beta} = \mathrm{E}(\mathbf{y})$ and $\boldsymbol{\mu}_* = \mathbf{X}_*\boldsymbol{\beta} = \mathrm{E}(\mathbf{y}_*)$ are known. Then the Best Linear Predictor of $\mathbf{y}_*$ on the basis of $\mathbf{y}$ is the following:

$$\mathrm{BLP}(\mathbf{y}_* \mid \mathbf{y}) = \boldsymbol{\mu}_* - \mathbf{V}_{21}\mathbf{V}^+(\mathbf{y} - \boldsymbol{\mu}). \tag{8.115}$$

The prediction error is

$$\mathbf{e}_{\mathbf{y}_*|\mathbf{y}} = \mathbf{y}_* - \mathrm{BLP}(\mathbf{y}_* \mid \mathbf{y}) = \mathbf{y}_* - [\boldsymbol{\mu}_* - \mathbf{V}_{21}\mathbf{V}^+(\mathbf{y} - \boldsymbol{\mu})], \tag{8.116}$$

with

$$\operatorname{cov}(\mathbf{e}_{\mathbf{y}_*|\mathbf{y}}) = \boldsymbol{\Sigma}_{22\cdot 1} = \mathbf{V}_{22} - \mathbf{V}_{21}\mathbf{V}^+\mathbf{V}_{12}\,. \tag{8.117}$$

Notice that $\boldsymbol{\Sigma}_{22\cdot 1} \leq_{\mathrm{L}} \operatorname{cov}(\mathbf{y}_* - \mathbf{N}\mathbf{y})$ for all $\mathbf{N}$. □

Next we consider the difference between the covariance matrices of the prediction errors. We have

$$\operatorname{cov}(\mathbf{y}_* - \tilde{\mathbf{y}}_*) = \boldsymbol{\Sigma}_{22\cdot 1} + \mathbf{SGVG}'\mathbf{S}' =: \mathbf{C}_{12}\,, \tag{8.118a}$$

$$\operatorname{cov}(\mathbf{y}_* - \tilde{\mathbf{y}}_{*1}) = \boldsymbol{\Sigma}_{22\cdot 1} + \mathbf{SG}_1\mathbf{VG}_1'\mathbf{S}' =: \mathbf{C}_1\,, \tag{8.118b}$$

and thereby, on account of (8.47),

$$\begin{aligned}
\mathbf{C}_{12} - \mathbf{C}_1 &= \mathbf{SGVG}'\mathbf{S}' - \mathbf{SG}_1\mathbf{VG}_1'\mathbf{S}' \\
&= \mathbf{SGVG}'\mathbf{S}' - \mathbf{SGG}_1\mathbf{VG}_1'\mathbf{G}'\mathbf{S}' \\
&= \mathbf{SGVM}_1(\mathbf{M}_1\mathbf{VM}_1)^-\mathbf{M}_1\mathbf{VG}'\mathbf{S}' \\
&= \operatorname{cov}(\tilde{\mathbf{y}}_* - \tilde{\mathbf{y}}_{*1})\,.
\end{aligned} \tag{8.119}$$

Obviously $\mathbf{C}_1 \leq_{\mathrm{L}} \mathbf{C}_{12}$ and $\mathbf{C}_{12} = \mathbf{C}_1$ if and only if $\mathbf{SGVM}_1 = \mathbf{0}$. Conclusion: If we add $\mathbf{X}_2$ into the model, observe the resulting $\mathbf{y}$, predict $\mathbf{y}_*$ on the basis of this particular $\mathbf{y}$, the resulting prediction error has a bigger covariance matrix (in the Löwner sense) than that error which is based on $\mathbf{X}_1$ only.

We omit the consideration of the inclusion of the type $\{\mathrm{BLUP}(\mathbf{y}_* \mid \mathscr{M}_1)\} \subset \{\mathrm{BLUP}(\mathbf{y}_* \mid \mathscr{M}_{12})\}$. Some related results appear in Haslett and Puntanen [6, 8].

The proposition below collects together the results obtained in this section.

Proposition 8.7 *Denote* $\tilde{\mathbf{y}}_* = \mathbf{Ty}$ *and* $\tilde{\mathbf{y}}_{*1} = \mathbf{T}_1\mathbf{y}$*, where* $\mathbf{T}$ *and* $\mathbf{T}_1$ *are defined as in* (8.100a)–(8.100b)*. Then the following statements hold:*

(a) $\operatorname{cov}(\tilde{\mathbf{y}}_* - \tilde{\mathbf{y}}_{*1}) = \mathbf{SGV}\dot{\mathbf{M}}_1\mathbf{VG}'\mathbf{S}'$,
(b) $\operatorname{cov}(\mathbf{y}_* - \tilde{\mathbf{y}}_*) - \operatorname{cov}(\mathbf{y}_* - \tilde{\mathbf{y}}_{*1}) = \operatorname{cov}(\tilde{\mathbf{y}}_* - \tilde{\mathbf{y}}_{*1})$,
(c) $\operatorname{cov}(\mathbf{y}_* - \tilde{\mathbf{y}}_{*1}) \leq_{\mathrm{L}} \operatorname{cov}(\mathbf{y}_* - \tilde{\mathbf{y}}_*)$.

The following statements are equivalent:

(d) $\operatorname{cov}(\tilde{\mathbf{y}}_* - \tilde{\mathbf{y}}_{*1}) = \mathbf{0}$,
(e) $\operatorname{cov}(\mathbf{y}_* - \tilde{\mathbf{y}}_{*1}) = \operatorname{cov}(\mathbf{y}_* - \tilde{\mathbf{y}}_*)$,
(f) $\tilde{\mathbf{y}}_* = \tilde{\mathbf{y}}_{*1}$ *for all* $\mathbf{y} \in \mathscr{C}(\mathbf{W}_1) = \mathscr{C}(\mathbf{X}_1 : \mathbf{VM}_1)$,
(g) $\mathbf{SGVM}_1 = \mathbf{0}$.

Moreover, the following statements are equivalent:

(h) $\operatorname{cov}(\tilde{\mathbf{y}}_* - \tilde{\mathbf{y}}_{*1}) = \operatorname{cov}(\tilde{\mathbf{y}}_*) - \operatorname{cov}(\tilde{\mathbf{y}}_{*1})$,
(i) $2(\mathbf{E}_1\mathbf{VE}_1' - \mathbf{EVE}') = \mathbf{LGVE}_1' + \mathbf{E}_1\mathbf{VG}'\mathbf{L}'$.

8.6 Conclusions

In this article we consider the partitioned linear model $\mathscr{M}_{12} = \{\mathbf{y}, \mathbf{X}_1\boldsymbol{\beta}_1 + \mathbf{X}_2\boldsymbol{\beta}_2, \mathbf{V}\}$ and the small model $\mathscr{M}_1 = \{\mathbf{y}, \mathbf{X}_1\boldsymbol{\beta}_1, \mathbf{V}\}$. Both models are supplemented with the new unobservable random vector $\mathbf{y}_*$, coming from $\mathbf{y}_* = \mathbf{K}\boldsymbol{\beta}_1 + \boldsymbol{\varepsilon}_*$, where $\mathbf{K}\boldsymbol{\beta}_1$ is estimable under both models. The covariance matrix of $\mathbf{y}_*$ is known as well as the cross-covariance matrix between $\mathbf{y}_*$ and $\mathbf{y}$.

Our aim to predict $\mathbf{y}_*$ on the basis of $\mathscr{M}_{12}$ and $\mathscr{M}_1$ and consider the resulting differences in the BLUEs and BLUPs. We consider the situation using given fixed multipliers of the response $\mathbf{y}$ yielding the BLUEs and BLUPs, and in addition, we characterize the whole class of multipliers in one model yielding the BLUEs and BLUPs that continue providing the BLUEs and BLUPs in the other model. Corresponding relations between the covariance matrices of the BLUEs, BLUPs and prediction errors are characterized. Particular attention is paid on the cases whether the response $\mathbf{y}$ lies in $\mathscr{C}(\mathbf{X}_1 : \mathbf{X}_2 : \mathbf{V})$ or in $\mathscr{C}(\mathbf{X}_1 : \mathbf{V})$.

We may mention, see, e.g., Isotalo et al. [10], that the results regarding the model $\mathscr{M}_{12}$ with new observations can be applied to the mixed linear model $\mathbf{y} = \mathbf{X}\boldsymbol{\beta} + \mathbf{Z}\mathbf{u} + \mathbf{e}$, where $\mathbf{X}_{n\times p}$ and $\mathbf{Z}_{n\times q}$ are known matrices, $\boldsymbol{\beta} \in \mathbb{R}^p$ is a vector of unknown fixed effects, $\mathbf{u}$ is an unobservable vector (q elements) of *random effects* with $\mathrm{E}(\mathbf{u}) = \mathbf{0}$, $\mathrm{cov}(\mathbf{u}) = \boldsymbol{\Delta}$, $\mathbf{e}$ is a random error vector with $\mathrm{E}(\mathbf{e}) = \mathbf{0}$, $\mathrm{cov}(\mathbf{e}) = \boldsymbol{\Phi}$, and $\mathrm{cov}(\mathbf{e}, \mathbf{u}) = \mathbf{0}$. Denoting $\mathbf{g} = \mathbf{X}\boldsymbol{\beta} + \mathbf{Z}\mathbf{u}$, we have $\mathrm{cov}(\mathbf{y}) = \mathrm{cov}(\mathbf{Z}\mathbf{u} + \mathbf{e}) = \mathbf{Z}\boldsymbol{\Delta}\mathbf{Z}' + \boldsymbol{\Phi}$, and the mixed linear model can be expressed as a version of the model with "new observations" corresponding $\mathbf{y}_*$ in (8.2) being in $\mathbf{g} = \mathbf{X}\boldsymbol{\beta} + \mathbf{Z}\mathbf{u}$.

Acknowledgments Part of this research has done during the MatTriad conference in Liblice, Czech Republic, September 2019, and during the meeting of an International Research Group on Multivariate and Mixed Linear Models in the Mathematical Research and Conference Center, Będlewo, Poland, in November 2019, supported by the Stefan Banach International Mathematical Center. Thanks go to Jarmo Niemelä for very helpful LaTeX assistance.

References

1. Baksalary, J.K., Puntanen, S., Styan, G.P.H.: A property of the dispersion matrix of the best linear unbiased estimator in the general Gauss–Markov model. Sankhyā Ser. A **52**, 279–296 (1990). http://www.jstor.org/stable/25050797
2. Baksalary, J.K., Rao, C.R., Markiewicz, A.: A study of the influence of the "natural restrictions" on estimation problems in the singular Gauss–Markov model. J. Stat. Plan. Inference **31**, 335–351 (1992). https://doi.org/10.1016/0378-3758(92)90141-E
3. Christensen, R.: Plane Answers to Complex Questions: the Theory of Linear Models, 4th edn. Springer, New York (2011). http://dx.doi.org/10.1007/978-1-4419-9816-3
4. Groß, J., Puntanen, S.: Estimation under a general partitioned linear model. Linear Algebra Appl. **321**, 131–144 (2000). http://dx.doi.org/10.1016/S0024-3795(00)00028-8

5. Haslett, S.J., Isotalo, J., Liu, Y., Puntanen, S.: Equalities between OLSE, BLUE and BLUP in the linear model. Stat. Pap. **55**, 543–561 (2014). http://dx.doi.org/10.1007/s00362-013-0500-7
6. Haslett, S.J., Puntanen, S.: A note on the equality of the BLUPs for new observations under two linear models. Acta Comment. Univ. Tartu. Math. **14**, 27–33 (2010a). http://math.ut.ee/acta/volume14.html
7. Haslett, S.J., Puntanen, S.: Effect of adding regressors on the equality of the BLUEs under two linear models. J. Stat. Plan. Inference **140**, 104–110 (2010b). http://dx.doi.org/10.1016/j.jspi.2009.06.010
8. Haslett, S.J., Puntanen, S.: A review of conditions under which BLUEs and/or BLUPs in one linear mixed model are also BLUEs and/or BLUPs in another. Calcutta Stat. Assoc. Bull. **65**, 25–41 (2013). http://dx.doi.org/10.1177/0008068320130103
9. Haslett, S.J., Puntanen, S.: Best linear unbiased prediction (BLUP). In: Davidian, M., Kenett, R., Longford, N.T., Molenberghs, G., Piegorsch, W.W., Ruggeri, F. (eds.) Wiley StatsRef: Statistics Reference Online. stat08120, 6 pp. Wiley, Chichester (2017) http://dx.doi.org/10.1002/9781118445112.stat08120
10. Isotalo, J., Markiewicz, A., Puntanen, S.: Some properties of linear prediction sufficiency in the linear model. In: Tez, M., von Rosen, D. (eds.) Trends and Perspectives in Linear Statistical Inference: LinStat, Istanbul, 2016, pp. 111–129. Springer, Cham (2018). https://doi.org/10.1007/978-3-319-73241-1_8
11. Isotalo, J., Puntanen, S.: Linear prediction sufficiency for new observations in the general Gauss–Markov model. Commun. Stat. Theory Methods **35**, 1011–1023 (2006). http://dx.doi.org/10.1080/03610920600672146
12. Isotalo, J., Puntanen, S., Styan, G.P.H.: A useful matrix decomposition and its statistical applications in linear regression. Commun. Stat. Theory Methods **37**, 1436–1457 (2008a). http://dx.doi.org/10.1080/03610920701666328
13. Isotalo, J., Puntanen, S., Styan, G.P.H.: The BLUE's covariance matrix revisited: a review. J. Stat. Plan. Inference **138**, 2722–2737 (2008b). http://dx.doi.org/10.1016/j.jspi.2008.03.029
14. Lu, C., Gan, S., Tian, Y.: Some remarks on general linear model with new regressors. Stat. Probab. Lett. **97**, 16–24 (2015). http://dx.doi.org/10.1016/j.spl.2014.10.015
15. Markiewicz, A., Puntanen, S.: Further properties of linear prediction sufficiency and the BLUPs in the linear model with new observations. Afrika Statistika **13**, 1511–1530 (2018). http://dx.doi.org/10.16929/as/1511.117
16. Markiewicz, A., Puntanen, S.: Further properties of the linear sufficiency in the partitioned linear model. In: Ahmed, S.E., Carvalho, F., Puntanen, S. (eds.) Matrices, Statistics and Big Data: Selected Contributions from IWMS 2016, pp. 1–22. Springer, Cham (2019). https://doi.org/10.1007/978-3-030-17519-1_1
17. Mitra, S.K., Moore, B.J.: Gauss–Markov estimation with an incorrect dispersion matrix. Sankhyā Ser. A **35**, 139–152 (1973). https://www.jstor.org/stable/25049864
18. Puntanen, S., Styan, G.P.H., Isotalo, J.: *Matrix Tricks for Linear Statistical Models: Our Personal Top Twenty*. Springer, Heidelberg (2011). http://dx.doi.org/10.1007/978-3-642-10473-2
19. Rao, C.R.: Representations of best linear estimators in the Gauss–Markoff model with a singular dispersion matrix. J. Multivariate Anal. **3**, 276–292 (1973). http://dx.doi.org/10.1016/0047-259X(73)90042-0
20. Rao, C.R.: Projectors, generalized inverses and the BLUE's, J. R. Stat. Soc. Ser. B **36**, 442–448 (1974). http://www.jstor.org/stable/2984930
21. Rao, C.R., Mitra, S.K.: *Generalized Inverse of Matrices and Its Applications*. Wiley, New York (1971)
22. Sengupta, D., Jammalamadaka, S.R.: *Linear Models: An Integrated Approach*. World Scientific, River Edge (2003)
23. Tian, Y.: A matrix handling of predictions of new observations under a general random-effects model. Electron. J. Linear Algebra **29**, 30–45 (2015a). http://dx.doi.org/10.13001/1081-3810.2895

24. Tian, Y.: A new derivation of BLUPs under random-effects model. Metrika **78**, 905–918 (2015b). https://doi.org/10.1007/s00184-015-0533-0
25. Tian, Y., Zhang, X.: On connections among OLSEs and BLUEs of whole and partial parameters under a general linear model. Stat. Probab. Lett. **112**, 105–112 (2016). http://dx.doi.org/10.1016/j.spl.2016.01.019

Chapter 9
A Collection of Moments of the Wishart Distribution

Thomas Holgersson and Jolanta Pielaszkiewicz

Abstract Moments of functions of Wishart distributed matrices appear frequently in multivariate analysis. Although a considerable number of such moments have long been available in the literature, they appear in rather dispersed sources and may sometimes be difficult to locate. This paper presents a collection of moments of the Wishart and inverse Wishart distribution, involving functions such as traces, determinants, Kronecker, and Hadamard products, etc. Moments of factors resulting from decompositions of Wishart matrices are also included.

9.1 Introduction

The Wishart distribution plays an important role in multivariate statistics and random matrix theory and is still subject to research and further development nearly 100 years after the original publication of J. Wishart [48]. The literature in the field is extensive, with findings appearing in books and journals across almost all fields of research involving statistics and probability. Introductions to Wishart matrices and their properties may be found in [8, 25], and [2]. The Wishart distribution is intimately connected with the multivariate normal distribution which in turn is one of the foundations stones in multivariate analysis and random matrices. Problems involving Wishart matrices thus arise frequently, and it is indeed difficult to find a textbook on these subjects that does not involve the Wishart distribution. Research in this field, however, usually does not require the full specification of the distribution but only its moments, typically the expectation, the dispersion, and perhaps third- or fourth-order moments. Although this only comprises a handful of moments, research problems are often concerned with functions of Wishart

T. Holgersson (✉)
Department of Economics and Statistics, Linnaeus University, Växjö, Sweden
e-mail: thomas.holgersson@lnu.se

J. Pielaszkiewicz
Department of Computer Science, Linköping University, Linköping, Sweden
e-mail: jolanta.pielaszkiewicz@liu.se

© Springer Nature Switzerland AG 2020

T. Holgersson, M. Singull (eds.), *Recent Developments in Multivariate and Random Matrix Analysis*, https://doi.org/10.1007/978-3-030-56773-6_9

matrices such as the trace, the determinant, or its submatrices. We will refer, somewhat imprecisely, to moments of such functions as *Wishart moments*. These may sometimes be calculated trivially. For example, the expected value of the trace of a Wishart matrix is obtained by simply taking the trace of the expected value. But the Wishart distribution is generally not closed under the transformation being conducted, and one may search the literature for available results on some specific moment. However, there presently does not seem to be any single source to consult in the matter, and it can be tedious to find the result sought for. The purpose of this paper is to list a number of moments of functions of Wishart moments in a convenient format for easy access.

When taking on such a task, there is an obvious need to constrain the types of moments included, in order to keep the size of the paper within reasonable bounds. We will not attempt to be comprehensive. Results that are cumbersome or involve the complex, the non-central or the singular Wishart distributions are thus not included. We will be content with only supplying references to these subjects for further reading.

Another concern is how to group the results. Should, for example, the results be categorized according to their dimension (scalars, vectors, etc.) or according to the type of transformation involved? In what follows we shall list the moments through classes of functions. This arrangement will naturally lead to some repetitions. For example, $\mathbb{E}[\text{Tr}\{\mathbf{W}^2\}]$ can be obtained by taking the trace of $\mathbb{E}[\mathbf{W}^2]$ or equivalently from $\mathbb{E}[1'(\mathbf{W}\odot\mathbf{W})1]$, and so on. However, since ease of access is given high priority, we consider such redundancies worthwhile.

The paper develops as follows: in Sect. 9.2 we give some preliminaries and definitions. Section 9.3 presents the promised Wishart moments, divided into separate sections for easy access. Finally, we give references to important properties not treated in the paper.

9.2 Preliminaries

9.2.1 Definitions

D.1 $(a)_k = a(a+1)\ldots(a+k-1) = (a+k-1)!/(a-1)!$

D.2 $\Gamma_m(a) = \pi^{m(m-1)/4}\prod_{j=1}^{m}\Gamma\left(a-\frac{j-1}{2}\right)$, $\mathscr{R}(a) > \frac{m-1}{2}$, where $\mathscr{R}(a)$ denotes the real part of a

D.3 Given a set $\Lambda = \{\lambda_1, \ldots, \lambda_p\}$ we define the elementary symmetric functions of Λ as follows:

- $\text{Tr}_1\{\Lambda\} = \text{Tr}\{\Lambda\} = \lambda_1 + \ldots + \lambda_p$
- $\text{Tr}_2\{\Lambda\} = \sum_{i<j}^{p}\lambda_i\lambda_j$
- $\ldots$
- $\text{Tr}_p\{\Lambda\} = det(\Lambda) = |\Lambda| = \lambda_1\lambda_2\cdots\lambda_p$

D.4 Some basic sets are defined as follows:

- $\mathbb{L}_p^+ = \{\mathrm{T} \in \mathbb{R}^{p \times p} : \mathrm{T} \text{ is lower triangular}, t_{jj} > 0, j = 1, \ldots, p\}$
- $\mathbb{U}_p^+ = \{\mathrm{T} \in \mathbb{R}^{p \times p} : \mathrm{T} \text{ is upper triangular}, t_{jj} > 0, j = 1, \ldots, p\}$
- $\mathbb{O}_p = \{\mathrm{H} \in \mathbb{R}^{p \times p} : \mathrm{H}'\mathrm{H} = \mathrm{H}\mathrm{H}' = \mathrm{I}\}$
- $\mathbb{O}_p^+ = \{\mathrm{H} \in \mathbb{O}_p : |\mathrm{H}| = 1\}$
- $\mathbb{S}_{p-1} = \{x \in \mathbb{R}^{p-1} : x'x = 1\}$

D.5 Consider some matrices $\mathrm{A} : m \times n$, $\mathrm{B} : p \times q$, $\mathrm{C} : p \times p$, $\mathrm{D} : p \times p$. The Kronecker product and the Hadamard product are defined by

$$\mathrm{A} \otimes \mathrm{B} = \begin{bmatrix} a_{11}\mathrm{B} \ldots a_{1n}\mathrm{B} \\ \ldots \;\; \ldots \;\; \ldots \\ a_{m1}\mathrm{B} \ldots a_{mn}\mathrm{B} \end{bmatrix} : pm \times qn,$$

$$\mathrm{C} \odot \mathrm{D} = \begin{bmatrix} c_{11}d_{11} \ldots c_{1p}d_{1p} \\ \ldots \;\; \ldots \;\; \ldots \\ c_{p1}d_{p1} \ldots c_{pp}d_{pp} \end{bmatrix} : p \times p.$$

D.6 The commutation matrix is defined as

$$K_{p,q} = \sum_{i=1}^{p} \sum_{j=1}^{q} e_i d_j' \otimes d_j e_i',$$

where e_i and d_j are unit basis vectors of size p and q, respectively.

D.7 Kronecker delta function: the scalar δ_{ij} is defined by

$$\delta_{ij} = \begin{cases} 1 : i = j \\ 0 : i \neq j. \end{cases}$$

We shall use the symbol $\sim$ to denote the distribution of a random quantity, e.g., $y \sim \chi^2(n)$, without distinguishing a formal distribution and a stochastic representation. For example, the two statements $y \sim \chi^2(n)$ and $y \sim (z_1^2 + z_2^2 + \ldots + z_n^2)$, where $z_i \overset{iid}{\sim} N(0, 1)$, will be used to denote the same thing.

9.2.2 *The Wishart Distribution*

The Wishart distribution appeared for the first time in [48], a publication which is sometimes considered as defining the birth of the theory of random matrices [28]. It describes the distributional properties of a sample covariance matrix calculated from a set of n (iid) normally distributed random vectors in $\mathbb{R}^p$. The density function is given by

$$f_{\mathbf{W}}(\mathbf{W}) = \frac{|\mathbf{W}|^{(n-p-1)/2}\exp\{-\frac{1}{2}\mathrm{Tr}\{\Sigma^{-1}\mathbf{W}\}\}}{2^{np/2}|\Sigma|^{n/2}\Gamma_p\left(\frac{n}{2}\right)}, \qquad n \geq p, \quad \Sigma > 0 \tag{9.1}$$

[2]. We will write $\mathbf{W} \sim W(\Sigma_p, n)$ for a random matrix $\mathbf{W}$ distributed according to (9.1), where n is the degrees of freedom parameter, which together with the matrix parameter Σ determine the distribution. The density $f(\mathbf{W})$ does not describe the joint density of all p^2 elements in $\mathbf{W}$ because it is symmetric, i.e., $W_{ij} = W_{ji}$ [48]. The density is usually defined through the unique $p(p+1)/2$ elements by specifying the density (9.1) with respect to the Lebesgue measure $\prod_{1\leq j}^{p} dw_{ij}$ in $\mathbb{R}^{p(p+1)/2}$ ([19, 28, 34]). The characteristic function is given by

$$\phi(\mathrm{T}) = \mathbb{E}[\exp\{\mathrm{iTr}\{\mathrm{T}'\mathbf{W}\}\}] = \mathbb{E}[\exp\{\frac{\mathrm{i}}{2}\mathrm{Tr}\{\mathrm{M}(\mathrm{T})\mathbf{W}\}\}] = |\mathrm{I} - \mathrm{iM}(\mathrm{T})\Sigma|^{-n/2},$$

where $M(\mathrm{T}) = \sum_{1\leq i\leq j}^{p} t_{ij}(e_i e_j' + e_j e_i')$, see [19, 25]. The inverse of a Wishart matrix, say $\mathbf{U} = \mathbf{W}^{-1}$, often appears in similar kinds of analyses as does $\mathbf{W}$. The density function is given by

$$f_{\mathbf{U}}(\mathbf{U}) = \frac{|\mathbf{U}|^{-(n+p+1)/2}\exp\{-\frac{1}{2}\mathrm{Tr}\{\Sigma^{-1}\mathbf{U}^{-1}\}\}}{2^{np/2}|\Sigma|^{n/2}\Gamma_p\left(\frac{n}{2}\right)}, \tag{9.2}$$

see [25]. An alternative parameterization for the density of the inverse Wishart is given in [19] and in [14].

There are several techniques available for obtaining moments arising from the Wishart distribution. Some results may be derived by algebraic treatments due to its close relation to the normal distribution. The Wishart class of distributions is closed under addition, i.e., for two independent Wishart matrices $\mathbf{W}_1$ and $\mathbf{W}_2$ having a common scale matrix, the sum $\mathbf{W}_1 + \mathbf{W}_2$ is again a Wishart matrix. Moreover, the class is closed under multiplication by some non-random B, which allows us to derive the moments of $\mathrm{B}\mathbf{W}\mathrm{B}'$ from the moments of $\mathbf{W}$ etc. More general moments of linear structure, on the other hand, require different tools. Differentiation of the characteristic function is a commonly used method. [19] developed differentials of symmetric matrices defined through a vectorizing operator that extracts the unique (non-repeated) elements only. Moments of more general transformations of Wishart matrices usually involve Jacobian transformations of the density function, see [15, 16, 25, 40] and [19] for general treatments. A comprehensive source of Jacobian transformations is available in [23].

Asymptotic results may be attractive when closed form results are unavailable or too complicated to be useful. A particularly important case involves high-dimensional settings, where $p/n \to c$ as $n, p \to \infty$. Recent developments in the theory of Free Probability have shown to be useful for deriving high-dimensional limiting behaviour of sample moments. Closed form recursive expressions for arbitrary moments in this line where proposed in [31] and further utilized by the authors of [32]. Some of the moments included in this paper have been derived from these formulas, which are also included in the appendix.

In this paper, we will not give any attention to the specific methods used for deriving Wishart moments. We shall merely specify the conditions necessary for the results to be valid (assumptions on n, p, Σ etc) and provide references for further readings.

9.3 Moments of Functions of the Wishart Distribution

9.3.1 Moments of the Type $\mathbb{E}[A\mathbf{W}B\mathbf{W}^{-1}]$

Let $\mathbf{W} \sim W(\Sigma_p, n)$, $p < n$ and $d = (n-p)(n-p-1)(n-p-3)$. Then

(9.3.1.a) $\mathbb{E}[\mathbf{W}] = n\Sigma$;

(9.3.1.b) $\mathbb{E}[\mathbf{W}A\mathbf{W}] = n\Sigma A'\Sigma + n^2\Sigma A\Sigma + n\mathrm{Tr}\{\Sigma A\}\Sigma$;

(9.3.1.c) $\mathbb{E}[\mathbf{W}^{-1}] = \frac{1}{n-p-1}\Sigma^{-1}$ for $n-p-1 > 0$;

(9.3.1.d) $\mathbb{E}[\mathbf{W}^{-1}A\mathbf{W}^{-1}] = \frac{1}{d}\mathrm{Tr}\{\Sigma^{-1}A\}\Sigma^{-1} + \frac{n-p-1}{d}\Sigma^{-1}A\Sigma^{-1}$ for $A' = A$;

(9.3.1.e) $\mathbb{E}[\mathbf{W}^{-1}A\mathbf{W}^{-1}] = \frac{1}{d}\mathrm{Tr}\{\Sigma^{-1}A\}\Sigma^{-1} + \frac{n-p-2}{d}\Sigma^{-1}A\Sigma^{-1} + \frac{1}{d}\Sigma^{-1}A'\Sigma^{-1}$;

(9.3.1.f) $\mathbb{E}[\mathbf{W}A\mathbf{W}^{-1}] = \frac{1}{n-p-1}\left[n\Sigma A\Sigma^{-1} - A' - \mathrm{Tr}\{A\}\mathrm{I}_p\right]$;

(9.3.1.g) $\mathbb{D}[\mathbf{W}] = \mathbb{E}[\mathrm{vec}\mathbf{W}\mathrm{vec}'\mathbf{W}] - \mathbb{E}[\mathrm{vec}\mathbf{W}]\mathbb{E}[\mathrm{vec}\mathbf{W}]' = n(\mathrm{I}_{p^2} + K_{p,p})(\Sigma \otimes \Sigma)$;

(9.3.1.h) $\mathbb{D}[\mathbf{W}^{-1}] = \frac{2}{d(n-p-1)}\mathrm{vec}\Sigma^{-1}\mathrm{vec}'\Sigma^{-1} + \frac{1}{d}(\mathrm{I}_{p^2} + K_{p,p})(\Sigma^{-1} \otimes \Sigma^{-1})$ for $n-p-3 > 0$;

(9.3.1.i) $\mathbb{E}[\mathbf{W}^2] = (n^2+n)\Sigma^2 + n\mathrm{Tr}\{\Sigma\}\Sigma$;

(9.3.1.j) $\mathbb{E}[\mathbf{W}^{-2}] = \frac{n-p-1}{d}\Sigma^{-2} + \frac{1}{d}\mathrm{Tr}\{\Sigma^{-1}\}\Sigma^{-1}$ for $n-p-3 > 0$;

(9.3.1.k) $\mathbb{E}[\mathbf{W}^3] = (n^3+3n^2+4n)\Sigma^3 + (2n^2+2n)(\mathrm{Tr}\{\Sigma\})\Sigma^2 + n(n+1)\mathrm{Tr}\{\Sigma^2\}\Sigma + (\mathrm{Tr}\{\Sigma\})^2\Sigma$;

(9.3.1.l) $\mathbb{E}[\mathbf{W}^{-3}] = \frac{n^2-2n(p+1)+p(p+2)+9}{(n-p-1)d_1}\Sigma^{-3} + \frac{4(n-p-2)}{(n-p-1)d_1}(\mathrm{Tr}\{\Sigma^{-1}\})\Sigma^{-2} - \frac{1}{d_1}(\mathrm{Tr}\{\Sigma^{-2}\})\Sigma^{-1} - \frac{2}{(n-p-1)d_1}(\mathrm{Tr}\{\Sigma^{-1}\})^2\Sigma^{-1}$, for $n-p-5 > 0$ and $d_1 = (n-p-5)(n-p-3)(n-p)(n-p+1)$;

Proof For (9.3.1.a), (9.3.1.b), (9.3.1.d) we refer to [14]. The results (9.3.1.c), (9.3.1.g), (9.3.1.h) and (9.3.1.j) can be found in [19]. (9.3.1.e) and (9.3.1.f) are due to [22] and [41], respectively. (9.3.1.i) and (9.3.1.k) are results of [7], while for (9.3.1.l) see [45]. □

9.3.2 Moments of the Type $\mathbb{E}[\mathrm{Tr}\{A\mathbf{W}\}B\mathbf{W}^{-1}]$

Let $\mathbf{W} \sim W(\Sigma_p, n)$, $p < n$ and let A be a non-random matrix of size $p \times p$. Then

(9.3.2.a) $\mathbb{E}[\mathrm{Tr}\{\mathbf{W}^{-1}\}\mathbf{W}^{-1}] = \frac{n-p-2}{(n-p-3)(n-p-1)(n-p)}(\mathrm{Tr}\{\Sigma^{-1}\})\Sigma^{-1} + \frac{2}{(n-p-3)(n-p-1)(n-p)}\Sigma^{-2}$ for $n-p-3>0$;

(9.3.2.b) $\mathbb{E}[\mathrm{Tr}\{\mathbf{W}^{-1}\}\mathbf{W}] = \frac{n}{n-p-1}(\mathrm{Tr}\{\Sigma^{-1}\})\Sigma - \frac{2}{n-p-1}\mathrm{I}_p$ for $n-p-1>0$;

(9.3.2.c) $\mathbb{E}[\mathrm{Tr}\{A\mathbf{W}^{-1}\}\mathbf{W}^{-1}] = \frac{1}{(n-p-3)(n-p-1)(n-p)}[\Sigma^{-1}A'\Sigma^{-1} + \Sigma^{-1}A\Sigma^{-1}] + \frac{n-p-2}{(n-p-3)(n-p-1)(n-p)}\mathrm{Tr}\{A\Sigma^{-1}\}\Sigma^{-1}$ for $n-p-3>0$;

(9.3.2.d) $\mathbb{E}[\mathrm{Tr}\{A\Sigma\}\mathbf{W}] = n\Sigma A\Sigma + n\Sigma A'\Sigma + n^2\mathrm{Tr}\{A\Sigma\}\Sigma$;

(9.3.2.e) $\mathbb{E}[\mathrm{Tr}\{\mathbf{W}^{-2}\}\mathbf{W}^{-1}] = \frac{2}{c_1}\Sigma^{-3} + \frac{3n-3p-7}{(n-p-1)c_1}(\mathrm{Tr}\{\Sigma^{-1}\})^2\Sigma^{-1} + \frac{n^2-2n(p+2)+p(p+4)+5}{(n-p-1)c_1}\mathrm{Tr}\{\Sigma^{-2}\}\Sigma^{-1} + \frac{4}{(n-p-1)c_1}\mathrm{Tr}\{\Sigma^{-1}\}\Sigma^{-2}$, where $n-p-5>0$ and $c_1 = (n-p-5)(n-p-3)(n-p)(n-p+1)$.

Proof For (9.3.2.a), (9.3.2.b), (9.3.2.e) see [45]. (9.3.2.c) was proven in [22]. (9.3.2.d) can be found in [14]. □

9.3.3 Moments of the Type $\mathbb{E}[A\mathbf{W} \otimes B\mathbf{W}^{-1}]$

Let $\mathbf{W} \sim W(\Sigma_p, n)$, $p < n$. Define $\mathrm{P} = \mathrm{I}_p \otimes K_{p,p}$, $\mathrm{Q} = K_{p,p} \otimes \mathrm{I}_p$, $\mathrm{V} = \left(\mathrm{vec}(\Sigma)(\mathrm{vec}(\Sigma))'\right) \otimes \Sigma$ and $\mathrm{U} = \left(\mathrm{vec}(\Sigma^{-1})(\mathrm{vec}(\Sigma^{-1}))'\right) \otimes \Sigma^{-1}$. Then

(9.3.3.a) $\mathbb{E}[\mathbf{W} \otimes \mathbf{W}] = n^2\Sigma \otimes \Sigma + n\mathrm{vec}\Sigma\mathrm{vec}'\Sigma + nK_{pg,p}(\Sigma \otimes \Sigma)$;

(9.3.3.b) $\mathbb{E}[\mathbf{W}^{-1} \otimes \mathbf{W}^{-1}] = \frac{n-p-2}{(n-p-3)(n-p-1)(n-p)}\Sigma^{-1} \otimes \Sigma^{-1} + \frac{1}{(n-p-3)(n-p-1)(n-p)}\left(\mathrm{vec}\Sigma^{-1}\mathrm{vec}'\Sigma^{-1} + K_{p,p}(\Sigma^{-1} \otimes \Sigma^{-1})\right)$ for $n-p-3>0$;

(9.3.3.c) $\mathbb{E}[\mathbf{W}^{-1} \otimes \mathbf{W}] = \frac{n}{n-p-1}\Sigma^{-1} \otimes \Sigma - \frac{1}{n-p-1}\left(\mathrm{vecI}_p(\mathrm{vecI}_p)' + K_{p,p}\right)$ for $n-p-1>0$;

(9.3.3.d) $\mathbb{E}[\mathbf{W}^{-1} \otimes \mathbf{W}^{-1} \otimes \mathbf{W}^{-1}] = \frac{(n-p-2)(n-p-3)}{(n-p-1)c_1}(\Sigma^{-1})^{\otimes 3} + \frac{3n-3p-7}{(n-p-1)c_1}\mathrm{U} + \frac{1}{c_1}\mathrm{Q}(\Sigma^{-1})^{\otimes 3} + \frac{n-p-3}{(n-p-1)c_1}\mathrm{PUP} + \frac{n-p-3}{(n-p-1)c_1}\mathrm{PQ}(\Sigma^{-1})^{\otimes 3}\mathrm{P} + \frac{1}{c_1}\mathrm{P}(\Sigma^{-1})^{\otimes 3} + \frac{2}{(n-p-1)c_1}\mathrm{PQ}(\Sigma^{-1})^{\otimes 3} + \frac{4}{(n-p-1)c_1}\mathrm{UPQ} - \frac{1}{c_1}K_{p,pp}\mathrm{U}K_{pp,p} - \frac{2}{(n-p-1)c_1}\mathrm{QPU}$, where $c_1 = (n-p-5)(n-p-3)(n-p)(n-p+1)$;

(9.3.3.e) $\mathbb{E}[\mathbf{W}^{\otimes 3}] = n^3\left(\Sigma^{\otimes 3}\right) + n^2\mathrm{V} + n^2\mathrm{Q}\left(\Sigma^{\otimes 3}\right) + n\mathrm{QPVP} + n\mathrm{PQ}\left(\Sigma^{\otimes 3}\right) + n^2\mathrm{QPVPQ} + n\mathrm{QPV} + n^2\mathrm{QP}\left(\Sigma^{\otimes 3}\right) + n\mathrm{VP} + n\mathrm{VPQ} + n^2\mathrm{P}\left(\Sigma^{\otimes 3}\right) + n\mathrm{PV} + n\mathrm{PVPQ} + n^2\mathrm{PVP} + n^2\mathrm{QPQ}\left(\Sigma^{\otimes 3}\right)$;

(9.3.3.f) $\mathbb{E}[(\mathbf{W} - \mathbb{E}[\mathbf{W}])^{\otimes 3}] = n\left(\mathrm{PV} + \mathrm{VP} + \mathrm{QPV} + \mathrm{VPQ} + \mathrm{PVPQ} + \mathrm{QPVP} + \mathrm{PQ}\left(\Sigma^{\otimes 3}\right)\right) + n^2\left(\mathrm{PVP} - \mathrm{PQVQP} + \mathrm{QP}\left(\Sigma^{\otimes 3}\right)\right) - 2n^3\left(\Sigma^{\otimes 3}\right)$.

Proof (9.3.3.b) can be found in [19], while for (9.3.3.d) and (9.3.3.c) see [14]. (9.3.3.e) is given by [43] and its alternative in [19]. (9.3.3.f) is result by [18]. □

9.3.4 Moments Involving $\left[\mathrm{Tr}\{A\mathbf{W}^{-1}\}^{a}, \ldots, \mathrm{Tr}\{(A\mathbf{W}^{-1})^{l}\}^{k}\right]$

Let $\mathbf{W} \sim W(\Sigma_p, n)$ such that Σ_p^{-1} exists and let $A : p \times p$ be a fixed symmetric matrix. Define the constants $d_1 = p-1$, $d_2 = p(p-1)(p-3)$ and $d_3 = (p+1)p(p-1)(p-3)(p-5)$. Then the following hold:

(9.3.4.a) $\mathbb{E}[\mathrm{Tr}\{A\mathbf{W}^{-1}\}] = d_1 \mathrm{Tr}\{A\Sigma^{-1}\}$;

(9.3.4.b) $\mathbb{E}\begin{bmatrix} (\mathrm{Tr}\{A\mathbf{W}^{-1}\})^2 \\ \mathrm{Tr}\{(A\mathbf{W}^{-1})^2\} \end{bmatrix} = d_2 \begin{pmatrix} n-p-2 & 2 \\ 1 & n-p-1 \end{pmatrix} \begin{pmatrix} (\mathrm{Tr}\{A\Sigma^{-1}\})^2 \\ \mathrm{Tr}\{(A\Sigma^{-1})^2\} \end{pmatrix}$;

(9.3.4.c) $\mathbb{E}\begin{bmatrix} (\mathrm{Tr}\{A\mathbf{W}^{-1}\})^3 \\ \mathrm{Tr}\{(A\mathbf{W}^{-1})^2\}\mathrm{Tr}\{A\mathbf{W}^{-1}\} \\ \mathrm{Tr}\{(A\mathbf{W}^{-1})^3\} \end{bmatrix} = d_3 \cdot$

$$\cdot \begin{pmatrix} (n-p)^2 - 5(n-p) + 2 & 6(n-p-3) & 16 \\ n-p-3 & (n-p)^2 - 4(n-p) + 7 & 4(n-p-1) \\ 2 & 3(n-p-1) & (n-p-1)^2 \end{pmatrix}$$

$$\cdot \begin{pmatrix} (\mathrm{Tr}\{A\Sigma^{-1}\})^3 \\ \mathrm{Tr}\{(A\Sigma^{-1})^2\}\mathrm{Tr}\{A\Sigma^{-1}\} \\ \mathrm{Tr}\{(A\Sigma^{-1})^3\} \end{pmatrix};$$

(9.3.4.d) $\mathbb{C}ov\begin{bmatrix} \frac{n}{p}\mathrm{Tr}\{(\Sigma^{-1}\mathbf{W})^{-1}\} \\ \frac{1}{np}\mathrm{Tr}\{\Sigma^{-1}\mathbf{W}\} \\ \frac{1}{pn^2}\mathrm{Tr}\{(\Sigma^{-1}\mathbf{W})^2\} \end{bmatrix} = \begin{bmatrix} \frac{2(n-1)n^2}{e_1} & \frac{2}{e_2} & \frac{4}{e_2} \\ -- & \frac{2}{np} & \frac{4(n+p+1)}{pn^2} \\ -- & -- & \frac{8(n^2+p^2)+20(n+1)(p+1)}{pn^3} \end{bmatrix}$,

where $e_1 = p(n-p-3)(n-p)(p-n+1)^2$ and $e_2 = (p-n+1)p$.

Proof [47] derived expressions for $\mathbb{E}\left[(\mathrm{Tr}\{A\mathbf{W}^{-1}\})^{v_1}(\mathrm{Tr}\{(A\mathbf{W}^{-1})^2\})^{v_2}\cdots\right]$ of degree k where $v = (1^{v_1}2^{v_2}3^{v_3}\cdots)$ is a partition of k consisting of v_1 ones, v_2 twos etc. The moments are expressed in terms of zonal polynomials where the scalars d_j and the matrices D_j are tabulated up to order 6. The first 3 are given in (9.3.4.a)–(9.3.4.c) above. For the result (9.3.4.d) see [30]. □

9.3.5 Moments Involving $a'(\mathbf{W} - \mathbf{W}^{-1})b$

Let $\mathbf{W} \sim W(\Sigma_p, n)$ and $a, b \in \mathbb{R}^p$ be non-random vectors. Let $D_a = diag(a)$ and $D_b = diag(b)$. Let $I\chi^2_{(df)}$ denote the inverse chi-square distribution and let $k \in \mathbb{Z}^+$. Then:

(9.3.5.a) $\mathbb{E}\left[\frac{a'\mathbf{W}^{-1}a}{a'\mathbf{W}^{-2}a}\right] = \frac{\alpha\gamma}{\alpha+\beta}$, where $\alpha = \frac{n-p+2}{2}$, $\beta = \frac{p-1}{2}$, $\gamma = n-p+1$;

(9.3.5.b) $\mathbb{E}\left[\left(\frac{a'\mathbf{W}^{-1}a}{a'\mathbf{W}^{-2}a}\right)^2\right] = \frac{\alpha\gamma(\gamma+2)(\beta+(\alpha+\beta+1)\alpha)}{(\alpha+\beta)^2(\alpha+\beta+1)}$;

(9.3.5.c) $\mathbb{E}\left[\frac{a'\mathbf{W}^{-1}aa'\mathbf{W}^{-2}a}{(a'a)^2}\right] = \frac{n-1}{(n-p)(n-p-1)(n-p-3)(n-p-5)}$, $n > p+5$ for $a'a \neq 0$.

(9.3.5.d) $\mathbb{E}\left[\left(\frac{a'\mathbf{W}a}{a'\Sigma a}\right)^k\right] = \mathbb{E}[(\chi^2_{(n)})^k]$;

(9.3.5.e) $\mathbb{E}\left[\left(\frac{a'\Sigma^{-1}a}{a'\mathbf{W}^{-1}a}\right)^k\right] = \mathbb{E}[(\chi^2_{(n-p+1)})^k]$, $n > p$;

(9.3.5.f) $\mathbb{E}\left[\left(a'(\mathbf{W} - \Sigma)a\right)^k\right] = (a'\Sigma a)^k \mathbb{E}\left[(\chi^2_{(n)} - 1)^k\right]$;

(9.3.5.g) $\mathbb{E}\left[\left(a'(\mathbf{W}^{-1} - \Sigma^{-1})a\right)^k\right] = (a'\Sigma^{-1}a)^k \mathbb{E}\left[I\chi^2_{(n-p+1)} - 1)^k\right]$, $n > p$;

(9.3.5.h) $\mathbb{E}\left[1'(\mathbf{W} - \Sigma)^{\odot 2}1\right] = np\text{Tr}\{\Sigma\} - (n-1)\text{Tr}\{\Sigma^2\}$;

(9.3.5.i) $\mathbb{E}\left[1'(\mathbf{W}^{-1} - \Sigma^{-1})^{\odot 2}1\right] = \left(\text{Tr}\{\Sigma^{-1}\}\right)^2 ((n-p-1)(n-p-3))^{-1} +$
$\text{Tr}\{\Sigma^{-2}\}\left(((n-p)(n-p-3))^{-1} - 2(n-p-1)^{-1} + 1\right)$
for $n - p - 3 > 0$;

(9.3.5.j) $\mathbb{E}\left[a'(\mathbf{W} \odot \mathbf{W})b\right] = n(1+n)a'(\Sigma \odot \Sigma)b + n\text{Tr}\{D_b\Sigma\}\text{Tr}\{D_a\Sigma\}$;

(9.3.5.k) $\mathbb{E}\left[a'(\mathbf{W}^{-1} \odot \mathbf{W}^{-1})b\right] = \frac{1}{(n-p)(n-p-3)} a'(\Sigma^{-1} \odot \Sigma^{-1})b$
$+ \frac{1}{(n-p)(n-p-1)(n-p-3)} \text{Tr}\{\Sigma^{-1}D_b\}\text{Tr}\{D_a\Sigma^{-1}\}$;

(9.3.5.l) $\mathbb{E}\left[a'(\mathbf{W}^{-1} \odot \mathbf{W})b\right] = \frac{1}{n-p-1}(na'(\Sigma^{-1} \odot \Sigma)b - a'(\mathrm{I} + 11')b)$.

Proof Let $r = \frac{a'\mathbf{W}^{-1}a}{a'\mathbf{W}^{-2}a}$. Gupta and Nagar [14] derived the stochastic representation $r \sim xy$, where $x \sim Beta_I(\frac{n-p+2}{2}, \frac{p-1}{2})$, $y \sim \chi^2_{(n-p+1)}$ and x and y are independent. Hence $\mathbb{E}[r^k] = \mathbb{E}[x^k]\mathbb{E}[y^k]$, $k = 1, 2, \ldots$. Using well-known moments of the beta and chi-square distribution (see [17]) we get (9.3.5.a)–(9.3.5.c). The expressions for $\mathbb{E}[r^{-k}]$ may be obtained similarly by using moments of inverse chi-square and inverse beta.

(9.3.5.d)–(9.3.5.g) are simple consequences of the facts that $a'\mathbf{W}a \sim a'\Sigma a\chi^2_{(n)}$ and $(a'\mathbf{W}^{-1}a)^{-1} \sim (a'\Sigma^{-1}a)^{-1}\chi^2_{(n-p+1)}$, [25]. Using $\mathbb{E}[\mathbf{W}]$ etc from Sect. 9.3.1 and the fact that for some symmetric matrices A and B there holds that $a'(A \odot B)b = \text{Tr}\{D_aAD_bB\}$, formulas (9.3.5.h)–(9.3.5.l) are obtained. □

9.3.6 Moments Involving $|\mathbf{W}|$, $\text{Tr}_j\{\mathbf{W}\}$, $\text{Tr}\{\mathbf{W}^k\}$

Let $\mathbf{W} \sim W(\Sigma_p, n)$. Then:

(9.3.6.a) $\mathbb{E}[|\mathbf{W}|] = |\Sigma|n(n-1)\cdots(n-(p-1))$;

(9.3.6.b) $\mathbb{E}[|\mathbf{W}|^2] = |\Sigma|^2 n(n-1)\cdots(n-(p-1))(n+2)(n+1)n\cdots(n-(p-1)+2) = |\Sigma|^2 \frac{\Gamma(n+1)\Gamma(n+3)}{\Gamma(n-p+1)\Gamma(n-p+3)}$;

(9.3.6.c) $\mathbb{E}[|\mathbf{W}|\text{Tr}\{\Sigma^{-1}\mathbf{W}\}] = 2|\Sigma|\frac{p(n+2)}{2} n(n-1)\cdots(n-(p-1))$;

(9.3.6.d) $\mathbb{E}[|\mathbf{W}|^2(\text{Tr}\{\Sigma^{-1}\mathbf{W}\})^2] = 2^2|\Sigma|^2\left(\frac{pn}{2} + 2p\right)_2 \frac{\Gamma(n+1)\Gamma(n+3)}{\Gamma(n-p+1)\Gamma(n-p+3)}$;

(9.3.6.e) $\mathbb{E}[\text{Tr}_j\{\mathbf{W}\}] = n(n-1)\ldots(n-j+1)\text{Tr}_j\{\Sigma\}$;

(9.3.6.f) $\mathbb{E}[\text{Tr}\{\mathbf{W}\}\text{Tr}_2\{\mathbf{W}\}] = n(n-1)(n+2)\text{Tr}\{\Sigma\}\text{Tr}_2\{\Sigma\} - 6n(n-1)\text{Tr}_3\{\Sigma\}$;

(9.3.6.g) $\mathbb{E}[(\text{Tr}_2\{\mathbf{W}\})^2] = n(n+2)(n-1)(n+1)(\text{Tr}_2\{\Sigma\})^2$
$- 4n(n-1)(n+2)\text{Tr}\{\Sigma\}\text{Tr}_3\{\Sigma\} - 4n(n-1)(2n-5)\text{Tr}_4\{\Sigma\}$;

(9.3.6.h) $\mathbb{E}[\mathrm{Tr}\{\mathbf{W}\}\mathrm{Tr}_3\{\mathbf{W}\}] = n(n-1)(n+2)(n-2)\mathrm{Tr}\{\Sigma\}\mathrm{Tr}_3\{\Sigma\} - 8n(n-1)(n-2)\mathrm{Tr}_4\{\Sigma\}$;

(9.3.6.i) $\mathbb{E}[(\mathrm{Tr}\{\mathbf{W}\})^2\mathrm{Tr}_2\{\mathbf{W}\}] = n(n+2)(n+4)(n-1)(\mathrm{Tr}\{\Sigma\})^2\mathrm{Tr}_2\{\Sigma\} - 4n(n-1)(n+2)(\mathrm{Tr}_2\{\Sigma\})^2 - 12n(n-1)(n+2)\mathrm{Tr}\{\Sigma\}\mathrm{Tr}_3\{\Sigma\} + 48n(n-1)\mathrm{Tr}_4\{\Sigma\}$;

(9.3.6.j) $\mathbb{E}[\mathrm{Tr}\{A\mathbf{W}\}\mathrm{Tr}\{B\mathbf{W}\}] = n\mathrm{Tr}\{A\Sigma B\Sigma\} + n\mathrm{Tr}\{A'\Sigma B\Sigma\} + n^2\mathrm{Tr}\{A\Sigma\}\mathrm{Tr}\{B\Sigma\}$;

(9.3.6.k) $\mathbb{E}[\mathrm{Tr}\{A\mathbf{W}^{-1}\}\mathrm{Tr}\{B\mathbf{W}^{-1}\}] = \frac{n-p-2}{(n-p-3)(n-p-1)(n-p)}\mathrm{Tr}\{A\Sigma^{-1}\}\mathrm{Tr}\{B\Sigma^{-1}\} + \frac{1}{(n-p-3)(n-p-1)(n-p)}\left[\mathrm{Tr}\{A'\Sigma^{-1}B\Sigma^{-1}\} + \mathrm{Tr}\{A\Sigma^{-1}B\Sigma^{-1}\}\right]$;

(9.3.6.l) $\mathbb{E}[\mathrm{Tr}\{\mathbf{W}^2\}] = (n+n^2)\mathrm{Tr}\{\Sigma^2\} + n(\mathrm{Tr}\{\Sigma\})^2 \overset{\Sigma=I}{=} np(1+n+p)$;

(9.3.6.m) $\mathbb{E}[(\mathrm{Tr}\{\Sigma^{-1}\mathbf{W}\})^k] = \sum_{(i)\in\mathscr{I}_k} \frac{k!}{i_1!\times\cdots\times i_k!1^{i_1}\times\cdots\times k^{i_k}}(np)^{i_1+\cdots+i_k}2^{k-\sum_{j=1}^k i_j}$, where the set $\mathscr{I}_k$ consists of k-tuples $(i) = (i_1,\ldots,i_k)$ such that $i_1 + 2i_2 + \ldots + ki_k = k$ and i_j, $j = 1,\ldots,k$ are non-negative integers;

(9.3.6.n) $\mathbb{E}[(\mathrm{Tr}\{\Sigma^{-1}\mathbf{W}\})^k] = (np + 2(k-1))\mathbb{E}[(\mathrm{Tr}\{\Sigma^{-1}\mathbf{W}\})^{k-1}]$;

(9.3.6.o) $\mathbb{E}[\mathrm{Tr}\{(\Sigma^{-1}\mathbf{W})^{-1}\}] = \frac{p}{n-p-1}$;

(9.3.6.p) $\mathbb{E}[(\mathrm{Tr}\{(\Sigma^{-1}\mathbf{W})^{-1}\})^2] = \frac{p(p(n-p-2)+2)}{(n-p-3)(n-p-1)(n-p)}$;

(9.3.6.q) $\mathbb{E}[\mathrm{Tr}\{(\Sigma^{-1}\mathbf{W})^{-2}\}] = \frac{(n-1)p}{(n-p-3)(n-p-1)(n-p)}$;

(9.3.6.r) $\mathbb{E}[\mathrm{Tr}\{\Sigma^{-1}\mathbf{W}\}\mathrm{Tr}\{(\Sigma^{-1}\mathbf{W})^k\}] = (np+2k)\mathbb{E}[\mathrm{Tr}\{(\Sigma^{-1}\mathbf{W})^k\}]$;

(9.3.6.s) $\mathbb{E}[\mathrm{Tr}\{\Sigma^{-1}\mathbf{W}\}\mathrm{Tr}\{(\Sigma^{-1}\mathbf{W})^2\}] = np(np+4)(n+p+1)$;

(9.3.6.t) $\mathbb{E}[(\mathrm{Tr}\{\Sigma^{-1}\mathbf{W}\})^2] = np(np+2)(n+p+1)$;

(9.3.6.u) $\mathbb{E}[\mathrm{Tr}\{(\Sigma^{-1}\mathbf{W})^3\}] = np(n^2+p^2+3np+3n+3p+4)$;

(9.3.6.v) $\mathbb{E}[\mathrm{Tr}\{(\Sigma^{-1}\mathbf{W})^4\}] = np(n^3+p^3+6n^2p+6np^2+6n^2+6p^2+17np+21n+21p+20)$;

(9.3.6.w) $\mathbb{E}[\mathrm{Tr}\{(\Sigma^{-1}\mathbf{W})^5\}] = np(p^4+n^4+10np^3+10n^3p+20n^2p^2+65n^2+65p^2+175np+10n^3+10p^3+55n^2p+55np^2+160n+160p+148)$;

(9.3.6.x) $\mathbb{E}[\mathrm{Tr}\{(\Sigma^{-1}\mathbf{W})^2\}\mathrm{Tr}\{(\Sigma^{-1}\mathbf{W})^k\}] = (n+p+1)(np+2k)\mathbb{E}[\mathrm{Tr}\{(\Sigma^{-1}\mathbf{W})^k\}] + 2k\mathbb{E}[\mathrm{Tr}\{(\Sigma^{-1}\mathbf{W})^{k+1}\}]$;

(9.3.6.y) $\mathbb{E}[\mathrm{Tr}\{(\Sigma^{-1}\mathbf{W})^2\}\mathrm{Tr}\{(\Sigma^{-1}\mathbf{W})^2\}] = (n+p+1)(np+4)np(1+p+n) + 4np(n^2+p^2+3np+3n+3p+4)$.

Proof [26] derived the expression:

$$\mathbb{E}[(\mathrm{Tr}\{\Sigma^{-1}\mathbf{W}\})^k|\mathbf{W}|^h] = 2^{ph+k}|\Sigma|^h\left(\frac{1}{2}np+ph\right)_k \frac{\Gamma_p\left(\frac{1}{2}n+h\right)}{\Gamma_p\left(\frac{1}{2}n\right)},$$

which gives (9.3.6.a)–(9.3.6.d). de Waal and Nel [7] gave a number of moments for $\mathbb{E}[\mathrm{Tr}\{\mathbf{W}\}\mathrm{Tr}_j\{\mathbf{W}\}]$, including those in (9.3.6.e)–(9.3.6.i). Results (9.3.6.j)–(9.3.6.k) can be found in [14]. For (9.3.6.l) see [10]. Letac and Massam [20] proved formula (9.3.6.m). Relation (9.3.6.n), (9.3.6.r), (9.3.6.x) as well as formulas (9.3.6.s)–(9.3.6.w), (9.3.6.y) are derived from **(A.i)**, see [31]. (9.3.6.o)–(9.3.6.q) are derived from **(A.iii)** and **(A.iv)** that were proven in [30]. □

9.3.7 Bartlett Decompositions

Let $\mathbf{W} \sim W(\mathrm{I}_p, n)$, $\mathbf{W} = \mathbf{TT}'$, $\mathbf{B} = (\mathbf{T}'\mathbf{T})^{-1}$ where $\mathbf{T} = \{t_{ij}\} \in \mathbb{L}_p^+$. Then the following holds:

(9.3.7.a) $\mathbb{E}[t_{ij}] = 0$, $\mathbb{V}ar[t_{ij}] = 1$, $1 \leq j < i \leq p$;
(9.3.7.b) $\mathbb{E}[(t_{ii})^2] = n - i + 1$, $\mathbb{V}ar[(t_{ii})^2] = 2(n - i + 1)$, $1 \leq i \leq p$;
(9.3.7.c) $\mathbb{E}[b_{jj}] = \frac{n-1}{(n-j-1)(n-j)}$, $b_{ij} = 0, i \neq j$, $1 \leq j \leq p$;

Proof (9.3.7.a) and (9.3.7.b) follows from the well-known property that $t_{ij} \sim N(0, 1)$, $1 \leq j < i \leq p$ and $t_{ii}^2 \sim \chi^2_{(n-i+1)}$, $1 \leq i \leq p$ (see [25] for details) while (9.3.7.c) is given in [9]. □

9.3.8 The Spectral Decomposition

Let $\mathbf{W} = \{W_{ij}\}$ where $\mathbf{W} \sim W(\mathrm{I}_p, n)$, $p > 2$. Let the spectral decomposition be defined by $\mathbf{V}'\mathbf{W}\mathbf{V} = \mathbf{L}$, where $\mathbf{V} \in \mathbb{O}_p$ and $\mathbf{L}$ is a diagonal matrix of eigenvalues of $\mathbf{W}$. Define two index sets by $\mathrm{i} = (i_1, \ldots, i_{2k})$, $\mathrm{j} = (j_1, \ldots, j_{2k})$. The following holds:

(9.3.8.a) $\mathbb{E}[V_{i_1 j_1} V_{i_2 j_2} \cdots V_{i_k j_k}] = 0$ for k odd;
(9.3.8.b) $\mathbb{E}[V_{1 j_1} V_{1 j_2} V_{2 j_3} V_{2 j_4}] = \frac{1}{p(p+2)(p-1)}((p + 1)\delta_{j_1 j_2}\delta_{j_3 j_4} - \delta_{j_1 j_3}\delta_{j_2 j_4} - \delta_{j_1 j_4}\delta_{j_2 j_3})$,
$p > 2$, $j_1, j_2, j_3, j_4 \leq p$;
(9.3.8.c) $\mathbb{E}[V_{11}^2 V_{22}^2] = \frac{p+1}{p(p+2)(p-1)}$;
(9.3.8.d) $\mathbb{E}[V_{11}^4] = \frac{3}{p(p+2)}$;

Let $\mathbf{W} \sim W(\Sigma_p, n)$, where $\Pi' \Sigma \Pi = \Lambda$ is a diagonal matrix of the eigenvalues $\lambda_1, \cdots, \lambda_p$ which are strictly unequal. Denote the vectors of eigenvalues of Σ and $\mathbf{W}$ by $\lambda = (\lambda_1, \ldots, \lambda_p)'$ and $\boldsymbol{l} = (l_1, \ldots, l_p)'$ respectively. Then, for some fixed finite p, there holds that $\sqrt{n}(\boldsymbol{l} - \lambda) \overset{law}{\to} N(0, 2\Lambda^2)$. In particular,

(9.3.8.e) $\lim_{n \to \infty} \mathbb{E}[\boldsymbol{l}] = \lambda$;
(9.3.8.f) $\lim_{n \to \infty} \mathbb{C}ov[\sqrt{n}\boldsymbol{l}] = 2\Lambda^2$.

Proof It is well known that $\mathbf{V} \sim h_p$, where h_p denotes the unique Haar measure on $\mathbb{O}_p$, [5]. Matsumoto [24] gave a general formula for $\mathbb{E}[V_{i_1 j_1} V_{i_2 j_2} \cdots V_{i_k j_k}]$ in terms of orthogonal Weingarten functions, of which (9.3.8.a)–(9.3.8.d) are special cases. For (9.3.8.e) and (9.3.8.f) see [1, 2]. □

Remark Most results on distributional properties of the eigenmatrix $\mathbf{V}$ involves the Haar distribution. Although it is well-known that that the eigenmatrix from a standard Wishart distribution (i.e., when $\Sigma = \mathrm{I}$) is distributed according to the Haar measure, much less is known for the case of general Σ [3, 36, 37].

9.3.9 Some Asymptotic Moments

Although exact moments are available in the above sections, they are sometimes lengthy and simpler expressions may be more attractive.

Let $\mathbf{W} \sim W(\frac{1}{n}\mathrm{I}_p, n)$. Assume $\frac{p}{n} \to c \in (0, 1)$ as $n, p \to \infty$. Then the following limits hold:

(9.3.9.a) $\mathbb{E}[p^{-1}\mathrm{Tr}\{\mathbf{W}\}] \to c$;
(9.3.9.b) $\mathbb{E}[p^{-1}\mathrm{Tr}\{\mathbf{W}^2\}] \to 1 + c$;
(9.3.9.c) $\mathbb{E}[p^{-1}\mathrm{Tr}\{\mathbf{W}^3\}] \to 1 + 3c + c^2$;
(9.3.9.d) $\mathbb{E}[p^{-1}\mathrm{Tr}\{\mathbf{W}^4\}] \to 1 + 6c + 6c^2 + c^3$;
(9.3.9.e) $\mathbb{E}[p^{-1}\mathrm{Tr}\{\mathbf{W}^5\}] \to 1 + 10c + 20c^2 + 10c^3 + c^4$;
(9.3.9.f) $\mathbb{E}[p^{-1}\mathrm{Tr}\{\mathbf{W}^{-1}\}] \to (1 - c)^{-1}$;
(9.3.9.g) $\mathbb{E}[p^{-1}\mathrm{Tr}\{\mathbf{W}^{-2}\}] \to (1 - c)^{-3}$;
(9.3.9.h) $\mathbb{E}[p^{-1}\mathrm{Tr}\{\mathbf{W}^{-3}\}] \to (1 + c)(1 - c)^{-5}$;
(9.3.9.i) $\mathbb{E}[p^{-1}\mathrm{Tr}\{\mathbf{W}^{-4}\}] \to (1 + 3c + c^2)(1 - c)^{-7}$;
(9.3.9.j) $\mathbb{E}[p^{-1}\mathrm{Tr}\{\mathbf{W}^{-5}\}] \to (1 + 6c + 6c^2 + c^3)(1 - c)^{-9}$;
(9.3.9.k) $\mathbb{E}[p^{-2}\mathrm{Tr}\{\mathbf{W}^k\}\mathrm{Tr}\{\mathbf{W}\}] \to \lim_{n,p\to\infty} \mathbb{E}[p^{-1}\mathrm{Tr}\{\mathbf{W}^k\}]$;
(9.3.9.l) $\mathbb{E}[p^{-2}\mathrm{Tr}\{\mathbf{W}^2\}\mathrm{Tr}\{\mathbf{W}\}] \to 1 + c$;
(9.3.9.m) $\mathbb{E}[p^{-2}\mathrm{Tr}\{\mathbf{W}^3\}\mathrm{Tr}\{\mathbf{W}\}] \to 1 + 3c + c^2$;
(9.3.9.n) $\mathbb{E}[p^{-2}\mathrm{Tr}\{\mathbf{W}^4\}\mathrm{Tr}\{\mathbf{W}\}] \to 1 + 6c + 6c^2 + c^3$;
(9.3.9.o) $\mathbb{E}[p^{-2}\mathrm{Tr}\{\mathbf{W}^k\}\mathrm{Tr}\{\mathbf{W}^2\}] \to (1 + c) \lim_{n,p\to\infty} \mathbb{E}[p^{-1}\mathrm{Tr}\{\mathbf{W}^k\}]$;
(9.3.9.p) $\mathbb{E}[p^{-2}\mathrm{Tr}\{\mathbf{W}^2\}\mathrm{Tr}\{\mathbf{W}^2\}] \to (1 + c)^2$;
(9.3.9.q) $\mathbb{E}[p^{-2}\mathrm{Tr}\{\mathbf{W}^2\}\mathrm{Tr}\{\mathbf{W}^3\}] \to (1 + c)(1 + 3c + c^2)$;
(9.3.9.r) $\mathbb{E}[p^{-2}\mathrm{Tr}\{\mathbf{W}^2\}\mathrm{Tr}\{\mathbf{W}^4\}] \to (1 + c)(1 + 6c + 6c^2 + c^3)$;
(9.3.9.s) $\mathbb{E}[p^{-3}\mathrm{Tr}\{\mathbf{W}\}\mathrm{Tr}\{\mathbf{W}^3\}\mathrm{Tr}\{\mathbf{W}^4\}] \to 1 + 9c + 25c^2 + 25c^3 + 9c^4 + c^5$;
(9.3.9.t) $\mathbb{E}[p^{-2}\mathrm{Tr}\{\mathbf{W}^k\}\mathrm{Tr}\{\mathbf{W}^{-1}\}] \to (1 - c)^{-1} \lim_{n,p\to\infty} \mathbb{E}[p^{-1}\mathrm{Tr}\{\mathbf{W}^k\}]$;
(9.3.9.u) $\mathbb{E}[p^{-2}\mathrm{Tr}\{\mathbf{W}^2\}\mathrm{Tr}\{\mathbf{W}^{-1}\}] \to (1 - c)^{-1}(1 + c)$;
(9.3.9.v) $\mathbb{E}[p^{-2}\mathrm{Tr}\{\mathbf{W}^3\}\mathrm{Tr}\{\mathbf{W}^{-1}\}] \to (1 - c)^{-1}(1 + 3c + c^2)$;
(9.3.9.w) $\mathbb{E}[p^{-2}\mathrm{Tr}\{\mathbf{W}^k\}\mathrm{Tr}\{\mathbf{W}^{-2}\}] \to \frac{1}{1-c} \lim_{n,p\to\infty} \mathbb{E}[p^{-2}\mathrm{Tr}\{\mathbf{W}^k\}\mathrm{Tr}\{\mathbf{W}^{-1}\}] + \frac{c}{1-c} \lim_{n,p\to\infty} \mathbb{E}[p^{-3}\mathrm{Tr}\{\mathbf{W}^k\}\mathrm{Tr}\{\mathbf{W}^{-1}\}\mathrm{Tr}\{\mathbf{W}^{-1}\}]$.

Proof The limiting moments in (9.3.9.a)–(9.3.9.j) are derived trough a general recursive expression in [35]. An alternative, non-recurrent formula is available in [29]. (9.3.9.k)–(9.3.9.s) are derived using the recursive formula in **(A.ii)**, which is due to [31]. The expressions (9.3.9.t)–(9.3.9.w) are derived from the general formula in **(A.iv)** which in turn is due to [30]. □

9.3.10 Moments for Non-central, Singular or Complex Wishart Distributions

The moments listed in Sects. 9.3.1–9.3.9 are only selections of results, restricted to matrices $\mathbf{W} \in \mathbb{R}^{p\times p}$ with $rank(\mathbf{W}) \geq p$ and zero non-centrality parameter. More general results for Wishart moments can often be obtained in a manner similar to the regular case. Linear structured moments for the non-central Wishart distribution (say, $\mathbf{W}_\Theta$) are fairly straightforward to obtain due to its close relation to the non-central multivariate normal distribution. $\mathbb{V}ar[vec(\mathbf{W}_\Theta)]$ has been derived by [21], and [12] derived the expected value and variance of the trace of a non-central Wishart. Gupta and Nagar [14] has derived $\mathbb{E}[|\mathbf{W}_\Theta|^h]$ as well as the distribution of $\mathbf{T}$, where $\mathbf{T}\mathbf{T}' = \mathbf{W}_\Theta$ is the Bartlett decomposition for the non-central Wishart. Expressions for $\mathbb{E}[\mathrm{Tr}_j\{\Sigma^{-1}\mathbf{W}_\Theta\}]$ are available in [33].

Another special case arises when the dimension of the Wishart matrix exceeds the degrees of freedom, i.e., when $p > n$ in the notation of (9.1). Singular Wishart matrices play an important role in typical low-dimensional settings such as in analysis of linear models but also arise in analysis of high-dimensional data, i.e., when $p/n \to c > 1$ as $n, p \to \infty$. The density for singular Wishart matrices has been derived by [44] and [39]. References [49] and [4] derived expressions for the density of the generalized inverse $\mathbf{W}^-$ of singular Wishart distribution. The first- and second- order moments for $\mathbf{W}^-$, $\mathrm{Tr}\{\mathbf{W}^-\}$, and $\mathrm{vec}\{\mathbf{W}^-\}$ have been derived by [6] in a manner analogous to [46].

The case of complex Wishart matrices, $\mathbf{W}_C \in \mathbb{C}^{p\times p}$, arises primarily from the complex normal distribution; some early references include [13] and [38]. Many kinds of moments of the complex Wishart have been derived since then. Sultan and Tracy [42] found the first four moments of the central and noncentral complex Wishart distributions. Nagar and Gupta [27] provided a number of results for moments of the complex and inverse complex Wishart including first- and second-order matrix moments, an expression for $\mathbb{E}[|\mathbf{W}_C|^h]$, moments of traces $\mathrm{Tr}\{\mathbf{W}_C\}$, $\mathrm{Tr}\{\mathbf{W}_C^{-1}\}$, $\mathrm{Tr}\{(\Sigma^{-1}\mathbf{W}_C)^h|\mathbf{W}_C|^k\}$, and the distribution of $\mathbf{T}$, where $\mathbf{T}\mathbf{T}^* = \mathbf{W}_C$.

Díaz-García and Gutiérrez-Jáimez [11] proposed a unified approach that enables the Wishart distribution to be studied simultaneously in the real, complex, quaternion, and octonion cases.

Appendix

(A.i) The following recursive expression is due to [31]. Let $\mathbf{W} \sim W(\mathrm{I}_p, n)$, $n > p-1$. Then the following recursive formula holds for all $k \in \mathbb{N}$ and all m_0, $m_1, \ldots, m_k$ such that $m_0 = 0$, $m_k \in \mathbb{N}$, $m_i \in \mathbb{N}_0$, $i = 1, \ldots, k-1$:

$$\mathbb{E}\Big[\prod_{i=0}^{k} \mathrm{Tr}\{(\Sigma^{-1}\mathbf{W})^{m_i}\}\Big] = (n-p+m_k-1)\mathbb{E}\Big[\mathrm{Tr}\{(\Sigma^{-1}\mathbf{W})^{m_k-1}\}\prod_{i=0}^{k-1}\mathrm{Tr}\{(\Sigma^{-1}\mathbf{W})^{m_i}\}\Big]$$
$$+2\sum_{i=0}^{k-1} m_i \mathbb{E}\Big[\mathrm{Tr}\{(\Sigma^{-1}\mathbf{W})^{m_k+m_i-1}\}\prod_{\substack{j=0\\ j\neq i}}^{k-1}\mathrm{Tr}\{(\Sigma^{-1}\mathbf{W})^{m_j}\}\Big]$$
$$+\sum_{i=0}^{m_k-1}\mathbb{E}\Big[\mathrm{Tr}\{(\Sigma^{-1}\mathbf{W})^{i}\}\mathrm{Tr}\{(\Sigma^{-1}\mathbf{W})^{m_k-1-i}\}\prod_{j=0}^{k-1}\mathrm{Tr}\{(\Sigma^{-1}\mathbf{W})^{m_j}\}\Big];$$

(A.ii) The following non-recursive asymptotic expression is due to [31]. Let $\mathbf{W} \sim W(\mathrm{I}_p, n)$, $n > p-1$. Define

$$_{(k)}Q(m_k, \overline{m}) := \lim_{n,p\to\infty} \frac{1}{p^{k+1}n^s}\mathbb{E}\Big[\prod_{i=0}^{k}\mathrm{Tr}\{\mathbf{W}^{m_i}\}\Big].$$

Then for all $k \in \mathbb{N}$ and all m_k, $\overline{m} = \{m_0, \ldots, m_{k-1}\}$ such that $m_0 = 0$, $m_k \in \mathbb{N}$, $m_i \in \mathbb{N}_0$ for $i = 1, \ldots, k-1$; and $s = \sum_{i=0}^{k} m_i$ the following holds:

$$_{(k)}Q(m_k,\overline{m}) = \begin{cases} (\frac{1}{c}+1)_{(k)}Q(m_k-1,\overline{m}) \\ \quad +\frac{1}{c}\sum_{i=1}^{m_k-2} {}_{(k+1)}Q(i,\{m_0,\ldots,m_{k-1},m_k-1-i\}), & m_k > 2, k \geq 1,\\ (\frac{1}{c}+1)_{(k)}Q(1,\overline{m}), & m_k = 2, k \geq 1,\\ _{(k-1)}Q(m_{k-1},\{m_0,\ldots,m_{k-2}\}), & m_k = 1, k > 1,\\ 1, & m_k = 1, k = 1. \end{cases}$$

(A.iii) The following recursive expression is due to [30]. Let $\mathbf{W} \sim W(\mathrm{I}_p, n)$. Then the following recursive formula holds for all $k \in \mathbb{N}$ and all $m_0, m_1, \ldots, m_k$ such that $m_0 = 0$, $m_k \in \mathbb{N}$, $m_i \in \mathbb{N}_0$, $i = 1, \ldots, k-1$:

$$(n-p-m_k)\mathbb{E}\Big[\prod_{i=0}^{k}\mathrm{Tr}\{(\Sigma^{-1}\mathbf{W})^{-m_i}\}\Big] = \mathbb{E}\Big[\mathrm{Tr}\{(\Sigma^{-1}\mathbf{W})^{-m_k+1}\}\prod_{i=0}^{k-1}\mathrm{Tr}\{(\Sigma^{-1}\mathbf{W})^{-m_i}\}\Big]$$
$$+2\sum_{i=0}^{k-1} m_i\mathbb{E}\Big[\mathrm{Tr}\{(\Sigma^{-1}\mathbf{W})^{-m_k-m_i}\}\prod_{\substack{j=0\\ j\neq i}}^{k-1}\mathrm{Tr}\{(\Sigma^{-1}\mathbf{W})^{-m_j}\}\Big]$$
$$+\sum_{i=0}^{m_k-2}\mathbb{E}\Big[\mathrm{Tr}\{(\Sigma^{-1}\mathbf{W})^{-i-1}\}\mathrm{Tr}\{(\Sigma^{-1}\mathbf{W})^{-m_k+1+i}\}\prod_{j=0}^{k-1}\mathrm{Tr}\{(\Sigma^{-1}\mathbf{W})^{-m_j}\}\Big];$$

(A.iv) Under same assumptions as in **(A.iii)** the following recursive formula holds:

$$(n-p-m_k)\mathbb{E}\Big[\mathrm{Tr}\{(\Sigma^{-1}\mathbf{W})^{-m_k}\}\prod_{i=0}^{k-1}\mathrm{Tr}\{(\Sigma^{-1}\mathbf{W})^{m_i}\}\Big]$$
$$=\mathbb{E}\Big[\mathrm{Tr}\{(\Sigma^{-1}\mathbf{W})^{-m_k+1}\}\prod_{i=0}^{k-1}\mathrm{Tr}\{(\Sigma^{-1}\mathbf{W})^{m_i}\}\Big]$$
$$-2\sum_{i=0}^{k-1}m_i\mathbb{E}\Big[\mathrm{Tr}\{(\Sigma^{-1}\mathbf{W})^{-m_k+m_i}\}\prod_{\substack{j=0\\ j\neq i}}^{k-1}\mathrm{Tr}\{(\Sigma^{-1}\mathbf{W})^{m_j}\}\Big]$$
$$+\sum_{i=0}^{m_k-2}\mathbb{E}\Big[\mathrm{Tr}\{(\Sigma^{-1}\mathbf{W})^{-i-1}\}\mathrm{Tr}\{(\Sigma^{-1}\mathbf{W})^{-m_k+1+i}\}\prod_{j=0}^{k-1}\mathrm{Tr}\{(\Sigma^{-1}\mathbf{W})^{m_j}\}\Big].$$

The result is due to [30].

References

1. Anderson, T.W.: Asymptotic theory for principal component analysis. Ann. Math. Stat. **34**, 203–224 (1963)
2. Anderson, T.W.: *An Introduction to Multivariate Statistical Analysis*, 3rd edn. Wiley-Interscience, Hoboken, NJ (2003)
3. Bai, Z.D., Miao, B.Q., Pan, G.M.: On asymptotics of eigenvectors of large sample covariance matrix. Ann. Probab. **35**(4), 1532–1572 (2007)
4. Bodnar, T., Okhrin, Y.: Properties of the singular, inverse and generalized inverse partitioned Wishart distributions. J. Multivariate Anal. **99**, 2389–2405 (2008)
5. Collins, B., Matsumoto, S., Saad, N.: Integration of invariant matrices and moments of inverses of Ginibre and Wishart matrices. J. Multivariate Anal. **126**, 1–13 (2014)
6. Cook, R.D., Forzani, L.: On the mean and variance of the generalized inverse of a singular Wishart matrix. Electron. J. Stat. **5**, 146–158 (2011)
7. de Waal, D.J., Nel, D.G.: On some expectations with respect to Wishart matrices. South African Stat. J. **7**, 61–67 (1973)
8. Eaton, M.L.: *Multivariate Statistics: A Vector Space Approach*. Wiley, New York, NY (1983)
9. Eaton, M.L., Olkin, I.: Best equivalent estimators for a Cholesky decomposition. Ann. Stat. **15**(4), 1639–1650 (1987)
10. Fujikoshi, Y., Ulyanov, V.V., Shimizu, R.: Multivariate Statistics: High-Dimensional and Large-Sample Approximations. Wiley, Hoboken, NJ (2011)
11. Díaz-García, J.A., Gutiérrez-Jáimez, R.: Matricvariate and matrix multivariate T distributions and associated distributions. Metrika **75**, 963–976 (2012)
12. Glueck, D.H., Muller K.E.: On the trace of a Wishart. Commun. Stat. Theory **27**, 2137–2141 (1998)
13. Goodman, N.R.: Distribution of the determinant of a complex Wishart distributed matrix. Ann. Math. Stat. **34**, 178–180 (1963)
14. Gupta, A.K., Nagar, D.K.: Matrix Variate Distributions. Chapman & Hall/CRC, Boca Raton, FL (2000)
15. Haff, L.R.: Further identities for the Wishart distribution with applications in regression. Can. J. Stat. **9**, 215–224 (1981)

16. Haff, L.R.: Identities for the inverse Wishart distribution with computational results in linear and quadratic discrimination. Sankhyā Ser. B **44**, 245–258 (1982)
17. Johnson, N.L., Kotz, S., Balakrishnan, N.: Continuous Univariate Distributions, vol. 2, 2nd edn. Wiley-Interscience, Hoboken, NJ (1995)
18. Kang, C., Kim, B.: A study for the moment of Wishart distribution. Appl. Math. Sci. **9**, 3643–3649 (2015)
19. Kollo, T., von Rosen, D.: Advanced Multivariate Statistics with Matrices. Springer, Dordrecht (2005)
20. Letac, G., Massam, H.: All invariant moments of the Wishart distribution. Scand. J. Stat. **31**, 295–318 (2004)
21. Magnus, J.R., Neudecker, H.: The commutation matrix: some properties and applications. Ann. Stat. **7**, 381–394 (1979)
22. Marx, D.G.: Aspects of the Matric t-Distribution. Ph.D. thesis, University of The Orange Free State, RSA (1981)
23. Mathai, A.M.: Jacobians of Matrix Transformations and Functions of Matrix Arguments. World Scientific, Singapore (1979)
24. Matsumoto, S.: General moments of the inverse real wishart distribution and orthogonal Weingarten functions. J. Theor. Probab. **25**, 798–822 (2012)
25. Muirhead, R.J.: Aspects of Multivariate Statistical Theory. Wiley, NY (1982)
26. Muirhead, R.J.: A note on some Wishart expectations. Metrika **33**, 247–251 (1986)
27. Nagar, D.K., Gupta, A.K.: Expectations of functions of complex Wishart matrix. Acta Appl. Math. **113**, 265–288 (2011)
28. Olkin, I.: The 70th anniversary of the distribution of random matrices. Linear Algebra Appl. **354**, 231–243 (2002)
29. Oravecz, F., Petz, D.: On the eigenvalue distribution of some symmetric random matrices. Acta Sci. Math. **63**, 383–395 (1997)
30. Pielaszkiewicz, J.M., Holgersson, T.: Mixtures of traces of Wishart and inverse Wishart matrices. Commun. Stat. Theory 1–17 (2019). Online
31. Pielaszkiewicz, J.M., von Rosen D., Singull M.: On $\mathbb{E}\big[\prod_{i=0}^{k} tr\{W^{m_i}\}\big]$, where $W \sim \mathcal{W}_p(I, n)$. Commun. Stat. Theory **46**, 2990–3005 (2017)
32. Pielaszkiewicz, J.M., von Rosen, D., Singull M.: On n/p-asymptotic distribution of vector of weighted traces of powers of Wishart matrices. Electron. J. Linear Al. **33**, 24–40 (2018)
33. Saw, J.G.: Expectation of elementary symmetric functions of a Wishart matrix. Ann. Stat. **1**, 580–582 (1973)
34. Seber, G.A.F.: A Matrix Handbook for Statisticians. Wiley, Hoboken, NJ (2008)
35. Serdobolskii, V.: Multivariate Statistical Analysis: A High-Dimensional Approach. Kluwer Academic Press, Dordrecht (2000)
36. Silverstein, J.W.: On the eigenvectors of large-dimensional sample covariance matrices. J. Multivariate Anal. **30**, 1–16 (1989)
37. Silverstein, J.W.: Weak convergence of random functions defined by the eigenvectors of sample covariance matrices. Ann. Probab. **18**, 1174–1194 (1990)
38. Srivastava, M.S.: On the complex Wishart distribution. Ann. Math. Stat. **36**, 313–315 (1965)
39. Srivastava, M.S.: Singular Wishart and multivariate beta distributions. Ann. Stat. **31**, 1537–1560 (2003)
40. Srivastava, M.S., Khatri, C.G.: An Introduction to Multivariate Statistics. North-Holland, New York, NY (1979)
41. Styan, G.P.H.: Three useful expressions for expectations involving Wishart matrices and its inverse. In: Dodge, Y. (ed.) Statistical Data Analysis and Inference, pp. 283–296. Elsevier Science, Amsterdam (1989)
42. Sultan, S.A., Tracy D.S.: Moments of the complex multivariate normal distribution. Linear Algebra Appl. **237/238**, 191–204 (1996)
43. Tracy, D.S., Sultan, S.A.: Third moment of matrix quadratic form. Stat. Prob. Lett. **16**, 71–76 (1993)

44. Uhlig, H.: On singular Wishart and singular multivariate beta distributions. Ann. Stat. **22**, 395–405 (1994)
45. von Rosen D.: Moments for the inverted Wishart distribution. Scand. J. Stat. **15**, 97–109 (1988)
46. von Rosen D.: On moments of the inverted Wishart distribution. Statistics **30**, 259–278 (1997)
47. Watamori, Y.: On the moments of traces of Wishart and inverted Wishart matrices. South African Stat. J. **24**, 153–176 (1990)
48. Wishart, J.: The generalised product moment distribution in samples from a normal multivariate population. Biometrika **20A**, 32–52 (1928)
49. Zhang, Z.: Pseudo-inverse multivariate/matrix-variate distributions. J. Multivariate Anal. **98**, 1684–1692 (2007)

Chapter 10
Risk and Bias in Portfolio Optimization

Thomas Holgersson and Martin Singull

Abstract In this paper we derive weighted squared risk measures for a commonly used Stein-type estimator of the global minimum variance portfolio. The risk functions are conveniently split in terms of variance and squared bias over different weight matrices. It is argued that the common out-of-sample variance criteria should be used with care and that a simple unweighted risk function may be more appropriate.

10.1 Introduction

A financial portfolio is defined by a vector, $\mathbf{w} = (w_1, \ldots, w_p)$, say, whose element w_j describes the fraction of the portfolio investment that is allocated to some asset x_j. The random return of the portfolio is obtained as $r = x_1 w_1 + \cdots + x_p w_p = \mathbf{w}'\mathbf{x}$, where $\mathbf{x}$ is a vector of random returns. An investor is interested in holding a portfolio that maximizes the expected return $E[r]$ at a given level of risk ($var[r]$) or, equivalently, minimizes the risk $var[r]$ at a given level of the expected return. A portfolio satisfying such a maximization/minimization is called an optimal portfolio. Markowitz [20] developed a theory for such mean-variance portfolio optimization that still plays a fundamental role. The theory considers a fixed (non-random) portfolio $\mathbf{w}$, which could be determined qualitatively, and gives closed-form expressions for the optimal portfolio as a function of $\mathbf{w}$, which in turn depends on the mean vector and covariance matrix of $\mathbf{x}$.

While Markowitz's theory is concerned with analyses and conclusions drawn from a fixed portfolio vector, there has been a growing interest during the last fifty years in how one can use statistical methods to estimate $\mathbf{w}$. Because the

T. Holgersson (✉)
Department of Economics and Statistics, Linnaeus University, Växjö, Sweden
e-mail: thomas.holgersson@lnu.se

M. Singull
Department of Mathematics, Linköping University, Linköping, Sweden
e-mail: martin.singull@liu.se

© Springer Nature Switzerland AG 2020

T. Holgersson, M. Singull (eds.), *Recent Developments in Multivariate and Random Matrix Analysis*, https://doi.org/10.1007/978-3-030-56773-6_10

optimal portfolio is a function of certain population parameters, one obtains the so-called standard estimator of $\mathbf{w}$ by substituting these unknown parameters by their sample counterparts. Some basic sampling properties such as low-order moments have long been known [6, 15], while the full normal-theory distribution of the standard estimator was derived by [22]. It has since been recognized, both from the theoretical distribution and from empirical findings, that the sampling variance of the standard estimator may be too large to be useful in investment strategies [4, 9, 17, 21].

As in most multivariate analyses, this applies in particular when the sample size n is close to the dimension p. A number of alternatives to the standard estimator have been derived recently. Jagannathan and Ma [13] noticed that imposing certain moment constraints leads to a reduction of sampling variance, while [4] proposed a generalized estimator that impose a threshold constraint on the portfolio vector. Bayesian methods have been proposed by [15] and [16], while inference solutions to the case of singular covariance have been derived by [1].

In this paper our interest lies in a particular family of estimators defined as a weighted mean between the standard estimator and a constant vector. This type of estimator, which is related to a family of estimators originally proposed by James and Stein [14], has been considered by [2, 8] and [12]. One of the main features of this weighted estimator is that it depends on a tuning coefficient, which also introduces a bias in the estimator. We derive some properties of this Stein-type estimator with respect to different kinds of weighted squared loss functions. Particular focus is set on the bias term since this has been given little attention in the literature previously. We will restrict ourselves to the specific case of the global minimum variance portfolio (GMVP), although many of the concerns in the paper also apply to more general portfolio optimizations. We will not attempt to derive any explicit estimators. Our primary interest lies in deriving and comparing different risk measures for a commonly used Stein-type estimator and discuss some of their differences.

10.2 Preliminaries

Consider a vector $\mathbf{x} : p \times 1$ of excess returns of p financial assets. A financial portfolio is defined as the weighted sum $\mathbf{w}'\mathbf{x}$, where $\mathbf{w} : p \times 1$ is called the portfolio weight. The vector $\mathbf{w}$ can be determined "qualitatively", i.e., based on expert knowledge about the market, or it could be estimated from historical data of returns, which is the objective of this article. An efficient portfolio is usually determined by minimizing the portfolio variance subject to a mean portfolio premium return and the additional constraint that investment proportions sum to one. According to [20], an efficient portfolio weight $\mathbf{w}$, assuming absence of short sale constraints, is determined by

$$\min_{\mathbf{w}\in\mathbb{R}^p} \{\mathbf{w}'\Sigma\mathbf{w} \mid \mathbf{w}'\mathbf{1} = 1\}, \tag{10.1}$$

where Σ is the covariance matrix of $\mathbf{x}$ and $\mathbf{1} : p \times 1$ is a vector of ones. The well-known solution to (10.1) is given by

$$\mathbf{w} = \frac{\Sigma^{-1}\mathbf{1}}{\mathbf{1}'\Sigma^{-1}\mathbf{1}}. \tag{10.2}$$

The vector $\mathbf{w}$ is known as the global minimum variance portfolio (GMVP). It is also possible to define portfolios under more general constraints than those of (10.1), see [15] and [4] for alternative formulations.

Since the quantity in (10.2) depends on an unknown parameter, it needs to be estimated from data. We are thus concerned with the problem of using a set of n observations on random returns, say $\mathbf{x}_1, \ldots, \mathbf{x}_n$, to develop an estimator of $\mathbf{w}$. We will assume a common setting where $\mathbf{x}_i \sim_{iid} N_p(\mu, \Sigma)$ under the assumptions $\max_j |\mu_j| \leq a_1 < \infty$, $\max_j \lambda_j(\Sigma) \leq a_2 < \infty$, and $0 < a_3 \leq \min_j \lambda_j(\Sigma)$, where $\mu = (\mu_1, \ldots, \mu_p)'$ and $\lambda_j(\Sigma)$ denote an eigenvalue of Σ. Although these assumptions are not strictly necessary for making inference of the GMPV weight, they simplify the technical treatments considerably.

An obvious estimator of $\mathbf{w}$ is obtained by replacing the unknown covariance matrix by its sample counterpart. The resulting estimate is commonly referred to as the *standard estimator*, henceforth denoted by $\widehat{\mathbf{w}}_0$. This estimator is central in the paper, and we state some basic properties for the sake of clarity.

Property 10.1 Assume $\mathbf{x}_i \sim_{iid} N_p(\mu, \Sigma)$, $i = 1, \ldots, n$, $p \geq 4$ and $n \geq p+2$. Let $S = n^{-1}\sum_{i=1}^{n}(\mathbf{x}_i - \bar{\mathbf{x}})(\mathbf{x}_i - \bar{\mathbf{x}})'$, where $\bar{\mathbf{x}} = n^{-1}\sum_{i=1}^{n}\mathbf{x}_i$. Define

$$(i) \quad \mathbf{w} = \frac{\Sigma^{-1}\mathbf{1}}{\mathbf{1}'\Sigma^{-1}\mathbf{1}},$$

$$(ii) \quad \sigma^2 = \frac{1}{\mathbf{1}'\Sigma^{-1}\mathbf{1}},$$

$$(iii) \quad \widehat{\mathbf{w}}_0 = (\widehat{w}_{0(1)}, \ldots, \widehat{w}_{0(p)})' = \frac{\mathbf{S}^{-1}\mathbf{1}}{\mathbf{1}'\mathbf{S}^{-1}\mathbf{1}}.$$

Then

$$\widehat{\mathbf{w}}_0 \sim t_p\left(\mathbf{w}, \frac{\sigma^2\Sigma^{-1} - \mathbf{w}\mathbf{w}'}{n-p+1}, n-p+1\right),$$

where $t_p(\cdot)$ denotes a p-dimensional singular t-distribution with $n - p + 1$ degrees of freedom, location vector $\mathbf{w}$ and dispersion matrix $\sigma^2\Sigma^{-1} - \mathbf{w}\mathbf{w}'$ with $rank(\sigma^2\Sigma^{-1} - \mathbf{w}\mathbf{w}') = p - 1$.

Proof ***([18, 22])*** See also [19, Chapter 1], for a definition of the multivariate t-distribution. □

It is well known that the sampling variance of $\widehat{\mathbf{w}}_0$ can be substantial when n is small relative to the dimension p and hence of limited relevance to an investor. A considerable amount of research has been concerned with development of improved estimators of $\mathbf{w}$ [2, 4, 8, 12]. A common approach is to first decide a family of estimators, which usually depends on some tuning coefficient, and then use a risk function to identify the appropriate value of this coefficient. One concern with this approach, however, is that two different risk functions usually produce two different values of the tuning coefficient and hence that the distributional properties of our portfolio weight estimator may strongly depend on which specific risk function that is being used. The next section will discuss this matter further.

10.3 Risk Measures and Portfolio Estimation

The original view of portfolio optimization and risk as stated by [20] is that *The w_j's are not random variables, but are fixed by the investor* and that *"Risk" is described by the variance of the return*. The "risk" an investor is facing is accordingly determined by $var[\mathbf{w}'\mathbf{x}] = \mathbf{w}'\Sigma\mathbf{w}$, where $\Sigma = cov[\mathbf{x}_i]$. However, the definition of "risk" is less obvious when $\mathbf{w}$ is estimated from data because of the additional uncertainty of sampling variance and bias. From a perspective of statistical inference the term "risk" refers to the sampling variance and bias of a parameter estimator (say $\widehat{\mathbf{w}}$), while in portfolio theory "risk" primarily refers to variance of the return vector $\mathbf{x}$. Since the estimated portfolio return is defined by $r_i = \widehat{\mathbf{w}}'\mathbf{x}_i$, it involves risk in both senses.

Following Markowitz's view of a fixed (non-random) portfolio, the Lagrangian of the optimization problem in (10.1) may be formulated

$$L(\mathbf{w}, \Sigma, \lambda_0) = \frac{1}{2}\mathbf{w}'\Sigma\mathbf{w} - \lambda_0(\mathbf{w}'\mathbf{1} - 1), \tag{10.3}$$

where λ_0 is a Lagrange multiplier. By taking derivatives of $L(\mathbf{w}, \Sigma, \lambda_0)$ w.r.t. $\mathbf{w}$ and equating at zero we get the condition

$$\mathbf{w} = \lambda_0\Sigma^{-1}\mathbf{1}. \tag{10.4}$$

Since $\mathbf{w}'\mathbf{1} = 1$ it follows that $\lambda_0 = (\mathbf{1}'\Sigma^{-1}\mathbf{1})^{-1}$ and we obtain (10.2). Note that the identity (10.4) is completely determined by $\Sigma^{-1}\mathbf{1}$ in the sense that if we define $\theta = \Sigma^{-1}\mathbf{1}$, then the solution to the optimization problem depends on θ only. When it comes to random (estimated) portfolio weights, the optimization problem has been formulated in different ways in the literature. Jagannathan and Ma [13] specify a constrained portfolio variance minimization problem as

$$\min_{\mathbf{w}\in\mathbb{R}^p}\{\mathbf{w}'\widehat{\Sigma}\mathbf{w}|\mathbf{w}'\mathbf{1} = 1, 0 \le w_j, w_j \le \varpi, j = 1, \ldots, p\}, \tag{10.5}$$

where $\widehat{\Sigma}$ is some estimator of Σ and ϖ is a finite constant such that $w_j \leq \varpi$ defines an upper bound constraint for the weights. Let $\lambda = \begin{pmatrix}\lambda_1 \dots \lambda_p\end{pmatrix}$ be the Lagrange multiplier for $0 \leq w_j$ and $\delta = \begin{pmatrix}\delta_1 \dots \delta_p\end{pmatrix}$ the Lagrange multiplier for $w_j \leq \varpi$ and define $\widetilde{\Sigma} = \widehat{\Sigma} + (\delta\mathbf{1}' + \mathbf{1}\delta') - (\lambda\mathbf{1}' + \mathbf{1}\lambda')$, where $\widehat{\Sigma}$ is the normal-theory unconstrained maximum likelihood (ML) estimate of Σ. Jagannathan and Ma [13] showed that $\widetilde{\Sigma}$ is positive semidefinite, and that constructing a constrained global minimum variance portfolio from $\widehat{\Sigma}$ (i.e., the solution to (10.5)) is equivalent to constructing the unconstrained minimum variance portfolio from $\widetilde{\Sigma}$. Thus the constraints can be imposed in the estimation stage instead of the optimization stage and the result would be the same. This is a fundamental result in portfolio theory because it connects Markowitz's theoretical portfolio theory with statistical inference theory, including both frequentistic and Bayesian treatments.

DeMiguel et al. [4] proposed a norm-constrained minimum variance portfolio as one that solves $\min_{\mathbf{w}\in\mathbb{R}^p}\{\mathbf{w}'\Sigma\mathbf{w} \mid \mathbf{w}'\mathbf{1} = 1, \|\mathbf{w}\| \leq \varpi\}$, where $\|\mathbf{w}\|$ denotes either the L_1 norm $\|\mathbf{w}\|_1 = \sum_{j=1}^n |w_j|$ or the L_2 norm $\|\mathbf{w}\|_2 = \sqrt{\mathbf{w}'Q\mathbf{w}}$, where Q is some positive definite matrix. They showed that the solution under the constraint $\|\mathbf{w}\|_1 = 1$ coincides with the shortsale constrained estimator.

[8] take a rather different view on the optimization problem and argue that, for some estimator $\widehat{\mathbf{w}}$ based on returns $\mathbf{x}_1, \dots, \mathbf{x}_n$, the quantity $var[\mathbf{x}_{n+1}'\widehat{\mathbf{w}}|\mathscr{F}_n] - \widehat{\mathbf{w}}'\Sigma\widehat{\mathbf{w}}$, where $\mathscr{F}_n$ is the information set up to time n, represents the actual variance of the return belonging to the portfolio $\widehat{\mathbf{w}}$. References [2] and [23] take on a similar approach.

There currently seems to be no consensus, or unified theory, on how to formulate a statistical version of Markowitz's fixed-portfolio optimization problem. In what follows we will discuss some consequences of the view one takes on the optimization problem and the measure used to evaluate the properties of a weight estimator.

Portfolio Weight Estimators The GMVP weight vector $\mathbf{w} = (\mathbf{1}'\Sigma^{-1}\mathbf{1})^{-1}\Sigma^{-1}$ only involves one unknown quantity, Σ^{-1}. One can therefore specify $\mathbf{w}$ as a function of Σ^{-1}, say $\mathbf{w} = f(\Sigma^{-1}) = (\mathbf{1}'\Sigma^{-1}\mathbf{1})^{-1}\Sigma^{-1}$. A portfolio estimator may thus be obtained by substituting an estimator $\widehat{\Sigma}^{-1}$ into $\mathbf{w}$ and obtain $\widehat{\mathbf{w}} = f(\widehat{\Sigma}^{-1}) = (\mathbf{1}'\widehat{\Sigma}^{-1}\mathbf{1})^{-1}\widehat{\Sigma}^{-1}$. Any estimator $f(\widehat{\Sigma}^{-1})$ will be a sufficient statistic for $\mathbf{w}$ as long as $\widehat{\Sigma}^{-1}$ is a sufficient statistic for Σ^{-1}. Such "plug-in" estimators are therefore completely legitimate from an inferential point of view. The literature on estimation of Σ^{-1} is, in turn, extensive. Some important references include [5, 7, 10–12, 24, 26]. A comprehensive survey of improved estimators of Σ^{-1}, including Bayesian frameworks, is given in [25]. We will, however, not proceed on this path but instead consider a more restricted class of estimators.

Let $\widehat{\mathbf{w}}_0$ denote the standard estimator defined in Property 10.1 and $\mathbf{w}_{ref}$ denote some pre-determined reference portfolio, which could be fixed or random (but independent of $\widehat{\mathbf{w}}_0$). We denote a weighted mean between these two quantities as follows

$$\widehat{\mathbf{w}}_\alpha = (1-\alpha)\widehat{\mathbf{w}}_0 + \alpha\mathbf{w}_{ref}, \quad 0 \leq \alpha \leq 1. \tag{10.6}$$

The estimator $\widehat{\mathbf{w}}_\alpha$, which is closely related to a family of estimators proposed by [14], has been used by [2, 8, 18]. In order to determine an appropriate value of the tuning coefficient α one usually applies a loss function. A common square loss is defined by $l = (\widehat{\mathbf{w}} - \mathbf{w})'Q(\widehat{\mathbf{w}} - \mathbf{w})$, where Q is a p.d. matrix. Taking the expected value of l we obtain a quadratic risk function for $\widehat{\mathbf{w}}_\alpha$ as

$$\mathfrak{R}(\widehat{\mathbf{w}}_\alpha, Q) = E[(\widehat{\mathbf{w}}_\alpha - \mathbf{w})'Q(\widehat{\mathbf{w}}_\alpha - \mathbf{w})]. \tag{10.7}$$

The value of α that minimizes $\mathfrak{R}$ is then defined as the *optimal* α. By different choices of Q we obtain several common risk functions as special cases of (10.7). For example, $\mathfrak{R}(\widehat{\mathbf{w}}_\alpha, I)$ is the common mean squared error (MSE), which consists of the variance plus squared bias. The particular risk function $\mathfrak{R}(\widehat{\mathbf{w}}_\alpha, \Sigma)$ has been used by [8] and [18] while [2] used $l = (\widehat{\mathbf{w}}_\alpha - \mathbf{w})'\Sigma(\widehat{\mathbf{w}}_\alpha - \mathbf{w})$ directly, rather than its expected value, to derive an optimal portfolio estimator. Yet another way to evaluate the properties of a portfolio weight estimator is through its predictive, or out-of-sample properties. The predicted return of some estimator $\widehat{\mathbf{w}}$ is obtained by $\widehat{\mathbf{w}}'\mathbf{x}_t$, where $\mathbf{x}_t$ is a vector of returns not used in $\widehat{\mathbf{w}}$, i.e., $\mathbf{x}_t \notin \{\mathbf{x}_1, \ldots, \mathbf{x}_n\}$. The mean squared error of prediction (MSEP) of $\widehat{\mathbf{w}}$ may then be evaluated by $E[(\widehat{\mathbf{w}}'\mathbf{x}_t - \mathbf{w}'\mathbf{x}_t)^2]$. Note that MSEP is also a special case of (10.7), for

$$\begin{aligned} E[(\widehat{\mathbf{w}}'\mathbf{x}_t - \mathbf{w}'\mathbf{x}_t)^2] &= E[(\widehat{\mathbf{w}} - \mathbf{w})'\mathbf{x}_t\mathbf{x}_t'(\widehat{\mathbf{w}} - \mathbf{w})] \\ &= E[(\widehat{\mathbf{w}} - \mathbf{w})'(\Sigma + \mu\mu')(\widehat{\mathbf{w}} - \mathbf{w})] = \mathfrak{R}(\widehat{\mathbf{w}}, \Sigma + \mu\mu'). \end{aligned}$$

In what follows we give explicit expressions of $\mathfrak{R}(\widehat{\mathbf{w}}_\alpha, Q)$ for some particular values of Q.

Proposition 10.1 *Let* $\widehat{\mathbf{w}}_\alpha = (1 - \alpha)\widehat{\mathbf{w}}_0 + \alpha p^{-1}\mathbf{1}$ *for some* $0 \leq \alpha \leq 1$, *and let* $E[\widehat{\mathbf{w}}_0] = \mathbf{w}$, $R = cov[\widehat{\mathbf{w}}_0] = \dfrac{\sigma^2\Sigma^{-1} - \mathbf{w}\mathbf{w}'}{n-p-1}$, *where* $\sigma^2 = (\mathbf{1}'\Sigma^{-1}\mathbf{1})^{-1}$. *Let* $\mathbf{x}_t$ *be an out-of-sample vector of returns, and define* $\mu = E[\mathbf{x}_t]$, $\mu_w = \mathbf{w}'\mu$ *and* $\bar{\mu} = p^{-1}\mathbf{1}'\mu$. *Then the following risk identities hold*

(a) $\mathfrak{R}(\widehat{\mathbf{w}}_\alpha, I) = (1-\alpha)^2\mathrm{tr}\{R\} + \alpha^2(\mathbf{w}'\mathbf{w} - p^{-1})$,

(b) $\mathfrak{R}(\widehat{\mathbf{w}}_\alpha, \Sigma) = \dfrac{(1-\alpha)^2}{n-p-1}\sigma^2(p-1) + \alpha^2(p^{-2}\mathbf{1}'\Sigma\mathbf{1} - \sigma^2)$,

(c) $\mathfrak{R}(\widehat{\mathbf{w}}_\alpha, \mu\mu') = (1-\alpha)^2\mu'R\mu + \alpha^2(\mu_w - \bar{\mu})^2$,

(d) $\mathfrak{R}(\widehat{\mathbf{w}}_\alpha, \Sigma + \mu\mu') = \dfrac{(1-\alpha)^2}{n-p-1}(\sigma^2(p-1) + \sigma^2\mu'\Sigma^{-1}\mu + \mu_w^2) + \alpha^2((p^{-2}\mathbf{1}'\Sigma\mathbf{1} - \sigma^2) + (\mu_w - \bar{\mu})^2)$.

Before the proof of Proposition 10.1 we will state a useful identity in the following lemma.

Lemma 10.1 *Let* $\widehat{\mathbf{w}}_\alpha = (1-\alpha)\widehat{\mathbf{w}}_0 + \alpha p^{-1}\mathbf{1}$ *for some* $0 \le \alpha \le 1$, *and let* $E[\widehat{\mathbf{w}}_0] = \mathbf{w}$, $cov[\widehat{\mathbf{w}}_0] = R$. *Then it holds that*

$$E[(\widehat{\mathbf{w}}_\alpha - \mathbf{w})(\widehat{\mathbf{w}}_\alpha - \mathbf{w})'] = (1-\alpha)^2 R + \alpha^2(\mathbf{w} - p^{-1}\mathbf{1})(\mathbf{w} - p^{-1}\mathbf{1})'.$$

Proof See Appendix. □

We are now ready to give the proof of Proposition 10.1.

Proof Proof of Proposition 10.1. From Lemma 10.1 we find that

$$E[(\widehat{\mathbf{w}}_\alpha - \mathbf{w})(\widehat{\mathbf{w}}_\alpha - \mathbf{w})'] = \mathrm{tr}\{E[(\widehat{\mathbf{w}}_\alpha - \mathbf{w})\widehat{\mathbf{w}}_\alpha - \mathbf{w}']\} = (1-\alpha)^2\mathrm{tr}\{R\} + \alpha^2(\mathbf{w}'\mathbf{w} - p^{-1}),$$

which establishes (a). Applying Lemma 10.1 again we obtain (b) since

$$\mathrm{tr}\{E[(\widehat{\mathbf{w}}_\alpha - \mathbf{w})(\widehat{\mathbf{w}}_\alpha - \mathbf{w})'\Sigma]\} = (1-\alpha)^2\mathrm{tr}\{R\Sigma\} + \alpha^2(p^{-2}\mathbf{1}'\Sigma\mathbf{1} - \sigma^2)$$

and

$$\mathrm{tr}\{R\Sigma\} = \frac{\mathrm{tr}\{(\sigma^2\Sigma^{-1} - \mathbf{w}\mathbf{w}')\Sigma\}}{n-p-1} = \frac{\mathrm{tr}\{\sigma^2 I - \mathbf{w}\mathbf{w}'\Sigma\}}{n-p-1} = \frac{p-1}{n-p-1}\sigma^2.$$

The identity (c) is similarly obtained by

$$\mathrm{tr}\{E[(\widehat{\mathbf{w}}_\alpha - \mathbf{w})(\widehat{\mathbf{w}}_\alpha - \mathbf{w})'\mu\mu']\} = (1-\alpha)^2\mu' R\mu + \alpha^2(\mathbf{w}'\mu - p^{-1}\mathbf{1}'\mu)^2,$$

while (d) is obtained by adding (b) and (c) and simplifying terms. □

Remark 10.1 The MSEP measure in (d), i.e., $E[(\widehat{\mathbf{w}}'_\alpha\mathbf{x}_t - \mathbf{w}'\mathbf{x}_t)^2]$, describes the variability of $\widehat{\mathbf{w}}'_\alpha\mathbf{x}_t$ around the random term $\mathbf{w}'\mathbf{x}_t$. An alternative way of defining MSEP is by $E[(\widehat{\mathbf{w}}'_\alpha\mathbf{x}_t - \mathbf{w}'\mu)^2]$ which describes the variability of $\widehat{\mathbf{w}}'_\alpha\mathbf{x}_t$ around the non-random point $\mathbf{w}'E[\mathbf{x}_t]$. These two differ in that

$$E[(\widehat{\mathbf{w}}'_\alpha\mathbf{x}_t - \mathbf{w}'\mu)^2] = E[(\widehat{\mathbf{w}}'_\alpha\mathbf{x}_t - \mathbf{w}'\mathbf{x}_t)^2] + (\mathbf{1}'\Sigma^{-1}\mathbf{1})^{-1}.$$

See Appendix for a proof of this identity. Note that the smallest possible return prediction variance for any of the risks (a)–(d) of Proposition 10.1 is 0, which is attained at $\alpha = 1$, whereas for $E[(\widehat{\mathbf{w}}'_\alpha\mathbf{x}_t - \mathbf{w}'\mu)^2]$ the minimum prediction variance is determined by $(\mathbf{1}'\Sigma^{-1}\mathbf{1})^{-1}$ regardless of the value of α.

For purposes of deriving an estimator of $\mathbf{w}$, the choice between $E[(\widehat{\mathbf{w}}'_\alpha\mathbf{x}_t - \mathbf{w}'\mathbf{x}_t)^2]$ and $E[(\widehat{\mathbf{w}}'_\alpha\mathbf{x}_t - \mathbf{w}'\mu)^2]$ makes no difference, but for calculation of prediction intervals our view of the centre point does matter.

Remark 10.2 The bias terms vanish trivially if $\alpha = 0$ and/or $\Sigma = I$. More generally, the bias terms for $\mathfrak{R}(\widehat{\mathbf{w}}_\alpha, I)$, $\mathfrak{R}(\widehat{\mathbf{w}}_\alpha, \Sigma)$, and $\mathfrak{R}(\widehat{\mathbf{w}}_\alpha, \mu\mu')$ vanish when $p\mathbf{1}'\Sigma^{-2}\mathbf{1} = (\mathbf{1}'\Sigma^{-1}\mathbf{1})^2$, $p^{-2}\mathbf{1}'\Sigma\mathbf{1} = \sigma^2$ and $\mu_w - \bar{\mu} = 0$, respectively. While the first two conditions are almost identical to the condition $\Sigma = I$, the identity $\mu_w = \bar{\mu}$, holds when the portfolio return equals the average asset mean.

Remark 10.3 The optimal value of the tuning coefficient α may be derived by equating $\frac{\partial \mathfrak{R}(\widehat{\mathbf{w}}_\alpha, Q)}{\partial \alpha}$ at zero and solving for α. The resulting value, say α_{opt}, then yields an adaptive estimator $\widehat{\mathbf{w}}_{\alpha_{opt}} = (1 - \alpha_{opt})\widehat{\mathbf{w}}_0 + \alpha_{opt} p^{-1}\mathbf{1}$. However, α_{opt} is by necessity a function of unknown population parameters. Operational, or "bona-fide", estimators are usually obtained by substituting these unknown parameters by some estimators to obtain an approximation $\widehat{\alpha}$. The adapted portfolio estimator is then defined by $\widehat{\mathbf{w}}_{\widehat{\alpha}} = (1 - \widehat{\alpha})\widehat{\mathbf{w}}_0 + \widehat{\alpha} p^{-1}\mathbf{1}$. The coefficient $\widehat{\alpha}$ is hence random, which in turn distorts the otherwise known distribution of $\widehat{\mathbf{w}}_\alpha$. It is still possible to conduct inferential analysis (interval estimation, etc.) through the theory of oracle inequalities but at the price of becoming more technically involved (see [3] for a comprehensive survey of the topic). Another option is to determine α qualitatively, for example based on expert knowledge of the financial market. This method has the appealing property of retaining the sampling distribution of $\widehat{\mathbf{w}}_\alpha$.

Remark 10.4 The original portfolio problem as posed by [20] considers the return of a non-random portfolio at data point t (e.g., a time point), determined by $\mathbf{w}'\mathbf{x}_t$ with variance $var[\mathbf{w}'\mathbf{x}_t] = \mathrm{tr}\{cov[\mathbf{x}_t]\mathbf{w}\mathbf{w}'\} = \mathbf{w}'\Sigma\mathbf{w}$, which is void of concerns about consistency and bias, etc. The fixed-portfolio theory uses Lagrange multipliers to derive the global minimum weight portfolio $\mathbf{w} = \frac{\Sigma^{-1}\mathbf{1}}{\mathbf{1}'\Sigma^{-1}\mathbf{1}}$, which is the minimum of $var[\mathbf{w}'\mathbf{x}_t]$ subject to the constraint $\sum_{j=1}^{p} w_j = 1$. How to proceed from there to statistical estimation is less obvious. It is customary to apply the criterion $\mathfrak{R}(\widehat{\mathbf{w}}_\alpha, \Sigma) = E[(\widehat{\mathbf{w}} - \mathbf{w})'\Sigma(\widehat{\mathbf{w}} - \mathbf{w})]$ to derive estimators of $\mathbf{w}$, but it is not clear what is gained by minimizing this risk instead of the unweighted risk $\mathfrak{R}(\widehat{\mathbf{w}}_\alpha, I) = E[(\widehat{\mathbf{w}} - \mathbf{w})'(\widehat{\mathbf{w}} - \mathbf{w})]$. For any consistent estimator, say $\widehat{\mathbf{w}} = \frac{\widehat{\Sigma}^{-1}\mathbf{1}}{\mathbf{1}'\widehat{\Sigma}^{-1}\mathbf{1}}$, where $\widehat{\Sigma}^{-1}$ some consistent estimator of Σ^{-1}, the asymptotic return variance is $\lim_{n\to\infty} var[\widehat{\mathbf{w}}'\mathbf{x}_t] = \mathbf{w}'\Sigma\mathbf{w}$ regardless of which risk function was used to obtain $\widehat{\mathbf{w}}$. The weight $Q = \Sigma$ has already been used in the minimization of $\mathbf{w}'\Sigma\mathbf{w}$ to obtain the functional form $\frac{\Sigma^{-1}\mathbf{1}}{\mathbf{1}'\Sigma^{-1}\mathbf{1}}$, and there is no apparent gain in using the weighting $(\widehat{\mathbf{w}} - \mathbf{w})'\Sigma(\widehat{\mathbf{w}} - \mathbf{w})$ once again in the estimation stage. In fact, Σ^{-1} is the only unknown parameter in $\frac{\Sigma^{-1}\mathbf{1}}{\mathbf{1}'\Sigma^{-1}\mathbf{1}}$. Knowing Σ (up to a scalar multiplication) is therefore equivalent to knowing $\mathbf{w}$. Since Σ is a one-to-one transformation of Σ^{-1}, the weighted loss $(\widehat{\mathbf{w}} - \mathbf{w})'\Sigma(\widehat{\mathbf{w}} - \mathbf{w})$ in a sense uses the true parameter to derive its own estimator. In this view the unweighted risk $\mathfrak{R}(\widehat{\mathbf{w}}_\alpha, I) = E[(\widehat{\mathbf{w}} - \mathbf{w})'(\widehat{\mathbf{w}} - \mathbf{w})]$ may better reflect the actual risk of an estimator $\widehat{\mathbf{w}}$.

Acknowledgments The authors would also like to thank the reviewer for several valuable and helpful suggestions and comments to improve the presentation of the paper.

Appendix

First we give the proof of Lemma 10.1.

Proof Note that

$$\widehat{\mathbf{w}}_\alpha - E[\widehat{\mathbf{w}}_\alpha] = \widehat{\mathbf{w}}_0 - \alpha(\widehat{\mathbf{w}}_0 - p^{-1}\mathbf{1}) - \mathbf{w} - \alpha(\mathbf{w} - p^{-1}\mathbf{1}) = (\alpha - 1)(\widehat{\mathbf{w}}_0 - \mathbf{w}).$$

Hence,

$$cov[\widehat{\mathbf{w}}_\alpha] = (\alpha - 1)^2 E[(\widehat{\mathbf{w}}_0 - \mathbf{w})(\widehat{\mathbf{w}}_0 - \mathbf{w})'] = (\alpha - 1)^2 cov[\widehat{\mathbf{w}}_0] = (\alpha - 1)^2 R.$$

Similarly, define the bias term $\mathbf{b} = E[\widehat{\mathbf{w}}_\alpha] - \mathbf{w}$. Then

$$E[\widehat{\mathbf{w}}_0] - \alpha(E[\widehat{\mathbf{w}}_0] - p^{-1}\mathbf{1}) - \mathbf{w} = -\alpha(\mathbf{w} - p^{-1}\mathbf{1}).$$

Hence it follows that $\mathbf{bb}' = \alpha^2(\mathbf{w} - p^{-1}\mathbf{1})(\mathbf{w} - p^{-1}\mathbf{1})'$ and

$$E[(\widehat{\mathbf{w}}_\alpha - \mathbf{w})(\widehat{\mathbf{w}}_\alpha - \mathbf{w})'] = cov[\widehat{\mathbf{w}}_\alpha] + \mathbf{bb}' = (1 - \alpha)^2 R + \alpha^2(\mathbf{w} - p^{-1}\mathbf{1})(\mathbf{w} - p^{-1}\mathbf{1})'.$$

□

Next we give the proof of the identity in Remark 10.1.

Proof We have

$$\begin{aligned} E[(\widehat{\mathbf{w}}_\alpha'\mathbf{x}_t - \mathbf{w}'\mu)^2] &= E[((\widehat{\mathbf{w}}_\alpha'\mathbf{x}_t - \mathbf{w}'\mathbf{x}_t) + (\mathbf{w}'\mathbf{x}_t - \mathbf{w}'\mu))^2] \\ &= \mathrm{tr}\{E[(\widehat{\mathbf{w}}_\alpha - \mathbf{w})\mathbf{x}_t'\mathbf{x}_t(\widehat{\mathbf{w}}_\alpha - \mathbf{w})]\} \\ &\quad + 2\mathrm{tr}\{E[(\widehat{\mathbf{w}}_\alpha'\mathbf{x}_t - \mathbf{w}'\mathbf{x}_t)(\mathbf{w}'\mathbf{x}_t - \mathbf{w}'\mu)]\} \\ &\quad + E[\mathbf{w}'(\mathbf{x}_t - \mu)(\mathbf{x}_t - \mu)'\mathbf{w}] \\ &= A + 2B + D, \end{aligned}$$

say. The identities $A = \mathrm{tr}\{(\Sigma + \mu\mu')E[(\widehat{\mathbf{w}}_\alpha - \mathbf{w})(\widehat{\mathbf{w}}_\alpha - \mathbf{w})']\}$ and $D = \mathrm{tr}\{E[(\mathbf{x}_t - \mu)(\mathbf{x}_t - \mu)'\mathbf{ww}']\} = \mathrm{tr}\{\Sigma\mathbf{ww}'\} = \mathbf{w}'\Sigma\mathbf{w}$ are immediate. For the middle term B we have

$$\begin{aligned} B &= \mathrm{tr}\{E[\mathbf{x}_t\mathbf{x}_t'\mathbf{w}\widehat{\mathbf{w}}_\alpha - \mathbf{x}_t\mathbf{x}_t'\mathbf{ww}' - \mu\mu'\mathbf{w}\widehat{\mathbf{w}}_\alpha' + \mu\mu'\mathbf{ww}']\} \\ &= \mathrm{tr}\{E[\mathbf{x}_t\mathbf{x}_t'(\mathbf{w}(\mathbf{w} + \alpha(\mathbf{w} - p^{-1}\mathbf{1}))' - \mathbf{ww}') - \mu\mu'(\mathbf{w}(\mathbf{w} + \alpha(\mathbf{w} - p^{-1}\mathbf{1}))' - \mathbf{ww}')]\} \\ &= \mathrm{tr}\{\alpha E[\mathbf{x}_t\mathbf{x}_t' - \mu\mu']\mathbf{w}(\mathbf{w} - p^{-1}\mathbf{1})'\} = \alpha(\mathbf{w} - p^{-1}\mathbf{1})'\Sigma\mathbf{w} = 0. \end{aligned}$$

□

References

1. Bodnar, T., Mazur, S., Podgórski, K.: Singular inverse Wishart distribution and its application to portfolio theory. J. Multivar. Anal. **143**, 314–326 (2016)
2. Bodnar, T., Parolya, N., Schmid, W.: Estimation of the global minimum variance portfolio in high dimensions. Eur. J. Oper. Res. **266**(1), 371–390 (2018)
3. CandŔes, E.J.: Modern statistical estimation via oracle inequalities. Acta Numer. **15**, 257–325 (2006)
4. DeMiguel, V., Garlappi, L., Nogales, F.J., Uppal, R.: A generalized approach to portfolio optimization: improving performance by constraining portfolio norms. Manage. Sci. **55**(5), 798–812 (2009)
5. Dey, D.K., Ghosh, M., Srinivasan, C.: A new class of improved estimators of a multinormal precision matrix. Stat. Decis. **8**(2), 141–152 (1990)
6. Dickinson, J.P.: The reliability of estimation procedures in portfolio analysis. J. Financ. Quant. Anal. **9**(3), 447–462 (1974)
7. Efron, B., Morris, C.: Multivariate empirical Bayes and estimation of covariance matrices. Ann. Stat. **4**(1), 22–32 (1976)
8. Frahm, G., Memmel, C.: Dominating estimators for minimum-variance portfolios. J. Econom. **159**(2), 289–302 (2010)
9. Frankfurter, G.M., Phillips H.E., Seagle, J.P.: Portfolio selection: the effects of uncertain means, variances, and covariances. J. Financ. Quant. Anal. **6**(5), 1251–1262 (1971)
10. Haff, L.R.: Estimation of the inverse covariance matrix: random mixtures of the inverse Wishart matrix and the identity. Ann. Stat. **7**(6), 1264–1276 (1979)
11. Haff, L.R.: The variational form of certain bayes estimators. Ann. Stat. **19**(3), 1163–1190 (1991)
12. Holgersson, T., Karlsson, P., Stephan, A.: A risk perspective of estimating portfolio weights of the global minimum variance portfolio. Adv. Stat. Anal. **104**(1), 59–80 (2020)
13. Jagannathan, R., Ma, T.: Risk reduction in large portfolios: why imposing the wrong constraints helps. J. Finance **58**(4), 1651–1684 (2003)
14. James, W., Stein, C.: Estimation with quadratic loss. In: Proceedings of the Fourth Berkeley Symposium on Mathematical Statistics and Probability, vol. 1, pp. 361–379 (1961)
15. Jobson, J.D., Korkie, B.: Estimation for Markowitz efficient portfolios. J. Am. Stat. Assoc. **75**(371), 544–554 (1980)
16. Jorion, P.: International portfolio diversification with estimation risk. J. Bus. **58**(3), 259–278 (1985)
17. Kan, R., Zhou, G.: Optimal portfolio choice with parameter uncertainty. J. Financ. Quant. Anal. **42**(3), 621–656 (1985)
18. Kempf, A., Memmel, C.: Estimating the global minimum variance portfolio. Schmalenbach Bus. Rev. **58**, 332–348 (2006)
19. Kotz, S., Nadarajah, S.: Multivariate T-Distributions and Their Applications, 1st edn. Cambridge University Press, Cambridge (2004)
20. Markowitz, H.: Portfolio selection. J. Financ. **7**(1), 77–91 (1952)
21. Michaud, R.O.: The Markowitz optimization enigma: is optimized optimal? ICFA Continuing Educ. Ser. **4**, 43–54 (1989)
22. Okhrin, Y., Schmidt, W.: Distributional properties of portfolio weights. J. Econom. **134**(1), 235–256 (2006)
23. Rubio, F., Mestre, X., Palomar, D.P.: Performance analysis and optimal selection of large minimum variance portfolios under estimation risk. IEEE J. Select. Top. Signal Process. **6**(4), 337–350 (2012)
24. Serdobolskii, V.I.: Estimation of high-dimensional inverse covariance matrices. In: Multivariate Statistical Analysis: A High-Dimensional Approach, pp. 87–101. Springer, New York, NY (2000)

25. Tsukuma, H., Konno, Y.: On improved estimation of normal precision matrix and discriminant coefficients. J. Multivar. Anal. **97**, 1477–1500 (2006)
26. Zhou, X., Sun, X., Wang, J.: Estimation of the multivariate normal precision matrix under the entropy loss. Ann. Inst. Stat. Math. **53**(4), 760–768 (2001)

Chapter 11
Approximating Noncentral Chi-Squared to the Moments and Distribution of the Likelihood Ratio Statistic for Multinomial Goodness of Fit

Björn Holmquist, Anna Sjöström, and Sultana Nasrin

Abstract The chi-square distribution is often assumed to hold for the asymptotic distribution of two times the log likelihood ratio statistic under the null hypothesis. Approximations are derived for the mean and variance of G^2, the likelihood ratio statistic for testing goodness of fit in a s category multinomial distribution. The first two moments of G^2 are used to fit the distribution of G^2 to a noncentral chi-square distribution. The fit is generally better than earlier attempts to fit to scaled versions of asymptotic central chi-square distributions. The results enlighten the complex role of the dimension of the multivariate variable in relation to the sample size, for asymptotic likelihood ratio distribution results to hold.

11.1 Introduction

Let us consider repeated independent observations, of a categorical distribution in s categories (or classes or cells) whose summation will end up in an observation from a multinomial distribution. The likelihood ratio test for goodness of fit in a multinomial distribution rejects if

$$G^2 = 2\sum_{i=1}^{s} Y_i \ln(Y_i/(Np_i)) \tag{11.1}$$

is large, where N is the number of trials, s is the number of category cells, p_i is the hypothesized probability of the ith cell. The null hypothesis is that the category probabilities are given by the quantities $\{p_i\}$ where $\sum_{i=1}^{s} p_i = 1$.

Our setup is thus a $N \times s$ matrix of zeros and ones and where each row sum up to one, and column j sum up to Y_j, for $j = 1, 2, \ldots, s$, and where $\sum_{j=1}^{s} Y_j = N$.

B. Holmquist (✉) · A. Sjöström · S. Nasrin
Department of Statistics, Lund University, Lund, Sweden
e-mail: Bjorn.Holmquist@stat.lu.se; Anna.Sjostrom@stat.lu.se; Sultana.Nasrin@stat.lu.se

© Springer Nature Switzerland AG 2020

T. Holgersson, M. Singull (eds.), *Recent Developments in Multivariate and Random Matrix Analysis*, https://doi.org/10.1007/978-3-030-56773-6_11

An often used approximation is that G^2, for large N, is chi-squared distributed with $s - 1$ degrees of freedom (df) when the null distribution is true, which follows from general theory, [15]. The standard asymptotic theory for likelihood ratio tests [12, p. 417] implies that as $N \to \infty$, $P[G^2 \leq x] \to P[\chi^2(s-1) \leq x]$.

Despite the fact that the log likelihood ratio statistic is often used in practice, in preference to the Pearson chi-square statistic, the distributions of these statistics for small, or moderate sample sizes are not very well known. The asymptotic (large sample) approximation is often used in practice even for small sample sizes. Chen et al. [2] made (among other things) a comparison of the asymptotic distribution and a Monte Carlo implementation of the maximum likelihood ratio test for homogeneity of the class probabilities (the equiprobable case) in a situation with moderate sample size, indicating deviations in the finite size distribution from the asymptotic one.

Studies of the distribution of this statistic appears to have started with [9], who gave cumulative probabilities for certain trinomial distributions.

Our study involves the investigation of the mean and variance of the statistic G^2 and approximations of the distribution of the statistic. Especially the influence on these characteristics from the relation between the 'dependence' dimension s and the 'independence' dimension N is emphasized, which is of importance in situations where there may be an increase in both these dimensions.

11.2 Mean and Variance of the G^2 Statistic

An approximate expression of the mean of G^2, given by Williams [16, p. 34], is in our notation

$$(s-1)\left(1+\frac{s^2-1}{6(s-1)N}\right).$$

For non-equiprobable cases this will be

$$\mathsf{e}^{(W)} := s - 1 + \left(\left(\sum_{i=1}^{s} \frac{1}{p_i}\right) - 1\right) / 6N\,. \tag{11.2}$$

These will be denoted $\mathsf{e}^{(W)}$ in Table 11.1 below.

Williams [16] suggested a multiplicative adjustment of the log-likelihood ratio statistic to achieve better fit to the chi-square distribution for small N. Using a multiplier

$$q^{(W)} = 1 + [6N(s-1)]^{-1}[(\sum_i p_i^{-1}) - 1] \tag{11.3}$$

Table 11.1 Mean and variance of the G^2 statistic

	E_{sim} [a]	V_{sim} [a]	E_{xct} [b]	V_{xct} [b]	$q^{(W)}$ (11.3)	$q^{(Sc)}$ (11.6)	$\mathsf{e}^{(W)}$ (11.2)	$\mathsf{e}^{(Sc)}$ (11.4)	$\mathsf{v}^{(Sc)}$ (11.5)	$2\mathsf{e}_1^{\text{KL}}$ (11.15)	$2\mathsf{e}_2^{\text{KL}}$ (11.15)	$2\,\mathsf{e}_3^{\text{KL}}$ (11.17)	$2\mathsf{e}_4^{\text{KL}}$ (11.18)	$2\mathsf{e}_5^{\text{KL}}$ (11.19)	$2\mathsf{e}_6^{\text{KL}}$ (11.20)	$4\mathsf{v}_1^{\text{KL}}$ (11.21)	$4\,\mathsf{v}_2^{\text{KL}}$ (11.22)	$\mathsf{e}^{(L)}$ (11.7)	$\mathsf{v}^{(L)}$ (11.8)	$\mathsf{e}^{(SH)}$ (11.9)	$\mathsf{v}^{(SH)}$ (11.10)
$N = 15, s = 2$																					
$\mathbf{p} = (1, 1)/s$, $^{(S)}$	1.044	2.197	$1.037^{(S)}$	$2.169^{(S)}$	1.03	1.04	1.03	1.04	2.16	1	1	1.03	1.03	1.036	1.036	1.87	1.87	1.037	2.161	1.036	2.157
$N = 30, s = 2$																					
$\mathbf{p} = (1, 1)/s$, $^{(S)}$	1.024	2.10	$1.017^{(S)}$	$2.074^{(S)}$	1.02	1.02	1.02	1.02	2.07	1	1	1.02	1.02	1.017	1.017	1.93	1.93	1.017	2.073	1.017	2.073
$N = 15, s = 5$																					
$\mathbf{p} = \mathbf{1}_s/s$, $^{(S)}$	4.43	10.07	$4.431^{(S)}$	$10.094^{(S)}$	1.07	1.09	4.27	4.34	9.66	4	3.73	4.27	4.06	4.38	4.115	7.47	5.21	4.570	10.153	4.341	9.659
$N = 30, s = 5$																					
$\mathbf{p} = \mathbf{1}_s/s$, $^{(S)}$	4.16	8.79	$4.164^{(S)}$	$8.863^{(S)}$	1.03	1.04	4.13	4.15	8.68	4	3.87	4.13	4.08	4.16	4.122	7.73	6.50	4.162	8.743	4.152	8.681
$N = 15, s = 10$																					
$\mathbf{p} = \mathbf{1}_s/s$, $^{(S)}$	10.62	18.14	$10.646^{(S)}$	$18.223^{(S)}$	1.12	1.20	10.1	10.77	27.73	9	7.4	10.41	7.54	12.83	3.374	16.8	6.25	61.941	37.412	10.767	27.733
$N = 30, s = 10$																					
$\mathbf{p} = \mathbf{1}_s/s$, $^{(S)}$	9.92	22.32	$9.912^{(S)}$	$22.155^{(S)}$	1.06	1.08	9.55	9.72	21.53	9	8.2	9.63	8.91	9.98	8.847	17.4	10.80	10.449	22.743	9.717	21.533
$N = 40, s = 3$																					
$\mathbf{p} = (0.3)^2, 0.4$, $^{(L)}$	2.035	4.165	$2.036^{(L)}$	$4.157^{(L)}$	1.017	1.018	2.034	2.04	4.15	2	1.98	2.03	2.03	2.04	2.035	3.90	3.72	2.036	4.155	2.036	4.152
$N = 40, s = 3$																					
$\mathbf{p} = (1, 1, 1)/s$	2.02	4.13	2.035	4.152	1.017	1.018	2.033	2.04	4.15	2	1.98	2.03	2.03	2.04	2.035	3.90	3.74	2.035	4.151	2.035	4.148
(0.25, 0.25, 0.5)	2.04	4.16	2.041	4.180	1.019	1.020	2.038	2.04	4.17	2	1.98	2.04	2.03	2.04	2.038	3.93	3.66	2.041	4.177	2.040	4.172
(0.1, 0.1, 0.8)	2.13	4.72	2.125	4.691	1.042	1.052	2.084	2.10	4.49	2	1.88	2.09	2.01	2.13	2.031	4.21	2.86	2.135	4.590	2.103	4.487

(continued)

Table 11.1 (continued)

	E_{sim} [a]	V_{sim} [a]	E_{xct} [b]	V_{xct} [b]	$q^{(W)}$ (11.3)	$q^{(Sc)}$ (11.6)	$\mathsf{e}^{(W)}$ (11.2)	$\mathsf{e}^{(Sc)}$ (11.4)	$\mathsf{v}^{(Sc)}$ (11.5)	$2\mathsf{e}_1^{\text{KL}}$ (11.15)	$2\mathsf{e}_2^{\text{KL}}$ (11.16)	$2\ \mathsf{e}_3^{\text{KL}}$ (11.17)	$2\mathsf{e}_4^{\text{KL}}$ (11.18)	$2\mathsf{e}_5^{\text{KL}}$ (11.19)	$2\mathsf{e}_6^{\text{KL}}$ (11.20)	$4\mathsf{v}_1^{\text{KL}}$ (11.21)	$4\ \mathsf{v}_2^{\text{KL}}$ (11.22)	$\mathsf{e}^{(L)}$ (11.7)	$\mathsf{v}^{(L)}$ (11.8)	$\mathsf{e}^{(SH)}$ (11.9)	$\mathsf{v}^{(SH)}$ (11.10)
$N = 40, s = 8$																					
$\mathbf{p} = \mathbf{1}_s/s$	7.35	16.01	7.352	15.980	1.038	1.044	7.263	7.31	15.42	7	6.65	7.28	7.10	7.35	7.182	13.65	10.51	7.351	15.623	7.309	15.423
$(0.1, 0.1, 0.3, (0.1)^5)$	7.44	16.42	7.444	16.442	1.043	1.053	7.301	7.37	15.74	7	6.57	7.33	7.05	7.45	7.115	13.88	9.92	7.481	16.096	7.368	15.737
$(0.1)^2, (0.3)^2, (0.05)^4$	7.71	15.99	7.709	15.872	1.063	1.088	7.440	7.62	17.19	7	6.29	7.56	6.65	8.19	5.838	14.72	8.31	11.658	19.088	7.619	17.191
$N = 300, s = 8$																					
$\mathbf{p} = \mathbf{1}_s/s$	7.05	14.14	7.036	14.147	1.005	1.005	7.035	7.04	14.15	7	6.95	7.04	7.03	7.04	7.036	13.95	13.49	7.036	14.147	7.036	14.147
$(0.1, 0.1, 0.3, (0.1)^5)$	7.06	14.23	7.041	14.171	1.006	1.006	7.040	7.04	14.17	7	6.94	7.04	7.04	7.04	7.041	13.98	13.39	7.041	14.171	7.041	14.170
$(0.1)^2, (0.3)^2, (0.05)^4$	7.05	14.18	7.062	14.266	1.008	1.009	7.059	7.06	14.26	7	6.91	7.06	7.04	7.06	7.058	14.10	13.04	7.062	14.265	7.062	14.260
$N = 500, s = 8$																					
$\mathbf{p} = \mathbf{1}_s/s$	7.02	14.14	7.021	14.087	1.003	1.003	7.021	7.02	14.09	7	6.97	7.02	7.02	7.02	7.021	13.97	13.69	7.021	14.086	7.021	14.086
$0.1, 0.1, 0.3, (0.1)^5$	7.04	14.13	7.025	14.100	1.003	1.004	7.024	7.03	14.10	7	6.97	7.02	7.02	7.02	7.024	13.99	13.63	7.025	14.100	7.025	14.100
$(0.1)^2, (0.3)^2, (0.05)^4$	7.05	14.19	7.036	14.151	1.005	1.005	7.035	7.03	14.14	7	6.94	7.04	7.03	7.04	7.036	14.06	13.42	7.036	14.151	7.034	14.137

[a] Simulation values of means (E_{sim}) and variances (V_{sim}) are based on 100,000 replications of G^2

[b] Exact values of mean and variance, (E_{xct}) and (V_{xct}), are based on (11.12) and (11.13), and approximated values of mean by $^{(W)}$ [16], Eq. (11.2), mean and variance by Smith et al. [14], Eq. ((11.4)–(11.5)), the Kullback-Leibler divergence equations (11.15)–(11.22), $^{(L)}$ [8], Eqs. (11.7)–(11.8) and [13], Eqs. (11.9)–(11.10). Multipliers $q^{(W)}$ and $q^{(Sc)}$ are also given using Eqs. (11.3) and (11.6), respectively. The category distributions for which exact expected values and variances were previous calculated are indicated by $^{(S)}$ [14] or $^{(L)}$ [8]

the suggestion of Williams implies that $P[G^2 \leq x]$ should be approximated by $P[q^{(W)}\chi^2(s-1) \leq x]$.

The multiplicative factor of Williams comes from the expansion of the expected value of G^2, shown to be $s - 1 + [6N]^{-1}[(\sum_i p_i^{-1}) - 1]$ when neglecting terms of order N^{-2}, and its ratio to the degrees of freedom $s-1$ in the asymptotic chi-square distribution.

Smith et al. [14] derived expansions of the mean and variance of G^2,

$$\mathsf{e}^{(Sc)} := s - 1 + [(\sum_i p_i^{-1}) - 1]/6N + [(\sum_i p_i^{-2}) - (\sum_i p_i^{-1})]/6N^2 \tag{11.4}$$

and

$$\mathsf{v}^{(Sc)} := 2(s-1) + (2/3N)[(\sum_i p_i^{-1}) - 1] + (4/3N^2)[(\sum_i p_i^{-2}) - (\sum_i p_i^{-1})] \tag{11.5}$$

([14], (2.1) and (2.2) when neglecting terms of order N^{-3} and corrected for the misprinted sign in the N^{-2} term), and suggested a multiplier

$$q^{(Sc)} = 1 + [6N(s-1)]^{-1} \times [(\sum_i p_i^{-1}) - 1 + N^{-1}[(\sum_i p_i^{-2}) - (\sum_i p_i^{-1})]] \tag{11.6}$$

(also corrected for misprinted sign in the N^{-2} term in their notation q') to be used in χ^2 fit to $G^2/q^{(Sc)}$. The expressions in (11.4) and (11.5) will be denoted $\mathsf{e}^{(Sc)}$ and $\mathsf{v}^{(Sc)}$ in Table 11.1.

Lyons et al. [8] list some moments of G^2 (denoted L in their paper) for some special cases. They also give some higher order approximations of the mean and variance of G^2:

$$\begin{aligned}
\mathsf{E}(G^2) \approx \mathsf{e}^{(L)} &:= (s-1) + (R_1 - 1)/6N + (R_2 - R_1)/6N^2 \\
&+ (19R_3/60 - R_2/2 + R_1/6 + 1/60)/N^3 \\
&+ (9R_4/10 - 19R_3/10 + 7R_2/6 - R_1/6)/N^4 \\
&+ (863R_5/252 - 9R_4 + 95R_3/12 - 5R_2 + R_1/6 - 1/126)/N^5 \\
&+ (1375R_6/84 - 4315R_5/84 + 117R_4/2 - 57R_3/2 + 31R_2/6 - R_1/6)/N^6 \\
&+ (\frac{33953R_7}{360} - \frac{1375R_6}{4} + \frac{431R_5}{9} - 315R_4 \\
&+ \frac{5719R_3}{60} - \frac{21R_2}{2} + \frac{R_1}{6} + \frac{1}{120})/N^7
\end{aligned} \tag{11.7}$$

and

$$V(G^2) \approx \mathsf{v}^{(L)} := 2(s-1) + 2(R_1-1)/3N + 4(R_2-R_1)/3N^2 \\ + (701R_3/180 - 6R_2 - R_1^2/36 + 2/15)/N^3, \quad (11.8)$$

where $R_j = \sum_{i=1}^{s} p_i^{-j}$. These approximating expressions will be denoted $\mathsf{e}^{(L)}$ and $\mathsf{v}^{(L)}$ in Table 11.1.

The accuracy in these expressions is not necessarily good. For small probabilities p_i, the terms R_j for large j can be quite large and compensate for the reduction due to N^j in the denominator. This is clearly seen in Table 11.1 for the case $N = 15, s = 10$ and $p_i = 1/s, i = 1, \ldots, s$ where the approximations for both mean and variance are very far away from the correct values. Another example is the case $N = 40, s = 8$, and $\mathbf{p} = (0.1, 0.1, 0.3, 0.3, 0.05, 0.05, 0.05, 0.05)$ where the approximations give considerably larger values than the ones obtained from the simulations which probably are closer to the true values. A term R_j is bounded from above by $s/p_{\min}^j$ and the magnitude of the ratio $(Np_{\min})^j/s$ determines very much the accuracy that can be obtained in the approximations. For the first case $(15 \cdot 0.1)^7/15 = 17.09/15 = 1.14$ which is not very large, and for the second case $(40 \cdot 0.05)^7/8 = 128/8 = 16$. In case of $N = 300, s = 8$ and $\mathbf{p} = (0.1, 0.1, 0.3, 0.3, 0.05, 0.05, 0.05, 0.05)$ we have $(300 \cdot 0.05)^7/8 \approx 171 \cdot 10^6/8 \approx 21 \cdot 10^6$, which is large enough to give good accuracy in the mean in a 7th order approximation.

Similar calculations can be performed for the variance approximations, for the first case $(15 \cdot 0.1)^3/15 = 3.375/15 = 0.225$ and for the second case $(40 \cdot 0.05)^3/8 = 8/8 = 1$ while for the last case $(300 \cdot 0.05)^3/8 = 3375/8 \approx 422$. It is in the last situation thus doubtful if the accuracy is more than a couple of units in the first decimal digit. In the remaining cases accuracy is considerably lower.

11.2.1 Exact Means and Variances of the G^2 Statistic

Hutcheson [6, p. 75 (5.3.2)] acknowledge [13], for the approximations

$$E(G^2) \approx \mathsf{e}^{(SH)} := (s-1) + (R_1-1)/6N + (R_2-R_1)/6N^2, \quad (11.9)$$

$$V(G^2) \approx \mathsf{v}^{(SH)} := 2(s-1) + 2(R_1-1)/3N + 4(R_2-R_1)/3N^2, \quad (11.10)$$

where $R_j = \sum_{i=1}^{s}(1/p_i)^j$ of the mean and variance of the G^2 statistic, i.e. L in original papers notation.

In [6] these approximations were compared with exact values of mean and variance of L derived in some special cases.

In mentioned paper $\bar{H} = -\sum_{i=1}^{s} \frac{Y_i}{N} \ln(Y_i/N)$ is defined that allows to rewrite statistic as

$$G^2 = 2\sum_{i=1}^{s} Y_i \ln(Y_i/Np_i) = -2N(\bar{H} + \sum_{i=1}^{s} \frac{Y_i}{N} \ln p_i) . \tag{11.11}$$

Explicit expressions are given [6, p. 78] for mean and variance of $\bar{H}$:

$$\mathrm{E}[\bar{H}] = \ln N - \left[\sum_{j=0}^{N-2} \binom{N-1}{j} \left(\sum_{i=1}^{s} p_i^{N-j} (1-p_i)^j \right) \ln(N-j) \right]$$

and

$$\mathrm{V}[\bar{H}] =$$

$$= \sum_{a=0}^{N-2} \binom{N-1}{a} \left(\sum_{i=1}^{s} p_i^{N-a} (1-p_i)^a \right)$$
$$\left[\sum_{b=a}^{N-1} \binom{N-1}{b} \left(\sum_{i=1}^{s} p_i^{N-b} (1-p_i)^b \right) \left(\ln \frac{N-b}{N-a} \right)^2 \right] -$$

$$\frac{N-1}{N} \sum_{b=0}^{N-3} \binom{N-2}{b} \left[\sum_{a=0}^{B} \binom{N-b-2}{a} \right.$$
$$\left. \sum_{i \neq j} p_i^{N-a-b-1} p_j^{a+1} (1-p_i-p_j)^b \left(\ln \frac{N-a-b-1}{a+1} \right)^2 \right]$$

where B is the integer part of $(N-b-2)/2$.

The expression for the mean can through (11.11) be directly applied to G^2, giving

$$\mathrm{E}(G^2) = 2N \left[\sum_{j=0}^{N-2} \binom{N-1}{j} \left(\sum_{i=1}^{s} p_i^{N-j} (1-p_i)^j \right) \ln(N-j) \right] - 2N \sum_{i=1}^{s} p_i \ln(Np_i) \tag{11.12}$$

which is then an exact expression for the mean of G^2.

Hutcheson [6, p. 78], also give an approximation for the expected value of $\bar{H}$

$$\mathrm{E}[\bar{H}] \approx -\sum p_i \ln p_i - (s-1)/2N - (\sum_i p_i^{-1} - 1)/12N^2 - (\sum_i p_i^{-2} - \sum_i p_i^{-1})/12N^3 + O(1/N^4)$$

which, when using (11.11), gives

$$\mathrm{E}[G^2] \approx (s-1) + (\sum_i p_i^{-1} - 1)/6N + (\sum_i p_i^{-2} - \sum_i p_i^{-1})/6N^2 + O(1/N^3)$$

which is the Schäffer approximation in (11.9).

Exact general expressions for the mean and variance of the G^2 statistic were derived by Hutcheson and Lyons [7]. The expression they give [7, below Eq. (9)] for the mean

$$-2N \ln N + 2N \left[\sum_{i=1}^{s} \sum_{a=0}^{N-1} \binom{N-1}{a} p_i^{a+1} (1-p_i)^{N-1-a} \ln(\frac{a+1}{p_i}) \right]$$

is expressed in another order of the summations than in (11.12), but they are equivalent.

The expression they give [7, Eq. below (10)] for the variance is

$$4N^2 \sum_{i=1}^{s} \sum_{j=1}^{s} \sum_{a=0}^{N-2} \sum_{b=a+1}^{N-1} \binom{N-1}{a}\binom{N-1}{b} p_i^{N-a}(1-p_i)^a p_j^{N-b}(1-p_i)^b \left(\ln(\frac{p_i(N-b)}{p_j(N-a)})\right)^2 +$$

$$+4N^2 \sum_{i \neq j} \sum_{a=0}^{N-1} \left(\binom{N-1}{a}\right)^2 (p_i p_j)^{N-a}((1-p_i)(1-p_j))^a \ln p_i \ln(\frac{p_i}{p_j})+$$

$$-4N(N-1) \sum_{i \neq j} \sum_{a=0}^{N-2} \binom{N-2}{a} (1-p_i-p_j)^a \sum_{b=0}^{A} \binom{A}{b} p_i^B p_j^{b+1} \ln(p_i(b+1)) \ln(\frac{p_i(b+1)}{p_j B}), \tag{11.13}$$

where $A = N - 2 - a$ and $B = N - 1 - a - b$.

Calculations using two presented exact formulas are rather computationally intensive for large N since they are of order N and N^2, respectively.

11.2.2 Comparison of Approximate Means and Variances of the G^2 Statistic

The values denoted by $\mathsf{e}^{(Sc)}$ and $\mathsf{v}^{(Sc)}$ in Table 11.1 are the corrected versions of equations (2.1) and (2.2) of [14]. These corrected versions also coincide with $\mathsf{e}^{(SH)}$ and $\mathsf{v}^{(SH)}$ given by Hutcheson [6, p. 75 (5.3.2)], and attributed to [13].

The exact values of the mean and the variance of the G^2 statistic given in Table 1 of [14] (also given in Table 11.1) have successfully been replicated with sufficient accuracy through simulation experiments based on 100,000 repetitions of the G^2 statistic. The differences obtained from these simulations are at most a number of digits in the first or the second decimal places. Moments obtained via simulations are often more accurate than any of these approximating expressions. Simulation experiments may be fairly easy to perform for moderate values of N, while for very large N the computation time may be quite extensive. On the other hand, for very large N, the approximate chi-square approximation may be valid with fairly good accuracy. It is hence for moderate N relative to s that the non-central approximation may give some advantage in determining critical values and p-values, as compared with the asymptotic (central) chi-square distribution. For such situations it may be quite easy (through simulation experiments) to obtain parameters for the non-central chi-square distribution approximation.

11.3 Approximate Mean and Variance of the G^2 Statistic Based on Central Moments

We give in this section a series of approximations of the mean and the variance of the G^2 statistic in (11.1). The derivation is through the relation to the Kullback-Leibler divergence statistic for the empirical distribution against the specific category distribution $\{p_i\}$ given by

$$\mathrm{KL} = \sum_{i=1}^{s} Y_i \ln(\frac{Y_i}{Np_i}) \tag{11.14}$$

for which then $G^2 = 2\mathrm{KL}$. (Strictly speaking, the Kullback-Leibler divergence would be the above given KL divided by N.)

By expanding the logarithmic function we can construct a series of approximations to the expectation ($\mathsf{e}_j^{\mathrm{KL}}$) and variance ($\mathsf{v}_j^{\mathrm{KL}}$) of the Kullback-Leibler divergence expressed in terms of central moments of various order of the multinomial distribution:

$$\mathsf{e}_1^{\mathrm{KL}} = \sum_{i=1}^{s} \frac{\nu_i^{(2)}}{Np_i}\frac{1}{2}, \tag{11.15}$$

$$\mathsf{e}_2^{\mathrm{KL}} = \sum_{i=1}^{s} \left[\frac{\nu_i^{(2)}}{Np_i}\frac{1}{2} - \frac{\nu_i^{(3)}}{(Np_i)^2}\frac{1}{6} \right], \tag{11.16}$$

$$\mathsf{e}_3^{\mathrm{KL}} = \sum_{i=1}^{s} \left[\frac{\nu_i^{(2)}}{Np_i}\frac{1}{2} - \frac{\nu_i^{(3)}}{(Np_i)^2}\frac{1}{6} + \frac{\nu_i^{(4)}}{(Np_i)^3}\frac{1}{12} \right], \tag{11.17}$$

$$\mathsf{e}_4^{\mathrm{KL}} = \sum_{i=1}^{s} \left[\frac{\nu_i^{(2)}}{Np_i}\frac{1}{2} - \frac{\nu_i^{(3)}}{(Np_i)^2}\frac{1}{6} + \frac{\nu_i^{(4)}}{(Np_i)^3}\frac{1}{12} - \frac{\nu_i^{(5)}}{(Np_i)^4}\frac{1}{20} \right], \tag{11.18}$$

$$\mathsf{e}_5^{\mathrm{KL}} = \sum_{i=1}^{s}\left[\frac{\nu_i^{(2)}}{Np_i}\frac{1}{2} - \frac{\nu_i^{(3)}}{(Np_i)^2}\frac{1}{6} + \frac{\nu_i^{(4)}}{(Np_i)^3}\frac{1}{12} - \frac{\nu_i^{(5)}}{(Np_i)^4}\frac{1}{20} + \frac{\nu_i^{(6)}}{(Np_i)^5}\frac{1}{30}\right], \tag{11.19}$$

$$\mathsf{e}_6^{\mathrm{KL}} = \sum_{i=1}^{s}\left[\frac{\nu_i^{(2)}}{Np_i}\frac{1}{2} - \frac{\nu_i^{(3)}}{(Np_i)^2}\frac{1}{6} + \frac{\nu_i^{(4)}}{(Np_i)^3}\frac{1}{12} - \frac{\nu_i^{(5)}}{(Np_i)^4}\frac{1}{20} + \frac{\nu_i^{(6)}}{(Np_i)^5}\frac{1}{30} - \frac{\nu_i^{(7)}}{(Np_i)^6}\frac{1}{42}\right], \tag{11.20}$$

$$\mathsf{v}_1^{\mathrm{KL}} = (\sum_{i=1}^{s}\sum_{j=1}^{s}\frac{\nu_{i,j}^{(2,2)}}{2Np_i 2Np_j}) - (\mathsf{e}_1^{\mathrm{KL}})^2, \tag{11.21}$$

$$\mathsf{v}_2^{\mathrm{KL}} = \sum_{i=1}^{s}\sum_{j=1}^{s}\frac{\nu_{i,j}^{(2,2)}}{2Np_i 2Np_j} - \frac{\nu_{i,j}^{(2,3)}}{2(Np_i)6(Np_j)^2} - \frac{\nu_{i,j}^{(3,2)}}{6(Np_i)^2 2(Np_j)} + \frac{\nu_{i,j}^{(3,3)}}{6(Np_i)^2 6(Np_j)^2} - (\,\mathsf{e}_2^{\mathrm{KL}})^2 . \tag{11.22}$$

The moments and central moments of the multinomial distribution are given in Appendix 1. The details of the derivation of the approximate expressions are deferred to Appendix 2.

Based on these approximations, we get approximations of the mean ($\mathsf{e}_j^{G^2}$) and of the variance ($\mathsf{v}_j^{G^2}$) of $G^2 = 2\mathrm{KL}$, through

$$\mathsf{e}_j^{G^2} = 2\mathsf{e}_j^{\mathrm{KL}}$$

and

$$\mathsf{v}_j^{G^2} = 4\mathsf{v}_j^{\mathrm{KL}}.$$

The expectation approximations $2e_j^{\mathrm{KL}}$ are presumably more accurate for larger j although the effect from the alternating sign in the additional term (see Appendix 2) is clearly seen, when turning from approximation j to approximation $j+1$. Especially for small or moderate N relative to s, very large j is needed to get a good approximation of the true mean. There are clear indications (from Table 11.1) that the mean of 2KL is larger than $s-1$. If this can be shown to be generally valid, based on (11.12), remains to be settled. Whether the difference decreases with larger N for any configuration $\mathbf{p}$, is also a question that remains to be settled. However the difference can be considerable when s is relative large relative to N for certain configurations $\mathbf{p}$, as is seen in Table 11.1.

From the explicit expression for the KL approximations in Appendix 2 we get from (11.28)

$$\mathsf{e}_1^{G^2} = s - 1$$

and from (11.29)

$$\mathrm{e}_2^{G^2} = s - 1 + (s-1)/N - (\sum_i p_i^{-1} - 1)/3N = (s-1)(1 + \frac{1}{N}) - (R_1 - 1)/3N$$

which is somewhat similar to (but not identical to) Williams' approximation (11.2).

We have also

$$\mathrm{e}_3^{G^2} = s - 1 + (s-1)/N - (\sum_i p_i^{-1} - 1)/3N + [3(2(s-1)$$
$$- (\sum_i p_i^{-1} - 1))(2N^{-2} - N^{-1}) + N^{-2}(\sum_i p_i^{-2} - \sum_i p_i^{-1})]/6$$

(see Appendix (11.30)), which by rearranging gives

$$\mathrm{e}_3^{G^2} = s - 1 + (s-1)/N - (\sum_i p_i^{-1} - 1)/3N + (s-1)(2N^{-2} - N^{-1})$$
$$+ [-3(\sum_i p_i^{-1} - 1))(2N^{-2} - N^{-1}) + N^{-2}(\sum_i p_i^{-2} - \sum_i p_i^{-1})]/6$$
$$= s - 1 + 2(s-1)/N^2 - (\sum_i p_i^{-1} - 1)/3N$$
$$+ [-3(\sum_i p_i^{-1} - 1)(2N^{-2} - N^{-1}) + N^{-2}(\sum_i p_i^{-2} - \sum_i p_i^{-1})]/6.$$

This expression bear similarities with both the approximation by Smith et al. (11.4) and Schäffer's approximation (11.9) but is not exactly the same.

The behaviour of $2\mathrm{e}_6^{KL}$ is almost the same as those given by $\mathrm{e}^{(L)}$ given by Lyons et al. [8], but is less influenced (see Table 11.1) by the small values of p_i which means to say that they are more accurate than $\mathrm{e}^{(L)}$ for a larger class of probabilities $\mathbf{p}$.

11.4 Approximation by Noncentral Chi-Square Distribution

The approximate distributions suggested for the G^2 statistic under the null hypothesis (i.e. that the category probabilities are those given by $\mathbf{p}$) involve, in addition to the asymptotic chi-square distribution, different scaling adjustments of the chi-square distribution as previously mentioned.

We suggest here the use of a non-central chi-square distribution as a (serious) competitor to the chi-square or scaled chi-square as an approximate distribution under the null hypothesis.

The suggestion comes primarily from the observation that simulation studies of the statistic G^2 tend to give means above the nominal $s - 1$ as well as variance above the nominal $2(s - 1)$, see E_{sim} and V_{sim} in Table 11.1. The same effects in the mean and variance are observed for many of the studied approximate expressions, see Table 11.1. This indicates that a non-central chi-square distribution would probably fit better to the statistics G^2 for relatively small N than a central chi-square distribution. Moreover, one can verify statement via calculation of exact values of the mean and variance for the studied cases.

We verify the appropriateness of this suggestion through analysis of improvement in upper tail quantiles for non-central chi-square distribution in comparison to ones obtained from the previously suggested chi-square and scaled chi-squared distributions.

Although the non-central distribution is often used when the alternative is true, it is to our knowledge a novelty approach to consider the non-central chi-squared also when the null hypothesis is true. The difference between non-central chi-square distribution under the alternative and under the null hypothesis is, of course, in the distributions' parameters.

11.4.1 Degrees of Freedom and Non-centrality Parameter

The non-central chi-squared distribution, here denoted $\chi^2(k, \lambda)$ depends on two parameters k and λ, referred to as degrees of freedom (df) and non-centrality parameter (nc) respectively. When $\lambda = 0$, this is the (ordinary) chi-squared distribution with k degrees of freedom. The non-central chi-square distribution has a probability density function at x which may be written as

$$f(x) = \frac{e^{-x/2}e^{-\lambda/2}}{2^{k/2}} \sum_{j=0}^{\infty} \frac{x^{k/2+j-1}\lambda^j}{2^{2j} j! \Gamma(k/2 + j)},$$

and the rth cumulant being $\kappa_r = 2^{r-1}(r - 1)!(k + r\lambda)$.

This distribution has been thoroughly investigated very early by Fisher [3], and tables of percentage points have been given by Fix [4]. Several methods of approximation have been given by Patnaik [10] and Abdel-Aty [1]. Here we will rely on the percentage points as given by the function `qchisq` in the R-package `stats` version 3.1.0 [11].

The expected value (E) and the variance (V) in a $\chi^2(k, \lambda)$ distribution are

$$E = k + \lambda,$$

$$V = 2(k + \lambda) + 2\lambda.$$

Hence the parameters are directly related to mean and variance through

$$\lambda = V/2 - E, \tag{11.23}$$

$$k = 2E - V/2. \tag{11.24}$$

11.4.2 Noncentral Chi-Squared Approximation of G^2

The values of the df (i.e. k) and the non-centrality parameter (i.e., λ) were, in Table 11.2, calculated on the basis of the mean and variance obtained from the Monte-Carlo simulations:

$$\lambda_0 = V_{\text{sim}}/2 - E_{\text{sim}},$$

$$k_0 = 2E_{\text{sim}} - V_{\text{sim}}/2.$$

The parameter estimates can be calculated from any estimates of the mean and variance. Especially, if simple and accurate estimates can be obtained (i.e., calculated) from the particular design, i.e., $N, s, \mathbf{p} = (p_1, \ldots, p_s)$, the appropriate values k and λ is then easily obtained.

As seen from Table 11.1, the approximations given there, are fairly good with respect to the mean, but the estimates of the variances may differ considerably from what is seen from the simulations (or the exact values given by Smith et al. [14] for certain cases).

Smith et al. [14] however claim that their multiplicative factor $q^{(Sc)}$ will make the mean of $G^2/q^{(Sc)}$ agree with the mean of the approximating chi-square, neglecting terms of order N^{-3}.

In Table 11.2 are listed quantiles from certain specific distributions, that may be used as approximating distribution for the G^2 statistic. The quantiles listed are the 90%, 95%, and 99% percentiles of the distributions which are denoted by x_{90}, x_{95} and x_{99} respectively.

The three quantiles of the scaled chi-square distribution, a gamma distribution, are given in Table 11.2 under the column header $q^{(Sc)}\chi^2(s-1)$, while the quantiles of the central chi-square distribution are given under the column header $\chi^2(s-1)$.

Under the column header 'Simulations' we list the empirical percentiles obtained from simulation studies in each certain case based on 100,000 replications.

As seen from Table 11.2, the quantiles based on the central chi-square distribution in many cases under-estimate those given by the simulations. The quantiles based on the scaled chi-square distribution, in many cases give improvements, but still often under-estimates those given by the simulations.

The quantiles, in Table 11.2, based on the $\chi^2(k_0, \lambda_0)$ distribution are generally much closer to the simulation quantiles, compared to those obtained from the scaled chi-square distribution. It may hence serve as a good approximation of the distribution of G^2.

Table 11.2 Quantiles, simulated and estimated through different distributions

					Simulations			$\chi^2(k_0, \lambda_0)$			$\chi^2(s-1, 0)$			$q^{(Sc)}\chi^2(s-1, 0)$								$\chi^2(k^{(Sc)}, \lambda^{(Sc)})$		
	E_{sim}	V_{sim}	k_0	λ_0	x_{90}	x_{95}	x_{99}	x_{90}	x_{95}	x_{99}	x_{90}	x_{95}	x_{99}	x_{90}	x_{95}	x_{99}	$E^{(Sc)}$	$V^{(Sc)}$	$q^{(Sc)}$	$k^{(Sc)}$	$\lambda^{(Sc)}$	x_{90}	x_{95}	x_{99}
$N = 15, s = 2$ $\mathbf{p} = (1, 1)/s$	1.044	2.197	0.990	0.054	3.397	3.397	5.782	2.832	4.024	6.95	2.71	3.84	6.63	2.787	3.957	6.834	1.036	2.147	1.036	0.999	0.037	2.804	3.980	6.872
$N = 30, s = 2$ $\mathbf{p} = (1, 1)/s$	1.024	2.10	0.998	0.026	2.16	3.40	6.79	2.77	3.94	6.80	2.71	3.84	6.63	2.752	3.907	6.748	1.017	2.069	1.017	1.000	0.017	2.752	3.907	6.747
$N = 15, s = 5$ $\mathbf{p} = \mathbf{1}_s/s$	4.43	10.07	3.83	0.60	9.00	10.72	14.18	8.68	10.60	14.83	7.78	9.49	13.28	8.441	10.294	14.405	4.340	9.418	1.085	3.971	0.370	8.447	10.297	14.393
$N = 30, s = 5$ $\mathbf{p} = \mathbf{1}_s/s$	4.16	8.79	3.93	0.24	8.15	9.75	14.14	8.13	9.92	13.90	7.78	9.49	13.28	8.075	9.848	13.781	4.152	8.620	1.038	3.994	0.158	8.076	9.849	13.778
$N = 15, s = 10$ $\mathbf{p} = \mathbf{1}_s/s$	10.62	18.14	12.17	−1.55[a]	16.06	17.85	21.61	16.79	19.16	24.15	14.68	16.92	21.67	17.562	20.235	25.913	10.764	25.747	1.196	8.653	2.114	17.576	20.227	25.826
$N = 30, s = 10$ $\mathbf{p} = \mathbf{1}_s/s$	9.92	22.32	8.69	1.24	16.41	18.67	23.39	16.26	18.74	24.02	14.68	16.92	21.67	15.858	18.273	23.399	9.720	20.995	1.080	8.943	0.775	15.857	18.267	23.377
$N = 30, s = 3$ $\mathbf{p} = (0.3, 0.3, 0.4)$	2.035	4.165	1.988	0.047	4.55	6.15	9.50	4.69	6.11	9.40	4.61	5.99	9.21	4.688	6.099	9.376	2.036	4.145	1.018	1.999	0.165	4.981	6.474	9.930

$N = 40, s = 3$																								
$\mathbf{p} = (1, 1, 1)/s$	2.02	4.13	1.981	0.042	4.71	6.25	9.48	4.67	6.08	9.36	4.61	5.99	9.21	4.688	6.099	9.376	2.036	4.145	1.018	1.999	0.036	4.686	6.097	9.372
(0.25, 0.25, 0.5)	2.04	4.16	1.998	0.040	4.54	6.16	9.23	4.69	6.11	9.39	4.61	5.99	9.21	4.697	6.111	9.395	2.040	4.162	1.020	1.999	0.041	4.698	6.112	9.394
(0.1, 0.1, 0.8)	2.13	4.72	1.90	0.227	4.87	6.35	9.91	4.96	6.48	9.99	4.61	5.99	9.21	4.845	6.303	9.689	2.104	4.427	1.052	1.995	0.108	4.844	6.301	9.680
$N = 40, s = 8$																								
$\mathbf{p} = \mathbf{1}_s/s$	7.35	16.01	6.70	0.65	12.80	15.03	19.70	12.72	14.92	19.65	12.02	14.07	18.48	12.546	14.686	19.288	7.308	15.259	1.044	6.985	0.322	12.547	14.686	19.284
$(0.1, 0.1, 0.3, (0.1)^5)$	7.44	16.42	6.67	0.77	13.03	15.26	19.70	12.88	15.10	19.89	12.02	14.07	18.48	12.654	14.813	19.455	7.371	15.523	1.053	6.981	0.388	12.652	14.808	19.442
$(0.1)^2, (0.3)^2, (0.05)^4$	7.71	15.99	7.42	0.29	13.05	15.12	19.41	13.07	15.24	19.89	12.02	14.07	18.48	13.075	15.305	20.101	7.616	16.572	1.088	6.945	0.674	13.083	15.310	20.089
$N = 300, s = 8$																								
$\mathbf{p} = \mathbf{1}_s/s$	7.05	14.14	7.02	0.025	12.08	14.09	18.61	12.09	14.15	18.58	12.02	14.07	18.48	12.077	14.137	18.568	7.035	14.140	1.005	7.000	0.036	12.079	14.139	18.570
$(0.1, 0.1, 0.3, (0.1)^5)$	7.06	14.23	7.00	0.057	12.14	14.14	18.61	12.12	14.18	18.63	12.02	14.07	18.48	12.089	14.152	18.586	7.042	14.169	1.006	7.000	0.042	12.089	14.151	18.586
$(0.1)^2, (0.3)^2, (0.05)^4$	7.05	14.18	7.01	0.039	12.07	14.15	18.66	12.10	14.15	18.60	12.02	14.07	18.48	12.125	14.194	18.642	7.063	14.253	1.009	6.999	0.063	12.124	14.192	18.639
$N = 500, s = 8$																								
$\mathbf{p} = \mathbf{1}_s/s$	7.02	14.14	6.97	0.048	12.06	14.12	18.61	12.06	14.13	18.56	12.02	14.07	18.48	12.053	14.109	18.531	7.021	14.084	1.003	7.000	0.021	12.053	14.109	18.531
$0.1, 0.1, 0.3, (0.1)^5$	7.04	14.13	7.00	0.031	12.07	14.11	18.58	12.08	14.14	18.57	12.02	14.07	18.48	12.065	14.123	18.549	7.028	14.112	1.004	7.000	0.025	12.060	14.117	18.541
$(0.1)^2, (0.3)^2, (0.05)^4$	7.05	14.19	6.996	0.050	12.09	14.16	18.61	12.10	14.16	18.60	12.02	14.07	18.48	12.077	14.137	18.568	7.035	14.140	1.005	7.000	0.035	12.077	14.137	18.567

Here x_{90}, x_{95} and x_{99} denote the upper 90%, 95% and 99% quantiles of the respective distribution. The simulation distribution is based on 100,000 replications

$E^{(Sc)} = q^{(Sc)}(s-1)$, $V^{(Sc)} = (q^{(Sc)})^2 2(s-1)$

[a]Since $\lambda_0 < 0$ we calculate the quantiles from $\chi^2(k_0 + \lambda_0, 0)$ instead of from $\chi^2(k_0, \lambda_0)$

It however rely on sufficiently accurate parameters k_0 and λ_0, here based on simulation estimates of mean and variance. One could hence argue that if a simulation is needed to obtain mean and variance of G, then the full distribution of G^2 (also the percentiles needed) could be obtained from this simulation in the particular case, so that the noncentral chi-square approximation would be superfluous.

Hence it would be worthwhile to use the values $E^{(Sc)} = q^{(Sc)}(s-1)$ and $V^{(Sc)} = (q^{(Sc)})^2 2(s-1)$ of mean and variance in the distribution of $q^{(Sc)}$ times a $\chi^2(s-1)$ distributed random variable, to see what quantiles are being obtained from the noncentral chi-squared distribution $\chi^2(k^{(Sc)}, \lambda^{(Sc)})$, where

$$\lambda^{(Sc)} = V^{(Sc)}/2 - E^{(Sc)} = ((q^{(Sc)})^2 - q^{(Sc)})(s-1),$$
$$k^{(Sc)} = 2E^{(Sc)} - V^{(Sc)}/2 = (2q^{(Sc)} - (q^{(Sc)})^2)(s-1).$$

It turns out, see Table 11.2, that these quantiles obtained with this choice of parameters are almost identical to those obtained by the $q^{(Sc)}$-scaled $\chi^2(s-1)$ distribution, a gamma distribution, and are thus no improvement compared to those obtained from the $\chi^2(k_0, \lambda_0)$ distribution. The reason for this can be seen from Table 11.2, in that $E^{(Sc)}$ and $V^{(Sc)}$ in almost all cases under-estimates the true values of the mean and variance of G^2 (which are found in Table 11.1). The multiplicative factor $q^{(Sc)}$ is not 'accurate enough' to give more accurate approximations of the mean and variance in the particular situations. These are instead more accurately estimated through the Monte Carlo simulations.

The possibility still remains to calculate the exact mean and variance of G^2 through (11.12) and (11.13) to obtain parameters λ and k in the noncentral chi-square approximation through (11.23) and (11.24).

11.5 Discussion

The poor fit of the G^2 statistic to central χ^2 distribution for very small sample sizes (i.e. small N in relation to s leading to many empty cells), have led to suggested improvements in 'smoothed' G^2 obtained by adding nonnegative constants to all class frequencies, cf. [5]. This effect may however not be so much connected with the log likelihood ratio statistic as being an asymmetric statistic (i.e. overly skewed and over-dispersed), which is the case considered here. Instead this may be related to that the distribution is more discrete in nature for small N since the averaging effect has not had any large impact for the very few combinations that are possible for this very few number of outcome possibilities.

In order to obtain appropriate critical values for the G^2 test statistic it is suggested that the non-central chi-square approximation should be used in preference to the standard chi-squared approximation or scaled chi-squared approximation. This suggested approximation requires accurate values of the expected value and variance of the G^2 statistic for the particular statistical design. Such values can be obtained from exact expressions, accurate approximating expressions, or accurate simulation estimates, whichever is most convenient to use in the particular situation.

This approach is also appropriate under an alternative distribution. If $\{q_i\}$ represent category probabilities under the alternative then

$$G^2 = 2\sum_{i=1}^{s} Y_i \ln(Y_i/(Nq_i)) + 2\sum_{i=1}^{s} Y_i \ln(q_i/p_i)$$

and the moments and the distribution of G^2 will essentially be the ones determined by the first term on the right hand side, which will depend on the $\{q_i\}$, and with the mean increased by $2N\sum_{i=1}^{s} q_i \ln(q_i/p_i)$, i.e., $2N$ times the Kullback-Leibler divergence from the null categorical distribution to the alternative categorical distribution.

Acknowledgments We thank the referee for the insightful suggestions that substantially improved the presentation of the results in this paper.

Appendix 1: Moments of a Multinomial Distribution

The multinomial distribution is given by the probability mass function

$$P(Y_1 = y_1, \ldots, Y_s = y_s) = \frac{N!}{y_1! \cdots y_s!} p_1^{y_1} \cdots p_s^{y_s}$$

for $y_1 + \cdots + y_s = N$ and 0 otherwise, for $p_i \geq 0, i = 1, \ldots, s$, and $p_1 + \cdots + p_s = 1$. We write $(Y_1, \ldots, Y_s) \in \text{Multinomial}(N, p_1, \ldots, p_s)$.

The moment generating function (mgf) is given by

$$\mathrm{E}[e^{t_1Y_1+t_2Y_2+\cdots+t_sY_s}] = (p_1e^{t_1} + p_2e^{t_2} + \cdots + p_se^{t_s})^N .$$

Raw Moments

The ith raw moment of order k, denoted $\mu_i^{(k)}$, is defined by $\mu_i^{(k)} = E[(Y_i)^k]$ for $i = 1, \ldots, s$. The raw moments are obtained as derivatives with respect to t_i of the moment generating function, and evaluated at $t_j = 0$ for a j, and are

$$\mu_i^{(1)} = Np_i$$

$$\mu_i^{(2)} = N^{(2)}p_i^2 + N^{(1)}p_i^1$$

$$\mu_i^{(3)} = N^{(3)} p_i^3 + 3N^{(2)} p_i^2 + N^{(1)} p_i^1$$

$$\mu_i^{(4)} = N^{(4)} p_i^4 + 6N^{(3)} p_i^3 + 7N^{(2)} p_i^2 + N^{(1)} p_i^1$$

$$\mu_i^{(5)} = N^{(5)} p_i^5 + 10N^{(4)} p_i^4 + 25N^{(3)} p_i^3 + 15N^{(2)} p_i^2 + N^{(1)} p_i^1$$

$$\mu_i^{(6)} = N^{(6)} p_i^6 + 15N^{(5)} p_i^5 + 65N^{(4)} p_i^4 + 90N^{(3)} p_i^3 + 31N^{(2)} p_i^2 + N^{(1)} p_i^1$$

$$\mu_i^{(7)} = N^{(7)} p_i^7 + 21N^{(6)} p_i^6 + 140N^{(5)} p_i^5 + 350N^{(4)} p_i^4 + 301N^{(3)} p_i^3 + 63N^{(2)} p_i^2 + N^{(1)} p_i^1$$

where $N^{(a)} = N(N-1)(N-2)\cdots(N-a+1)$.

A general raw moment is given by

$$\mu_i^{(k)} = \sum_{j=0}^{k-1} a_{k,j} N^{(k-j)} p_i^{k-j}$$

where

$$a_{k,j} = (k-j)a_{k-1,j-1} + a_{k-1,j} \,.$$

The coefficients $a_{k,j}$ for the first values of k, up to 12 are given in Table 11.3.

Central Moments

The ith central moment of order k, denoted $\nu_i^{(k)}$, is defined by $\nu_i^{(k)} = \mathrm{E}[(Y_i - Np_i)^k]$ for $i = 1, \ldots, s$. The central moments are most easily expressed in terms of raw moments, using binomial expansions.

We have $\nu_i^{(1)} = 0$, and

$$\nu_i^{(2)} = \mu_i^{(2)} - 2\mu_i^{(1)}\mu_i^{(1)} + (\mu_i^{(1)})^2$$

$$\nu_i^{(3)} = \mu_i^{(3)} - 3\mu_i^{(2)}\mu_i^{(1)} + 3\mu_i^{(1)}(\mu_i^{(1)})^2 - (\mu_i^{(1)})^3$$

$$\nu_i^{(4)} = \mu_i^{(4)} - 4\mu_i^{(3)}\mu_i^{(1)} + 6\mu_i^{(2)}(\mu_i^{(1)})^2 - 4\mu_i^{(1)}(\mu_i^{(1)})^3 + (\mu_i^{(1)})^4$$

$$\nu_i^{(5)} = \mu_i^{(5)} - 5\mu_i^{(4)}\mu_i^{(1)} + 10\mu_i^{(3)}(\mu_i^{(1)})^2 - 10\mu_i^{(2)}(\mu_i^{(1)})^3 + 5\mu_i^{(1)}(\mu_i^{(1)})^4 - (\mu_i^{(1)})^5$$

$$\nu_i^{(6)} = \mu_i^{(6)} - 6\mu_i^{(5)}\mu_i^{(1)} + 15\mu_i^{(4)}(\mu_i^{(1)})^2 - 20\mu_i^{(3)}(\mu_i^{(1)})^3 + 15\mu_i^{(2)}(\mu_i^{(1)})^4 - 6\mu_i^{(1)}(\mu_i^{(1)})^5 + (\mu_i^{(1)})^6$$

Table 11.3 Coefficients $a_{k,j}$ for indices $0 \le j \le 11$ and $j \le k \le 12$

$a_{k,j}$	$j =$											
k	0	1	2	3	4	5	6	7	8	9	10	11
0	0											
1	1	0										
2	1	1	0									
3	1	3	1	0								
4	1	6	7	1	0							
5	1	10	25	15	1	0						
6	1	15	65	90	31	1	0					
7	1	21	140	350	301	63	1	0				
8	1	28	266	1050	1701	966	127	1	0			
9	1	36	462	2646	6951	7770	3025	255	1	0		
10	1	45	750	5880	22,827	42,525	34,105	9330	511	1	0	
11	1	55	1155	11,880	63,987	179,487	246,730	145,750	28,501	1023	1	0
12	1	66	1705	22,275	159,027	627,396	1,323,652	1,379,400	611,501	86,526	2047	1

$$\nu_i^{(7)} = \mu_i^{(7)} - 7\mu_i^{(6)}\mu_i^{(1)} + 21\mu_i^{(5)}(\mu_i^{(1)})^2 - 35\mu_i^{(4)}(\mu_i^{(1)})^3 \\ + 35\mu_i^{(3)}(\mu_i^{(1)})^4 - 21\mu_i^{(2)}(\mu_i^{(1)})^5 + 6(\mu_i^{(1)})^7 .$$

A general central moment is

$$\nu_i^{(k)} = \sum_{j=0}^{k} \binom{k}{j} (-1)^j \mu_i^{(k-j)} (\mu_i^{(1)})^j , \tag{11.25}$$

where $\mu_i^{(0)} = 1$.

Appendix 2: Approximate Moments of the Kullback-Leibler Statistic

The Kullback-Leibler statistic for the empirical distribution against the category distribution $\{p_i\}$ is given by

$$\mathrm{KL} = \sum_{i=1}^{s} Y_i \ln(\frac{Y_i}{Np_i}) . \tag{11.26}$$

By expansion of the logarithmic function we may write

$$\sum_{i=1}^{s} Y_i \ln(\frac{Y_i}{Np_i}) = \sum_{i=1}^{s} \sum_{k=2}^{\infty} \frac{(-1)^k}{k(k-1)} \frac{(Y_i - Np_i)^k}{(Np_i)^{k-1}} . \tag{11.27}$$

The alternating power series on the right hand side converges, as is seen using Leibnitz Criterion ($\lim_{k\to\infty} \frac{1}{k(k-1)} = 0$) and the radius of convergence is

$$r = \lim_{k\to\infty} |\frac{a_k}{a_{k+1}}| = \lim_{k\to\infty} \frac{(k+1)k}{k(k-1)} = 1$$

where $a_k = (k(k-1))^{-1}$.

From the expansion (11.27) we immediately have

$$E[\mathrm{KL}] = \sum_{i=1}^{s} \sum_{k=2}^{\infty} \frac{(-1)^k}{k(k-1)} \frac{E[(Y_i - Np_i)^k]}{(Np_i)^{k-1}} \\ = \sum_{i=1}^{s} \mu_i^{(1)} \sum_{k=2}^{\infty} \frac{(-1)^k}{k(k-1)} \frac{\nu_i^{(k)}}{(\mu_i^{(1)})^k} ,$$

where $\nu_i^{(k)} = E[(Y_i - Np_i)^k] = E[(Y_i - \mu_i^{(1)})^k]$.

By stopping at different specific orders of the central moment we get a series of different approximations of the expectation of KL:

$$\mathsf{e}_1^{\mathrm{KL}} = \sum_{i=1}^{s} \frac{\nu_i^{(2)}}{Np_i}\frac{1}{2}$$

$$\mathsf{e}_2^{\mathrm{KL}} = \sum_i \left[\frac{\nu_i^{(2)}}{Np_i}\frac{1}{2} - \frac{\nu_i^{(3)}}{(Np_i)^2}\frac{1}{6}\right]$$

$$\mathsf{e}_3^{\mathrm{KL}} = \sum_i \left[\frac{\nu_i^{(2)}}{Np_i}\frac{1}{2} - \frac{\nu_i^{(3)}}{(Np_i)^2}\frac{1}{6} + \frac{\nu_i^{(4)}}{(Np_i)^3}\frac{1}{12}\right]$$

$$\mathsf{e}_4^{\mathrm{KL}} = \sum_i \left[\frac{\nu_i^{(2)}}{Np_i}\frac{1}{2} - \frac{\nu_i^{(3)}}{(Np_i)^2}\frac{1}{6} + \frac{\nu_i^{(4)}}{(Np_i)^3}\frac{1}{12} - \frac{\nu_i^{(5)}}{(Np_i)^4}\frac{1}{20}\right]$$

$$\mathsf{e}_5^{\mathrm{KL}} = \sum_i \left[\frac{\nu_i^{(2)}}{Np_i}\frac{1}{2} - \frac{\nu_i^{(3)}}{(Np_i)^2}\frac{1}{6} + \frac{\nu_i^{(4)}}{(Np_i)^3}\frac{1}{12} - \frac{\nu_i^{(5)}}{(Np_i)^4}\frac{1}{20} + \frac{\nu_i^{(6)}}{(Np_i)^5}\frac{1}{30}\right]$$

$$\mathsf{e}_6^{\mathrm{KL}} = \sum_i \left[\frac{\nu_i^{(2)}}{Np_i}\frac{1}{2} - \frac{\nu_i^{(3)}}{(Np_i)^2}\frac{1}{6} + \frac{\nu_i^{(4)}}{(Np_i)^3}\frac{1}{12} - \frac{\nu_i^{(5)}}{(Np_i)^4}\frac{1}{20} + \frac{\nu_i^{(6)}}{(Np_i)^5}\frac{1}{30} - \frac{\nu_i^{(7)}}{(Np_i)^6}\frac{1}{42}\right]$$

with

$$\mathsf{e}_k^{\mathrm{KL}} = \mathsf{e}_{k-1}^{\mathrm{KL}} + \sum_i (-1)^{k+1} \frac{\nu_i^{(k+1)}}{(Np_i)^k}\frac{1}{(k+1)k}.$$

Using (11.25) we may write

$$E[\mathrm{KL}] = \sum_{i=1}^{s} \mu_i^{(1)} \sum_{k=2}^{\infty} \frac{(-1)^k}{k(k-1)} \sum_{j=0}^{k} \binom{k}{j} (-1)^j \mu_i^{(k-j)} (\mu_i^{(1)})^{j-k}$$

in terms of the raw moments $\{\mu_i^{(k)}\}$.

An alternative form is obtained by reversing the order of summation,

$$E[\mathrm{KL}] = \sum_{i=1}^{s} \mu_i^{(1)} [\sum_{k=2}^{\infty} \frac{(-1)^k}{k(k-1)} \mu_i^{(k)} (\mu_i^{(1)})^{-k} - \sum_{k=2}^{\infty} \frac{(-1)^k}{(k-1)} \mu_i^{(k-1)} (\mu_i^{(1)})^{1-k}$$

$$+ \sum_{j=2}^{\infty} \sum_{k=j}^{\infty} \frac{(-1)^k}{k(k-1)} \binom{k}{j} (-1)^j \mu_i^{(k-j)} (\mu_i^{(1)})^{j-k}].$$

Different approximations are obtained stopping the summation at various stages in the infinite series expressions.

Some Explicit Expressions

1. Here it generally holds for any $\{p_i\}$, that

$$\mathsf{e}_1^{\text{KL}} = \sum_{i=1}^{s} \frac{\nu_i^{(2)}}{Np_i}\frac{1}{2} = \sum_{i=1}^{s} \frac{\mu_i^{(2)} - (\mu_i^{(2)})^2}{Np_i}\frac{1}{2} = \sum_{i=1}^{s} \frac{N^{(2)}p_i^2 + Np_i - (Np_i)^2}{Np_i}\frac{1}{2}$$

$$= \sum_{i=1}^{s} \frac{Np_i(1-p_i)}{Np_i}\frac{1}{2} = (s-1)\frac{1}{2}\,. \tag{11.28}$$

2. Looking at the additional term in e_2^{KL} we have $\nu_i^{(3)} = \mu_i^{(3)} - 3\mu_i^{(2)}\mu_i^{(1)} + 2(\mu_i^{(1)})^3 = (N^{(3)}p_i^3 + 3N^{(2)}p_i^2 + Np_i) - 3(N^{(2)}p_i^2 + Np_i)(Np_i) + 2(Np_i)^3 = (N^{(3)} - 3NN^{(2)} + 2N^3)p_i^3 + (3N^{(2)} - 3N^2)p_i^2 + Np_i = (N(N-1)(N-2) - 3NN(N-1) + 2N^3)p_i^3 - 3Np_i^2 = (N(N^2-3N+2) - 3N(N^2-N) + 2N^3)p_i^3 - 3Np_i^2 + Np_i = 2Np_i^3 - 3Np_i^2 + Np_i = -Np_i^2(3-2p_i) + Np_i$ and thus

$$-\sum_i \frac{\nu_i^{(3)}}{(Np_i)^2}\frac{1}{6} = \sum_i (3-2p_i)\frac{1}{N\cdot 6} - \sum_i \frac{1}{Np_i}\frac{1}{6} = (6N)^{-1}(3s-2) - \sum_i \frac{1}{Np_i}\frac{1}{6}$$

$$= \frac{1}{6N}(3s - 3 - (\sum_i \frac{1}{p_i} - 1)) = \frac{s-1}{2N} - (\sum_i \frac{1}{p_i} - 1)/6N\,. \tag{11.29}$$

If $p_i = 1/s$, the last of these two terms is $-(s^2-1)/6N$ and the whole expression is $\frac{s-1}{2}(1 - \frac{s+1}{3})/N$ which is negative (for $s > 2$), and the magnitude of the term is hence strongly depending on the ratio s/N. For non-homogeneous category probabilities $\{p_i\}$, the magnitude of the term is even more accentuated. Hence

$$\mathsf{e}_2^{\text{KL}} = (s-1)\frac{1}{2} + \frac{s-1}{2N} - (\sum_i \frac{1}{p_i} - 1)/6N$$

which is strongly influenced by $\min_i p_i$.

3. For the additional term in e_3^{KL} we have

$$\nu_i^{(4)}6 = \mu_i^{(4)} - 4\mu_i^{(3)}\mu_i^{(1)} + 6\mu_i^{(2)}(\mu_i^{(1)})^2 - 3(\mu_i^{(1)})^4$$

$$= (N^{(4)}p_i^4 + 6N^{(3)}p_i^3 + 7N^{(2)}p_i^2 + N^{(1)}p_i^1) - 4(N^{(3)}p_i^3 + 3N^{(2)}p_i^2 + Np_i)Np_i$$

$$+ 6(N^{(2)}p_i^2 + Np_i)(Np_i)^2 - 3(Np_i)^4$$

$$= (N^{(4)} - 4NN^{(3)} + 6N^2N^{(2)} - 3N^4)p_i^4 + (6N^{(3)} - 4 \cdot 3NN^{(2)} + 6N^3)p_i^3$$
$$+ (7N^{(2)} - 4N^2)p_i^2 + N^{(1)}p_i^1$$
$$= (3N^2 - 6N)p_i^4 + (12N - 6N^2)p_i^3 + (3N^2 - 7N)p_i^2 + Np_i^1 .$$

Hence,

$$\frac{\nu_i^{(4)}}{(Np_i)^3}\frac{1}{12} = \frac{(3N^{-1} - 6N^{-2})p_i + (12N^{-2} - 6N^{-1}) + (3N^{-1} - 7N^{-2})p_i^{-1} + N^{-2}p_i^{-2}}{12}$$

and

$$\sum_i \frac{\nu_i^{(4)}}{(Np_i)^3}\frac{1}{12}$$
$$= \frac{(3N^{-1} - 6N^{-2}) + (12N^{-2} - 6N^{-1})s + (3N^{-1} - 7N^{-2})\sum_i p_i^{-1} + N^{-2}\sum_i p_i^{-2}}{12}$$
$$= \frac{(2N^{-2} - N^{-1})3\left(2(s-1) - (\sum_i p_i^{-1} - 1)\right) + N^{-2}(\sum_i p_i^{-2} - \sum_i p_i^{-1})}{12} . \tag{11.30}$$

When $p_i = 1/s$ the first of these two terms is $-(2N^{-2} - N^{-1})3(s-1)^2/12$ which is positive for $N > 3$. The second term is also positive and for $p_i = 1/s$ it equals $s^2(s-1)/12N^2$ and the whole expression in (11.30) is

$$\frac{(s-1)^2}{4N} + \frac{(s-1)}{12N^2}(s^2 - 6(s-1)) .$$

This expression depends on the two dimensions s and N, but not in a simple way of the ratio s/N only.

References

1. Abdel-Aty, S.H.: Approximate formulae for the percentage points and the probability integral of the non-central χ^2 distribution. Biometrika **41**, 538–540 (1954)
2. Chen, J., Glaz, J., Sison, C.P.: Testing homogeneity of the multinomial proportions. Commun. Stat. Theory Methods **45**, 1747–1777 (2016)
3. Fisher, R.A.: The general sampling distribution of the multiple correlation coefficient. Proc. R. Soc. A **121**, 654–673 (1928)
4. Fix, E.: Tables of non-central chi-square. Univ. Calif. Publ. Stat. **1**, 15–19 (1949)
5. Hosmane, B.S.: Smoothing of likelihood ratio statistic for equiprobable multinomial goodness-of-fit. Ann. Inst. Stat. Math. **42**, 133–147 (1990)
6. Hutcheson, K.: The moments and distribution for an estimate of the Shannon information measure and its application to ecology. PhD Thesis, Virginia Polytechnic Institute (1969)

7. Hutcheson, K., Lyons, N.I.: The mean and variance of the likelihood ratio criterion for the multinomial distribution. Metrika **26**, 57–61 (1979)
8. Lyons, N.I., Hutcheson, K., Sugg, M.N.: A small sample study of the moments and distribution of the log likelihood ratio statistic. Metrika **36**, 195–207 (1989)
9. Neyman, J., Pearson, E.S.: Further notes on the X^2 distribution. Biometrika **22**, 296–305 (1931)
10. Patnaik, P.B.: The non-central x2- and F-distributions and their applications. Biometrika **36**, 202–232 (1949)
11. R Core Team: R: A Language and Environment for Statistical Computing. R Foundation for Statistical Computing, Vienna (2014). http://www.R-project.org/
12. Rao, C.R.: Linear Statistical Inference and Its Applications, 2nd edn. Wiley, New York (1973)
13. Schäffer, K.A.: The likelihood test for fit. Mittbl. Math. Stat. **9**, 27–54 (1957)
14. Smith, P.J., Rae, D.S. Manderscheid, R.W., Silbergeld, S.: Approximating the moments and distribution of the likelihood ratio statistic for multinomial goodness of fit. J. Am. Stat. Assoc. **76**, 737–740 (1981)
15. Wilks, S.S.: The large-sample distribution of the likelihood ratio for testing composite hypotheses. Ann. Math. Stat. **9**, 60–62 (1938)
16. Williams, D.A.: Improved likelihood ratio tests for complete contingency tables. Biometrika **63** (1), 33–37 (1976)

Chapter 12
Covariance Structure Tests for t-distribution

Tõnu Kollo and Marju Valge

Abstract We derive expressions of statistics for testing covariance structures when the population is t-distributed. The likelihood ratio test, Rao's score test and Wald's score test are derived for basic covariance structures. Expressions of all three statistics are obtained under the general null-hypothesis $H_{01} : \boldsymbol{\Sigma} = \boldsymbol{\Sigma}_0$, using matrix derivative technique. Here $p \times p$-matrix $\boldsymbol{\Sigma}$ is a dispersion/scale parameter. The special cases $H_{02} : \boldsymbol{\Sigma} = \mathbf{I}_p$ and $H_{03} : \boldsymbol{\Sigma} = \gamma_0 \mathbf{I}_p$ where $\gamma_0 > 0$ is a known constant are also considered. Expressions of the statistics are obtained as approximations using first terms from Taylor expansions. The method can be carried over to other continuous multivariate elliptical distributions which have power function in the expression of the density function.

12.1 Introduction

In data analysis and modelling one is often interested in testing hypotheses about mean or presence of a specific covariance structure. In a simple case the hypothized covariance matrix is an identity matrix, whereas in more complex situations e.g. when analysing spacial-temporal data the Kronecker product structure is present [15, for example]. When population distribution is elliptical the covariance matrix is a product of a univariate multiplier which characterizes the distribution and a scale matrix, denoted by $\boldsymbol{\Sigma}$. The test statistics are derived for the parameter $\boldsymbol{\Sigma}$ which determines the structure of the corresponding covariance matrix. Probably the most commonly used test is the likelihood ratio test (LRT) under assumption of normality of the population [1, 3, for example]. Profound study of basic likelihood ratio tests about covariance structures under normality is presented in [12, Ch. 8]. However, it is known that when the number of parameters to be tested is large the likelihood ratio test will almost always reject the null hypothesis. In order to overcome this

T. Kollo (✉) · M. Valge
Institute of Mathematics and Statistics, University of Tartu, Tartu, Estonia
e-mail: tonu.kollo@ut.ee

© Springer Nature Switzerland AG 2020

T. Holgersson, M. Singull (eds.), *Recent Developments in Multivariate and Random Matrix Analysis*, https://doi.org/10.1007/978-3-030-56773-6_12

problem, corrections to the test have been made so that it could be used in a high-dimensional setup as well, see [2], for example. In practice one still carries on using the uncorrected test. In [9] it is shown that instead of the likelihood ratio test or Wald's score test (WST) more consistent Rao's score test (RST) should be used to test a particular covariance structure under normality. In practice empirical distribution with heavier than normal tail area is quite common. In this paper we develop these three tests for a multivariate t-distributed population.

In Sect. 12.2 notation and necessary notions are presented. In Sect. 12.3 we will find the likelihood ratio test statistic, in Sect. 12.4 Rao's score test statistic and in Sect. 12.5 Wald's score test statistic are derived for testing covariance structures under a multivariate t-distribution. Both, the general formula for testing $H_{01} : \boldsymbol{\Sigma} = \boldsymbol{\Sigma}_0$ and specific formulae for the sphericity test and for testing if the covariance matrix is an identity matrix, are derived.

Finally, in Sect. 12.6 we summarize the results and discuss directions of further research. Expressions of the test statistics for t-distributed observations are more complicated than in the normal case and found as approximations using first terms from Taylor expansions.

12.2 Notation and Notions

The derivations in the paper utilize a matrix technique which is based on the vec-operator, the Kronecker product and matrix derivatives. For deeper insight into this technique an interested reader is referred to [6, 11] or [8]. The following two vec-operator rules are frequently used

$$\mathrm{vec}\,(\mathbf{ABC}) = (\mathbf{C}' \otimes \mathbf{A})\mathrm{vec}\,\mathbf{B},$$
$$\mathrm{tr}(\mathbf{AB}) = \mathrm{vec}\,'\mathbf{A}'\mathrm{vec}\,\mathbf{B},$$

where tr denotes the trace function.

Later on we shall use matrix derivatives repeatedly, and the definition of [8], p. 127, is applied.

Definition 12.1 Let the elements of $\mathbf{Y} \in \mathbb{R}^{r\times s}$ be functions of $\mathbf{X} \in \mathbb{R}^{p\times q}$. The matrix $\frac{d\mathbf{Y}}{d\mathbf{X}} \in \mathbb{R}^{pq\times rs}$ is called matrix derivative of $\mathbf{Y}$ by $\mathbf{X}$ in a set A, if the partial derivatives $\frac{\partial y_{kl}}{\partial x_{ij}}$ exist, are continuous in A, and

$$\frac{d\mathbf{Y}}{d\mathbf{X}} = \frac{d}{d\mathrm{vec}\,\mathbf{X}}\mathrm{vec}\,'\mathbf{Y}$$

where

$$\frac{d}{d\mathrm{vec}\,\mathbf{X}} = \left(\frac{\partial}{\partial x_{11}}, \ldots, \frac{\partial}{\partial x_{p1}}, \frac{\partial}{\partial x_{12}}, \ldots, \frac{\partial}{\partial x_{p2}}, \ldots, \frac{\partial}{\partial x_{1q}}, \ldots, \frac{\partial}{\partial x_{pq}}\right)'$$

and vec $(\cdot)$ is the usual vectorization operator.

Further, the following properties of the matrix derivative are used [8, p. 149]

(a) $\frac{d\mathbf{X}}{d\mathbf{X}} = \mathbf{I}_{pq}$;
(b) $\frac{d\mathbf{Y}+\mathbf{Z}}{d\mathbf{X}} = \frac{d\mathbf{Y}}{d\mathbf{X}} + \frac{d\mathbf{Z}}{d\mathbf{X}}$;
(c) $\frac{d\mathbf{AXB}}{d\mathbf{X}} = \mathbf{B} \otimes \mathbf{A}'$;
(d) When $\mathbf{Z} = \mathbf{Z}(\mathbf{Y})$, $\mathbf{Y} = \mathbf{Y}(\mathbf{X})$ then $\frac{d\mathbf{Z}}{d\mathbf{X}} = \frac{d\mathbf{Y}}{d\mathbf{X}}\frac{d\mathbf{Z}}{d\mathbf{Y}}$;
(e) When $\mathbf{W} = \mathbf{YZ}$, $\mathbf{Z} \in \mathbb{R}^{s\times t}$ then $\frac{d\mathbf{W}}{d\mathbf{X}} = \frac{d\mathbf{Y}}{d\mathbf{X}}(\mathbf{Z} \otimes \mathbf{I}_r) + \frac{d\mathbf{Z}}{d\mathbf{X}}(\mathbf{I}_t \otimes \mathbf{Y}')$;
(f) When $\mathbf{X} \in \mathbb{R}^{p\times p}$ then $\frac{d\mathbf{X}^{-1}}{d\mathbf{X}} = -\mathbf{X}^{-1} \otimes (\mathbf{X}')^{-1}$;
(g) When $\mathbf{X} \in \mathbb{R}^{p\times p}$ then $\frac{d|\mathbf{X}|}{d\mathbf{X}} = |\mathbf{X}|\mathrm{vec}\,(\mathbf{X}^{-1})'$, where $|\cdot|$ denotes the determinant;
(h) $\frac{d\mathrm{tr}(\mathbf{A}'\mathbf{X})}{d\mathbf{X}} = \mathrm{vec}\,\mathbf{A}$.

Let $\mathbf{x}$ be a continuous random vector with distribution $P_{\mathbf{x}}(\boldsymbol{\theta})$ and density function $f(\mathbf{x}, \boldsymbol{\theta})$, where $\boldsymbol{\theta}$ is a vector of unknown parameters.

Definition 12.2 The score vector of random vector $\mathbf{x}$ is given by the matrix derivative

$$\mathbf{u}(\mathbf{x}, \boldsymbol{\theta}) = \frac{d}{d\boldsymbol{\theta}} \ln f(\mathbf{x}, \boldsymbol{\theta}).$$

Definition 12.3 The information matrix of random vector $\mathbf{x}$ is the covariance matrix of the score vector $\mathbf{u}(\mathbf{x}, \boldsymbol{\theta})$:

$$\mathrm{I}(\mathbf{x}, \boldsymbol{\theta}) = \mathrm{D}(\mathbf{u}(\mathbf{x}, \boldsymbol{\theta})) = \mathrm{E}(\mathbf{u}(\mathbf{x}, \boldsymbol{\theta})\mathbf{u}'(\mathbf{x}, \boldsymbol{\theta})).$$

Definition 12.4 The Hessian matrix of random vector $\mathbf{x}$ is given by the second order matrix derivative

$$\mathrm{H}(\mathbf{x}, \boldsymbol{\theta}) = \frac{d^2}{d\boldsymbol{\theta}^2} \ln f(\mathbf{x}, \boldsymbol{\theta}) = \frac{d\mathbf{u}(\mathbf{x}, \boldsymbol{\theta})}{d\boldsymbol{\theta}}.$$

Note that when the distribution of the random vector $\mathbf{x}$ is regular, then the information matrix of $\mathbf{x}$ can be calculated as an expectation:

$$\mathrm{I}(\mathbf{x}, \boldsymbol{\theta}) = -\mathrm{E}(\mathbf{H}(\mathbf{x}, \boldsymbol{\theta})).$$

Let $\mathscr{X} = (\mathbf{x}_1, \ldots, \mathbf{x}_n)$ denote the theoretical sample from the distribution $P_{\mathbf{x}}(\boldsymbol{\theta})$. Then the log-likelihood function of the sample is given by $l(\mathscr{X}, \boldsymbol{\theta}) = \sum_{i=1}^{n} \ln f(\mathbf{x}_i, \boldsymbol{\theta})$ and the score function of the sample, $\mathbf{u}(\mathscr{X}, \boldsymbol{\theta})$, is given by $\mathbf{u}(\mathscr{X}, \boldsymbol{\theta}) = \sum_{i=1}^{n} \mathbf{u}(\mathbf{x}_i, \boldsymbol{\theta})$. The information matrix of the sample is given by $\mathrm{I}(\mathscr{X}, \boldsymbol{\theta}) = n \cdot \mathrm{I}(\mathbf{x}, \boldsymbol{\theta})$.

The convergence in distribution is denoted by $\xrightarrow{D}$.

In this paper the random vector $\mathbf{x}$ has multivariate t-distribution.

Definition 12.5 Let random p-vector $\mathbf{y}$ be normally distributed, $\mathbf{y} \sim N_p(\mathbf{0}, \boldsymbol{\Sigma})$ and $Z^2 \sim \chi^2_\nu$ independent of $\mathbf{y}$. Then $\mathbf{x} = \sqrt{\nu}\frac{\mathbf{y}}{Z} + \boldsymbol{\mu}$ is multivariate t-distributed with parameters $\boldsymbol{\mu}$ and $\boldsymbol{\Sigma}$, $\mathbf{x} \sim t_{\nu,p}(\boldsymbol{\mu}, \boldsymbol{\Sigma})$, with ν degrees of freedom.

Remark When $\boldsymbol{\mu} = \mathbf{0}$ and instead of $\boldsymbol{\Sigma}$ we have correlation matrix $\mathbf{R}$ as the parameter of the normal distribution, we get in the univariate case standard t-distribution. From Definition 12.5 we get a member of the location-scale family of the univariate t-distribution.

The density function of $\mathbf{x}$ is given by Kotz and Nadarajah [10, Ch. 5]

$$f_\nu(\mathbf{x}, \boldsymbol{\mu}, \boldsymbol{\Sigma}) = c_p|\boldsymbol{\Sigma}|^{-\frac{1}{2}}\left[1 + \frac{1}{\nu}(\mathbf{x} - \boldsymbol{\mu})'\boldsymbol{\Sigma}^{-1}(\mathbf{x} - \boldsymbol{\mu})\right]^{-\frac{\nu+p}{2}} \tag{12.1}$$

where $c_p = \dfrac{\Gamma\left(\frac{\nu+p}{2}\right)}{(\pi\nu)^{\frac{p}{2}}\Gamma\left(\frac{\nu}{2}\right)}$.

The first moments of $\mathbf{x} \sim t_{\nu,p}(\boldsymbol{\mu}, \boldsymbol{\Sigma})$ are [10, p. 104]

$$\mathrm{E}\mathbf{x} = \boldsymbol{\mu},$$

$$\mathrm{D}\mathbf{x} = \frac{\nu}{\nu - 2}\boldsymbol{\Sigma}, \quad \nu > 2.$$

12.3 Likelihood Ratio Test

For testing the hypothesis

$$\begin{cases} \mathrm{H}_0 : \boldsymbol{\theta} = \boldsymbol{\theta}_0, \\ \mathrm{H}_1 : \boldsymbol{\theta} \neq \boldsymbol{\theta}_0 \end{cases} \tag{12.2}$$

we use the likelihood ratio test (LRT) in logarithmic form

$$LRT(\mathscr{X}, \boldsymbol{\theta}_0) = -2\ln\left(\frac{L(\mathscr{X}, \boldsymbol{\theta}_0)}{\max_{\boldsymbol{\theta}} L(\mathscr{X}, \boldsymbol{\theta})}\right) = -2[\ln L(\mathscr{X}, \boldsymbol{\theta}_0) - \ln L(\mathscr{X}, T_{\boldsymbol{\theta}}(\mathscr{X}))],$$

where $L(\cdot)$ is the likelihood function of $\mathbf{x}$, $\mathscr{X} = (\mathbf{x}_1, \ldots, \mathbf{x}_n)$ is a random sample from $P_{\mathbf{x}}(\boldsymbol{\theta})$, $\boldsymbol{\theta} = (\theta_1, \ldots, \theta_r)'$ and $T_{\boldsymbol{\theta}}$ is the maximum likelihood estimator of $\boldsymbol{\theta}$. When sample size $n \to \infty$ and H_0 holds then $LRT(\mathscr{X}, \boldsymbol{\theta}) \xrightarrow{D} \chi^2_r$.

When under the null hypothesis there are k parameters not fixed then the LRT in logarithmic form has the representation

$$LRT(\mathscr{X},\boldsymbol{\theta}_0) = -2\ln\left(\frac{\max_{\boldsymbol{\theta}_1} L(\mathscr{X},\boldsymbol{\theta}_0)}{\max_{\boldsymbol{\theta}} L(\mathscr{X},\boldsymbol{\theta})}\right)$$
$$= -2[\ln L(\mathscr{X}, T_{\boldsymbol{\theta}_1}(\mathscr{X})) - \ln L(\mathscr{X}, T_{\boldsymbol{\theta}}(\mathscr{X}))],$$

where $\boldsymbol{\theta}_1$ is the set of $(r-k)$ fixed parameters and $LRT(\mathscr{X},\boldsymbol{\theta}) \xrightarrow{D} \chi^2_{r-k}$ when the sample size $n \to \infty$ and H_0 holds (see [14, §6e, for instance]).

Let us find the likelihood ratio test statistic for testing hypothesis (12.2) when the population is multivariate t-distributed with ν degrees of freedom. Let $\mathbf{x} \sim t_{\nu,p}(\boldsymbol{\mu}, \boldsymbol{\Sigma})$, with $\boldsymbol{\mu} = (\mu_1, \ldots, \mu_p)'$, the scale parameter $\boldsymbol{\Sigma} > 0 : p \times p$ and let $\boldsymbol{\theta} = (\boldsymbol{\mu}', \mathrm{vec}'\boldsymbol{\Sigma})'$. The likelihood function $L(\mathscr{X},\boldsymbol{\theta})$ is

$$L(\mathscr{X},\boldsymbol{\theta}) = c_p^n|\boldsymbol{\Sigma}|^{-\frac{n}{2}}\prod_{i=1}^{n}\left[1+\frac{1}{\nu}(\mathbf{x}_i-\boldsymbol{\mu})'\boldsymbol{\Sigma}^{-1}(\mathbf{x}_i-\boldsymbol{\mu})\right]^{-\frac{\nu+p}{2}}$$

where $c_p = \dfrac{\Gamma\left(\frac{\nu+p}{2}\right)}{(\pi\nu)^{\frac{p}{2}}\Gamma\left(\frac{\nu}{2}\right)}$. The log-likelihood function

$$l(\mathscr{X},\boldsymbol{\theta}) = n\ln c_p - \frac{n}{2}\ln|\boldsymbol{\Sigma}| - \frac{\nu+p}{2}\sum_{I=1}^{n}\ln\left[1+\frac{1}{\nu}(\mathbf{x}_i-\boldsymbol{\mu})'\boldsymbol{\Sigma}^{-1}(\mathbf{x}_i-\boldsymbol{\mu})\right].$$

The logarithmic likelihood ratio test statistic for testing $H_{01} : \boldsymbol{\Sigma} = \boldsymbol{\Sigma}_0$ when no constraints are imposed on $\boldsymbol{\mu}$ has the following form

$$LRT(\mathscr{X},\boldsymbol{\Sigma}_0) = -2\ln\left[\left(\frac{|\widehat{\boldsymbol{\Sigma}}|}{|\boldsymbol{\Sigma}_0|}\right)^{\frac{n}{2}}\prod_{i=1}^{n}\left(\frac{1+\frac{1}{\nu}(\mathbf{x}_i-\overline{\mathbf{x}})'\boldsymbol{\Sigma}_0^{-1}(\mathbf{x}_i-\overline{\mathbf{x}})}{1+\frac{1}{\nu}(\mathbf{x}_i-\overline{\mathbf{x}})'\widehat{\boldsymbol{\Sigma}}^{-1}(\mathbf{x}_i-\overline{\mathbf{x}})}\right)^{-\frac{\nu+p}{2}}\right], \tag{12.3}$$

where $\overline{\mathbf{x}} = \frac{1}{n}\sum_{i=1}^{n}\mathbf{x}_i$ and $\widehat{\boldsymbol{\Sigma}}$ is the maximum likelihood estimator of $\boldsymbol{\Sigma}$: $\widehat{\boldsymbol{\Sigma}} = \frac{\nu-2}{\nu}\mathbf{S}$ where $\mathbf{S} = \frac{1}{n}\sum_{i=1}^{n}(\mathbf{x}_i-\overline{\mathbf{x}})(\mathbf{x}_i-\overline{\mathbf{x}})'$ is the maximum likelihood estimator of $D\mathbf{x}$. Then we get from (12.3) the equality

$$LRT(\mathscr{X},\boldsymbol{\Sigma}_0) = -n\ln\left(\frac{\nu-2}{\nu}\frac{|\mathbf{S}|}{|\boldsymbol{\Sigma}_0|}\right)$$
$$+(\nu+p)\left[\sum_{i=1}^{n}\left(\ln\left(1+\frac{1}{\nu}(\mathbf{x}_i-\overline{\mathbf{x}})'\boldsymbol{\Sigma}_0^{-1}(\mathbf{x}_i-\overline{\mathbf{x}})\right)\right.\right.$$
$$\left.\left.-\ln\left(1+\frac{1}{\nu-2}(\mathbf{x}_i-\overline{\mathbf{x}})'\mathbf{S}^{-1}(\mathbf{x}_i-\overline{\mathbf{x}})\right)\right)\right].$$

We can formulate the following result.

Proposition 12.1 *The likelihood ratio test statistic for testing* $\mathrm{H}_{01}: \boldsymbol{\Sigma} = \boldsymbol{\Sigma}_0$ *when no constraints are imposed on* $\boldsymbol{\mu}$ *is given by*

$$
\begin{aligned}
LRT(\mathscr{X}, \boldsymbol{\Sigma}_0) = & -n \ln\Big(\frac{\nu-2}{\nu}\frac{|\mathbf{S}|}{|\boldsymbol{\Sigma}_0|}\Big) \\
& + (\nu+p)\Big[\sum_{i=1}^{n}\Big(\ln\big(1+\frac{1}{\nu}(\mathbf{x}_i-\overline{\mathbf{x}})'\boldsymbol{\Sigma}_0^{-1}(\mathbf{x}_i-\overline{\mathbf{x}})\big) \\
& - \ln\big(1+\frac{1}{\nu-2}(\mathbf{x}_i-\overline{\mathbf{x}})'\mathbf{S}^{-1}(\mathbf{x}_i-\overline{\mathbf{x}})\big)\Big)\Big] \qquad (12.4)
\end{aligned}
$$

where $\mathscr{X} = (\mathbf{x}_1, \ldots, \mathbf{x}_n)$ *is a sample from* $t_{\nu,p}(\boldsymbol{\mu}, \boldsymbol{\Sigma})$*. When* $n \to \infty$ *and* H_{01} *holds then* $LRT(\mathscr{X}, \boldsymbol{\Sigma}_0) \xrightarrow{D} \chi^2$ *with degrees of freedom* $df = \frac{p(p+1)}{2}$

The proof of convergence to the chi-square distribution follows from the general theory of LRT, see [4, §8.4.1], for example.

Corollary 12.1 *Under the assumptions of Proposition 12.1 an approximation of the likelihood ratio test statistic is of the form*

$$
LRT(\mathscr{X}, \boldsymbol{\Sigma}_0) \approx -n \ln\Big(\frac{\nu-2}{\nu}\frac{|\mathbf{S}|}{|\boldsymbol{\Sigma}_0|}\Big) + n(\nu+p)\Big[\frac{1}{\nu}\mathrm{tr}(\boldsymbol{\Sigma}_0^{-1}\mathbf{S}) - \frac{p}{\nu-2}\Big].
$$

Proof Let us expand logarithmic function of the sum in (12.4) into power series and take the first term from the expansion

$$
\begin{aligned}
LRT(\mathscr{X}, \boldsymbol{\Sigma}_0) \approx & -n \ln\Big(\frac{\nu-2}{\nu}\frac{|\mathbf{S}|}{|\boldsymbol{\Sigma}_0|}\Big) + (\nu+p)\Big[\sum_{i=1}^{n}\Big(\frac{1}{\nu}\mathrm{tr}(\boldsymbol{\Sigma}_0^{-1}(\mathbf{x}_i-\overline{\mathbf{x}})(\mathbf{x}_i-\overline{\mathbf{x}})') \\
& - \frac{1}{\nu-2}\mathrm{tr}(\mathbf{S}^{-1}(\mathbf{x}_i-\overline{\mathbf{x}})(\mathbf{x}_i-\overline{\mathbf{x}})')\Big)\Big] \\
= & -n \ln\Big(\frac{\nu-2}{\nu}\frac{|\mathbf{S}|}{|\boldsymbol{\Sigma}_0|}\Big) + n(\nu+p)\Big[\frac{1}{\nu}\mathrm{tr}(\boldsymbol{\Sigma}_0^{-1}\mathbf{S}) - \frac{1}{\nu-2}\mathrm{tr}(\mathbf{I}_p)\Big)\Big].
\end{aligned}
$$

As trace of the identity matrix equals p, the statement is proved. □

Remark When $\nu \to \infty$, then $t_{\nu,p}(\boldsymbol{\mu}, \boldsymbol{\Sigma})$-distribution tends to the normal distribution $N_p(\boldsymbol{\mu}, \boldsymbol{\Sigma})$ and the approximation of the LRT statistic in Corollary 12.1 converges to the corresponding LRT statistic for the normal population [9, for example]

$$
LRT = n[\mathrm{tr}(\boldsymbol{\Sigma}_0^{-1}\mathbf{S}) - p - \ln(|\boldsymbol{\Sigma}_0^{-1}||\mathbf{S}|)].
$$

Corollary 12.2 *The likelihood ratio test statistic for testing* $\mathrm{H}_{02}: \boldsymbol{\Sigma} = \mathbf{I}_p$ *when no constraints are imposed on* $\boldsymbol{\mu}$ *has the following form*

$$\begin{aligned} LRT(\mathscr{X}, \mathbf{I}_p) = &-n \ln\Big(\frac{\nu-2}{\nu}|\mathbf{S}|\Big) \\ &+ (\nu+p)\Big[\sum_{i=1}^{n}\Big(\ln\big(1+\frac{1}{\nu}(\mathbf{x}_i-\bar{\mathbf{x}})'(\mathbf{x}_i-\bar{\mathbf{x}})\big) \\ &- \ln\big(1+\frac{1}{\nu-2}(\mathbf{x}_i-\bar{\mathbf{x}})'\mathbf{S}^{-1}(\mathbf{x}_i-\bar{\mathbf{x}})\big)\Big)\Big] \end{aligned} \tag{12.5}$$

where $\mathscr{X} = (\mathbf{x}_1, \ldots, \mathbf{x}_n)$ *is a sample from* $t_{\nu,p}(\boldsymbol{\mu}, \boldsymbol{\Sigma})$. *When* $n \to \infty$ *and* H_{02} *holds then the* $LRT(\mathscr{X}, \mathbf{I}_p) \sim \chi^2$ *with degrees of freedom* $df = \frac{p(p+1)}{2}$. *An approximation of the test statistic is of the form*

$$LRT(\mathscr{X}, \mathbf{I}_p) \approx -n \ln\Big(\frac{\nu-2}{\nu}|\mathbf{S}|\Big) + n(\nu+p)\Big[\frac{1}{\nu}\mathrm{tr}(\mathbf{S}) - \frac{p}{\nu-2}\Big].$$

Proof Expression (12.5) we get directly from (12.4) by replacing $\boldsymbol{\Sigma}_0$ with $\mathbf{I}_p$. The approximation comes straightforwardly from Corollary 12.1. □

Next, the LRT statistic for testing sphericity is given in Corollary 12.3 in the case of specified $\gamma = \gamma_0$.

Corollary 12.3 *In the assumptions of Proposition 12.1 the likelihood ratio test statistic for testing* $\mathrm{H}_{03}: \boldsymbol{\Sigma} = \gamma_0\mathbf{I}_p$ *for a fixed value* $\gamma_0 > 0$, *when no constraints are imposed on* $\boldsymbol{\mu}$, *has the following form*

$$\begin{aligned} LRT(\mathscr{X}, \gamma_0\mathbf{I}_p) = &-n \ln\Big(\frac{\nu-2}{\nu\gamma_0^p}|\mathbf{S}|\Big) \\ &+ (\nu+p)\Big[\sum_{i=1}^{n}\Big(\ln\big(1+\frac{1}{\nu\gamma_0}(\mathbf{x}_i-\bar{\mathbf{x}})'(\mathbf{x}_i-\bar{\mathbf{x}})\big) \\ &- \ln\big(1+\frac{1}{\nu-2}(\mathbf{x}_i-\bar{\mathbf{x}})'\mathbf{S}^{-1}(\mathbf{x}_i-\bar{\mathbf{x}})\big)\Big)\Big]. \end{aligned} \tag{12.6}$$

When $n \to \infty$ *and* H_{03} *holds then the* $LRT(\mathscr{X}, \gamma_0\mathbf{I}_p) \xrightarrow{D} \chi^2$ *with degrees of freedom* $df = \frac{p(p+1)}{2}$. *An approximation of the test statistic is of the form*

$$LRT(\mathscr{X}, \gamma_0\mathbf{I}_p) \approx -n \ln\Big(\frac{\nu-2}{\nu\gamma_0^p}|\mathbf{S}|\Big) + n(\nu+p)\Big[\frac{1}{\nu\gamma_0}\mathrm{tr}(\mathbf{S}) - \frac{p}{\nu-2}\Big].$$

Proof Statement (12.6) follows directly from (12.4) by replacing $\boldsymbol{\Sigma}_0$ with $\gamma_0\mathbf{I}_p$. The approximation we get from Corollary 12.1. □

Remark For the general sphericity test $\boldsymbol{\Sigma} = \gamma \mathbf{I}_p$ explicit expression of the statistic is not available as maximum likelihood estimate for γ has to be found iteratively.

12.4 Rao's Score Test

Definition 12.6 Rao's score test statistic (RST) for testing hypothesis (12.2) is of the form

$$RST(\mathscr{X}, \boldsymbol{\theta}_0) = \mathbf{u}'(\mathscr{X}, \boldsymbol{\theta}_0)\mathrm{I}(\mathscr{X}, \boldsymbol{\theta}_0)^{-1}\mathbf{u}(\mathscr{X}, \boldsymbol{\theta}_0),$$

where $\mathscr{X} = (\mathbf{x}_1, \ldots, \mathbf{x}_n)$ is a theoretical sample from $P_{\mathbf{x}}(\boldsymbol{\theta})$ and $\boldsymbol{\theta} = (\theta_1, \ldots, \theta_r)'$.

When the sample size $n \to \infty$ and H_0 holds then $RST(\mathscr{X}, \boldsymbol{\theta}) \xrightarrow{D} \chi_r^2$ [13].

Let us derive Rao's score test statistic for testing (12.2) for the multivariate t-distributed population. The logarithm of the density (12.1) is given by

$$l(\mathbf{x}, \boldsymbol{\theta}) = \ln c_p - \frac{1}{2}\ln|\boldsymbol{\Sigma}| - \frac{\nu + p}{2}\ln\left[1 + \frac{1}{\nu}(\mathbf{x} - \boldsymbol{\mu})'\boldsymbol{\Sigma}^{-1}(\mathbf{x} - \boldsymbol{\mu})\right],$$

where c_p is a constant that does not depend on the parameter $\boldsymbol{\theta} = (\boldsymbol{\mu}', \mathrm{vec}'\boldsymbol{\Sigma})'$.

According to Definition 12.2 the score vector can be derived as the matrix derivative

$$\mathbf{u}(\mathbf{x}, \boldsymbol{\theta}) = \frac{d}{d\boldsymbol{\theta}} l(\mathbf{x}, \boldsymbol{\theta}),$$

where $\frac{d}{d\boldsymbol{\theta}} = \begin{pmatrix} \frac{d}{d\boldsymbol{\mu}} \\ \frac{d}{d\mathrm{vec}\,\boldsymbol{\Sigma}} \end{pmatrix} : (p + p^2) \times 1.$

Using the differentiation rules in Sect. 12.2 the derivative $\frac{dl}{d\boldsymbol{\mu}}$ equals

$$\begin{aligned}
\frac{dl}{d\boldsymbol{\mu}} &= -\frac{\nu + p}{2}\frac{d\left(1 + \frac{1}{\nu}(\mathbf{x} - \boldsymbol{\mu})'\boldsymbol{\Sigma}^{-1}(\mathbf{x} - \boldsymbol{\mu})\right)}{d\boldsymbol{\mu}}\frac{d\ln\left(1 + \frac{1}{\nu}(\mathbf{x} - \boldsymbol{\mu})'\boldsymbol{\Sigma}^{-1}(\mathbf{x} - \boldsymbol{\mu})\right)}{d\left(1 + \frac{1}{\nu}(\mathbf{x} - \boldsymbol{\mu})'\boldsymbol{\Sigma}^{-1}(\mathbf{x} - \boldsymbol{\mu})\right)} \\
&= \frac{\nu + p}{2}\frac{1}{\nu}2\boldsymbol{\Sigma}^{-1}(\mathbf{x} - \boldsymbol{\mu})\frac{1}{1 + \frac{1}{\nu}(\mathbf{x} - \boldsymbol{\mu})'\boldsymbol{\Sigma}^{-1}(\mathbf{x} - \boldsymbol{\mu})} \\
&= \frac{\nu + p}{\nu}\frac{\boldsymbol{\Sigma}^{-1}(\mathbf{x} - \boldsymbol{\mu})}{1 + \frac{1}{\nu}(\mathbf{x} - \boldsymbol{\mu})'\boldsymbol{\Sigma}^{-1}(\mathbf{x} - \boldsymbol{\mu})}.
\end{aligned} \tag{12.7}$$

The derivative of $l(\cdot)$ by $\boldsymbol{\Sigma}$ we get from the following row of equalities

$$\begin{aligned}
\frac{d}{d\text{vec}\,\boldsymbol{\Sigma}}l(\mathbf{x},\boldsymbol{\theta}) &= -\frac{1}{2}\text{vec}\,\boldsymbol{\Sigma}^{-1} - \frac{\nu+p}{2}\frac{d\boldsymbol{\Sigma}^{-1}}{d\boldsymbol{\Sigma}}\frac{d\big(1+\frac{1}{\nu}(\mathbf{x}-\boldsymbol{\mu})'\boldsymbol{\Sigma}^{-1}(\mathbf{x}-\boldsymbol{\mu})\big)}{d\boldsymbol{\Sigma}^{-1}} \\
&\quad\times\frac{d\ln\big(1+\frac{1}{\nu}(\mathbf{x}-\boldsymbol{\mu})'\boldsymbol{\Sigma}^{-1}(\mathbf{x}-\boldsymbol{\mu})\big)}{d\big(1+\frac{1}{\nu}(\mathbf{x}-\boldsymbol{\mu})'\boldsymbol{\Sigma}^{-1}(\mathbf{x}-\boldsymbol{\mu})\big)} \\
&= -\frac{1}{2}\text{vec}\,\boldsymbol{\Sigma}^{-1} + \frac{\nu+p}{2}(\boldsymbol{\Sigma}^{-1}\otimes\boldsymbol{\Sigma}^{-1})\frac{1}{\nu}\text{vec}\,\mathbf{M} \\
&\quad\times\frac{1}{1+\frac{1}{\nu}(\mathbf{x}-\boldsymbol{\mu})'\boldsymbol{\Sigma}^{-1}(\mathbf{x}-\boldsymbol{\mu})} \\
&= -\frac{1}{2}\text{vec}\,\boldsymbol{\Sigma}^{-1} + \frac{\nu+p}{2\nu}\frac{\text{vec}\,(\boldsymbol{\Sigma}^{-1}\mathbf{M}\boldsymbol{\Sigma}^{-1})}{1+\frac{1}{\nu}\text{tr}(\boldsymbol{\Sigma}^{-1}\mathbf{M})},
\end{aligned} \tag{12.8}$$

where

$$\mathbf{M} = (\mathbf{x}-\boldsymbol{\mu})(\mathbf{x}-\boldsymbol{\mu})'. \tag{12.9}$$

Hence the score vector we get from (12.7)–(12.9) as a partitioned matrix

$$\mathbf{u}(\mathbf{x},\boldsymbol{\theta}) = \begin{pmatrix}\mathbf{u}_1\\ \mathbf{u}_2\end{pmatrix} = \begin{pmatrix}\frac{\nu+p}{\nu}\frac{\boldsymbol{\Sigma}^{-1}(\mathbf{x}-\boldsymbol{\mu})}{1+\frac{1}{\nu}\text{tr}(\boldsymbol{\Sigma}^{-1}\mathbf{M})}\\ -\frac{1}{2}\text{vec}\,\boldsymbol{\Sigma}^{-1}+\frac{\nu+p}{2\nu}\frac{\text{vec}\,(\boldsymbol{\Sigma}^{-1}\mathbf{M}\boldsymbol{\Sigma}^{-1})}{1+\frac{1}{\nu}\text{tr}(\boldsymbol{\Sigma}^{-1}\mathbf{M})}\end{pmatrix}.$$

The score vector for the sample equals

$$\mathbf{u}(\mathcal{X},\boldsymbol{\theta}) = \begin{pmatrix}\frac{(\nu+p)}{\nu}\sum_{i=1}^{n}\frac{\boldsymbol{\Sigma}^{-1}(\mathbf{x}_i-\boldsymbol{\mu})}{1+\frac{1}{\nu}\text{tr}(\boldsymbol{\Sigma}^{-1}\mathbf{M}_i)}\\ -\frac{n}{2}\text{vec}\,\boldsymbol{\Sigma}^{-1}+\frac{(\nu+p)}{2\nu}\sum_{i=1}^{n}\frac{\text{vec}\,(\boldsymbol{\Sigma}^{-1}\mathbf{M}_i\boldsymbol{\Sigma}^{-1})}{1+\frac{1}{\nu}\text{tr}(\boldsymbol{\Sigma}^{-1}\mathbf{M}_i)}\end{pmatrix}, \tag{12.10}$$

where

$$\mathbf{M}_i = (\mathbf{x}_i-\boldsymbol{\mu})(\mathbf{x}_i-\boldsymbol{\mu})'. \tag{12.11}$$

Next, the Hessian matrix $\mathbf{H}(\mathbf{x},\boldsymbol{\theta})$ is derived following to Definition 12.4

$$\mathbf{H}(\mathbf{x},\boldsymbol{\theta}) = \frac{d}{d\boldsymbol{\theta}}\mathbf{u}(\mathbf{x},\boldsymbol{\theta}) = \begin{pmatrix}\frac{d\mathbf{u}_1}{d\boldsymbol{\mu}} & \frac{d\mathbf{u}_2}{d\boldsymbol{\mu}}\\ \frac{d\mathbf{u}_1}{d\boldsymbol{\Sigma}} & \frac{d\mathbf{u}_2}{d\boldsymbol{\Sigma}}\end{pmatrix} =: \begin{pmatrix}\mathbf{H}_{1,1} & \mathbf{H}_{1,2}\\ \mathbf{H}_{2,1} & \mathbf{H}_{2,2}\end{pmatrix} : (p+p^2)\times(p+p^2).$$

First, let us derive $\mathbf{H}_{1,1}$

$$\mathbf{H}_{1,1} = \frac{(\nu+p)}{\nu}\frac{d}{d\boldsymbol{\mu}}\Big(\frac{\boldsymbol{\Sigma}^{-1}(\mathbf{x}-\boldsymbol{\mu})}{1+\frac{1}{\nu}\mathrm{tr}(\boldsymbol{\Sigma}^{-1}\mathbf{M})}\Big)$$

$$= \frac{(\nu+p)}{\nu}\Big(\frac{d\boldsymbol{\Sigma}^{-1}(\mathbf{x}-\boldsymbol{\mu})}{d\boldsymbol{\mu}}\frac{1}{1+\frac{1}{\nu}\mathrm{tr}(\boldsymbol{\Sigma}^{-1}\mathbf{M})}$$

$$+\frac{d(1+\frac{1}{\nu}(\mathbf{x}-\boldsymbol{\mu})'\boldsymbol{\Sigma}^{-1}(\mathbf{x}-\boldsymbol{\mu}))^{-1}}{d\boldsymbol{\mu}}(\mathbf{x}-\boldsymbol{\mu})'\boldsymbol{\Sigma}^{-1}\Big)$$

$$= \frac{(\nu+p)}{\nu}\Big(\frac{-\boldsymbol{\Sigma}^{-1}}{1+\frac{1}{\nu}\mathrm{tr}(\boldsymbol{\Sigma}^{-1}\mathbf{M})}+\frac{2}{\nu}\frac{\boldsymbol{\Sigma}^{-1}\mathbf{M}\boldsymbol{\Sigma}^{-1}}{(1+\frac{1}{\nu}\mathrm{tr}(\boldsymbol{\Sigma}^{-1}\mathbf{M}))^{2}}\Big)$$

$$= -\frac{(\nu+p)}{\nu(1+\frac{1}{\nu}\mathrm{tr}(\boldsymbol{\Sigma}^{-1}\mathbf{M}))}\Big(\boldsymbol{\Sigma}^{-1}-\frac{2}{\nu}\frac{\boldsymbol{\Sigma}^{-1}\mathbf{M}\boldsymbol{\Sigma}^{-1}}{(1+\frac{1}{\nu}\mathrm{tr}(\boldsymbol{\Sigma}^{-1}\mathbf{M}))}\Big). \quad (12.12)$$

Next we derive $\mathbf{H}_{2,1}$

$$\mathbf{H}_{2,1} = \frac{d\mathbf{u}_1}{d\boldsymbol{\Sigma}} = \frac{(\nu+p)}{\nu}\frac{d}{d\boldsymbol{\Sigma}}\Big(\frac{\boldsymbol{\Sigma}^{-1}(\mathbf{x}-\boldsymbol{\mu})}{1+\frac{1}{\nu}\mathrm{tr}(\boldsymbol{\Sigma}^{-1}\mathbf{M})}\Big)$$

$$= -(\boldsymbol{\Sigma}^{-1}\otimes\boldsymbol{\Sigma}^{-1})((\mathbf{x}-\boldsymbol{\mu})\otimes\mathbf{I}_p)\frac{1}{1+\frac{1}{\nu}\mathrm{tr}(\boldsymbol{\Sigma}^{-1}\mathbf{M})}$$

$$+\frac{1}{(1+\frac{1}{\nu}\mathrm{tr}(\boldsymbol{\Sigma}^{-1}\mathbf{M}))^{2}}\frac{1}{\nu}(\boldsymbol{\Sigma}^{-1}\otimes\boldsymbol{\Sigma}^{-1})(\mathbf{M}\otimes\mathbf{I}_p)\mathrm{vec}\,\mathbf{I}_p(\mathbf{x}-\boldsymbol{\mu})'\boldsymbol{\Sigma}^{-1}$$

$$= -(\boldsymbol{\Sigma}^{-1}(\mathbf{x}-\boldsymbol{\mu})\otimes\boldsymbol{\Sigma}^{-1})\frac{1}{1+\frac{1}{\nu}\mathrm{tr}(\boldsymbol{\Sigma}^{-1}\mathbf{M})}+\frac{\mathrm{vec}\,(\boldsymbol{\Sigma}^{-1}\mathbf{M}\boldsymbol{\Sigma}^{-1})(\mathbf{x}-\boldsymbol{\mu})'\boldsymbol{\Sigma}^{-1}}{\nu(1+\frac{1}{\nu}\mathrm{tr}(\boldsymbol{\Sigma}^{-1}\mathbf{M}))^{2}}. \quad (12.13)$$

Because of symmetry of matrix H we have equality $\mathbf{H}_{1,2} = \mathbf{H}'_{2,1}$.

Finally, we derive $\mathbf{H}_{2,2}$

$$\mathbf{H}_{2,2} = \frac{d\mathbf{u}_2}{d\boldsymbol{\Sigma}} = \frac{1}{2}(\boldsymbol{\Sigma}^{-1}\otimes\boldsymbol{\Sigma}^{-1})+\frac{\nu+p}{2\nu}\frac{d}{d\boldsymbol{\Sigma}}\Big(\frac{\mathrm{vec}\,(\boldsymbol{\Sigma}^{-1}\mathbf{M}\boldsymbol{\Sigma}^{-1})}{1+\frac{1}{\nu}\mathrm{tr}(\boldsymbol{\Sigma}^{-1}\mathbf{M})}\Big)$$

$$= \frac{1}{2}(\boldsymbol{\Sigma}^{-1}\otimes\boldsymbol{\Sigma}^{-1})+\frac{\nu+p}{2\nu}\Big(\frac{-(\boldsymbol{\Sigma}^{-1}\otimes\boldsymbol{\Sigma}^{-1}\mathbf{M}\boldsymbol{\Sigma}^{-1})-(\boldsymbol{\Sigma}^{-1}\mathbf{M}\boldsymbol{\Sigma}^{-1}\otimes\boldsymbol{\Sigma}^{-1})}{(1+\frac{1}{\nu}\mathrm{tr}(\boldsymbol{\Sigma}^{-1}\mathbf{M}))}$$

$$+\mathrm{vec}\,(\boldsymbol{\Sigma}^{-1}\mathbf{M}\boldsymbol{\Sigma}^{-1})\mathrm{vec}'(\boldsymbol{\Sigma}^{-1}\mathbf{M}\boldsymbol{\Sigma}^{-1})\frac{1}{\nu(1+\frac{1}{\nu}\mathrm{tr}(\boldsymbol{\Sigma}^{-1}\mathbf{M}))^{2}}\Big)$$

$$= \frac{1}{2}(\boldsymbol{\Sigma}^{-1} \otimes \boldsymbol{\Sigma}^{-1}) + \frac{(\nu + p)}{2\nu(1 + \frac{1}{\nu}\mathrm{tr}(\boldsymbol{\Sigma}^{-1}\mathbf{M}))}\Big(\frac{\mathrm{vec}\,(\boldsymbol{\Sigma}^{-1}\mathbf{M}\boldsymbol{\Sigma}^{-1})\mathrm{vec}\,'(\boldsymbol{\Sigma}^{-1}\mathbf{M}\boldsymbol{\Sigma}^{-1})}{\nu(1 + \frac{1}{\nu}\mathrm{tr}(\boldsymbol{\Sigma}^{-1}\mathbf{M}))}$$

$$- (\boldsymbol{\Sigma}^{-1} \otimes \boldsymbol{\Sigma}^{-1}\mathbf{M}\boldsymbol{\Sigma}^{-1}) - (\boldsymbol{\Sigma}^{-1}\mathbf{M}\boldsymbol{\Sigma}^{-1} \otimes \boldsymbol{\Sigma}^{-1})\Big). \tag{12.14}$$

The information matrix $\mathbf{I}(\mathbf{x}, \boldsymbol{\theta})$ is of the form

$$\mathbf{I}(\mathbf{x}, \boldsymbol{\theta}) = -\mathrm{E}(\mathbf{H}(\mathbf{x}, \boldsymbol{\theta})) = \begin{pmatrix} \mathbf{I}_{11}(\boldsymbol{\mu}) & \mathbf{I}_{12}(\boldsymbol{\mu}, \boldsymbol{\Sigma}) \\ \mathbf{I}_{21}(\boldsymbol{\mu}, \boldsymbol{\Sigma}) & \mathbf{I}_{22}(\boldsymbol{\Sigma}) \end{pmatrix},$$

where

$$\mathbf{I}_{i,j} = -\mathrm{E}(\mathbf{H}_{i,j})$$

and $\mathbf{H}_{i,j}$ are given in (12.12)–(12.14). Expressions of $\mathbf{H}_{i,j}$ are complicated and explicit expressions of the expectations can not be found. Instead we use the first term from the Taylor expansion of $\mathbf{H}_{i,j}$ at $\mathrm{E}\mathbf{x} = \boldsymbol{\mu}$, $\mathrm{E}[(\mathbf{x} - \boldsymbol{\mu})(\mathbf{x} - \boldsymbol{\mu})'] = \mathrm{D}\mathbf{x} = \frac{\nu}{\nu-2}\boldsymbol{\Sigma}$ as an approximation of the expectation. We keep the notation of the expectation $\mathbf{I}_{i,j}$ also for the obtained approximation

$$\mathbf{I}_{1,1}(\boldsymbol{\mu}) = \frac{\nu + p}{\nu}\frac{1}{1 + \frac{1}{\nu}\mathrm{tr}(\frac{\nu}{\nu-2}\mathbf{I}_p)}\left(\boldsymbol{\Sigma}^{-1} - \frac{2}{\nu}\frac{\frac{\nu}{\nu-2}\boldsymbol{\Sigma}^{-1}}{\left(1 + \frac{1}{\nu}\mathrm{tr}(\frac{\nu}{\nu-2}\mathbf{I}_p)\right)}\right) = a_1\boldsymbol{\Sigma}^{-1}, \tag{12.15}$$

where

$$a_1 = \frac{(\nu - 2)}{\nu}\frac{(\nu + p)(\nu + p - 4)}{(\nu + p - 2)^2}. \tag{12.16}$$

At $\mathbf{x} = \boldsymbol{\mu}$ the matrix $\mathbf{H}_{1,2} = \mathbf{H}'_{2,1} = \mathbf{0}$ and therefore the matrix $\mathbf{I}(\mathbf{x}, \boldsymbol{\theta})$ is block-diagonal. It remains to find $\mathbf{I}_{2,2}$

$$\mathbf{I}_{2,2}(\boldsymbol{\Sigma}) = -\frac{\nu + p}{2\nu(1 + \frac{1}{\nu}\mathrm{tr}(\frac{\nu}{\nu-2}\mathbf{I}_p))}\Bigg[\frac{\mathrm{vec}\,(\frac{\nu}{\nu-2}\boldsymbol{\Sigma}^{-1})\mathrm{vec}\,'(\frac{\nu}{\nu-2}\boldsymbol{\Sigma}^{-1})}{\nu(1 + \frac{1}{\nu}\mathrm{tr}(\frac{\nu}{\nu-2}\mathbf{I}_p))}$$

$$- \frac{2\nu}{\nu - 2}(\boldsymbol{\Sigma}^{-1} \otimes \boldsymbol{\Sigma}^{-1})\Bigg] - \frac{1}{2}(\boldsymbol{\Sigma}^{-1} \otimes \boldsymbol{\Sigma}^{-1})$$

$$= \frac{(\nu + p)}{(\nu + p - 2)}\left[(\boldsymbol{\Sigma}^{-1} \otimes \boldsymbol{\Sigma}^{-1}) - \frac{1}{2(\nu + p - 2)}\mathrm{vec}\,\boldsymbol{\Sigma}^{-1}\mathrm{vec}\,'\boldsymbol{\Sigma}^{-1}\right]$$

$$- \frac{1}{2}(\boldsymbol{\Sigma}^{-1} \otimes \boldsymbol{\Sigma}^{-1})$$

$$= \Big(\frac{(\nu+p)}{\nu+p-2} - \frac{1}{2}\Big)(\boldsymbol{\Sigma}^{-1} \otimes \boldsymbol{\Sigma}^{-1}) - \frac{(\nu+p)}{2(\nu+p-2)^2}\text{vec}\,\boldsymbol{\Sigma}^{-1}\text{vec}'\boldsymbol{\Sigma}^{-1}$$

$$= \frac{1}{2}(\boldsymbol{\Sigma}^{-1} \otimes \boldsymbol{\Sigma}^{-1}) - a_2\text{vec}\,\boldsymbol{\Sigma}^{-1}\text{vec}'\boldsymbol{\Sigma}^{-1}, \qquad (12.17)$$

where

$$a_2 = \frac{(\nu+p)}{2(\nu+p-2)^2}. \qquad (12.18)$$

Remark When $\nu \to \infty$ then $\mathbf{I}_{1,1} \to \boldsymbol{\Sigma}^{-1}$ and $\mathbf{I}_{2,2} \to \frac{1}{2}(\boldsymbol{\Sigma}^{-1} \otimes \boldsymbol{\Sigma}^{-1})$ which are the corresponding expressions of the blocks of the information matrix when we have a normal population $N_p(\boldsymbol{\mu}, \boldsymbol{\Sigma})$ [9, for instance]. This indicates that we have got a reasonable approximation of the information matrix.

For a sample of size n

$$\mathbf{I}(\mathcal{X}, \boldsymbol{\theta}) = n\mathbf{I}(\mathbf{x}, \boldsymbol{\theta})$$

and we have $\mathbf{I}(\mathcal{X}, \boldsymbol{\theta})$ in the block-diagonal form. Summarizing we can formulate the following result.

Proposition 12.2 *Let* $\mathcal{X} = (\mathbf{x}_1, \ldots, \mathbf{x}_n)$ *be a sample from* $t_{\nu,p}$*-distribution with density* (12.1)*. Under the null hypothesis* $\mathrm{H}_{01} : \boldsymbol{\Sigma} = \boldsymbol{\Sigma}_0$ *approximation of the information matrix of the sample has the form*

$$\mathbf{I}(\mathcal{X}, \boldsymbol{\Sigma}_0) = n\begin{pmatrix} a_1\boldsymbol{\Sigma}_0^{-1} & \mathbf{0} \\ \mathbf{0} & \frac{1}{2}(\boldsymbol{\Sigma}_0^{-1} \otimes \boldsymbol{\Sigma}_0^{-1}) - a_2\text{vec}\,\boldsymbol{\Sigma}_0^{-1}\text{vec}'\boldsymbol{\Sigma}_0^{-1} \end{pmatrix}.$$

Under the hypothesis $\mathrm{H}_{02} : \boldsymbol{\Sigma} = \mathbf{I}_p$ *we have*

$$\mathbf{I}(\mathcal{X}, \mathbf{I}_p) = n\begin{pmatrix} a_1\mathbf{I}_p & \mathbf{0} \\ \mathbf{0} & \frac{1}{2}\mathbf{I}_{p^2} + a_2\text{vec}\,\mathbf{I}_p\text{vec}'\mathbf{I}_p \end{pmatrix}.$$

Under $\mathrm{H}_{03} : \boldsymbol{\Sigma} = \gamma_0\mathbf{I}_p$ *where* $\gamma_0 > 0$ *is a known constant, we have*

$$\mathbf{I}(\mathcal{X}, \gamma_0\mathbf{I}_p) = n\begin{pmatrix} \frac{a_1}{\gamma_0}\mathbf{I}_p & \mathbf{0} \\ \mathbf{0} & \frac{1}{\gamma_0^2}(\frac{1}{2}\mathbf{I}_{p^2} - a_2\text{vec}\,\mathbf{I}_p\text{vec}'\mathbf{I}_p) \end{pmatrix},$$

where a_1 *and* a_2 *are defined by* (12.16) *and* (12.18)*, respectively.*

For deriving Rao's statistic we need expression of the score function $\mathbf{u}(\mathcal{X}, \boldsymbol{\theta}_0)$. We shall find an approximation of $\mathbf{u}(\mathcal{X}, \boldsymbol{\theta}_0)$ by using the first two terms from the Taylor expansion of its expression in (12.10) by substituting $\boldsymbol{\Sigma}$ with $\boldsymbol{\Sigma}_0$. We use notation $\mathbf{u}(\mathscr{X}, \boldsymbol{\Sigma}_0)$ also for the approximation. The result is presented in the next proposition.

Proposition 12.3 *The score vector* $\mathbf{u}(\mathscr{X}, \boldsymbol{\theta}_0)$ *when no constraints have been imposed to* $\boldsymbol{\mu}$*, equals*

$$\mathbf{u}(\mathscr{X}, \boldsymbol{\theta}_0) = \begin{pmatrix} \mathbf{u}_1(\mathscr{X}, \boldsymbol{\mu}) \\ \mathbf{u}_2(\mathscr{X}, \boldsymbol{\Sigma}_0) \end{pmatrix}$$

$$\approx \frac{n}{2}\begin{pmatrix} \mathbf{0} \\ (a_3 \mathrm{vec}\, \boldsymbol{\Sigma}_0^{-1} + a_4 \mathrm{tr}(\boldsymbol{\Sigma}_0^{-1}\mathbf{S}))\mathrm{vec}\,(\boldsymbol{\Sigma}_0^{-1}) + a_5 \mathrm{vec}\,(\boldsymbol{\Sigma}_0^{-1}\mathbf{S}\boldsymbol{\Sigma}_0^{-1}) \end{pmatrix}, \qquad (12.19)$$

where $a_3 = \frac{p(\nu+p)}{(\nu+p-2)^2} - 1$, $a_4 = \frac{(\nu-2)(\nu+p)}{\nu(\nu+p-2)^2}$ *and* $a_5 = \frac{(\nu+p)(\nu-2)}{\nu(\nu+p-2)}$.

Proof We expand $\mathbf{u}(\mathscr{X}, \boldsymbol{\theta}_0)$ at $(\boldsymbol{\mu}', \mathrm{vec}'(\frac{\nu}{\nu-2}\boldsymbol{\Sigma}_0))'$. When there are no restrictions put on $\boldsymbol{\mu}$ then maximum likelihood estimate of $\boldsymbol{\mu}$ is inserted into score vector to replace $\boldsymbol{\mu}$ (see [5]). Then the part of the score vector corresponding to $\boldsymbol{\mu}$ turns to zero and hence only the second part of the score vector $\mathbf{u}_2(\mathscr{X}, \boldsymbol{\Sigma}_0)$ contributes to the calculation of the test statistic. Taylor expansion of the second block is

$$\mathbf{u}_2(\mathscr{X}, \boldsymbol{\Sigma}_0) = \mathbf{u}_2(\mathscr{X}, \boldsymbol{\Sigma}_0)\Big|_{\mathbf{M}_i = \frac{\nu}{\nu-2}\boldsymbol{\Sigma}_0}$$

$$+ \frac{\nu+p}{2\nu}\sum_{i=1}^{n} \frac{d}{d\mathbf{M}_i}\Big(\frac{\mathrm{vec}\,(\boldsymbol{\Sigma}_0^{-1}\mathbf{M}_i\boldsymbol{\Sigma}_0^{-1})}{1+\frac{1}{\nu}\mathrm{tr}(\boldsymbol{\Sigma}_0^{-1}\mathbf{M}_i)}\Big)\Big|_{\mathbf{M}_i = \frac{\nu}{\nu-2}\boldsymbol{\Sigma}_0} \mathrm{vec}\,(\mathbf{M}_i - \frac{\nu}{\nu-2}\boldsymbol{\Sigma}_0) + \ldots$$

where $\mathbf{M}_i$ is defined in (12.11).

Let us find the terms step-by-step. The first term equals

$$\mathbf{u}_2(\mathscr{X}, \boldsymbol{\Sigma}_0)\Big|_{\mathbf{M}_i = \frac{\nu}{\nu-2}\boldsymbol{\Sigma}_0} = -\frac{n}{2}\mathrm{vec}\,\boldsymbol{\Sigma}_0^{-1} + \frac{\nu+p}{2\nu}\sum_{i=1}^{n}\frac{\mathrm{vec}\,(\frac{\nu}{\nu-2}\boldsymbol{\Sigma}_0^{-1})}{1+\frac{1}{\nu}\mathrm{tr}(\frac{\nu}{\nu-2}\mathbf{I}_p)}$$

$$= \frac{n}{2}\mathrm{vec}\,\boldsymbol{\Sigma}_0^{-1}\Big(\frac{\nu+p}{\nu+p-2} - 1\Big) = \frac{n}{\nu+p-2}\mathrm{vec}\,\boldsymbol{\Sigma}_0^{-1}.$$

Let us find the derivative below the sum

$$\frac{d}{d\mathbf{M}_i}\Big(\frac{\mathrm{vec}\,(\boldsymbol{\Sigma}_0^{-1}\mathbf{M}_i\boldsymbol{\Sigma}_0^{-1})}{1+\frac{1}{\nu}\mathrm{tr}(\boldsymbol{\Sigma}_0^{-1}\mathbf{M}_i)}\Big)\Big|_{\mathbf{M}_i = \frac{\nu}{\nu-2}\boldsymbol{\Sigma}_0}$$

$$= \frac{d}{d\mathbf{M}_i}\Big(\mathrm{vec}\,(\boldsymbol{\Sigma}_0^{-1}\mathbf{M}_i\boldsymbol{\Sigma}_0^{-1})\frac{1}{1+\frac{1}{\nu}\frac{\nu}{\nu-2}\mathrm{tr}(\boldsymbol{\Sigma}_0^{-1}\mathbf{M}_i)}\Big)\Big|_{\mathbf{M}_i = \frac{\nu}{\nu-2}\boldsymbol{\Sigma}_0}$$

$$+ \frac{d(1+\frac{1}{\nu}\mathrm{tr}(\boldsymbol{\Sigma}_0^{-1}\mathbf{M}_i))}{d\mathbf{M}_i}\frac{d(1+\frac{1}{\nu}\mathrm{tr}(\boldsymbol{\Sigma}_0^{-1}\mathbf{M}_i))^{-1}}{d(1+\frac{1}{\nu}\mathrm{tr}(\boldsymbol{\Sigma}_0^{-1}\mathbf{M}_i))}\mathrm{vec}'(\boldsymbol{\Sigma}_0^{-1}\mathbf{M}_i\boldsymbol{\Sigma}_0^{-1})\Big|_{\mathbf{M}_i = \frac{\nu}{\nu-2}\boldsymbol{\Sigma}_0}$$

$$= \frac{1}{1+\frac{p}{\nu-2}}(\boldsymbol{\Sigma}_0^{-1} \otimes \boldsymbol{\Sigma}_0^{-1}) - \frac{1}{\nu}\frac{1}{(1+\frac{p}{\nu-2})^2}(\mathbf{I}_p \otimes \boldsymbol{\Sigma}_0^{-1})\mathrm{vec}\,\mathbf{I}_p\mathrm{vec}'(\frac{\nu}{\nu-2}\boldsymbol{\Sigma}_0^{-1})$$

$$= \frac{\nu-2}{\nu+p-2}(\boldsymbol{\Sigma}_0^{-1} \otimes \boldsymbol{\Sigma}_0^{-1}) - \frac{\nu-2}{(\nu+p-2)^2}\mathrm{vec}\,\boldsymbol{\Sigma}_0^{-1}\mathrm{vec}'\boldsymbol{\Sigma}_0^{-1}.$$

When we plug-in the obtained expressions into the Taylor expansion, we get

$$\mathbf{u}_2(\mathscr{X}, \boldsymbol{\Sigma}_0) = \frac{n}{\nu+p-2}\mathrm{vec}\,\boldsymbol{\Sigma}_0^{-1} + \frac{\nu+p}{2\nu}n\Big[\frac{\nu-2}{\nu+p-2}(\boldsymbol{\Sigma}_0^{-1} \otimes \boldsymbol{\Sigma}_0^{-1})$$

$$- \frac{(\nu-2)}{(\nu+p-2)^2}\mathrm{vec}\,\boldsymbol{\Sigma}_0^{-1}\mathrm{vec}'\boldsymbol{\Sigma}_0^{-1}\Big]\mathrm{vec}\Big(\mathbf{S} - \frac{\nu}{\nu-2}\boldsymbol{\Sigma}_0\Big) + \ldots$$

$$= \frac{n}{\nu+p-2}\Big[\mathrm{vec}\,\boldsymbol{\Sigma}_0^{-1} + \frac{(\nu+p)(\nu-2)}{2\nu}(\boldsymbol{\Sigma}_0^{-1} \otimes \boldsymbol{\Sigma}_0^{-1})\mathrm{vec}\Big(\mathbf{S} - \frac{\nu}{\nu-2}\boldsymbol{\Sigma}_0\Big)$$

$$- \frac{(\nu+p)(\nu-2)}{2\nu(\nu+p-2)}\mathrm{vec}\,\boldsymbol{\Sigma}_0^{-1}\mathrm{vec}'\boldsymbol{\Sigma}_0^{-1}\mathrm{vec}\Big(\mathbf{S} - \frac{\nu}{\nu-2}\boldsymbol{\Sigma}_0\Big)\Big] + \ldots$$

$$= \frac{n}{\nu+p-2}\Big[\mathrm{vec}\,\boldsymbol{\Sigma}_0^{-1} + \frac{(\nu+p)(\nu-2)}{2\nu}\mathrm{vec}\,(\boldsymbol{\Sigma}_0^{-1}\mathbf{S}\boldsymbol{\Sigma}_0^{-1}) - \frac{\nu+p}{2}\mathrm{vec}\,\boldsymbol{\Sigma}_0^{-1}$$

$$- \frac{(\nu+p)(\nu-2)}{2\nu(\nu+p-2)}\mathrm{vec}\,\boldsymbol{\Sigma}_0^{-1}\cdot\mathrm{tr}(\boldsymbol{\Sigma}_0^{-1}\mathbf{S}) + \frac{(\nu+p)(p)}{2(\nu+p-2)}\mathrm{vec}\,\boldsymbol{\Sigma}_0^{-1}\Big] + \ldots$$

$$= \frac{n}{\nu+p-2}\Big[\Big(1 - \frac{\nu+p}{2} + \frac{(\nu+p)(p)}{2(\nu+p-2)}\Big)\mathrm{vec}\,\boldsymbol{\Sigma}_0^{-1}$$

$$+ \frac{(\nu+p)(\nu-2)}{2\nu}\mathrm{vec}\,(\boldsymbol{\Sigma}_0^{-1}\mathbf{S}\boldsymbol{\Sigma}_0^{-1}) - \frac{(\nu+p)(\nu-2)}{2\nu(\nu+p-2)}\mathrm{tr}(\boldsymbol{\Sigma}_0^{-1}\mathbf{S})\mathrm{vec}\,\boldsymbol{\Sigma}_0^{-1}\Big] + \ldots$$

We have obtained $\mathbf{u}_2(\mathscr{X}, \boldsymbol{\Sigma}_0)$ of the form

$$\mathbf{u}_2(\mathscr{X}, \boldsymbol{\Sigma}_0) = \frac{n}{2}\Big[\Big(\frac{p(\nu+p)}{(\nu+p-2)^2} - 1\Big)\mathrm{vec}\,\boldsymbol{\Sigma}_0^{-1} - \frac{(\nu-2)(\nu+p)}{\nu(\nu+p-2)^2}\mathrm{tr}(\boldsymbol{\Sigma}_0^{-1}\mathbf{S})\mathrm{vec}\,\boldsymbol{\Sigma}_0^{-1}$$

$$+ \frac{(\nu-2)(\nu+p)}{\nu(\nu+p-2)}\mathrm{vec}\,(\boldsymbol{\Sigma}_0^{-1}\mathbf{S}\boldsymbol{\Sigma}_0^{-1})\Big] + \ldots$$

where $\mathbf{S}$ is the maximum likelihood estimator of $D\,\mathbf{x}$

$$\mathbf{S} = \frac{1}{n}\sum_{i=1}^{n}(\mathbf{x}_i - \overline{\mathbf{x}})(\mathbf{x}_i - \overline{\mathbf{x}})'.$$

We have got the expression (12.19) where multipliers of $\mathrm{vec}\,\boldsymbol{\Sigma}_0^{-1}$ and $\mathrm{vec}\,(\boldsymbol{\Sigma}_0^{-1}\mathbf{S}\boldsymbol{\Sigma}_0^{-1})$ are equal to the constants defined in proposition. The proposition is proved. □

Remark The obtained approximate expression of $\mathbf{u}_2(\mathscr{X}, \boldsymbol{\Sigma}_0)$ is in concordance with the expression of the score vector for the normal population. When $\nu \to \infty$, then

$$\mathbf{u}_2(\mathscr{X}, \boldsymbol{\Sigma}_0) \to \frac{n}{2}\text{vec}\,((\boldsymbol{\Sigma}_0^{-1}\mathbf{S} + \mathbf{I}_p)\boldsymbol{\Sigma}_0^{-1}).$$

This is the expression of the score vector when population distribution is $N_p(\boldsymbol{\mu}, \boldsymbol{\Sigma})$ [9].

Corollary 12.4 *When* $\text{H}_{02} : \boldsymbol{\Sigma} = \mathbf{I}_p$*, approximation of the score vector is of the form*

$$\mathbf{u}_2(\mathscr{X}, \mathbf{I}_p) \approx \frac{n}{2}(a_3 + a_4\text{tr}\mathbf{S})\text{vec}\,\mathbf{I}_p + a_5\text{vec}\,\mathbf{S},$$

where a_i, $i = 3, 4, 5$ *are given in Proposition 12.3.*

Corollary 12.5 *Under the null hypothesis* $\text{H}_{03} : \boldsymbol{\Sigma} = \gamma_0\mathbf{I}_p$ *where* $\gamma_0 > 0$ *is a known constant, approximation of the score vector is of the form*

$$\mathbf{u}_2(\mathscr{X}, \gamma_0\mathbf{I}_p) \approx \frac{n}{2}(a_3 + \frac{a_4}{\gamma_0^2}\text{tr}\mathbf{S})\text{vec}\,\mathbf{I}_p + \frac{a_5}{\gamma_0^2}\text{vec}\,\mathbf{S},$$

where a_i, $i = 3, 4, 5$ *are given in Proposition 12.3.*

Now, Rao's score test (RST) statistic for testing $\text{H}_0 : \boldsymbol{\theta} = \boldsymbol{\theta}_0$ can be calculated using the expressions of $\mathbf{u}(\mathscr{X}, \boldsymbol{\theta}_0)$ and $\text{I}(\mathscr{X}, \boldsymbol{\theta})$, substituting $\boldsymbol{\Sigma}$ with $\boldsymbol{\Sigma}_0$, and noting that the inverse of a block-diagonal matrix is a block-diagonal matrix with the inverses of the diagonal blocks.

Proposition 12.4 *Rao's score test statistic for testing* $\text{H}_{01} : \boldsymbol{\Sigma} = \boldsymbol{\Sigma}_0$ *when no constraints are imposed on* $\boldsymbol{\mu}$ *is approximated by the following equality*

$$RST(\mathscr{X}, \boldsymbol{\Sigma}_0) = \mathbf{u}_2(\mathscr{X}, \boldsymbol{\Sigma}_0)'(\mathbf{I}_{2.2}(\boldsymbol{\Sigma}_0))^{-1}\mathbf{u}_2(\mathscr{X}, \boldsymbol{\Sigma}_0), \tag{12.20}$$

where $\mathbf{u}_2(\mathscr{X}, \boldsymbol{\Sigma}_0)$ *is given in Proposition 12.3 and* $\mathbf{I}_{2.2}$ *by equalities* (12.17) *and* (12.18).

Remark Note that while deriving Rao's score test statistic we have taken derivatives by vec $\boldsymbol{\Sigma}$, not by vech $\boldsymbol{\Sigma}$ which consists of the non-repeated elements of the lower triangle of $\boldsymbol{\Sigma}$. A transformation via the duplication matrix which is used in the literature (see [17, for example]) complicates derivation and is not necessary. It turns out that both approaches lead to the same expression of the test statistic. This statement is proved in [9] and is a corollary of a more general result in [7] for patterned matrices.

Corollary 12.6 *An approximation of the Rao's score test statistic for testing* H_{02} : $\boldsymbol{\Sigma} = \mathbf{I}_p$ *when no constraints are imposed on* $\boldsymbol{\mu}$ *has the following form*

$$RST(\mathscr{X}, \mathbf{I}_p) = \mathbf{u}_2(\mathscr{X}, \mathbf{I}_p)'(\mathbf{I}_{2.2}(\mathbf{I}_p))^{-1}\mathbf{u}_2(\mathscr{X}, \mathbf{I}_p), \tag{12.21}$$

where $\mathscr{X} = (\mathbf{x}_1, \dots, \mathbf{x}_n)$ *is a sample from* $t_{\nu,p}(\boldsymbol{\mu}, \boldsymbol{\Sigma})$, $\mathbf{u}_2(\mathscr{X}, \mathbf{I}_p)$ *is given in Corollary 12.4 and* $\mathbf{I}_{2.2}(\mathbf{I}_p)$ *in Proposition 12.2.*

Proof Equality (12.21) follows directly from (12.20). □

Corollary 12.7 *An approximation of the Rao's score test statistic for testing* H_{03} : $\boldsymbol{\Sigma} = \gamma_0\mathbf{I}_p$ *when no constraints are imposed on* $\boldsymbol{\mu}$ *has the following form*

$$RST(\mathscr{X}, \gamma_0\mathbf{I}_p) = \mathbf{u}_2(\mathscr{X}, \gamma_0\mathbf{I}_p)'(\mathbf{I}_{2.2}(\gamma_0\mathbf{I}_p))^{-1}\mathbf{u}_2(\mathscr{X}, \gamma_0\mathbf{I}_p), \tag{12.22}$$

where $\gamma_0 > 0$ *is a known constant,* $\mathscr{X} = (\mathbf{x}_1, \dots, \mathbf{x}_n)$ *is a sample from* $t_{\nu,p}(\boldsymbol{\mu}, \boldsymbol{\Sigma})$, *score vector* $\mathbf{u}_2(\mathscr{X}, \gamma_0\mathbf{I}_p)$ *is given in Corollary 12.5 and* $\mathbf{I}_{2.2}(\gamma_0\mathbf{I}_p)$ *in Proposition 12.2.*

Proof Equality (12.22) we get from (12.20) when substituting $\boldsymbol{\Sigma}_0$ by $\gamma_0\mathbf{I}_p$. □

12.5 Wald's Score Statistic

Definition 12.7 Wald's score test statistic for testing hypothesis (12.2) is given by

$$WST(\mathscr{X}, \boldsymbol{\theta}_0) = (T_{\boldsymbol{\theta}}(\mathscr{X}) - \boldsymbol{\theta}_0)' \cdot \boldsymbol{I}(\mathscr{X}, T_{\boldsymbol{\theta}}(\mathscr{X})) \cdot (T_{\boldsymbol{\theta}}(\mathscr{X}) - \boldsymbol{\theta}_0), \tag{12.23}$$

where $\boldsymbol{\theta} = (\theta_1, \dots, \theta_r)'$, $T_{\boldsymbol{\theta}}$ is the maximum likelihood estimator of $\boldsymbol{\theta}$ and $\mathscr{X}$ is a random sample.

When sample size $n \to \infty$ and H_0 holds then $WST(\mathscr{X}, \boldsymbol{\theta}_0)$ in (12.23) converges to χ_r^2-distribution [16].

Compared with the likelihood ratio test LRT and Rao's score test RST the Wald's score test WST has simpler form for the multivariate $t_{\nu,p}$-distributed observations. As $D\,\mathbf{x} = \frac{\nu}{\nu-2}\boldsymbol{\Sigma}$, the maximum likelihood estimator of $\boldsymbol{\Sigma}$ is

$$\widehat{\boldsymbol{\Sigma}} = \frac{\nu-2}{\nu}\mathbf{S}.$$

Using the results obtained above in Sects. 12.3 and 12.4 one can calculate Wald's score test statistic for $\mathrm{H}_0 : \boldsymbol{\Sigma} = \boldsymbol{\Sigma}_0$ by substituting $\boldsymbol{\mu}$ with maximum likelihood estimator $\overline{\mathbf{x}}$ and $\boldsymbol{\Sigma}$ with maximum likelihood estimator $\widehat{\boldsymbol{\Sigma}}$ in the expression of $\mathbf{I}_{2,2}(\mathscr{X}, T_{\boldsymbol{\theta}}(\mathscr{X}))$.

Proposition 12.5 *Wald's score test statistic for testing* $\mathrm{H}_{01} : \boldsymbol{\Sigma} = \boldsymbol{\Sigma}_0$ *when no constraints are imposed on* $\boldsymbol{\mu}$ *has the following form*

$$WST(\mathscr{X}, \boldsymbol{\Sigma}_0) = \frac{n}{2}\Big(\mathrm{tr}(\boldsymbol{\Sigma}_0\widehat{\boldsymbol{\Sigma}}^{-1})^2 - 2a_2\big(\mathrm{tr}(\boldsymbol{\Sigma}_0\widehat{\boldsymbol{\Sigma}}^{-1})\big)^2 + (1-2a_2p)(p-2\mathrm{tr}(\boldsymbol{\Sigma}_0\widehat{\boldsymbol{\Sigma}}^{-1})\Big), \tag{12.24}$$

where $\mathscr{X} = (\mathbf{x}_1, \ldots, \mathbf{x}_n)$ *is a sample from* $t_{\nu,p}(\boldsymbol{\mu}, \boldsymbol{\Sigma})$ *and* $\widehat{\boldsymbol{\Sigma}}^{-1} = \frac{\nu}{\nu-2}\mathbf{S}^{-1}$.

Proof When there are no restrictions put on $\boldsymbol{\mu}$ then $T_{\boldsymbol{\theta}}(\mathscr{X})$ is replaced by $\widehat{\boldsymbol{\Sigma}}$, $\boldsymbol{\theta}_0$ by $\boldsymbol{\Sigma}_0$ and $\mathrm{I}(\mathscr{X}, T_{\boldsymbol{\theta}}(\mathscr{X}))$ by $\mathrm{I}_{2,2}(\mathscr{X}, \widehat{\boldsymbol{\Sigma}})$ in (12.17) and (12.18). Then

$$WST(\mathscr{X}, \boldsymbol{\Sigma}_0) = \mathrm{vec}'(\widehat{\boldsymbol{\Sigma}} - \boldsymbol{\Sigma}_0)\mathbf{I}_{2,2}(\widehat{\boldsymbol{\Sigma}})\mathrm{vec}\,(\widehat{\boldsymbol{\Sigma}} - \boldsymbol{\Sigma}_0),$$

where

$$\mathbf{I}_{2,2}(\widehat{\boldsymbol{\Sigma}}) = \frac{n}{2}(\widehat{\boldsymbol{\Sigma}}^{-1} \otimes \widehat{\boldsymbol{\Sigma}}^{-1}) - na_2\mathrm{vec}\,\widehat{\boldsymbol{\Sigma}}^{-1}\mathrm{vec}'\widehat{\boldsymbol{\Sigma}}^{-1},$$

a_2 is given in (12.18).

Using properties of vec -operator the Wald's statistic can be simplified

$$\begin{aligned}WST(\mathscr{X}, \boldsymbol{\Sigma}_0) &= \mathrm{vec}'(\widehat{\boldsymbol{\Sigma}} - \boldsymbol{\Sigma}_0)\Big[\frac{n}{2}\mathrm{vec}\,(\widehat{\boldsymbol{\Sigma}}^{-1}(\widehat{\boldsymbol{\Sigma}} - \boldsymbol{\Sigma}_0)\widehat{\boldsymbol{\Sigma}}^{-1}) \\ &\quad - na_2\mathrm{vec}\,\widehat{\boldsymbol{\Sigma}}^{-1}\mathrm{tr}(\widehat{\boldsymbol{\Sigma}}^{-1}(\widehat{\boldsymbol{\Sigma}} - \boldsymbol{\Sigma}_0))\Big] \\ &= \frac{n}{2}\Big[\mathrm{tr}(\widehat{\boldsymbol{\Sigma}}^{-1}(\widehat{\boldsymbol{\Sigma}} - \boldsymbol{\Sigma}_0))^2 - 2a_2(\mathrm{tr}(\widehat{\boldsymbol{\Sigma}}^{-1}(\widehat{\boldsymbol{\Sigma}} - \boldsymbol{\Sigma}_0)))^2\Big].\end{aligned}$$

When we open the brackets around $\widehat{\boldsymbol{\Sigma}} - \boldsymbol{\Sigma}_0$ the statistic takes the following form:

$$WST(\mathscr{X}, \boldsymbol{\Sigma}_0) = \frac{n}{2}\Big(\mathrm{tr}(\boldsymbol{\Sigma}_0\widehat{\boldsymbol{\Sigma}}^{-1})^2 - 2a_2\big(\mathrm{tr}(\boldsymbol{\Sigma}_0\widehat{\boldsymbol{\Sigma}}^{-1})\big)^2 + (1-2a_2p)(p-2\mathrm{tr}(\boldsymbol{\Sigma}_0\widehat{\boldsymbol{\Sigma}}^{-1}))\Big).$$

The statement is proved. □

Remark When $\nu \to \infty$, coefficient $a_2 \to 0$ and WST statistic tends to the corresponding expression for the normal population [9].

Corollary 12.8 *Wald's score test statistic for testing* $\mathrm{H}_{02} : \boldsymbol{\Sigma} = \mathbf{I}_p$ *when no constraints are imposed on* $\boldsymbol{\mu}$ *has the following form*

$$WST(\mathscr{X}, \mathbf{I}_p) = \frac{n}{2}\Big(\mathrm{tr}(\widehat{\boldsymbol{\Sigma}}^{-1})^2 - 2a_2\big(\mathrm{tr}\widehat{\boldsymbol{\Sigma}}^{-1}\big)^2 + (1-2a_2p)(p-2\mathrm{tr}\widehat{\boldsymbol{\Sigma}}^{-1}\Big), \tag{12.25}$$

where $\mathscr{X} = (\mathbf{x}_1, \ldots, \mathbf{x}_n)$ *is a sample from* $t_{\nu,p}(\boldsymbol{\mu}, \boldsymbol{\Sigma})$ *and* $\widehat{\boldsymbol{\Sigma}}^{-1} = \frac{\nu}{\nu-2}\mathbf{S}^{-1}$.

Proof The expression of WST in (12.25) follows directly from (12.24). □

Corollary 12.9 *Wald's score test statistic for testing* $\mathrm{H}_{03} : \boldsymbol{\Sigma} = \gamma_0 \mathbf{I}_p$, $\gamma_0 > 0$ *is a known constant, when no constraints are imposed on* $\boldsymbol{\mu}$, *has the following form*

$$WST(\mathscr{X}, \gamma \mathbf{I}_p) = \frac{n\gamma_0}{2}\Big(\gamma_0 \mathrm{tr}(\widehat{\boldsymbol{\Sigma}}^{-1})^2 - 2a_2\gamma_0\big(\mathrm{tr}\widehat{\boldsymbol{\Sigma}}^{-1}\big)^2 + (1 - 2a_2 p)(\frac{p}{\gamma_0} - 2\mathrm{tr}\widehat{\boldsymbol{\Sigma}}^{-1})\Big), \tag{12.26}$$

where $\mathscr{X} = (\mathbf{x}_1, \ldots, \mathbf{x}_n)$ *is a sample from* $t_{\nu,p}(\boldsymbol{\mu}, \boldsymbol{\Sigma})$ *and* $\widehat{\boldsymbol{\Sigma}}^{-1} = \frac{\nu}{\nu-2}\mathbf{S}^{-1}$.

Proof The expression of WST in (12.26) follows directly from (12.24) when substituting $\boldsymbol{\Sigma}_0$ by $\gamma_0 \mathbf{I}_p$. □

12.6 Summary and Discussion

We have derived test statistics for classical tests about covariance structure for the $t_{\nu,p}$-distributed population. In Sect. 12.3 we derived the likelihood ratio test statistic LRT, Rao's score test statistic RST in Sect. 12.4 and Wald's score test statistic WST in Sect. 12.5 under the null hypothesis $\mathrm{H}_{01} : \boldsymbol{\Sigma} = \boldsymbol{\Sigma}_0$ and got expressions of these statistics in the special cases $\mathrm{H}_{02} : \boldsymbol{\Sigma} = \mathbf{I}_p$ and $\mathrm{H}_{03} : \boldsymbol{\Sigma} = \gamma_0 \mathbf{I}_p$ where $\gamma_0 > 0$ is known. In the derivation of the LRT and RST statistics we used first terms from Taylor expansions and the expressions were obtained as approximations of the statistics. The obtained results are in concordance with the expressions of these test statistics in the normal case. When the number of degrees of freedom $\nu \to \infty$ the $t_{\nu,p}$-distribution tends to the normal distribution $N_p(\boldsymbol{\mu}, \boldsymbol{\Sigma})$ and the expressions of all derived statistics converge to the corresponding expressions of the statistics for normal observations. The used approximation technique can be applied also to other continuous elliptical distributions which have power function in the expression of the density function (Kotz type distributions, for instance).

The behaviour of the derived test statistics will be examined in a later paper. It is necessary to compare speed of convergence of the statistics to the chi-square distribution depending on the ratio of the dimension p and the sample size n. Also the power properties of the test statistics have to be examined for the $t_{\nu,p}$-distributed population. At the same time it will be interesting to see how robust the corresponding normality-based statistics in [9] are, when the data comes from a t-distribution, for instance. It is also necessary to get information about the behaviour of the statistics in an increasing dimension situation when both the sample size n and the dimensionality p grow simultaneously. In a simulation experiment [9] Rao's score test was more consistent that LRT and Wald's score test. It would be interesting to know if Rao's score test still performs better than the other two tests in the case of $t_{\nu,p}$-distributed sample. In the paper the sample size $n \to \infty$, when

$n > p$. A challenging but complicated problem is to test covariance structures in high-dimensional set-up when the number of variables p can be greater than the sample size n.

Acknowledgments The authors are thankful to the Referee for valuable comments which helped to avoid some misprints and improved presentation of the paper. Tõnu Kollo is grateful to Professor Dietrich von Rosen for discussions on different topics of multivariate statistics and inspiring ideas during cooperation over 30 years. T. Kollo and M. Valge are grateful for financial support from institutional target financed project IUT34-5. M. Valge is also thankful for financial support from Estonian Mathematics and Statistics Doctoral School.

References

1. Anderson, T.W.: An Introduction to Multivariate Statistical Analysis, 3rd edn. Wiley, New York (2003)
2. Bai, Z., Jiang, D., Yao, J., Zheng, S.: Corrections to LRT on large-dimensional covariance matrix by RMT. Ann. Statist. **37**, 3822–3840 (2009)
3. Bilodeau, M., Brenner, D.: Theory of Multivariate Statistics. Springer, New York (1999)
4. Casella, G., Berger, R.L.: Statistical Inference. Duxbury Press, Belmont (1990)
5. Gombay, E.: Parametric sequential tests in the presence of nuisance parameters. Theory Stoch. Process. **8**, 106–118 (2002)
6. Harville D.A.: Matrix Algebra from a Statistician's Perspective. Springer, New York (1997)
7. Kollo, T.: A note on patterned matrices with applications in multivariate analysis. Acta Comment. Univ. Tartu. Math. **968**, 17–27 (1994)
8. Kollo, T., von Rosen, D.: Advanced Multivariate Statistics with Matrices. Springer, Dordrecht (2005)
9. Kollo, T., von Rosen, D., Valge, M.: Hypotheses testing on covariance structures: comparison of likelihood ratio test, Rao's score test and Wald's score test. In: Bozeman, J.R., Oliveira, T., Skiadas, C.H. (eds.) Stochastic and Data Analysis Methods and Applications in Statistics and Demography (ISAST), pp. 423–435 (2016)
10. Kotz, S., Nadarajah, S.: Multivariate t Distributions and their Applications. Cambridge University, Cambridge (2004)
11. Magnus, J., Neudecker, H.: Matrix Differential Calculus with Applications in Statistics and Econometrics, Revised edn. Wiley, New York (1999)
12. Muirhead, R.J.: Aspects of Multivariate Statistical Theory. 2nd edn. Wiley, Hoboken (2005)
13. Rao, C.R.: Large sample tests of statistical hypotheses concerning several parameters with applications to problems of estimation. Math. Proc. Cambridge Philos. Soc. **44**, 50–57 (1948)
14. Rao, C.R.: Linear Statistical Inference and Its Applications. 2nd edn. Wiley, New York (1973)
15. Srivastava, M.S., von Rosen, T., von Rosen, D.: Estimation and testing in general multivariate linear models with Kronecker product covariance structure. Sankhyā Ser. A **71**, 137–163 (2009)
16. Wald, A.: Tests of statistical hypotheses concerning several parameters, when the number of observations is large. Trans. Amer. Math. Soc. **54**, 426–482 (1943)
17. Yuan, K.-H., Bentler, P.M.: On normal theory and associated test statistics in covariance structure analysis under two classes of nonnormal distributions. Statist. Sinica **9**, 831–853 (1999)

Chapter 13
Variable Selection in Joint Mean and Covariance Models

Chaofeng Kou and Jianxin Pan

Abstract In this paper, we propose a penalized maximum likelihood method for variable selection in joint mean and covariance models for longitudinal data. Under certain regularity conditions, we establish the consistency and asymptotic normality of the penalized maximum likelihood estimators of parameters in the models. We further show that the proposed estimation method can correctly identify the true models, as if the true models would be known in advance. We also carry out real data analysis and simulation studies to assess the small sample performance of the new procedure, showing that the proposed variable selection method works satisfactorily.

13.1 Introduction

In longitudinal studies, one of the main objectives is to find out how the average value of the response varies over time and how the average response profile is affected by different treatments or various explanatory variables of interest. Traditionally the within-subject covariance matrices are treated as nuisance parameters or assumed to have a very simple parsimonious structure, which inevitably leads to a misspecification of the covariance structure. Although the misspecification need not affect the consistency of the estimators of the parameters in the mean, it can lead to a great loss of efficiency of the estimators. In some circumstances, for example, when missing data are present, the estimators of the mean parameters can be severely biased if the covariance structure is misspecified. Therefore, correct specification of the covariance structure is really important.

On the other hand, the within-subject covariance structure itself may be of scientific interest, for example, in prediction problems arising in econometrics and finance. Moreover, like the mean, the covariances may be dependent on various explanatory variables. A natural constraint for modelling of covariance structures

C. Kou · J. Pan (✉)
Department of Mathematics, University of Manchester, Manchester, UK
e-mail: ckou@maths.man.ac.uk; jianxin.pan@manchester.ac.uk

© Springer Nature Switzerland AG 2020
T. Holgersson, M. Singull (eds.), *Recent Developments in Multivariate and Random Matrix Analysis*, https://doi.org/10.1007/978-3-030-56773-6_13

is that the estimated covariance matrices must be positive definite, making the covariance modelling rather challenging. Chiu et al. [2] proposed to solve this problem by using a matrix logarithmic transformation, defined as the inverse of the matrix exponential transformation by taking the spectral decomposition of the covariance matrix. Since there are no constraints on the upper triangular elements of the matrix logarithm, any structures of interest may be imposed on the elements of the matrix logarithm. But the limitation of this approach is that the matrix logarithm is lack a clear statistical interpretation. An alternative method to deal with the positive definite constraint of covariance matrices is to work on the modified Cholesky decomposition advocated by Pourahmadi [9, 10], and use regression formulations to model the unconstrained elements in the decomposition. The key idea is that any covariance matrix can be diagonalized by a unique lower triangular matrix with 1's as its diagonal elements. The elements of the lower triangular matrix and the diagonal matrix enjoy a very clear statistical interpretation in terms of autoregressive coefficients and innovation variances, see, e.g., Pan and MacKenzie [8]. Ye and Pan [13] proposed an approach for joint modelling of mean and covariance structures for longitudinal data within the framework of generalized estimation equations, which does not require any distribution assumptions and only assumes the existence of the first four moments of the responses. However, a challenging issue for modelling joint mean and covariance structures is the high-dimensional problem, which arises frequently in many fields such as genomics, gene expression, signal processing, image analysis and finance. For example, the number of explanatory variables may be very large. Intuitively, all these variables should be included in the initial model in order to reduce the modelling bias. But it is very likely that only a small number of these explanatory variables contribute to the model fitting and the majority of them do not. Accordingly, these insignificant variables should be excluded from the initial model to increase prediction accuracy and avoid overfitting problem. Variable selection thus can improve estimation accuracy by effectively identifying the important subset of the explanatory variables, which may be just tens out of several thousands of predictors with a sample size being in tens or hundreds.

There are many variable selection criteria existing in the literature. Traditional variable selection criteria such as Mallow's C_p criteria, Akaike Information Criterion (AIC) and Bayes Information Criterion (BIC) all involve a combinatorial optimization problem, with computational loads increasing exponentially with the number of explanatory variables. This intensive computation problem hampers the use of traditional procedures. Fan and Li [4] discussed a class of penalized likelihood based methods for variable selection, including the bridge regression by Frank and Friedman [6], Lasso by Tibshirani [11] and smoothly clipped absolute deviation by Fan and Li [4]. In the setting of finite parameters, [4] further studied oracle properties for non-concave penalized likelihood estimators in the sense that the penalized maximum likelihood estimator can correctly identify the true model as if we would know it in advance. Fan and Peng [5] extended the results by letting the number of parameters have the order $o(n^{1/3})$ and showed that the oracle properties still hold in this case. Zou [14] proposed an adaptive Lasso in a finite parameter setting and showed that the Lasso does not have oracle properties as conjectured by Fan and Li [4], but the adaptive Lasso does.

In this paper we aim to develop an efficient penalized likelihood based method to select important explanatory variables that make a significant contribution to the joint modelling of mean and covariance structures for longitudinal data. We show that the proposed approach produces good estimation results and can correctly identify zero regression coefficients for the joint mean and covariance models, simultaneously. The rest of the paper is organized as follows. In Sect. 13.2, we first describe a reparameterisation of covariance matrix through the modified Cholesky decomposition and introduce the joint mean and covariance models for longitudinal data. We then propose a variable selection method for the joint models via penalized likelihood function. Asymptotic properties of the resulting estimators are considered. The standard error formula of the parameter estimators and the choice of the tuning parameters are provided. In Sect. 13.3, we study the variable selection method and its sample properties when the number of explanatory variables tends to infinity with the sample size. In Sect. 13.4, we illustrate the proposed method via a real data analysis. In Sect. 13.5, we carry out simulation studies to assess the small sample performance of the method. In Sect. 13.6, we give a further discussion on the proposed variable selection method. Technical details on calculating the penalized likelihood estimators of parameters are given in Appendix A, and theoretical proofs of the theorems that summarize the asymptotic results are presented in Appendix B.

13.2 Variable Selection via Penalized Maximum Likelihood

13.2.1 *Joint Mean and Covariance Models*

Suppose that there are n independent subjects and the ith subject has m_i repeated measurements. Let y_{ij} be the jth measurement of the ith subject and t_{ij} be the time at which the measurement y_{ij} is made. Throughout this paper we assume that $\mathbf{y}_i = (y_{i1}, \dots, y_{im_i})^T$ is a random sample of the ith subject from the multivariate normal distribution with the mean μ_i and covariance matrix Σ_i, where $\mu_i = (\mu_{i1}, \dots, \mu_{im_i})^T$ is an $(m_i \times 1)$ vector and Σ_i is an $(m_i \times m_i)$ positive definite matrix $(i = 1, \dots, n)$. We consider the simultaneous variable selection procedure for the mean and covariance structures using penalized maximum likelihood estimation methods.

To deal with the positive definite constraint of the covariance matrices, we design an effective regularization approach to gain statistical efficiency and overcome the high dimensionality problem in the covariance matrices. We actually use a statistically meaningful representation that reparameterizes the covariance matrices by the modified Cholesky decomposition advocated by Pourahmadi [9, 10]. Specifically, any covariance matrix Σ_i $(1 \leq i \leq n)$ can be diagonalized by a unique lower triangular matrix T_i with 1's as its diagonal elements. In other words,

$$T_i \Sigma_i T_i^T = D_i, \tag{13.1}$$

where D_i is a unique diagonal matrix with positive diagonal elements. The elements of T_i and D_i have a very clear statistical interpretation in terms of autoregressive least square regressions. More precisely, the below-diagonal entries of $T_i = (-\phi_{ijk})$ are the negatives of the regression coefficients of $\widehat{y}_{ij} = \mu_{ij} + \sum_{k=1}^{j-1} \phi_{ijk}(y_{ik} - \mu_{ik})$, the linear least square predictor of y_{ij} based on its predecessors $y_{i1}, \ldots, y_{i(j-1)}$, and the diagonal entries of $D_i = \text{diag}(\sigma_{i1}^2, \ldots, \sigma_{im_i}^2)$ are the prediction error variances $\sigma_{ij}^2 = \text{var}(y_{ij} - \widehat{y}_{ij})$ $(1 \leq i \leq n, 1 \leq j \leq m_i)$. The new parameters ϕ_{ijk}'s and σ_{ij}^2's are called generalized autoregressive parameters and innovation variances, respectively. By taking log transformation to the innovation variances, the decomposition (13.1) converts the constrained entries of $\{\Sigma_i : i = 1, \ldots, n\}$ into two groups of unconstrained autoregressive regression parameters and innovation variances, given by $\{\phi_{ijk} : i = 1, \ldots, n; j = 2, \ldots, m_i; k = 1, \ldots, (j-1)\}$ and $\{\log \sigma_{ij}^2 : i = 1, \ldots, n, j = 1, \ldots, m_i\}$, respectively.

Based on the modified Cholesky decomposition, the unconstrained parameters μ_{ij}, ϕ_{ijk} and $\log \sigma_{ij}^2$ are modelled in terms of the linear regression models

$$\mu_{ij} = \mathbf{x}_{ij}^T \boldsymbol{\beta}, \quad \phi_{ijk} = \mathbf{z}_{ijk}^T \boldsymbol{\gamma} \quad \text{and} \quad \log \sigma_{ij}^2 = \mathbf{h}_{ij}^T \boldsymbol{\lambda}, \tag{13.2}$$

where $\mathbf{x}_{ij}$, $\mathbf{z}_{ijk}$ and $\mathbf{h}_{ij}$ are $(p \times 1)$, $(q \times 1)$ and $(d \times 1)$ covariates vectors, and $\boldsymbol{\beta}$, $\boldsymbol{\gamma}$ and $\boldsymbol{\lambda}$ are the associated regression coefficients. The covariates $\mathbf{x}_{ij}$, $\mathbf{z}_{ijk}$ and $\mathbf{h}_{ij}$ may contain baseline covariates, polynomials in time and their interactions, etc. For example, when modelling stationary growth curve data using polynomials in time, the explanatory variables may take the forms $\mathbf{x}_{ij} = (1, t_{ij}, t_{ij}^2, \ldots, t_{ij}^{p-1})^T$, $\mathbf{z}_{ijk} = (1, (t_{ij} - t_{ik}), (t_{ij} - t_{ik})^2, \ldots, (t_{ij} - t_{ik})^{q-1})^T$ and $\mathbf{h}_{ij} = (1, t_{ij}, t_{ij}^2, \ldots, t_{ij}^{d-1})^T$. An advantage of the model (13.2) is that the resulting estimators of the covariance matrices can be guaranteed to be positive definite. In this paper we assume that the covariates $\mathbf{x}_{ij}$, $\mathbf{z}_{ijk}$ and $\mathbf{h}_{ij}$ may be of high dimension and we would select the important subsets of the covariates $\mathbf{x}_{ij}$, $\mathbf{z}_{ijk}$ and $\mathbf{h}_{ij}$, simultaneously. We first assume all the explanatory variables of interest, and perhaps their interactions as well, are already included into the initial models. We then aim to remove the unnecessary explanatory variables from the models.

13.2.2 Penalized Maximum Likelihood

Many traditional variable selection criteria can be considered as a penalized likelihood which balances modelling biases and estimation variances [4]. Let $\ell(\boldsymbol{\theta})$

denote the log-likelihood function. For the joint mean and covariance models (13.2), we propose the penalized likelihood function

$$Q(\boldsymbol{\theta}) = \ell(\boldsymbol{\theta}) - n\sum_{i=1}^{p} p_{\tau^{(1)}}(|\beta_i|) - n\sum_{j=1}^{q} p_{\tau^{(2)}}(|\gamma_j|) - n\sum_{k=1}^{d} p_{\tau^{(3)}}(|\lambda_k|), \quad (13.3)$$

where $\boldsymbol{\theta} = (\theta_1, \ldots, \theta_s)^T = (\beta_1, \ldots, \beta_p; \gamma_1, \ldots, \gamma_q; \lambda_1, \ldots, \lambda_d)^T$ with $s = p+q+d$ and $p_{\tau^{(l)}}(\cdot)$ is a given penalty function with the tuning parameter $\tau^{(l)}$ $(l = 1, 2, 3)$. Here we use the same penalty function $p(\cdot)$ for all the regression coefficients but with different tuning parameters $\tau^{(1)}$, $\tau^{(2)}$ and $\tau^{(3)}$ for the mean parameters, generalized autoregressive parameters and log-innovation variances, respectively. The function form of $p_\tau(\cdot)$ determines the general behavior of the estimators. Antoniadis [1] defined the hard thresholding rule for variable selection by taking the hard thresholding penalty function as $P_\tau(|t|) = \tau^2 - (|t| - \tau)^2 I(|t| < \tau)$, where $I(.)$ is the indicator function. The penality function $p_\tau(\cdot)$ may also be chosen as L_p penalty. Especially, the use of L_1 penalty, defined by $p_\tau(t) = \tau|t|$, leads to the least absolute shrinkage and selection operator (Lasso) proposed by Tibshirani [11]. Fan and Li [4] suggested using the smoothly clipped absolute deviation (SCAD) penalty function, which is defined by

$$p_\tau(|t|) = \begin{cases} \tau|t| & \text{if } 0 \le |t| < \tau \\ -(|t|^2 - 2a\tau|t| + \tau^2)/\{2(a-1)\} & \text{if } \tau \le |t| < a\tau \\ (a+1)\tau^2/2 & \text{if } |t| \ge a\tau \end{cases} \quad (13.4)$$

for some $a > 2$. This penalty function is continuous, symmetric and convex on $(0, \infty)$ but singular at the origin. It improves the Lasso by avoiding excessive estimation biases. Details of penalty functions can be found in [4].

The penalized maximum likelihood estimator of $\boldsymbol{\theta}$, denoted by $\widehat{\boldsymbol{\theta}}$, maximizes the function $Q(\boldsymbol{\theta})$ in (13.3). With appropriate penalty functions, maximizing $Q(\boldsymbol{\theta})$ with respect to $\boldsymbol{\theta}$ leads to certain parameter estimators vanishing from the initial models so that the corresponding explanatory variables are automatically removed. Hence, through maximizing $Q(\boldsymbol{\theta})$ we achieve the goal of selecting important variables and obtaining the parameter estimators, simultaneously. In Appendix A, we provide the technical details and an algorithm for calculating the penalized maximum likelihood estimator $\widehat{\boldsymbol{\theta}}$.

13.2.3 Asymptotic Properties

In this subsection we consider the consistency and asymptotic normality of the penalized maximum likelihood estimator $\widehat{\boldsymbol{\theta}}$. To emphasize its dependence on the subject number n, we also denote it by $\widehat{\boldsymbol{\theta}}_n$. We assume that the number of the parameters, $s = p + q + d$, is fixed in the first instance. In the next section we will consider the case when s is a variable tending to infinity with n. Denote the true value of $\boldsymbol{\theta}$ by $\boldsymbol{\theta}_0$. Without loss of generality, we assume that $\boldsymbol{\theta}_0 = ((\boldsymbol{\theta}_0^{(1)})^T, (\boldsymbol{\theta}_0^{(2)})^T)^T$ where $\boldsymbol{\theta}_0^{(1)}$ and $\boldsymbol{\theta}_0^{(2)}$ are the nonzero and zero components of $\boldsymbol{\theta}_0$, respectively. Otherwise the components of $\boldsymbol{\theta}_0$ can be reordered. Denote the dimension of $\boldsymbol{\theta}_0^{(1)}$ by s_1. In what follows we first show that the penalized maximum likelihood estimator $\widehat{\boldsymbol{\theta}}_n$ exists and converges to $\boldsymbol{\theta}_0$ at the rate $O_p(n^{-1/2})$, implying that it has the same consistency rate as the ordinary maximum likelihood estimator. We then prove that the $\sqrt{n}$-consistent estimator $\widehat{\boldsymbol{\theta}}_n$ has the asymptotic normal distribution and possesses the oracle property under certain regularity conditions. The results are summarized in the following two theorems and the detailed proofs are provided in Appendix B. To prove the theorems in this paper, we require the following regularity conditions:

(A1) The covariates $\mathbf{x}_{ij}$, $\mathbf{z}_{ijk}$ and $\mathbf{h}_{ij}$ are fixed. Also, for each subject the number of repeated measurements, m_i, is fixed ($i = 1, \ldots, n$; $j = 1, \ldots, m_i$; $k = 1, \ldots, j-1$).
(A2) The parameter space is compact and the true value $\boldsymbol{\theta}_0$ is in the interior of the parameter space.
(A3) The design matrices $\mathbf{x}_i$, $\mathbf{z}_i$ and $\mathbf{h}_i$ in the joint models are all bounded, meaning that all the elements of the matrices are bounded by a single finite real number.
(A4) The dimensions of the parameter vectors $\boldsymbol{\beta}$, $\boldsymbol{\gamma}$, and $\boldsymbol{\lambda}$, that is, p_n, q_n and d_n, have the same order as s_n.
(A5) The nonzero components of the true parameters $\theta_{01}^{(1)}, \ldots, \theta_{0s_1}^{(1)}$ satisfy

$$\min_{1 \le j \le s_1} \left\{ \frac{|\theta_{0j}^{(1)}|}{\tau_n} \right\} \to \infty \ (\text{as } n \to \infty),$$

where τ_n is equal to either $\tau_n^{(1)}$, $\tau_n^{(2)}$ or $\tau_n^{(3)}$, depending on whether $\theta_{0j}^{(1)}$ is a component of $\boldsymbol{\beta}_0$, $\boldsymbol{\gamma}_0$, and $\boldsymbol{\lambda}_0$ ($j = 1, \ldots, s_1$).

Theorem 13.1 *Let*

$$a_n = \max_{1 \le j \le s} \{p'_{\tau_n}(|\theta_{0j}|) : \theta_{0j} \neq 0\} \textit{ and } b_n = \max_{1 \le j \le s} \{|p''_{\tau_n}(|\theta_{0j}|)| : \theta_{0j} \neq 0\},$$

where $\boldsymbol{\theta}_0 = (\theta_{01}, \ldots, \theta_{0s})^T$ is the true value of $\boldsymbol{\theta}$, and τ_n is equal to either $\tau_n^{(1)}$, $\tau_n^{(2)}$ or $\tau_n^{(3)}$, depending on whether θ_{0j} is a component of $\boldsymbol{\beta}_0$, $\boldsymbol{\gamma}_0$ or $\boldsymbol{\lambda}_0$ ($1 \le j \le s$). Assume $a_n = O_p(n^{-1/2})$, $b_n \to 0$ and $\tau_n \to 0$ as $n \to \infty$. Under the conditions (A1)–(A3) above, with probability tending to 1 there must exist a local maximizer $\widehat{\boldsymbol{\theta}}_n$ of the penalized likelihood function $Q(\boldsymbol{\theta})$ in (13.3) such that $\widehat{\boldsymbol{\theta}}_n$ is a $\sqrt{n}$-consistent estimator of $\boldsymbol{\theta}_0$.

We now consider the asymptotic normality property of $\widehat{\theta}_n$. Let

$$A_n = \mathrm{diag}(p''_{\tau_n}(|\theta_{01}^{(1)}|), \ldots, p''_{\tau_n}(|\theta_{0s_1}^{(1)}|)),$$
$$\boldsymbol{c}_n = (p'_{\tau_n}(|\theta_{01}^{(1)}|)\mathrm{sgn}(\theta_{01}^{(1)}), \ldots, p'_{\tau_n}(|\theta_{0s_1}^{(1)}|)\mathrm{sgn}(\theta_{0s_1}^{(1)}))^T,$$

where τ_n has the same definition as that in Theorem 13.1, and $\theta_{0j}^{(1)}$ is the jth component of $\boldsymbol{\theta}_0^{(1)}$ $(1 \leq j \leq s_1)$. Denote the Fisher information matrix of $\boldsymbol{\theta}$ by $\mathscr{I}_n(\boldsymbol{\theta})$.

Theorem 13.2 *Assume that the penalty function $p_{\tau_n}(t)$ satisfies*

$$\liminf_{n\to\infty} \liminf_{t\to 0+} \frac{p'_{\tau_n}(t)}{\tau_n} > 0$$

and $\bar{\mathscr{I}}_n = \mathscr{I}_n(\boldsymbol{\theta}_0)/n$ converges to a finite and positive definite matrix $\mathscr{I}(\boldsymbol{\theta}_0)$ as $n \to \infty$. Under the same mild conditions as these given in Theorem 13.1, if $\tau_n \to 0$ and $\sqrt{n}\tau_n \to \infty$ as $n \to \infty$, then the $\sqrt{n}$-consistent estimator $\widehat{\boldsymbol{\theta}}_n = (\widehat{\boldsymbol{\theta}}_n^{(1)T}, \widehat{\boldsymbol{\theta}}_n^{(2)T})^T$ in Theorem 13.1 must satisfy $\widehat{\boldsymbol{\theta}}_n^{(2)} = 0$ and

$$\sqrt{n}(\bar{\mathscr{I}}_n^{(1)})^{-1/2}(\bar{\mathscr{I}}_n^{(1)} + A_n)\left\{(\widehat{\boldsymbol{\theta}}_n^{(1)} - \boldsymbol{\theta}_0^{(1)}) + (\bar{\mathscr{I}}_n^{(1)} + A_n)^{-1}\boldsymbol{c}_n\right\} \to \mathscr{N}_{s_1}(\mathbf{0}, I_{s_1})$$

in distribution, where $\bar{\mathscr{I}}_n^{(1)}$ is the $(s_1 \times s_1)$ submatrix of $\bar{\mathscr{I}}_n$ corresponding to the nonzero components $\boldsymbol{\theta}_0^{(1)}$ and I_{s_1} is the $(s_1 \times s_1)$ identity matrix.

Note for the SCAD penalty we can show

$$p'_{\tau_n}(t) = \tau_n\left\{I(t \leq \tau_n) + \frac{(a\tau_n - t)_+}{(a-1)\tau_n} I(t > \tau_n)\right\},$$
$$p''_{\tau_n}(t) = \frac{1}{1-a} I(\tau_n < t \leq a\tau_n)$$

for $t > 0$, where $a > 2$ and $(x)_+ = xI(x > 0)$. Since $\tau_n \to 0$ as $n \to \infty$, we then have $a_n = 0$ and $b_n = 0$ so that $\boldsymbol{c}_n = \mathbf{0}$ and $A_n = \mathbf{0}$ when the sample size n is large enough. It can be verified that in this case the conditions in Theorems 13.1 and 13.2 are all satisfied. Accordingly, we must have

$$\sqrt{n}(\bar{\mathscr{I}}_n^{(1)})^{1/2}(\widehat{\boldsymbol{\theta}}_n^{(1)} - \boldsymbol{\theta}_0^{(1)}) \to \mathscr{N}_{s_1}(\mathbf{0}, I_{s_1})$$

in distribution. This means that the estimator $\widehat{\boldsymbol{\theta}}_n^{(1)}$ shares the same sampling property as if we would know $\boldsymbol{\theta}_0^{(2)} = \mathbf{0}$ in advance. In other words, the penalized maximum likelihood estimator of $\boldsymbol{\theta}$ based on the SCAD penalty can correctly identify the true model as if we would know it in advance. This property is the so-called oracle property by Fan and Li [4]. Similarly, the parameter estimator based on the hard thresholding penalty also possesses the oracle property. For the Lasso penalty, however, the parameter estimator does not have the oracle property. A brief explanation for this is given as follows. Since $p_{\tau_n}(t) = \tau_n t$ for $t > 0$ and then $p'_{\tau_n}(t) = \tau_n$, the assumption of $a_n = O_p(n^{-1/2})$ in Theorem 13.1 implies $\tau_n = O_p(n^{-1/2})$, leading to $\sqrt{n}\tau_n = O_p(1)$. On the other hand, one of the conditions in Theorem 13.2 is $\sqrt{n}\tau_n \to \infty$ as $n \to \infty$, which conflicts the assumption of $\sqrt{n}\tau_n = O_p(1)$. Hence the oracle property cannot be guaranteed in this case.

13.2.4 Standard Error Formula

As a consequence of Theorem 13.2, the asymptotic covariance matrix of $\widehat{\boldsymbol{\theta}}_n^{(1)}$ is

$$\mathrm{Cov}(\widehat{\boldsymbol{\theta}}_n^{(1)}) = \frac{1}{n}(\bar{\mathscr{I}}_n^{(1)} + A_n)^{-1}\bar{\mathscr{I}}_n^{(1)}(\bar{\mathscr{I}}_n^{(1)} + A_n)^{-1} \tag{13.5}$$

so that the asymptotic standard error for $\widehat{\boldsymbol{\theta}}_n^{(1)}$ is straightforward. However, $\bar{\mathscr{I}}_n^{(1)}$ and A_n are evaluated at the true value $\boldsymbol{\theta}_0^{(1)}$, which is unknown. A natural choice is to evaluate $\bar{\mathscr{I}}_n^{(1)}$ and A_n at the estimator $\widehat{\boldsymbol{\theta}}_n^{(1)}$ so that the estimator of the asymptotic covariance matrix of $\widehat{\boldsymbol{\theta}}_n^{(1)}$ is obtained through (13.5).

Corresponding to the partition of $\boldsymbol{\theta}_0$, we assume $\boldsymbol{\theta} = (\boldsymbol{\theta}^{(1)T}, \boldsymbol{\theta}^{(2)T})^T$. Denote

$$\ell'(\boldsymbol{\theta}_0^{(1)}) = \left[\frac{\partial \ell(\boldsymbol{\theta})}{\partial \boldsymbol{\theta}^{(1)}}\right]_{\boldsymbol{\theta}=\boldsymbol{\theta}_0} \quad \text{and} \quad \ell''(\boldsymbol{\theta}_0^{(1)}) = \left[\frac{\partial^2 \ell(\boldsymbol{\theta})}{\partial \boldsymbol{\theta}^{(1)}\partial \boldsymbol{\theta}^{(1)T}}\right]_{\boldsymbol{\theta}=\boldsymbol{\theta}_0},$$

where $\boldsymbol{\theta}_0 = (\boldsymbol{\theta}_0^{(1)T}, \mathbf{0})^T$. Also, let

$$\Sigma_{\tau_n}(\boldsymbol{\theta}_0^{(1)}) = \mathrm{diag}\left\{\frac{p'_{\tau_n^{(1)}}(|\theta_{01}^{(1)}|)}{|\theta_{01}^{(1)}|}, \ldots, \frac{p'_{\tau_n^{(s_1)}}(|\theta_{0s_1}^{(1)}|)}{|\theta_{0s_1}^{(1)}|}\right\}.$$

Using the observed information matrix to approximate the Fisher information matrix, the covariance matrix of $\widehat{\boldsymbol{\theta}}_n^{(1)}$ can be estimated through

$$\widehat{\mathrm{Cov}}(\widehat{\boldsymbol{\theta}}_n^{(1)}) = \left\{\ell''(\widehat{\boldsymbol{\theta}}_n^{(1)}) - n\Sigma_{\tau_n}(\widehat{\boldsymbol{\theta}}_n^{(1)})\right\}^{-1}\widehat{\mathrm{Cov}}\left\{\ell'(\widehat{\boldsymbol{\theta}}_n^{(1)})\right\}\left\{\ell''(\widehat{\boldsymbol{\theta}}_n^{(1)}) - n\Sigma_{\tau_n}(\widehat{\boldsymbol{\theta}}_n^{(1)})\right\}^{-1},$$

where $\widehat{\mathrm{Cov}}\{\ell'(\widehat{\boldsymbol{\theta}}_n^{(1)})\}$ is the covariance of $\ell'(\boldsymbol{\theta}^{(1)})$ evaluated at $\boldsymbol{\theta}^{(1)} = \widehat{\boldsymbol{\theta}}_n^{(1)}$.

13.2.5 Choosing the Tuning Parameters

The penalty function $p_{\tau^{(l)}}(\cdot)$ involves the tuning parameter $\tau^{(l)}$ $(l = 1, 2, 3)$ that controls the amount of penalty. We may use K-fold cross-validation or generalized cross-validation [4, 11] to choose the most appropriate tuning parameters τ's. For the purpose of fast computation, we prefer the K-fold cross-validation approach, which is described briefly as follows. First, we randomly split the full dataset $\mathscr{D}$ into K subsets which are of about the same sample size, denoted by $\mathscr{D}^v$ $(v = 1, \ldots, K)$. For each v, we use the data in $\mathscr{D} - \mathscr{D}^v$ to estimate the parameters and $\mathscr{D}^v$ to validate the model. We also use the log-likelihood function to measure the performance of the cross-validation method. For each $\boldsymbol{\tau} = (\tau^{(1)}, \tau^{(2)}, \tau^{(3)})^T$, the K-fold likelihood based cross-validation criterion is defined by

$$\mathrm{CV}(\boldsymbol{\tau}) = \frac{1}{K}\sum_{v=1}^{K}\left\{\sum_{i\in I_v}\log(|\widehat{\Sigma}_i^{-v}|) + \sum_{i\in I_v}(\mathbf{y}_i - \mathbf{x}_i\widehat{\boldsymbol{\beta}}^{-v})^T(\widehat{\Sigma}_i^{-v})^{-1}(\mathbf{y}_i - \mathbf{x}_i\widehat{\boldsymbol{\beta}}^{-v})\right\},$$

where I_v is the index set of the data in $\mathscr{D}^v$, and $\widehat{\boldsymbol{\beta}}^{-v}$ and $\widehat{\Sigma}_i^{-v}$ are the estimators of the mean parameter $\boldsymbol{\beta}$ and the covariance matrix Σ_i obtained by using the training dataset $\mathscr{D} - \mathscr{D}^v$. We then choose the most appropriate tuning parameter τ by minimizing $\mathrm{CV}(\tau)$. In general, we may choose the number of data subsets as $K = 5$ or $K = 10$.

13.3 Variable Selection when the Number of Parameters $s = s_n \to \infty$

In the previous section, we assume that the numbers of the parameters $\boldsymbol{\beta}$, $\boldsymbol{\gamma}$, and $\boldsymbol{\lambda}$, i.e., p, q and d and therefore s, are fixed. In some circumstances, it is not uncommon that the number of explanatory variables increase with the sample size. In this section we consider the case where the number of parameters s_n is a variable, which goes to infinity as the sample size n tends to infinity. In what follows, we study the asymptotic properties of the penalized maximum likelihood estimator in this case.

As before, we assume that $\boldsymbol{\theta}_0 = (\boldsymbol{\theta}_0^{(1)T}, \boldsymbol{\theta}_0^{(2)T})^T$ is the true value of $\boldsymbol{\theta}$ where $\boldsymbol{\theta}_0^{(1)}$ and $\boldsymbol{\theta}_0^{(2)}$ are the nonzero and zero components of $\boldsymbol{\theta}_0$, respectively. Also, we denote the dimension of $\boldsymbol{\theta}_0$ by s_n, which increases with the sample size n this time. Similar to the previous section, we first show that there exists a consistent penalized maximum likelihood estimator $\widehat{\boldsymbol{\theta}}_n$ that converges to $\boldsymbol{\theta}_0$ at the rate $O_p(\sqrt{s_n/n})$. We then show that the $\sqrt{n/s_n}$-consistent estimator $\widehat{\boldsymbol{\theta}}_n$ has an asymptotic normal distribution and possesses the oracle property.

Theorem 13.3 *Let*

$$a_n^* = \max_{1\le j\le s_n}\{p'_{\tau_n}(|\theta_{0j}|) : \theta_{0j} \ne 0\} \text{ and } b_n^* = \max_{1\le j\le s_n}\{|p''_{\tau_n}(|\theta_{0j}|)| : \theta_{0j} \ne 0\},$$

where $\boldsymbol{\theta}_0 = (\theta_{01}, \ldots, \theta_{0s_n})^T$ is the true value of $\boldsymbol{\theta}$, and τ_n is equal to either $\tau_n^{(1)}$, $\tau_n^{(2)}$ or $\tau_n^{(3)}$, depending on whether θ_{0j} is a component of β_0, γ_0 or λ_0 $(1 \le j \le s)$. Assume $a_n^* = O_p(n^{-1/2})$, $b_n^* \to 0$, $\tau_n \to 0$ and $s_n^4/n \to 0$ as $n \to \infty$. Under the conditions (A1)–(A5) above, with probability tending to one there exists a local maximizer $\widehat{\boldsymbol{\theta}}_n$ of the penalized likelihood function $Q(\boldsymbol{\theta})$ in (13.3) such that $\widehat{\boldsymbol{\theta}}_n$ is a $\sqrt{n/s_n}$-consistent estimator of $\boldsymbol{\theta}_0$.

In what follows we consider the asymptotic normality property of the estimator $\widehat{\boldsymbol{\theta}}_n$. Denote the number of nonzero components of $\boldsymbol{\theta}_0$ by $s_{1n}(\le s_n)$. Let

$$A_n^* = \text{diag}(p''_{\tau_n}(|\theta_{01}^{(1)}|), \ldots, p''_{\tau_n}(|\theta_{0s_{1n}}^{(1)}|)),$$

$$\boldsymbol{c}_n^* = (p'_{\tau_n}(|\theta_{01}^{(1)}|)\text{sgn}(\theta_{01}^{(1)}), \ldots, p'_{\tau_n}(|\theta_{0s_{1n}}^{(1)}|)\text{sgn}(\theta_{0s_{1n}}^{(1)}))^T,$$

where τ_n is equal to either $\tau_n^{(1)}$, $\tau_n^{(2)}$ or $\tau_n^{(3)}$, depending on whether θ_{0j} is a component of $\boldsymbol{\beta}_0$, $\boldsymbol{\gamma}_0$ or $\boldsymbol{\lambda}_0$ $(1 \le j \le s)$, and $\theta_{0j}^{(1)}$ is the jth component of $\boldsymbol{\theta}_0^{(1)}$ $(1 \le j \le s_{1n})$. Denote the Fisher information matrix of $\boldsymbol{\theta}$ by $\mathscr{I}_n(\boldsymbol{\theta})$.

Theorem 13.4 *Assume that the penalty function $p_{\tau_n}(t)$ satisfies*

$$\liminf_{n\to\infty}\liminf_{t\to 0+}\frac{p'_{\tau_n}(t)}{\tau_n} > 0$$

and $\bar{\mathscr{I}}_n = \mathscr{I}_n(\boldsymbol{\theta}_0)/n$ converges to a finite and positive definite matrix $\mathscr{I}(\boldsymbol{\theta}_0)$ as $n \to \infty$. Under the same mild conditions as these in Theorem 13.3, if $\tau_n \to 0$, $s_n^5/n \to 0$ and $\tau_n\sqrt{n/s_n} \to \infty$ as $n \to \infty$, then the $\sqrt{n/s_n}$-consistent estimator $\widehat{\boldsymbol{\theta}}_n = (\widehat{\boldsymbol{\theta}}_n^{(1)T}, \widehat{\boldsymbol{\theta}}_n^{(2)T})^T$ in Theorem 13.3 must satisfy $\widehat{\boldsymbol{\theta}}_n^{(2)} = 0$ and

$$\sqrt{n}M_n(\bar{\mathscr{I}}_n^{(1)})^{-1/2}(\bar{\mathscr{I}}_n^{(1)} + A_n^*)\left\{(\widehat{\boldsymbol{\theta}}_n^{(1)} - \boldsymbol{\theta}_0^{(1)}) + (\bar{\mathscr{I}}_n^{(1)} + A_n^*)^{-1}\boldsymbol{c}_n^*\right\} \to \mathscr{N}_k(\mathbf{0}, \boldsymbol{G})$$

in distribution, where $\bar{\mathscr{I}}_n^{(1)}$ is the $(s_{1n} \times s_{1n})$ submatrix of $\bar{\mathscr{I}}_n$ corresponding to the nonzero components $\boldsymbol{\theta}_0^{(1)}$, M_n is an $(k \times s_{1n})$ matrix satisfying $M_n M_n^T \to \boldsymbol{G}$ as $n \to \infty$, $\boldsymbol{G}$ is an $(k \times k)$ positive definite matrix and $k(\le s_{1n})$ is a constant.

The technical proofs of Theorems 13.3 and 13.4 are provided in Appendix B. Similar to the finite parameters setting, for the SCAD penalty and hard thresholding penalty functions, it can be verified that the conditions in Theorems 13.3 and 13.4 are all satisfied. In this case, we have

$$\sqrt{n} M_n (\bar{\mathscr{I}}_n^{(1)})^{1/2} (\widehat{\boldsymbol{\theta}}_n^{(1)} - \boldsymbol{\theta}_0^{(1)}) \to \mathscr{N}_k(\mathbf{0}, \boldsymbol{G})$$

in distribution. That means the estimator $\widehat{\boldsymbol{\theta}}_n^{(1)}$ shares the same sampling property as if we would know $\boldsymbol{\theta}_0^{(2)} = 0$ in advance. In other words, the estimation procedures based on the SCAD and hard thresholding penalty have the oracle property. However, the L_1-based penalized maximum likelihood estimator like Lasso does not have this property. Based on Theorem 13.4, similar to the finite parameters case the asymptotic covariance estimator of $\widehat{\boldsymbol{\theta}}^{(1)}$ can also be constructed but the details are omitted.

13.4 Real Data Analysis

In this section, we apply the proposed procedure to the well known CD4+ cell data analysis, of which the data details can be found in [3]. The human immune deficiency virus (HIV) causes AIDS by reducing a person's ability to fight infection. The HIV attacks an immune cell called the CD4+ cell which orchestrates the body's immunoresponse to infectious agents. An uninfected individual usually has around 1100 cells per millilitre of blood. When infected, the CD4+ cells decrease in number with time and an infected person's CD4+ cell number can be used to monitor the disease progression. The data set we analyzed consists of 369 HIV-infected men. Altogether there are 2376 values of CD4+ cell numbers, with several repeated measurements being made for each individual at different times covering a period of approximately eight and a half years.

For this unbalanced longitudinal data set, information from several explanatory variables is recorded, including X_1 =time, X_2 =age, X_3 =smoking habit (the number of packs of cigarettes smoked per day), X_4 =recreational drug use (1, yes; 0, no), X_5 =number of sexual partners, and X_6 =score on center for epidemiological studies of depression scale. The objectives of our analysis are: (a) to identify covariates that really affect the CD4+ cell numbers in the sense that they are statistically significant in either the mean or covariance models, and (b) to estimate the average time course for the HIV-infected men by taking account of measurement errors in the CD4+ cell collection. Ye and Pan [13] analyzed the CD4+ count data with a focus on the second objective and did not include the explanatory variables except the time. Following [13], we propose to use three polynomials in time, one of degree 6 and two cubics, to model the mean μ_{ij}, the generalized autoregressive parameters ϕ_{ijk} and the log-innovation variances $\log \sigma_{ij}^2$. In the meantime, the explanatory variables $X_2, \ldots, X_6$ above and the intercept X_0 are also included in the initial models for the selection purpose. The ordinary maximum likelihood estimation and the penalized maximum likelihood estimation methods using the

Table 13.1 Estimated tuning parameters

Parameters	SCAD	LASSO	Hard-thresholding
$\tau^{(1)}$	0.42	0.01	0.79
$\tau^{(2)}$	0.21	0.01	0.46
$\tau^{(3)}$	0.84	0.04	0.88

Table 13.2 Estimators of the mean parameters $\boldsymbol{\beta}$

Coefficient	MLE	SCAD	LASSO	Hard-thresholding
β_1 (X_0)	776.60(20.96)	776.68(20.31)	775.35 (20.96)	776.60(20.96)
β_2 (X_1)	−209.05(14.24)	−209.10(9.40)	−209.04(14.25)	−209.05(14.24)
β_3 (X_1^2)	−14.47(8.36)	−14.49(8.04)	−14.51(8.37)	−14.47(8.36)
β_4 (X_1^3)	32.68(5.93)	32.74(2.17)	32.72 (5.93)	32.68(5.93)
β_5 (X_1^4)	−1.97(1.05)	−1.97(1.02)	−1.96(1.05)	−1.97(1.05)
β_6 (X_1^5)	−1.84(0.57)	−1.84(0.21)	−1.85(0.55)	−1.84(0.57)
β_7 (X_1^6)	0.25(0.08)	0.26(0.02)	0.26 (0.08)	0.25(0.08)
β_8 (X_2)	0.88(1.34)	0.88(0.007)	0.88 (1.35)	0.88(1.34)
β_9 (X_3)	61.27(5.36)	61.32(6.35)	61.04 (6.30)	61.27(6.36)
β_{10} (X_4)	45.70(18.84)	45.71(18.71)	45.61 (18.84)	45.70(18.84)
β_{11} (X_5)	−3.61(2.09)	−3.60(2.09)	−3.64(2.09)	−3.61(2.09)
β_{12} (X_6)	−2.24(0.80)	−2.30(0.82)	0(−)	−2.24(0.80)

SCAD, Lasso and Hard-thresholding penalty functions are all considered. The unknown tuning parameters $\tau^{(l)}$ $(l = 1, 2, 3)$ of the penalty functions are estimated through using the 5-fold cross-validation principle described in Sect. 13.2.5, and the resulting estimators are summarized in Table 13.1. It is noted that the SCAD penalty function given in (13.4) also involves another parameter a. Here we choose $a = 3.7$ as suggested by Fan and Li [4].

For the mean, generalized autoregressive parameters and log-innovation variances, the estimated regression coefficients and their associated standard errors, in parentheses, by different penalty estimation methods, are presented in Tables 13.2, 13.3, and 13.4. It is noted that in Table 13.3 $\gamma_1, \ldots, \gamma_4$ correspond to the coefficients of the cubic polynomial in time lag, γ_5 is associated with the time-independent covariate X_2, and the other coefficients $\gamma_6, \ldots, \gamma_{13}$ correspond to the time-dependent covariates $X_3, \ldots, X_6$ measured at two different time points, denoted by $(X_{31}, X_{32}), \ldots, (X_{61}, X_{62})$.

From Tables 13.2, 13.3, and 13.4, it is clear that for the mean structure the estimated regression coefficients of the sixth power polynomial in time are statistically significant. For the generalized autoregressive parameters and the innovation variances, the estimated regression coefficients of cubic polynomials in time are significant. This confirms the conclusion drawn by Ye and Pan [13]. Furthermore, Table 13.2 shows that there is little evidence for the association between age and immune response, but the smoking habit and the use of recreational drug have significant positive effects on the CD4+ numbers. In addition, the number of sexual partners seems to have little effect on the immune response, although it shows

Table 13.3 Estimators of the generalized autoregressive parameters $\boldsymbol{\gamma}$

Coefficient	MLE	SCAD	LASSO	Hard-thresholding
γ_1 (X_0)	0.29(0.06)	0.29 (0.02)	0.29 (0.06)	0.29 (0.06)
γ_2 (X_1)	−0.33(0.09)	−0.33(0.02)	−0.33(0.09)	−0.33(0.09)
γ_3 (X_1^2)	0.20(0.04)	0.20 (0.01)	0.20 (0.04)	0.20 (0.04)
γ_4 (X_1^3)	−0.03(0.004)	−0.03(0.002)	−0.03(0.003)	−0.03(0.004)
γ_5 (X_2)	−0.001(0.0008)	0(−)	0(−)	0(−)
γ_6 (X_{31})	−0.01(0.008)	−0.01(0.005)	−0.01(0.007)	−0.01(0.007)
γ_7 (X_{32})	0.007(0.008)	0(−)	0(−)	0(−)
γ_8 (X_{41})	−0.01(0.02)	0.01 (0.06)	0.01 (0.01)	0.01 (0.02)
γ_9 (X_{42})	0.02(0.02)	0.02 (0.07)	0.02 (0.01)	0.02 (0.02)
γ_{10} (X_{51})	0.001(0.002)	0(−)	0(−)	0(−)
γ_{11} (X_{52})	−0.005(0.003)	0(−)	0(−)	0(−)
γ_{12} (X_{61})	0.004(0.0009)	0(−)	0(−)	0(−)
γ_{13} (X_{62})	0.006(0.001)	0(−)	0(−)	0(−)

Table 13.4 Estimators of the log-innovation variance parameters $\boldsymbol{\lambda}$

Coefficient	MLE	SCAD	LASSO	Hard-thresholding
λ_1 (X_0)	11.64(0.07)	11.63 (0.04)	11.63 (0.08)	11.64 (0.07)
λ_2 (X_1)	−0.22(0.03)	−0.22(0.01)	−0.22(0.03)	−0.22(0.03)
λ_3 (X_1^2)	−0.03(0.01)	−0.03(0.04)	−0.03(0.01)	−0.03(0.01)
λ_4 (X_1^3)	−0.02(0.003)	−0.02(0.001)	−0.02(0.004)	−0.02(0.003)
λ_5 (X_2)	−0.005(0.004)	0(−)	0(−)	0(−)
λ_6 (X_3)	0.21(0.02)	0.21 (0.01)	0.21 (0.02)	0.21 (0.02)
λ_7 (X_4)	−0.12(0.07)	−0.12(0.005)	−0.12(0.06)	−0.12(0.07)
λ_8 (X_5)	−0.02(0.008)	−0.02(0.004)	−0.02(0.008)	−0.02(0.009)
λ_9 (X_6)	−0.006(0.003)	0(−)	0(−)	0(−)

some evidence of negative association. Also, there is a negative association between depression symptoms (score) and immune response.

Interestingly, Table 13.3 clearly indicates that except the cubic polynomial in time lag all other covariates do not have significant influences to the generalized autoregressive parameters, implying that the generalized autoregressive parameters are characterized only by the cubic polynomial in time lag. For the log-innovation variances, however, Table 13.4 shows that in addition to the cubic polynomial in time, the smoking habit, the use of recreational drug, and the number of sexual partners do have significant effects, implying that the innovation variances and therefore the within-subject covariances are not homogeneous and are actually dependent on the covariates of interests. Finally, we notice that in this data example the SCAD, Lasso and Hard thresholding penalty based methods perform very similarly in terms of the selected variables.

13.5 Simulation Study

In this section we conduct a simulation study to assess the small sample performance of the proposed procedures. We simulate 100 subjects, each of which has five observations drawn from the multivariate normal distribution $\mathcal{N}_5(\mu_i, \Sigma_i)$, where the mean μ_i and the within-subject covariance matrix Σ_i are formed by the joint models (13.2) in the framework of the modified Cholesky decomposition. We choose the true values of the parameters in the mean, generalized autoregressive parameters and log-innovation variances to be $\boldsymbol{\beta} = (3, 0, 0, -2, 1, 0, 0, 0, 0, -4)^T$, $\boldsymbol{\gamma} = (-4, 0, 0, 2, 0, 0, 0)^T$ and $\boldsymbol{\lambda} = (0, 1, 0, 0, 0, -2, 0)^T$, respectively. We form the mean covariates $\mathbf{x}_{ij} = (x_{ijt})_{t=1}^{10}$ by drawing random samples from the multivariate normal distribution with mean 0 and covariance matrix of AR(1) structure with $\sigma^2 = 1$ and $\rho = 0.5$ ($i = 1, 2, \ldots, 100$; $j = 1, 2, \ldots, 5$). We then form the covariates $\mathbf{z}_{ijk} = (x_{ijt} - x_{ikt})_{t=1}^{7}$ and $\mathbf{h}_{ij} = (x_{ijt})_{t=1}^{7}$ for the generalized autoregressive parameters and the log-innovation variances. Using these values, the mean μ_i and covariance matrix Σ_i are constructed through the modified Cholesky decomposition. The responses $\mathbf{y}_i$ are then drawn from the multivariate normal distribution $\mathcal{N}(\mu_i, \Sigma_i)$ ($i = 1, 2, \ldots, 100$).

In the simulation study, 1000 repetitions of random samples are generated by using the above data generation procedure. For each simulated data set, the proposed estimation procedures for finding out the ordinary maximum likelihood estimators and penalized maximum likelihood estimators with SCAD, Lasso and Hard-thresholding penalty functions are considered. The unknown tuning parameters $\tau^{(l)}, l = 1, 2, 3$ for the penalty functions are chosen by a 5-fold cross-validation criterion in the simulation. For each of these methods, the average of zero coefficients over the 1000 simulated data sets is reported in Table 13.5. Note that 'True' in Table 13.5 means the average number of zero regression coefficients that are correctly estimated as zero, and 'Wrong' depicts the average number of non-zero regression coefficients that are erroneously set to zero. In addition, the non-zero parameter estimators, and their associated standard errors as well, are provided in Table 13.6. From those simulation results, it is clear that the SCAD penalty method outperforms the Lasso and Hard thresholding penalty approaches in the sense of correct variable selection rate, which significantly reduces the model uncertainty and complexity.

Table 13.5 Average number of zero regression coefficients

	SCAD		LASSO		Hard-thresholding	
Parameter	True	Wrong	True	Wrong	True	Wrong
β	5.42	0.00	4.76	0.00	4.92	0.00
γ	4.18	0.06	3.28	0.08	3.55	0.21
λ	4.53	0.00	3.70	0.00	4.06	0.00

Table 13.6 Estimators of non-zero regression coefficients

Coefficient	True value	SCAD	LASSO	Hard-thresholding
β_1	3	3.08(0.95)	3.08(0.95)	3.09(0.93)
β_4	−2	−1.94(0.68)	−1.93(0.63)	−1.95(0.65)
β_5	1	0.95(0.32)	0.96(0.39)	0.97(0.39)
β_{10}	−4	−4.12(1.65)	−4.13(1.74)	−4.14(1.75)
γ_1	−4	−4.13(1.88)	−4.07(2.14)	−4.10(2.14)
γ_4	2	1.77(0.79)	1.71(0.85)	1.75(0.85)
λ_2	1	1.05(0.05)	1.03(0.06)	1.03(0.06)
λ_6	−2	−2.20(0.83)	−2.11(0.81)	−2.11(0.82)

13.6 Discussion

Within the framework of joint modelling of mean and covariance structures for longitudinal data, we proposed a variable selection method based on penalized likelihood approaches. Like the mean, the covariance structures may be dependent on various explanatory variables of interest so that simultaneous variable selection to the mean and covariance structures becomes fundamental to avoid the modelling biases and reduce the model complexities.

We have shown that under mild conditions the proposed penalized maximum likelihood estimators of the parameters in the mean and covariance models are asymptotically consistent and normally distributed. Also, we have shown that the SCAD and Hard thresholding penalty based estimation approaches have the oracle property. In other words, they can correctly identify the true models as if the true models would be known in advance. In contrast, the Lasso penalty based estimation method does not share the oracle property. We also considered the case when the number of explanatory variables goes to infinity with the sample size and obtained similar results to the case with finite number of variables.

The proposed method differs from [7] where they only addressed the issue of variable selection in the mean model without modelling the generalized autoregressive parameters and innovation variances. It is also different from [12] where a different decomposition of the covariance matrix, namely moving average coefficient based model, was employed, and the variable selection issue was discussed under the decomposition but with the number of explanatory variables fixed. In contrast, the proposed models and methods in this paper are more flexible, interpretable and practicable.

Appendix A: Penalized Maximum Likelihood Estimation

Firstly, note the first two derivatives of the log-likelihood function $\ell(\boldsymbol{\theta})$ are continuous. Around a given point $\boldsymbol{\theta}_0$, the log-likelihood function can be approximated by

$$\ell(\boldsymbol{\theta}) \approx \ell(\boldsymbol{\theta}_0) + \left[\frac{\partial \ell(\boldsymbol{\theta}_0)}{\partial \boldsymbol{\theta}}\right]^T (\boldsymbol{\theta} - \boldsymbol{\theta}_0) + \frac{1}{2}(\boldsymbol{\theta} - \boldsymbol{\theta}_0)^T \left[\frac{\partial^2 \ell(\boldsymbol{\theta}_0)}{\partial \boldsymbol{\theta} \partial \boldsymbol{\theta}^T}\right] (\boldsymbol{\theta} - \boldsymbol{\theta}_0).$$

Also, given an initial value t_0 we can approximate the penalty function $p'_r(t)$ by a quadratic function [4]

$$[p_\tau(|t|)]' = p'_\tau(|t|)\text{sgn}(t) \approx \frac{p'_\tau(|t_0|)t}{t_0}, \quad \text{for } t \approx t_0.$$

In other words,

$$p_\tau(|t|) \approx p_\tau(|t_0|) + \frac{1}{2} p'_\tau(|t_0|) \frac{t^2 - t_0^2}{|t_0|}, \quad \text{for } t \approx t_0.$$

Therefore, the penalized likelihood function (13.3) can be locally approximated, apart from a constant term, by

$$Q(\boldsymbol{\theta}) \approx \ell(\boldsymbol{\theta}_0) + \left[\frac{\partial \ell(\boldsymbol{\theta}_0)}{\partial \boldsymbol{\theta}}\right]^T (\boldsymbol{\theta} - \boldsymbol{\theta}_0)$$
$$+ \frac{1}{2}(\boldsymbol{\theta} - \boldsymbol{\theta}_0)^T \left[\frac{\partial^2 \ell(\boldsymbol{\theta}_0)}{\partial \boldsymbol{\theta} \partial \boldsymbol{\theta}^T}\right] (\boldsymbol{\theta} - \boldsymbol{\theta}_0) - \frac{n}{2} \boldsymbol{\theta}^T \Sigma_\tau(\boldsymbol{\theta}_0) \boldsymbol{\theta},$$

where

$$\Sigma_\tau(\boldsymbol{\theta}_0) = \text{diag}\Big\{\frac{p'_{\tau(1)}(|\beta_{01}|)}{|\beta_{01}|}, \ldots, \frac{p'_{\tau(1)}(|\beta_{0p}|)}{|\beta_{0p}|}, \frac{p'_{\tau(2)}(|\gamma_{01}|)}{|\gamma_{01}|}, \ldots, \frac{p'_{\tau(2)}(|\gamma_{0q}|)}{|\gamma_{0q}|},$$
$$\frac{p'_{\tau(3)}(|\lambda_{01}|)}{|\lambda_{01}|}, \ldots, \frac{p'_{\tau(3)}(|\lambda_{0d}|)}{|\lambda_{0d}|}\Big\},$$

where $\boldsymbol{\theta} = (\theta_1, \ldots, \theta_s)^T = (\beta_1, \ldots, \beta_p, \gamma_1, \ldots, \gamma_q, \lambda_1, \ldots, \lambda_d)^T$ and $\boldsymbol{\theta}_0 = (\theta_{01}, \ldots, \theta_{0s})^T = (\beta_{01}, \ldots, \beta_{0p}, \gamma_{01}, \ldots, \gamma_{0q}, \lambda_{01}, \ldots, \lambda_{0d})^T$. Accordingly, the quadratic maximization problem for $Q(\boldsymbol{\theta})$ leads to a solution iterated by

$$\boldsymbol{\theta}_1 \approx \boldsymbol{\theta}_0 + \left\{\frac{\partial^2 \ell(\boldsymbol{\theta}_0)}{\partial \boldsymbol{\theta} \partial \boldsymbol{\theta}^T} - n\Sigma_\tau(\boldsymbol{\theta}_0)\right\}^{-1} \left\{n\Sigma_\tau(\boldsymbol{\theta}_0)\boldsymbol{\theta}_0 - \frac{\partial \ell(\boldsymbol{\theta}_0)}{\partial \boldsymbol{\theta}}\right\}.$$

Secondly, as the data are normally distributed the log-likelihood function $\ell(\boldsymbol{\theta})$ can be written as

$$
\begin{aligned}
-2\ell(\boldsymbol{\theta}) &= \sum_{i=1}^{n} \log|\Sigma_i| + \sum_{i=1}^{n} (\mathbf{y}_i - \mathbf{x}_i\boldsymbol{\beta})^T \Sigma_i^{-1} (\mathbf{y}_i - \mathbf{x}_i\boldsymbol{\beta}), \\
&= \sum_{i=1}^{n} \log|D_i| + \sum_{i=1}^{n} (\boldsymbol{r}_i - \mathbf{z}_i\boldsymbol{\gamma})^T D_i^{-1} (\boldsymbol{r}_i - \mathbf{z}_i\boldsymbol{\gamma}), \\
&= \sum_{i=1}^{n}\sum_{j=1}^{m_i} \log\sigma_{ij}^2 + \sum_{i=1}^{n}\sum_{j=1}^{m_i} \frac{(r_{ij} - \widehat{r}_{ij})^2}{\sigma_{ij}^2},
\end{aligned}
$$

where

$$
\begin{aligned}
\boldsymbol{r}_i &= \mathbf{y}_i - \mathbf{x}_i\boldsymbol{\beta} = (r_{i1}, \ldots, r_{im_i})^T, \\
\widehat{r}_{ij} &= \sum_{k-1}^{j-1} \phi_{ijk} r_{ik}, \quad (j = 2, \ldots, m_i) \\
\mathbf{z}_i &= (\mathbf{z}_{i1}, \ldots, \mathbf{z}_{im_i})^T, \\
\mathbf{z}_{ij} &= \sum_{k=1}^{j-1} r_{ik}\mathbf{z}_{ijk}, \quad (j = 2, \ldots, m_i) \\
\mathbf{x}_i &= (\mathbf{x}_{i1}, \ldots, \mathbf{x}_{im_i})^T, \quad (i = 1, \ldots, n).
\end{aligned}
$$

Therefore, the resulting score functions are

$$
U(\boldsymbol{\theta}) = \frac{\partial \ell(\boldsymbol{\theta})}{\partial \boldsymbol{\theta}} = (U_1^T(\boldsymbol{\beta}), U_2^T(\boldsymbol{\gamma}), U_3^T(\boldsymbol{\lambda}))^T
$$

where

$$
\begin{aligned}
U_1(\boldsymbol{\beta}) &= \sum_{i=1}^{n} \mathbf{x}_i^T \Sigma_i^{-1} (\mathbf{y}_i - \mathbf{x}_i\boldsymbol{\beta}), \\
U_2(\boldsymbol{\gamma}) &= \sum_{i=1}^{n} \mathbf{z}_i^T D_i^{-1} (\boldsymbol{r}_i - \mathbf{z}_i\boldsymbol{\gamma}), \\
U_3(\boldsymbol{\lambda}) &= \frac{1}{2}\sum_{i=1}^{n} \mathbf{h}_i^T D_i^{-1} (\boldsymbol{\varepsilon}_i^2 - \Sigma_i^2),
\end{aligned}
$$

where

$$\begin{aligned}
\mathbf{h}_i &= (\mathbf{h}_{i1}, \ldots, \mathbf{h}_{im_i})^T, \\
\boldsymbol{\varepsilon}_i^2 &= (\varepsilon_{i1}^2, \ldots, \varepsilon_{im_i}^2)^T, \\
\varepsilon_{ij}^2 &= (r_{ij} - \widehat{r}_{ij})^2, \quad (j = 1, \ldots, m_i) \\
\Sigma_i^2 &= (\sigma_{i1}^2, \ldots, \sigma_{im_i}^2)^T.
\end{aligned}$$

According to [13], the Fisher information matrix $\mathscr{I}_n(\boldsymbol{\theta})$ must be block diagonal. In other words, $\mathscr{I}_n(\boldsymbol{\theta}) = \mathrm{diag}(\mathscr{I}_{11}, \mathscr{I}_{22}, \mathscr{I}_{33})$, where

$$\begin{aligned}
\mathscr{I}_{11} &= \sum_{i=1}^{n} \mathbf{x}_i^T \Sigma_i^{-1} \mathbf{x}_i, \\
\mathscr{I}_{22} &= \sum_{i=1}^{n} E(\mathbf{z}_i^T D_i^{-1} \mathbf{z}_i), \\
\mathscr{I}_{33} &= \frac{1}{2} \sum_{i=1}^{n} \mathbf{h}_i^T \mathbf{h}_i.
\end{aligned}$$

By using the Fisher information matrix to approximate the observed information matrix, we obtain the following iteration solution

$$\begin{aligned}
\boldsymbol{\theta}_1 &\approx \boldsymbol{\theta}_0 + \left\{ \frac{\partial^2 \ell(\boldsymbol{\theta}_0)}{\partial \boldsymbol{\theta} \partial \boldsymbol{\theta}^T} - n\Sigma_\tau(\boldsymbol{\theta}_0) \right\}^{-1} \left\{ n\Sigma_\tau(\boldsymbol{\theta}_0)\boldsymbol{\theta}_0 - \frac{\partial \ell(\boldsymbol{\theta}_0)}{\partial \boldsymbol{\theta}} \right\} \\
&\approx \boldsymbol{\theta}_0 + \{\mathscr{I}_n(\boldsymbol{\theta}_0) + n\Sigma_\tau(\boldsymbol{\theta}_0)\}^{-1} \{U(\boldsymbol{\theta}_0) - n\Sigma_\tau(\boldsymbol{\theta}_0)\boldsymbol{\theta}_0\} \\
&= \{\mathscr{I}_n(\boldsymbol{\theta}_0) + n\Sigma_\tau(\boldsymbol{\theta}_0)\}^{-1} \{U(\boldsymbol{\theta}_0) + \mathscr{I}_n(\boldsymbol{\theta}_0)\boldsymbol{\theta}_0\}.
\end{aligned}$$

Since $\mathscr{I}_n(\boldsymbol{\theta})$ is block diagonal, the above iteration solution is equivalent to

$$\begin{aligned}
\boldsymbol{\beta}_1 &= \left\{ \sum_{i=1}^{n} \mathbf{x}_i^T \Sigma_i^{-1} \mathbf{x}_i + n\Sigma_{\tau(1)}(\boldsymbol{\beta}_0) \right\}^{-1} \left\{ \sum_{i=1}^{n} \mathbf{x}_i^T \Sigma_i^{-1} \mathbf{y}_i \right\}, \\
\boldsymbol{\gamma}_1 &= \left\{ \sum_{i=1}^{n} \mathbf{z}_i^T D_i^{-1} \mathbf{z}_i + n\Sigma_{\tau(2)}(\boldsymbol{\gamma}_0) \right\}^{-1} \left\{ \sum_{i=1}^{n} \mathbf{z}_i^T D_i^{-1} \boldsymbol{r}_i \right\}, \\
\boldsymbol{\lambda}_1 &= \left\{ \sum_{i=1}^{n} \mathbf{h}_i^T \mathbf{h}_i + 2n\Sigma_{\tau(3)}(\boldsymbol{\lambda}_0) \right\}^{-1} \left\{ \sum_{i=1}^{n} \mathbf{h}_i^T D_i^{-1} (\boldsymbol{\varepsilon}_i^2 - \Sigma_i^2 + D_i \log \Sigma_i^2) \right\},
\end{aligned}$$

where all the relevant quantities on the right hand side are evaluated at $\boldsymbol{\theta} = \boldsymbol{\theta}_0$, and

$$\Sigma_{\tau^{(1)}}(\boldsymbol{\beta}_0) = \text{diag}\{\frac{p'_{\tau^{(1)}}(|\beta_{01}|)}{|\beta_{01}|}, \ldots, \frac{p'_{\tau^{(1)}}(|\beta_{0p}|)}{|\beta_{0p}|}\},$$

$$\Sigma_{\tau^{(2)}}(\boldsymbol{\gamma}_0) = \text{diag}\{\frac{p'_{\tau^{(2)}}(|\gamma_{01}|)}{|\gamma_{01}|}, \ldots, \frac{p'_{\tau^{(2)}}(|\gamma_{0q}|)}{|\gamma_{0q}|}\},$$

$$\Sigma_{\tau^{(3)}}(\boldsymbol{\lambda}_0) = \text{diag}\{\frac{p'_{\tau^{(3)}}(|\lambda_{01}|)}{|\lambda_{01}|}, \ldots, \frac{p'_{\tau^{(3)}}(|\lambda_{0d}|)}{|\lambda_{0d}|}\}.$$

Finally, the following algorithm summarizes the computation of the penalized maximum likelihood estimators of the parameters in the joint mean and covariance models.

Algorithm

0. Take the ordinary least squares estimators (without penalty) $\boldsymbol{\beta}^{(0)}$, $\boldsymbol{\gamma}^{(0)}$ and $\boldsymbol{\lambda}^{(0)}$ of $\boldsymbol{\beta}$, $\boldsymbol{\gamma}$ and $\boldsymbol{\lambda}$ as their initial values.
1. Given the current values $\{\boldsymbol{\beta}^{(s)}, \boldsymbol{\gamma}^{(s)}, \boldsymbol{\lambda}^{(s)}\}$, update

$$\boldsymbol{r}_i^{(s)} = \mathbf{y}_i - \mathbf{x}_i\boldsymbol{\beta}^{(s)}, \ \phi_{ijk}^{(s)} = \mathbf{z}_{ijk}^T\boldsymbol{\gamma}^{(s)}, \ \log[(\sigma_{ij}^2)^{(s)}] = \mathbf{h}_{ij}^T\boldsymbol{\lambda}^{(s)},$$

and then use the above iteration solutions to update $\boldsymbol{\gamma}$ and $\boldsymbol{\lambda}$ until convergence. Denote the updated results by $\boldsymbol{\gamma}^{(s+1)}$ and $\boldsymbol{\lambda}^{(s+1)}$.
2. For the updated values $\boldsymbol{\gamma}^{(s+1)}$ and $\boldsymbol{\lambda}^{(s+1)}$, form

$$\phi_{ijk}^{(s+1)} = \mathbf{z}_{ijk}^T\boldsymbol{\gamma}^{(s+1)}, \text{ and } \log[(\sigma_{ij}^2)^{(s+1)}] = \mathbf{h}_{ij}^T\boldsymbol{\lambda}^{(s+1)},$$

and construct

$$\Sigma_i^{(s+1)} = (T_i^{(s+1)})^{-1}D_i^{(s+1)}[(T_i^{(s+1)})^T]^{-1}.$$

Then update $\boldsymbol{\beta}$ according to

$$\boldsymbol{\beta}^{(s+1)} = \left\{\sum_{i=1}^n \mathbf{x}_i^T(\Sigma_i^{(s+1)})^{-1}\mathbf{x}_i + n\Sigma_{\tau^{(1)}}(\boldsymbol{\beta}^{(s)})\right\}^{-1}\left\{\sum_{i=1}^n \mathbf{x}_i^T(\Sigma_i^{(s+1)})^{-1}\mathbf{y}_i\right\}.$$

3. Repeat Step 1 and Step 2 above until certain convergence criteria are satisfied. For example, it can be considered as convergence if the L_2-norm of the difference of the parameter vectors between two adjacent iterations is sufficiently small.

Appendix B: Proofs of Theorems

Proof of Theorem 13.1 Note that $p_{\tau_n}(0) = 0$ and $p_{\tau_n}(\cdot) > 0$. Obviously, we have

$$\begin{aligned}
&Q(\boldsymbol{\theta}_0+n^{-1/2}\mathbf{u}) - Q(\boldsymbol{\theta}_0)\\
&\quad \leq [\ell(\boldsymbol{\theta}_0 + n^{-1/2}\mathbf{u}) - \ell(\boldsymbol{\theta}_0)] - n\sum_{j=1}^{s_1}[p_{\tau_n}(|\theta_{0j} + n^{-1/2}u_j|) - p_{\tau_n}(|\theta_{0j}|)]\\
&\quad = K_1 + K_2.
\end{aligned}$$

We consider K_1 first. By using Taylor expansion, we know

$$\begin{aligned}
K_1 &= \ell(\boldsymbol{\theta}_0 + n^{-1/2}\mathbf{u}) - \ell(\boldsymbol{\theta}_0)\\
&= n^{-1/2}\mathbf{u}^T\ell'(\boldsymbol{\theta}_0) + \frac{1}{2}n^{-1}\mathbf{u}^T\ell''(\boldsymbol{\theta}^*)\mathbf{u}\\
&= K_{11} + K_{12},
\end{aligned}$$

where $\boldsymbol{\theta}^*$ lies between $\boldsymbol{\theta}_0$ and $\boldsymbol{\theta}_0 + n^{-1/2}u$. Note the fact that $n^{-1/2}\|\ell'(\boldsymbol{\theta}_0)\| = O_p(1)$. By applying Cauchy-Schwartz inequality, we obtain

$$K_{11} = n^{-1/2}\mathbf{u}^T\ell'(\boldsymbol{\theta}_0) \leq n^{-1/2}\|\ell'(\boldsymbol{\theta}_0)\|\|\mathbf{u}\| = O_p(1).$$

According to Chebyshev's inequality, we know that for any $\varepsilon > 0$,

$$\begin{aligned}
P\left\{\frac{1}{n}\|\ell''(\boldsymbol{\theta}_0) - E\ell''(\boldsymbol{\theta}_0)\| \geq \varepsilon\right\} &\leq \frac{1}{n^2\varepsilon^2}E\left\{\sum_{j=1}^{s}\sum_{l=1}^{s}\left(\frac{\partial^2\ell(\boldsymbol{\theta}_0)}{\partial\theta_j\partial\theta_l} - E\frac{\partial^2\ell(\boldsymbol{\theta}_0)}{\partial\theta_j\partial\theta_l}\right)^2\right\}\\
&\leq \frac{Cs^2}{n\varepsilon^2} = o(1)
\end{aligned}$$

so that $n^{-1}\|\ell''(\boldsymbol{\theta}_0) - E\ell''(\boldsymbol{\theta}_0)\| = o_p(1)$. It then follows directly that

$$\begin{aligned}
K_{12} &= \frac{1}{2}n^{-1}\mathbf{u}^T\ell''(\boldsymbol{\theta}^*)\mathbf{u} = \frac{1}{2}\mathbf{u}^T\{n^{-1}[\ell''(\boldsymbol{\theta}_0) - E\ell''(\boldsymbol{\theta}_0) - \mathscr{I}_n(\boldsymbol{\theta}_0)]\}\mathbf{u}[1 + o_p(1)]\\
&= -\frac{1}{2}\mathbf{u}^T\mathscr{I}(\boldsymbol{\theta}_0)\mathbf{u}[1 + o_p(1)].
\end{aligned}$$

Therefore we conclude that K_{12} dominates K_{11} uniformly in $\|\mathbf{u}\| = C$ if the constant C is sufficiently large.

We then study the term K_2. It follows from Taylor expansion and Cauchy-Schwartz inequality that

$$
\begin{aligned}
K_2 &= -n \sum_{j=1}^{s_1} [p_{\tau_n}(|\theta_{0j} + n^{-1/2} u_j|) - p_{\tau_n}(|\theta_{0j}|)] \\
&= -\sum_{j=1}^{s_1} \{n^{1/2} p'_{\tau_n}(|\theta_{0j}|)\text{sgn}(\theta_{0j}) u_j + \frac{1}{2} p''_{\tau_n}(|\boldsymbol{\theta}_{0j}|) u_j^2 [1 + o_p(1)]\} \\
&\leq \sqrt{s_1} n^{1/2} \|\mathbf{u}\| \max_{1 \leq j \leq s} \{p'_{\tau_n}(|\theta_{0j}|) : \theta_{0j} \neq 0\} + 2\|\mathbf{u}\|^2 \max_{1 \leq j \leq s} \{|p''_{\tau_n}(|\theta_{0j}|)| : \theta_{0j} \neq 0\} \\
&= \sqrt{s_1} n^{1/2} \|\mathbf{u}\| a_n + 2\|\mathbf{u}\|^2 b_n.
\end{aligned}
$$

Since it is assumed that $a_n = O_p(n^{-1/2})$ and $b_n \to 0$, we conclude that K_{12} dominates K_2 if we choose a sufficiently large C. Therefore for any given $\varepsilon > 0$, there exists a large constant C such that

$$
P\left\{ \sup_{\|\mathbf{u}\|=C} Q(\boldsymbol{\theta}_0 + n^{-1/2}\mathbf{u}) < Q(\boldsymbol{\theta}_0) \right\} \geq 1 - \varepsilon,
$$

implying that there exists a local maximizer $\widehat{\boldsymbol{\theta}}_n$ such that $\widehat{\boldsymbol{\theta}}_n$ is a $\sqrt{n}$-consistent estimator of $\boldsymbol{\theta}_0$. The proof of Theorem 13.1 is completed. □

Proof of Theorem 13.2 First, we prove that under the conditions of Theorem 13.2, for any given $\boldsymbol{\theta}^{(1)}$ satisfying $\boldsymbol{\theta}^{(1)} - \boldsymbol{\theta}_0^{(1)} = O_p(n^{-1/2})$ and any constant $C > 0$, we have

$$
Q\{((\boldsymbol{\theta}^{(1)})^T, \mathbf{0}^T)^T\} = \max_{\|\boldsymbol{\theta}^{(2)}\| \leq Cn^{-1/2}} Q\{((\boldsymbol{\theta}^{(1)})^T, (\boldsymbol{\theta}^{(2)})^T)^T\}.
$$

In fact, for any θ_j ($j = s_1 + 1, \ldots, s$), using Taylor's expansion we obtain

$$
\begin{aligned}
\frac{\partial Q(\boldsymbol{\theta})}{\partial \theta_j} &= \frac{\partial \ell(\boldsymbol{\theta})}{\partial \theta_j} - n p'_{\tau_n}(|\theta_j|)\text{sgn}(\theta_j) \\
&= \frac{\partial \ell(\boldsymbol{\theta}_0)}{\partial \theta_j} + \sum_{l=1}^{s} \frac{\partial^2 \ell(\boldsymbol{\theta}^*)}{\partial \theta_j \partial \theta_l} (\theta_l - \theta_{0l}) - n p'_{\tau_n}(|\theta_j|)\text{sgn}(\theta_j)
\end{aligned}
$$

where $\boldsymbol{\theta}^*$ lies between $\boldsymbol{\theta}$ and $\boldsymbol{\theta}_0$. By using the standard argument, we know

$$
\frac{1}{n} \frac{\partial \ell(\boldsymbol{\theta}_0)}{\partial \theta_j} = O_p(n^{-1/2}) \text{ and } \frac{1}{n} \left\{ \frac{\partial^2 \ell(\boldsymbol{\theta}_0)}{\partial \theta_j \partial \theta_l} - E\left(\frac{\partial^2 \ell(\boldsymbol{\theta}_0)}{\partial \theta_j \partial \theta_l} \right) \right\} = o_p(1).
$$

Note $\|\boldsymbol{\theta}-\boldsymbol{\theta}_0\| = O_p(n^{-1/2})$. We then have

$$\frac{\partial Q(\boldsymbol{\theta})}{\partial \theta_j} = n\tau_n\{-\tau_n^{-1} p'_{\tau_n}(|\theta_j|)\mathrm{sgn}(\theta_j) + O_p(n^{-1/2}\tau_n^{-1})\}.$$

According to the assumption in Theorem 13.2, we obtain

$$\liminf_{n\to\infty}\liminf_{t\to 0+} \frac{p'_{\tau_n}(t)}{\tau_n} > 0 \text{ and } n^{-1/2}\tau_n^{-1} = (\sqrt{n}\tau_n)^{-1} \to 0,$$

so that

$$\frac{\partial Q(\boldsymbol{\theta})}{\partial \theta_j} \begin{cases} < 0, \text{ for } \ 0 < \theta_j < Cn^{-1/2}; \\ > 0, \text{ for } -Cn^{-1/2} < \theta_j < 0. \end{cases}$$

Therefore $Q(\boldsymbol{\theta})$ achieves its maximum at $\boldsymbol{\theta} = ((\boldsymbol{\theta}^{(1)})^T, \mathbf{0}^T)^T$ and the first part of Theorem 13.2 has been proved. □

Second, we discuss the asymptotic normality of $\widehat{\boldsymbol{\theta}}_n^{(1)}$. From Theorem 13.1 and the first part of Theorem 13.2, there exists a penalized maximum likelihood estimator $\widehat{\boldsymbol{\theta}}_n^{(1)}$ that is the $\sqrt{n}$-consistent local maximizer of the function $Q\{((\boldsymbol{\theta}^{(1)})^T, \mathbf{0}^T)^T\}$. The estimator $\widehat{\boldsymbol{\theta}}_n^{(1)}$ must satisfy

$$\begin{aligned} 0 = \left.\frac{\partial Q(\boldsymbol{\theta})}{\partial \theta_j}\right|_{\boldsymbol{\theta}=\binom{\widehat{\boldsymbol{\theta}}_n^{(1)}}{\mathbf{0}}} &= \left.\frac{\partial \ell(\boldsymbol{\theta})}{\partial \theta_j}\right|_{\boldsymbol{\theta}=\binom{\widehat{\boldsymbol{\theta}}_n^{(1)}}{\mathbf{0}}} - np'_{\tau_n}(|\widehat{\theta}_{nj}^{(1)}|)\mathrm{sgn}(\widehat{\theta}_{nj}^{(1)}) \\ &= \frac{\partial \ell(\boldsymbol{\theta}_0)}{\partial \theta_j} + \sum_{l=1}^{s_1}\left\{\frac{\partial^2 \ell(\boldsymbol{\theta}_0)}{\partial\theta_j\partial\theta_l} + o_p(1)\right\}(\widehat{\theta}_{nl}^{(1)} - \theta_{0l}^{(1)}) \\ &\quad -np'_{\tau_n}(|\theta_{0j}^{(1)}|)\mathrm{sgn}(\theta_{0j}^{(1)}) - n\{p''_{\tau_n}(|\theta_{0j}^{(1)}|) + o_p(1)\}(\widehat{\theta}_{nj}^{(1)} - \theta_{0j}^{(1)}). \end{aligned}$$

In other words, we have

$$\left\{-\frac{\partial^2 \ell(\boldsymbol{\theta}_0)}{\partial\boldsymbol{\theta}^{(1)}\partial(\boldsymbol{\theta}^{(1)})^T} + nA_n + o_p(1)\right\}(\widehat{\boldsymbol{\theta}}_n^{(1)} - \boldsymbol{\theta}_0^{(1)}) + \mathbf{c}_n = \frac{\partial \ell(\boldsymbol{\theta}_0)}{\partial\boldsymbol{\theta}^{(1)}}.$$

Using the Liapounov form of the multivariate central limit theorem, we obtain

$$\frac{1}{\sqrt{n}}\frac{\partial \ell(\boldsymbol{\theta}_0)}{\partial\boldsymbol{\theta}^{(1)}} \to \mathscr{N}_{s_1}(\mathbf{0}, \mathscr{I}^{(1)})$$

in distribution. Note that

$$\frac{1}{n}\left\{\frac{\partial^2 \ell(\boldsymbol{\theta}_0)}{\partial \boldsymbol{\theta}^{(1)} \partial (\boldsymbol{\theta}^{(1)})^T} - E\left(\frac{\partial^2 \ell(\boldsymbol{\theta}_0)}{\partial \boldsymbol{\theta}^{(1)} \partial (\boldsymbol{\theta}^{(1)})^T}\right)\right\} = o_p(1),$$

it follows immediately by using Slustsky's theorem that

$$\sqrt{n}(\bar{\mathscr{I}}_n^{(1)})^{-1/2}(\bar{\mathscr{I}}_n^{(1)} + A_n)\left\{(\widehat{\boldsymbol{\theta}}_n^{(1)} - \boldsymbol{\theta}_0^{(1)}) + (\bar{\mathscr{I}}_n^{(1)} + A_n)^{-1}\mathbf{c}_n\right\} \to \mathscr{N}_{s_1}(\mathbf{0}, I_{s_1})$$

in distribution. The proof of Theorem 13.2 is complete. □

Proof of Theorem 13.3 Let $\alpha_n = (n/s_n)^{-1/2}$. Note $p_{\tau_n}(0) = 0$ and $p_{\tau_n}(\cdot) > 0$. We then have

$$\begin{aligned} Q(\boldsymbol{\theta}_0 + \alpha_n \mathbf{u}) - Q(\boldsymbol{\theta}_0) &\le [\ell(\boldsymbol{\theta}_0 + \alpha_n \mathbf{u}) - \ell(\boldsymbol{\theta}_0)] - n\sum_{j=1}^{s_{1n}}[p_{\tau_n}(|\theta_{0j} + \alpha_n u_j|) - p_{\tau_n}(|\theta_{0j}|)] \\ &= K_1 + K_2. \end{aligned}$$

Using Taylor's expansion, we obtain

$$\begin{aligned} K_1 &= \ell(\boldsymbol{\theta}_0 + \alpha_n \mathbf{u}) - \ell(\boldsymbol{\theta}_0) \\ &= \alpha_n \mathbf{u}^T \ell'(\boldsymbol{\theta}_0) + \frac{1}{2}\alpha_n^2 \mathbf{u}^T \ell''(\boldsymbol{\theta}_0^*)\mathbf{u} \\ &= K_{11} + K_{12}, \end{aligned}$$

where $\boldsymbol{\theta}_0^*$ lies between $\boldsymbol{\theta}_0$ and $\boldsymbol{\theta}_0 + \alpha_n \mathbf{u}$. Note that $\|\ell'(\boldsymbol{\theta}_0)\| = O_p(\sqrt{ns_n})$. By using Cauchy-Schwartz inequality, we conclude that

$$|K_{11}| = |\alpha_n \mathbf{u}^T \ell'(\boldsymbol{\theta}_0)| \le \alpha_n \|\ell'(\boldsymbol{\theta}_0)\| \|\mathbf{u}\| = O_p(\alpha_n (ns_n)^{1/2})\|\mathbf{u}\| = O_p(n\alpha_n^2)\|\mathbf{u}\|.$$

According to Chebyshev's inequality, for any $\varepsilon > 0$ we have

$$\begin{aligned} P\{\left\|\frac{s_n}{n}\left(\ell''(\boldsymbol{\theta}_0) - E\ell''(\boldsymbol{\theta}_0)\right)\right\| \ge \varepsilon\} &\le \frac{1}{\varepsilon^2}E\left(\left\|\frac{s_n}{n}\left(\ell''(\boldsymbol{\theta}_0) - E\ell''(\boldsymbol{\theta}_0)\right)\right\|^2\right) \\ &= \frac{s_n^2}{n^2\varepsilon^2}E\left\{\sum_{j=1}^{s_n}\sum_{l=1}^{s_n}\left(\frac{\partial^2 \ell(\boldsymbol{\theta}_0)}{\partial \theta_j \partial \theta_l} - E\frac{\partial^2 \ell(\boldsymbol{\theta}_0)}{\partial \theta_j \partial \theta_l}\right)^2\right\} \\ &\le \frac{Cs_n^4}{n\varepsilon^2} = o(1), \end{aligned}$$

which implies that $\frac{s_n}{n}\left\|\ell''(\boldsymbol{\theta}_0)-E\ell''(\boldsymbol{\theta}_0)\right\|=o_p(1)$. It then follows that

$$K_{12}=\frac{1}{2}\alpha_n^2\mathbf{u}^T\ell''(\boldsymbol{\theta}_0^*)\mathbf{u}=\frac{1}{2}n\alpha_n^2\mathbf{u}^T\left\{[\frac{1}{n}(\ell''(\boldsymbol{\theta}_0)-E\ell''(\boldsymbol{\theta}_0))-\mathscr{I}_n(\boldsymbol{\theta}_0)]\right\}\mathbf{u}[1+o_p(1)]$$
$$=-\frac{1}{2}n\alpha_n^2\mathbf{u}^T\mathscr{I}(\boldsymbol{\theta}_0)\mathbf{u}[1+o_p(1)].$$

Therefore we know that K_{12} dominates K_{11} uniformly in $\|\mathbf{u}\|=C$ for a sufficiently large constant C.

We now turn to K_2. It follows from Taylor's expansion that

$$K_2=-n\sum_{j=1}^{s_{1n}}[p_{\tau_n}(|\theta_{0j}+\alpha_n u_j|)-p_{\tau_n}(|\theta_{0j}|)]$$
$$=-\sum_{j=1}^{s_{1n}}\{n\alpha_n p'_{\tau_n}(|\theta_{0j}|)\mathrm{sgn}(\theta_{0j})u_j+\frac{1}{2}n\alpha_n^2 p''_{\tau_n}(|\theta_{0j}|)u_j^2[1+o_p(1)]\}$$
$$\leq\sqrt{s_{1n}}n\alpha_n\|\mathbf{u}\|\max_{1\leq j\leq s_n}\{p'_{\tau_n}(|\theta_{0j}|):\theta_{0j}\neq 0\}$$
$$+2n\alpha_n^2\|\mathbf{u}\|^2\max_{1\leq j\leq s_n}\{|p''_{\tau_n^{(j)}}(|\theta_{0j}|)|:\theta_{0j}\neq 0\}$$
$$\leq\sqrt{s_n}n\alpha_n\|\mathbf{u}\|a_n^*+2n\alpha_n^2\|\mathbf{u}\|^2 b_n^*$$
$$=n\alpha_n^2\|\mathbf{u}\|O_p(1)+2n\alpha_n^2\|\mathbf{u}\|^2 b_n^*.$$

Since $b_n^*\to 0$ as $n\longrightarrow 0$, it is clear that K_{12} dominates K_2 if a sufficiently large constant C is chosen. In other words, for any given $\varepsilon>0$ there exists a large constant C such that

$$P\left\{\sup_{\|\mathbf{u}\|=C}Q(\boldsymbol{\theta}_0+\alpha_n\mathbf{u})<Q(\boldsymbol{\theta}_0)\right\}\geq 1-\varepsilon$$

as long as n is large enough. This implies that there exists a local maximizer $\widehat{\boldsymbol{\theta}}_n$ in the ball $\{\boldsymbol{\theta}_0+\alpha_n\mathbf{u}:\|\mathbf{u}\|\leq C\}$ such that $\widehat{\boldsymbol{\theta}}_n$ is a $\sqrt{n/s_n}$-consistent estimator of $\boldsymbol{\theta}_0$. The proof of Theorem 13.3 is completed. □

Proof of Theorem 13.4 The proof of Theorem 13.4 is similar to that of Theorem 13.2. In what follows we only give a very brief proof. First, it is easy to show that under the conditions of Theorem 13.4, for any given $\boldsymbol{\theta}^{(1)}$ satisfying $\|\boldsymbol{\theta}^{(1)}-\boldsymbol{\theta}_0^{(1)}\|=O_p((n/s_n)^{-1/2})$ and any constant C, the following equality holds

$$Q\{((\boldsymbol{\theta}^{(1)})^T,\mathbf{0}^T)^T\}=\max_{\|\boldsymbol{\theta}^{(2)}\|\leq C(n/s_n)^{-1/2}}Q\{((\boldsymbol{\theta}^{(1)})^T,(\boldsymbol{\theta}^{(2)})^T)^T\}.$$

Based on this fact and Theorem 13.3, there exists an $\sqrt{n/s_n}$-consistent estimator $\widehat{\boldsymbol{\theta}}_n^{(1)}$ that is the local maximizer of $Q\{((\boldsymbol{\theta}^{(1)})^T, \mathbf{0}^T)^T\}$. Let $\bar{\mathscr{I}}_n^{(1)} = \mathscr{I}_n^{(1)}/n$. Similar to the proof of Theorem 13.2, we can show that

$$(\bar{\mathscr{I}}_n^{(1)} + A_n^*)(\widehat{\boldsymbol{\theta}}_n^{(1)} - \boldsymbol{\theta}_0^{(1)}) + \boldsymbol{c}_n^* = \frac{1}{n}\frac{\partial \ell(\boldsymbol{\theta}_0)}{\partial \boldsymbol{\theta}^{(1)}} + o_p(\frac{1}{\sqrt{n}}),$$

so that

$$\sqrt{n}M_n(\bar{\mathscr{I}}_n^{(1)})^{-1/2}(\bar{\mathscr{I}}_n^{(1)} + A_n^*)\{(\widehat{\boldsymbol{\theta}}_n^{(1)} - \boldsymbol{\theta}_0^{(1)}) + (\bar{\mathscr{I}}_n^{(1)} + A_n^*)^{-1}\boldsymbol{c}_n^*\}$$
$$= \frac{1}{\sqrt{n}}M_n(\bar{\mathscr{I}}_n^{(1)})^{-1/2}\frac{\partial \ell(\boldsymbol{\theta}_0)}{\partial \boldsymbol{\theta}^{(1)}} + o_p(M_n(\bar{\mathscr{I}}_n^{(1)})^{-1/2}).$$

By using Lindeberg-Feller central limit theorem, we can show that

$$\frac{1}{\sqrt{n}}M_n(\bar{\mathscr{I}}_n^{(1)})^{-1/2}\frac{\partial \ell(\boldsymbol{\theta}_0)}{\partial \boldsymbol{\theta}^{(1)}}$$

has an asymptotic multivariate normal distribution. The result in Theorem 13.4 follows immediately according to Slustsky's theorem. The proof of Theorem 13.4 is complete. □

Acknowledgments We would like to thank the editors and an anonymous referee for their very constructive comments and suggestions, which makes the paper significantly improved. This research is supported by a research grant from the Royal Society of the UK (R124683).

References

1. Antoniadis, A.: Wavelets in statistics: a review (with discussion). J. Ital. Stat. Assoc. **6**, 97–144 (1997)
2. Chiu, T.Y.M., Leonard, T., Tsui, K.W.: The matrix-logarithm covariance model. J. Am. Stat. Assoc. **91**, 198–210 (1996)
3. Diggle, P.J., Heagerty, P.J., Liang, K.Y., Zeger, S.L.: Analysis of Longitudinal Data, 2nd edn. Oxford University, Oxford (2002)
4. Fan, J., Li, R.: Variable selection via nonconcave penalized likelihood and its oracle properties. J. Am. Stat. Assoc. **96**, 1348–60 (2001)
5. Fan, J., Peng, H.: Nonconcave Penalized likelihood with a diverging number of parameters. Ann. Stat. **32**, 928–61 (2004)
6. Frank, I.E., Friedman, J.H.: A statistical view of some chemometrics regression tools. Technometrics **35**, 109–148 (1993)
7. Lee, J., Kim, S., Jhong, J., Koo, J.: Variable selection and joint estimation of mean and covariance models with an application to eQTL data. Comput. Math. Methods Med. **2018**, 13 (2018)
8. Pan, J., MacKenzie, G.: Model selection for joint mean-covariance structures in longitudinal studies. Biometrika **90**, 239–44 (2003)

9. Pourahmadi, M.: Joint mean-covariance models with applications to lontidinal data: unconstrained parameterisation. Biometrika **86**, 677–90 (1999)
10. Pourahmadi, M.: Maximum likelihood estimation fo generalised linear models for multivariate normal covariance matrix. Biometrika **87**, 425–35 (2000)
11. Tibshirani, R.: Regression shrinkage and selection via the Lasso. J. R. Stat. Soc. B **58**, 267–88 (1996)
12. Xu, D., Zhang, Z., Wu, L.: Joint variable selection of mean-covariance model for longitudinal data. Open J. Stat. **3**, 27–35 (2013)
13. Ye, H.J., Pan, J.: Modelling of covariance structures in generalized estimating equations for longitudinal data. Biometrika **93**, 927–41 (2006)
14. Zou, H.: The adaptive Lasso and its oracle properties. J. Am. Stat. Assoc. **101**, 1418–29 (2006)

Chapter 14
On Shrinkage Estimators and "Effective Degrees of Freedom"

Lynn R. LaMotte, Julia Volaufova, and Simo Puntanen

Abstract Explicit expressions for the estimated mean $\tilde{\mathbf{y}}_k = X\tilde{\boldsymbol{\beta}}_k = H_k\mathbf{y}$ and *effective degrees of freedom* $\nu_k = \mathrm{tr}(H_k)$ by penalized least squares, with penalty $k||D\boldsymbol{\beta}||^2$, can be found readily when $X'X + D'D$ is nonsingular. We establish them here in general under only the condition that X be a non-zero matrix, and we show that the monotonicity properties that are known when $X'X$ is nonsingular also hold in general, but that they are affected by estimability of $D\boldsymbol{\beta}$. We establish the relation between these penalized least squares estimators and least squares under the restriction that $D\boldsymbol{\beta} = \mathbf{0}$.

14.1 Introduction

For an n-vector response $\mathbf{y}$ with mean vector $\boldsymbol{\mu}$, variance-covariance matrix $\sigma^2\mathrm{I}_n$, and a model $X\boldsymbol{\beta}$, the least-squares (OLS) estimate of $\boldsymbol{\mu}$ is

$$\hat{\mathbf{y}} = X\hat{\boldsymbol{\beta}} = X(X'X)^{-}X'\mathbf{y} = \mathbf{P}_X\mathbf{y} = H_0\mathbf{y}. \tag{14.1}$$

X is a fixed, known $n \times p$ matrix, $\boldsymbol{\beta}$ is a p-vector of unknown parameters, and $\hat{\boldsymbol{\beta}}$ satisfies $X\hat{\boldsymbol{\beta}} = H_0\mathbf{y}$. The dimension of the column space of X is $\nu_0 = \mathrm{rank}(X) = \mathrm{tr}(H_0)$: that is, the model provides ν_0 *degrees of freedom* to fit $\boldsymbol{\mu}$ to $\mathbf{y}$. To state the link directly, the dimension of the model space $\mathbf{R}(X) = \mathbf{R}(H_0)$ is its degrees of freedom.

The term *degrees of freedom* (used here as a singular noun) has been in use since the beginnings of applied statistics. It has also been used in linear algebra

L. R. LaMotte (✉) · J. Volaufova
Biostatistics Program, LSU Health-NO, School of Public Health, New Orleans, LA, USA
e-mail: llamot@lsuhsc.edu; jvolau@lsuhsc.edu

S. Puntanen
Faculty of Information Technology and Communication Sciences, Tampere University, Tampere, Finland
e-mail: simo.puntanen@tuni.fi

© Springer Nature Switzerland AG 2020

T. Holgersson, M. Singull (eds.), *Recent Developments in Multivariate and Random Matrix Analysis*, https://doi.org/10.1007/978-3-030-56773-6_14

for the dimension of the solution set for a system of linear equations. Thus $SS = \sum_{i=1}^{n}(y_i - \bar{y})^2$ has $n - 1$ degrees of freedom because the n terms squared, $y_i - \bar{y}$, which satisfy the single equation $\sum_{i=1}^{n}(y_i - \bar{y}) = 0$, can be represented linearly in terms of $n - 1$ independent linear functions of $y_1, \ldots, y_n$.

Degrees of freedom shows up wherever a chi-squared random variable is involved, where it is the number of independent normal random variables squared and summed. Consequently it shows up in Student's t and F statistics. In all these uses, it can be directly defined as the dimension of a linear subspace, and hence also as the trace of an orthogonal projection matrix, that is, of a symmetric idempotent matrix.

Given a matrix D and constant $k \geq 0$, penalties for departures of $D\boldsymbol{\beta}$ from $\mathbf{0}$ can be introduced by regressing $\begin{pmatrix} \mathbf{y} \\ \mathbf{0} \end{pmatrix}$ on $\begin{pmatrix} X \\ \sqrt{k}D \end{pmatrix}$ (see [2]). The penalized least squares estimate of $\boldsymbol{\mu}$ is then

$$\tilde{\mathbf{y}}_k = X\tilde{\boldsymbol{\beta}}_k = X(X'X + kD'D)^{-}X'\mathbf{y} = H_k\mathbf{y}. \tag{14.2}$$

Increasing the *smoothing parameter k* increases the penalty $k\boldsymbol{\beta}'D'D\boldsymbol{\beta}$ and so pushes $D\tilde{\boldsymbol{\beta}}_k$ toward zero. Other possibilities include replacing $\sqrt{k}D$ with $\sqrt{K}D$, with diagonal matrix $\sqrt{K} = \text{Diag}(\sqrt{k_i} \geq 0)$, permitting separate controls in multiple dimensions.

Often $\nu_k = \text{tr}(H_k)$ is called the *effective degrees of freedom* of the penalized model: see [3, p. 153], for example. It is supposed to index, somehow, the complexity and flexibility of the model. Ruppert et al. [6] say that it "has the rough interpretation as the *equivalent number of parameters* and can be calibrated with polynomial fits."

Estimates like (14.2) have been called *shrinkage* estimates, because $||H_k\mathbf{y}||^2 \leq ||H_0\mathbf{y}||^2$. The notion of shrinkage dates back at least to Stein's seminal paper, [7]. A later version, *ridge regression*, received widespread attention (see [1]). It is (14.2) with $D = \text{I}$. In that case, $X'X + k\text{I}$ is nonsingular when $k > 0$, and its shrinkage property is apparent. It can be shown that ridge regression shrinkage is monotone: for $k_2 > k_1$, both $H_{k_1} - H_{k_2}$ and $H_{k_1}^2 - H_{k_2}^2$ are nnd, and hence $\nu_{k_2} < \nu_{k_1}$ and $||\tilde{\mathbf{y}}_{k_2}||^2 < ||\tilde{\mathbf{y}}_{k_1}||^2$. In the limit as $k \to \infty$, $\tilde{\mathbf{y}}_k \to \mathbf{0}$.

Extant accounts ([3, 6], for example) implicitly assume that $X'X + kD'D$ is nonsingular in order to establish these properties of H_k, H_k^2, and ν_k, and their limits. Models widely used in practice entail X and D matrices such that $X'X + D'D$ is singular. Models that include effects of one or more treatment factors often are formulated with dummy, or indicator, variables, in which case the columns of the X matrix are linearly dependent. Further, main effects and interaction effects are formulated correspondingly, and imposing conditions like "no AB interaction effects" takes the form $D\boldsymbol{\beta} = \mathbf{0}$ with non-full-column-rank D. Similarly, differences (first order, second order) can also take the form $D\boldsymbol{\beta}$ with linearly dependent columns of D. Questions of estimability and effects arise when X has less than full column rank. For example, increasing penalties on $D\boldsymbol{\beta}$ in some directions may have no effect at all, as if D or k were zero.

Our objective here is to assess these same properties in general, when $X'X + D'D$ is not necessarily positive definite. We will show that ν_k and H_k are monotone decreasing functions of k, and we will show that $\tilde{\mathbf{y}}_k$ shrinks toward the least-squares estimate under the restriction that $D\tilde{\boldsymbol{\beta}} = \mathbf{0}$. Furthermore, we will show that the dimension of the space over which $\tilde{\mathbf{y}}_k$ ranges is the same for all k, and hence that ν_k has no relation to the dimension of the model space.

14.2 Propositions

Assume throughout that matrices (all real here) are conformable for the operations and expressions in which they appear. For matrices A and B, matrix sum, product, transpose, trace, generalized inverse, Moore-Penrose pseudoinverse, and inverse are denoted by $A+B$, AB, A', $\mathrm{tr}(A)$, A^-, A^+, and A^{-1}, respectively. For a symmetric, non-negative definite (nnd) matrix A, $A^{1/2}$ denotes a symmetric matrix such that $A^{1/2}A^{1/2} = A$. $\mathrm{Diag}(\alpha_i)$ denotes an $r \times r$ diagonal matrix with diagonal entries α_i, $i = 1, \ldots, r$. The orthogonal complement of a set $\mathscr{S}$ of vectors in $\Re^n$ is denoted $\mathscr{S}^\perp$. Denote the column space of M by $\mathbf{R}(M) = \{M\mathbf{x} : \mathbf{x} \in \Re^c\}$ and the null space of M by $\mathbf{N}(M) = \{\mathbf{x} \in \Re^c : M\mathbf{x} = \mathbf{0}\} = \mathbf{R}(M')^\perp$. $\mathbf{P}_A$ denotes the orthogonal projection matrix onto the column space of A.

Let X and D be matrices, both with $p \geq 1$ columns. We shall use D without $\sqrt{k}$ to simplify notation in the following propositions. As the results hold for any D, they hold for $\sqrt{k}D$ too.

Proposition 14.1 *$H = X(X'X + D'D)^-X'$ is invariant to the choice of generalized inverse, and $\mathbf{R}(H) = \mathbf{R}(X)$.*

Proof Clearly $\mathbf{R}(X') \subset \mathbf{R}(X', D') = \mathbf{R}(X'X + D'D)$, so there exists a matrix M such that $X' = (X'X + D'D)M$. Then, with $(X'X + D'D)^\dagger$ an arbitrary generalized inverse of $(X'X + D'D)$,

$$
\begin{aligned}
H &= X(X'X + D'D)^-X' \\
&= M'(X'X + D'D)(X'X + D'D)^-(X'X + D'D)M \\
&= M'(X'X + D'D)M \\
&= M'(X'X + D'D)(X'X + D'D)^\dagger(X'X + D'D)M \\
&= X(X'X + D'D)^\dagger X', \text{ hence the invariance.}
\end{aligned}
$$

It is clear that $\mathbf{R}(H) \subset \mathbf{R}(X)$. Because $H = M'(X'X + D'D)M$ is nnd, $H\mathbf{z} = \mathbf{0} \Longrightarrow (X'X + D'D)M\mathbf{z} = X'\mathbf{z} = \mathbf{0}$, and so it follows that $\mathbf{R}(H)^\perp \subset \mathbf{R}(X)^\perp$, and hence $\mathbf{R}(X) \subset \mathbf{R}(H)$. □

Proposition 14.8 in Sect. 14.4 is a re-statement of Theorem 6.2.3, p. 122, in [5]. Here, with both $A = X'X$ and $B = D'D$ nnd, the condition that

$\mathbf{R}(N'A) \subset \mathbf{R}(N'AN)$ is satisfied. Then there exists a nonsingular matrix T such that $T'X'XT = \Delta_1 = \text{Diag}(\delta_{1i})$ and $T'D'DT = \Delta_2 = \text{Diag}(\delta_{2i})$. Note that the diagonal entries of both Δ_1 and Δ_2 are nonnegative. Let $\Delta = \Delta_1 + \Delta_2$. Note that $T\Delta^+T'$ is a generalized inverse of $T'^{-1}\Delta T^{-1} = X'X + D'D$. It follows that

$$\begin{aligned} H &= X(T'^{-1}\Delta T^{-1})^- X' \\ &= XT\Delta^+T'X' \\ &= XT(\Delta_1 + \Delta_2)^+T'X'. \end{aligned}$$

Keep in mind that when we replace D by $\sqrt{k}D$, only Δ_2 is affected, and it is replaced by $k\Delta_2$.

Rearrange the columns of T as $T = (T_{++}, T_{+0}, T_{0+}, T_{00})$ to correspond to column numbers $J_{++} = \{j : \delta_{1j} > 0 \text{ and } \delta_{2j} > 0\}$, $J_{+0} = \{j : \delta_{1j} > 0 \text{ and } \delta_{2j} = 0\}$, $J_{0+} = \{j : \delta_{1j} = 0 \text{ and } \delta_{2j} > 0\}$, and $J_{00} = \{j : \delta_{1j} = 0 \text{ and } \delta_{2j} = 0\}$, respectively. With $X \neq 0$, not both J_{++} and J_{+0} are empty; if $D \neq 0$, not both J_{++} and J_{0+} are empty. All other configurations are possible. With these conventions,

$$T'X'XT = \begin{pmatrix} T'_{++}X'XT_{++} & T'_{++}X'XT_{+0} & T'_{++}X'XT_{0+} & T'_{++}X'XT_{00} \\ T'_{+0}X'XT_{++} & T'_{+0}X'XT_{+0} & T'_{+0}X'XT_{0+} & T'_{+0}X'XT_{00} \\ T'_{0+}X'XT_{++} & T'_{0+}X'XT_{+0} & T'_{0+}X'XT_{0+} & T'_{0+}X'XT_{00} \\ T'_{00}X'XT_{++} & T'_{00}X'XT_{+0} & T'_{00}X'XT_{0+} & T'_{00}X'XT_{00} \end{pmatrix}$$

$$= \begin{pmatrix} \Delta_{1++} & 0 & 0 & 0 \\ 0 & \Delta_{1+0} & 0 & 0 \\ 0 & 0 & 0 & 0 \\ 0 & 0 & 0 & 0 \end{pmatrix}. \tag{14.3}$$

Similar notation will be applied to submatrices of Δ_2, that is,

$$T'D'DT = \Delta_2 = \begin{pmatrix} \Delta_{2++} & 0 & 0 & 0 \\ 0 & 0 & 0 & 0 \\ 0 & 0 & \Delta_{20+} & 0 \\ 0 & 0 & 0 & 0 \end{pmatrix}. \tag{14.4}$$

From (14.3) we get immediately that $XT_{0+} = 0$ and $XT_{00} = 0$. Analogously, $DT_{+0} = 0$ and $DT_{00} = 0$.

Proposition 14.2

$$X(X'X + D'D)^{-}X' = XT(\Delta_1 + \Delta_2)^{+}T'X'$$
$$= XT \begin{pmatrix} (\Delta_{1++} + \Delta_{2++})^{-1} & 0 & 0 & 0 \\ 0 & \Delta_{1+0}^{-1} & 0 & 0 \\ 0 & 0 & \Delta_{20+}^{-1} & 0 \\ 0 & 0 & 0 & 0 \end{pmatrix} T'X'$$
$$= XT_{++}(\Delta_{1++} + \Delta_{2++})^{-1}T'_{++}X'$$
$$+XT_{+0}\Delta_{1+0}^{-1}T'_{+0}X'. \tag{14.5}$$

Proposition 14.3 $\mathbf{N}(D'D) = \mathbf{R}(T_{+0}, T_{00})$.

Proof Note that $(T_{+0}, T_{00})'D'D(T_{+0}, T_{00}) = 0 \Longrightarrow D'D(T_{+0}, T_{00}) = 0 \Longrightarrow \mathbf{R}(T_{+0}, T_{00}) \subset \mathbf{N}(D'D)$.

Let $\mathbf{u} \in \mathbf{N}(D'D)$. Since T is nonsingular, $\mathbf{u} = T\mathbf{v}$, with $\mathbf{v} = T^{-1}\mathbf{u}$; and $D'D\mathbf{u} = D'DT\mathbf{v} = \mathbf{0} \Longrightarrow \mathbf{v}'T'D'DT\mathbf{v} = \mathbf{v}'\Delta_2\mathbf{v} = \mathbf{v}'_{++}\Delta_{2++}\mathbf{v}_{++} + \mathbf{v}'_{0+}\Delta_{20+}\mathbf{v}_{0+} = 0 \Longrightarrow \mathbf{v}_{++} = \mathbf{0}$ and $\mathbf{v}_{0+} = \mathbf{0}$. Therefore $\mathbf{u} = T\mathbf{v} = T(\mathbf{0}', \mathbf{v}'_{+0}, \mathbf{0}', \mathbf{v}'_{00})' = T_{+0}\mathbf{v}_{+0} + T_{00}\mathbf{v}_{00} \in \mathbf{R}(T_{+0}, T_{00})$. □

The linear subspace $\{X\boldsymbol{\beta} : \boldsymbol{\beta} \in \Re^p$ and $D\boldsymbol{\beta} = \mathbf{0}\}$ is the restricted model $X\boldsymbol{\beta}$ under the condition that $D\boldsymbol{\beta} = \mathbf{0}$. The following proposition establishes that it is the same as $\mathbf{R}(XT_{+0})$.

Proposition 14.4 $\{X\boldsymbol{\beta} : \boldsymbol{\beta} \in \Re^p$ *and* $D\boldsymbol{\beta} = \mathbf{0}\} = \mathbf{R}(XT_{+0})$.

Proof If $\boldsymbol{\mu} = X\boldsymbol{\beta}_0$ and $D\boldsymbol{\beta}_0 = \mathbf{0}$ then $\boldsymbol{\beta}_0 \in \mathbf{N}(D'D)$, and so there exist $\mathbf{v}_{+0}$ and $\mathbf{v}_{00}$ such that $\boldsymbol{\beta}_0 = T_{+0}\mathbf{v}_{+0} + T_{00}\mathbf{v}_{00}$, and hence $X\boldsymbol{\beta}_0 = XT_{+0}\mathbf{v}_{+0}$ because $XT_{00} = 0$.

If $\boldsymbol{\mu} \in \mathbf{R}(XT_{+0})$ then $\exists$ $\mathbf{v}_{+0}$ such that $\boldsymbol{\mu} = XT_{+0}\mathbf{v}_{+0}$; and $DT_{+0} = \mathbf{0}$, so $\boldsymbol{\mu} \in \{X\boldsymbol{\beta} : \boldsymbol{\beta} \in \Re^p$ and $D\boldsymbol{\beta} = \mathbf{0}\}$. □

Now replace D by $\sqrt{k}D$, $k \geq 0$, so that

$$H_k = XT_{++}(\Delta_{1++} + k\Delta_{2++})^{-1}T'_{++}X'$$
$$+XT_{+0}\Delta_{1+0}^{-1}T'_{+0}X', \tag{14.6}$$

and

$$\nu_k = \mathrm{tr}(H_k) = p_{+0} + \mathrm{tr}[(\Delta_{1++} + k\Delta_{2++})^{-1}\Delta_{1++}], \tag{14.7}$$

where p_{+0} is the column dimension of T_{+0}.

Proposition 14.5

$$\lim_{k\to\infty} H_k = XT_{+0}\Delta_{1+0}^{-1}T_{+0}'X' = \mathbf{P}_{XT_{+0}} \textit{ and}$$

$$\lim_{k\to\infty} \nu_k = p_{+0}.$$

Proposition 14.6

$$\frac{d}{dk}H_k = -XT_{++}(\Delta_{1++} + k\Delta_{2++})^{-2}\Delta_{2++}T_{++}'X',$$

and

$$\frac{d\nu_k}{dk} = \frac{d}{dk}\mathrm{tr}(H_k) = -\mathrm{tr}[(\Delta_{1++} + k\Delta_{2++})^{-2}\Delta_{1++}\Delta_{2++}].$$

Proposition 14.7 *If $\Delta_{2++} \neq 0$, then for any $k_2 > k_1 \geq 0$, both $H_{k_1} - H_{k_2}$ and $H_{k_1}^2 - H_{k_2}^2$ are non-zero and nnd.*

14.3 Discussion

Some parts of the partition of T can be absent, depending on X and D. If columns of X are linearly independent, then both J_{0+} and J_{00} are void, and so $T = (T_{++}, T_{+0})$. If $D'D$ is pd, then both J_{+0} and J_{00} are void, and $T = (T_{++}, T_{0+})$. In general, whether or not X has full column rank, if $\mathbf{R}(D') \subset \mathbf{R}(X')$, it can be seen that $\delta_{1j} = 0 \Longrightarrow \delta_{2j} = 0$, and so J_{0+} is empty, and $T = (T_{++}, T_{+0}, T_{00})$. Then positive entries in Δ_2 occur only with positive entries of Δ_1. On the other hand, if none of $D\boldsymbol{\beta}$ is estimable, so that $\mathbf{R}(D') \cap \mathbf{R}(X') = \{\mathbf{0}\}$, it can be shown that $H_k = \mathbf{P}_X$ and $\tilde{\mathbf{y}}_k = \hat{\mathbf{y}}$ for all $k \geq 0$; positive entries in Δ_2 occur only with 0 entries in Δ_1. More generally, it can be shown that H_k is affected only by the *estimable part* of D, the part of $\mathbf{R}(D')$ that is contained in $\mathbf{R}(X')$.

Proposition 14.1 establishes that, for any $k \geq 0$, $\mathbf{R}(H_k) = \mathbf{R}(X)$, and hence the set of possibilities for the shrinkage estimator $\tilde{\mathbf{y}}_k$ is the same as the set of possibilities for the ordinary least squares estimator. The dimension of $\mathbf{R}(H_k)$ is $\nu_0 = \mathrm{rank}(X)$ for all k. Only at $k = 0$, when $\nu_k = \nu_0 = \mathrm{rank}(X)$, is ν_k the dimension of any linear subspace that appears in this setting.

The partitioning of T provides a full-column-rank reparametrization of the model $X\boldsymbol{\beta}$ that provides restricted model and full model estimates and sums of squares for the restriction $D\boldsymbol{\beta} = \mathbf{0}$. This follows from the fact that $X(T_{++}, T_{+0})$ has full column rank, $\mathbf{R}(X) = \mathbf{R}[X(T_{++}, T_{+0})]$ is the orthogonal sum of $\mathbf{R}(XT_{++})$ and $\mathbf{R}(XT_{+0})$, and that the latter is the restricted model $\{X\boldsymbol{\beta} : \boldsymbol{\beta} \in \Re^p \text{ and } D\boldsymbol{\beta} = \mathbf{0}\}$. It follows that $\mathbf{y}'\mathbf{P}_{XT_{++}}\mathbf{y}$ is the numerator sum of squares, with p_{++} degrees of freedom, for the F-statistic that tests the testable part of $\mathrm{H}_0 : D\boldsymbol{\beta} = \mathbf{0}$.

The expression (14.6) and Propositions 14.5 and 14.7 establish that $\tilde{\mathbf{y}}_k$ shrinks monotonically toward $\mathbf{P}_{XT_{+0}}\mathbf{y}$, which is the least-squares estimate of $X\boldsymbol{\beta}$ under the restriction $D\boldsymbol{\beta} = \mathbf{0}$, by Proposition 14.4.

Both Propositions 14.6 and 14.7 establish that ν_k is a strictly decreasing function of k if $\Delta_{2++} \neq 0$. At $k = 0$ it is $\nu_0 = \text{rank}(X)$, and it decreases with k to approach its limit from above, which is p_{+0}. Proposition 14.7 establishes the shrinkage property, that the squared norm of the estimate $\tilde{\mathbf{y}}_k$ is also strictly decreasing in k if $\Delta_{2++} \neq 0$.

It has been suggested in some settings (e.g., [3, p. 158]) that the value of k be chosen so that ν_k is some chosen value, say p_*. The expression for the derivative of ν_k with respect to k in Proposition 14.6 can be used in Newton's method to solve the nonlinear equation $\nu_k = p_*$.

The limiting behavior of H_k may seem abrupt: for all $k \geq 0$, $\mathbf{R}(H_k) = \mathbf{R}(X)$ and its rank stays undiminished as ν_0, but it approaches $\mathbf{P}_{XT_{+0}}$, which has rank p_{+0}. However, this reflects a fundamental property of closed convex cones of nnd matrices: in the relative interior (as here with finite k) of such a cone, all matrices have the same rank, and the rank can decrease only at the relative boundary. (See Lemma 2 in [4].) Here, H_k follows a path within the relative interior, and it approaches a matrix on the relative boundary that has lesser rank.

14.4 Construction of *T*

Proposition 14.8 is a slight re-statement of Theorem 6.2.3 in [5, p. 122]. The proof we have included shows in detail how to construct T.

Proposition 14.8 *Let A and B be symmetric $n \times n$ matrices, B nnd, and let columns of N comprise an orthonormal basis of $\mathbf{R}(N) = \mathbf{R}(B)^{\perp} = \{\mathbf{x} \in \Re^n : B\mathbf{x} = \mathbf{0}\}$. Let r denote the rank of B.*

(a) If $\mathbf{R}(N'A) = \mathbf{R}(N'AN)$ then there exists a nonsingular matrix T such that $T'AT = \Delta_1 = \text{Diag}(\delta_{1i})$ and $T'BT = \Delta_2 = \text{Diag}(\delta_{2i})$.

(b) If there exists a nonsingular matrix T such that $T'AT$ and $T'BT$ are both diagonal, then $\mathbf{R}(N'A) = \mathbf{R}(N'AN)$.

Proof *(a)* By spectral decomposition of B, there exists a matrix L such that $L'BL = \text{I}_r$. (With $B = P\Lambda P' = (P_+, P_0)\begin{pmatrix} \Lambda_+ & 0 \\ 0 & 0 \end{pmatrix} P' = P_+\Lambda_+P_+'$, with $P'P = PP' = \text{I}$, let $L = P_+\Lambda_+^{-1/2}$. P_+ has r columns and P_0 has $n-r$ columns.) Let $N = P_0$.

Let $S = ((\text{I} - N(N'AN)^{-}N'A)L, N)$. Then, with M such that $N'A = N'ANM$ because $\mathbf{R}(N'A) = \mathbf{R}(N'AN)$,

$$N'A(\text{I} - N(N'AN)^{-}N'A) = N'A - N'AN(N'AN)^{-}N'ANM) = N'A - N'ANM = 0.$$

Then

$$S'AS = \begin{pmatrix} L'(\mathrm{I} - AN(N'AN)^{-'}N') \\ N' \end{pmatrix} A((\mathrm{I} - N(N'AN)^{-}N'A)L, N),$$

and

$$\begin{aligned} &L'(\mathrm{I} - AN(N'AN)^{-'}N')A(\mathrm{I} - N(N'AN)^{-}N'A)L \\ &\qquad = L'AL - L'AN(N'AN)^{-'}N'AL - L'AN(N'AN)^{-}N'AL \\ &\qquad\quad + L'AN(N'AN)^{-'}N'AN(N'AN)^{-}N'AL \\ &\qquad = L'AL - L'AN(N'AN)^{-}N'AL \end{aligned}$$

because all the terms involving $AN(N'AN)^{-}N'A$ are invariant to the choice of generalized inverse, and $(N'AN)^{-'}$ is a generalized inverse of $N'AN$. Then

$$S'AS = \begin{pmatrix} L'(A - AN(N'AN)^{-}N'A)L & 0 \\ 0 & N'AN \end{pmatrix}.$$

Because $BN = 0$, $B(\mathrm{I} - N(N'AN)^{-}N'A) = B$, so that

$$L'(\mathrm{I} - AN(N'AN)^{-'}N')B((\mathrm{I} - N(N'AN)^{-}N'A)L = L'BL = \mathrm{I}_r.$$

Then

$$S'BS = \begin{pmatrix} L'BL & 0 \\ 0 & 0 \end{pmatrix} = \begin{pmatrix} \mathrm{I}_r & 0 \\ 0 & 0 \end{pmatrix}.$$

Turning back to $S'AS$, the two matrices on the diagonal are symmetric, and so there exist orthonormal matrices M and Q such that

$$\begin{aligned} M'L'(A - AN(N'AN)^{-}N'A)LM &= \Phi_1 \text{ and} \\ Q'N'ANQ &= \Phi_2, \end{aligned}$$

where both Φ_1 and Φ_2 are diagonal. Let

$$T = S\begin{pmatrix} M & 0 \\ 0 & Q \end{pmatrix} = ((\mathrm{I} - N(N'AN)^{-}N'A)LM, NQ).$$

Then

$$T'AT = \begin{pmatrix} M' & 0 \\ 0 & Q' \end{pmatrix} S'AS \begin{pmatrix} M & 0 \\ 0 & Q \end{pmatrix} = \begin{pmatrix} \Phi_1 & 0 \\ 0 & \Phi_2 \end{pmatrix} = \Delta_1,$$

and

$$T'BT = \begin{pmatrix} M' & 0 \\ 0 & Q' \end{pmatrix} S'BS \begin{pmatrix} M & 0 \\ 0 & Q \end{pmatrix} = \begin{pmatrix} M' & 0 \\ 0 & Q' \end{pmatrix} \begin{pmatrix} \mathrm{I}_r & 0 \\ 0 & 0 \end{pmatrix} \begin{pmatrix} M & 0 \\ 0 & Q \end{pmatrix}$$

$$= \begin{pmatrix} M'M & 0 \\ 0 & 0 \end{pmatrix} = \begin{pmatrix} \mathrm{I} & 0 \\ 0 & 0 \end{pmatrix} = \Delta_2.$$

To show that the columns of T are linearly independent, suppose that $T\begin{pmatrix} \mathbf{u} \\ \mathbf{v} \end{pmatrix} = \mathbf{0}$. Then

$$T\begin{pmatrix} \mathbf{u} \\ \mathbf{v} \end{pmatrix} = (\mathrm{I} - N(N'AN)^{-}N'A)LM\mathbf{u} + NQ\mathbf{v} = \mathbf{0}$$

implies that

$$BT\begin{pmatrix} \mathbf{u} \\ \mathbf{v} \end{pmatrix} = B(\mathrm{I} - N(N'AN)^{-}N'A)LM\mathbf{u} + BNQ\mathbf{v} = BLM\mathbf{u} = \mathbf{0}$$

because $BN = 0$. That implies that $\mathbf{u}'M'L'BLM\mathbf{u} = 0 \Longrightarrow \mathbf{u}'\mathbf{u} = 0$ because $M'L'BLM = \mathrm{I}$, hence $\mathbf{u} = \mathbf{0}$. That leaves $NQ\mathbf{v} = \mathbf{0}$, which implies that $\mathbf{v}'Q'N'NQ\mathbf{v} = \mathbf{v}'Q'Q\mathbf{v} = \mathbf{v}'\mathbf{v} = 0$, which implies that $\mathbf{v} = \mathbf{0}$. □

For the proof of (b), let T be a nonsingular matrix such that $T'AT = \Delta_1$ and $T'BT = \Delta_2$, where both Δ_1 and Δ_2 are diagonal. Order columns of $T = (T_+, T_0)$ so that

$$T'BT = \begin{pmatrix} T_+'BT_+ & T_+'BT_0 \\ T_0'BT_+ & T_0'BT_0 \end{pmatrix} = \begin{pmatrix} \Delta_{2+} & 0 \\ 0 & 0 \end{pmatrix}$$

and

$$T'AT = \begin{pmatrix} T_+'AT_+ & T_+'AT_0 \\ T_0'AT_+ & T_0'AT_0 \end{pmatrix} = \begin{pmatrix} \Delta_{1+} & 0 \\ 0 & \Delta_{10} \end{pmatrix}.$$

Also, since B is symmetric and nnd $\Longrightarrow T_0'BT_0 = \mathbf{0} \Longrightarrow BT_0 = 0$. Therefore $\mathbf{R}(T_0) \subset \mathbf{R}(B)^{\perp}$. With T nonsingular, $\mathbf{z} \in \mathbf{R}(B)^{\perp} \Longrightarrow \exists\, \mathbf{x}, \boldsymbol{w}$ such that $B\mathbf{z} = \mathbf{0} = B(T_0\mathbf{x} + T_+\boldsymbol{w}) = BT_+\boldsymbol{w} \Longrightarrow \boldsymbol{w}'T_+'BT_+\boldsymbol{w} = 0 \Longrightarrow \boldsymbol{w}'\Delta_{2+}\boldsymbol{w} = 0 \Longrightarrow \boldsymbol{w} = \mathbf{0} \Longrightarrow \mathbf{z} = T_0\mathbf{x} \in \mathbf{R}(T_0)$. Hence, $\mathbf{R}(T_0) = \mathbf{R}(B)^{\perp}$ and $\mathbf{R}(B) = \mathbf{R}(BT_+)$. Define N to be $T_0(T_0'T_0)^{-1/2}$, so that $N'N = \mathrm{I}$ and $\mathbf{R}(N) = \mathbf{R}(B)^{\perp}$.

Proof *(b)* With A, B, $T = (T_+, T_0)$, and N as defined above, proposition (b) is equivalent to

$$\mathbf{R}(N'AN)^{\perp} = \mathbf{R}(N'A)^{\perp}.$$

Clearly $\mathbf{R}(N'AN) \subset \mathbf{R}(N'A)$, and so it remains to prove that, for any $\mathbf{x} \in \Re^{n-r}$, $N'AN\mathbf{x} = 0 \Longrightarrow AN\mathbf{x} = \mathbf{0}$.

Suppose $N'AN\mathbf{x} = \mathbf{0}$. Then $AN\mathbf{x} \in \mathbf{R}(N)^{\perp} = \mathbf{R}(B) = \mathbf{R}(BT_{+}) \Longrightarrow \exists\ \boldsymbol{w}$ such that

$$AN\mathbf{x} = BT_{+}\boldsymbol{w}.$$

Then

$$\boldsymbol{w}'T_{+}'BT_{+}\boldsymbol{w} = \boldsymbol{w}'T_{+}'AN\mathbf{x} = 0,$$

because $T_{+}'AN = 0$; and this implies that $BT_{+}\boldsymbol{w} = \mathbf{0}$ because B is nnd. Therefore $AN\mathbf{x} = \mathbf{0}$. □

Given A and B, defining T entails the following steps.

(a) Spectrally decompose B to get P_{+}, Λ_{+}, $L = P_{+}\Lambda_{+}^{-1/2}$, $N = P_0$, and $(N'AN)^{-}$. Check whether $\mathbf{R}(N'A) \subset \mathbf{R}(N'AN)$ (one way is whether $(N'AN)(N'AN)^{-}N'A = N'A$).
(b) Spectrally decompose $L'(A - AN(N'AN)^{-}N'A)L$ and $N'AN$ to get M and Q.
(c) Compute

$$T = ((\mathrm{I} - N(N'AN)^{-}N'A)LM, NQ).$$

(d) For $A = X'X$ and $B = D'D$, both nnd, compute $T'AT = \Delta_1$ and $T'BT = \Delta_2$; permute the columns of T to correspond to the sets J_{++}, J_{+0}, J_{0+}, and J_{00} that are non-void.

References

1. Alldredge, J.R., Gilb, N.S.: Ridge regression: an annotated bibliography. Int. Stat. Rev. **44**(3), 355–360 (1976)
2. Allen, D.M.: The relationship between variable selection and data augmentation and a method for prediction. Technometrics **16**(1), 125–127 (1974)
3. Hastie, T., Tibshirani, R., Friedman, J.: The Elements of Statistical Learning: Data Mining, Inference, and Prediction, 2nd edn. Springer Science+Business Media, New York (2009)
4. LaMotte, L.R.: On admissibility and completeness of linear unbiased estimators in a general linear model. J. Am. Stat. Assoc. **72**(358), 438–441 (1977)
5. Rao, C.R., Mitra, S.K.: Generalized Inverse of Matrices and Its Applications. Wiley, New York (2015)
6. Ruppert, D., Wand, M.P., Carroll, R.J.: Semiparametric Regression. Cambridge University Press, Cambridge (2003)
7. Stein, C.: Inadmissibility of the usual estimator for the mean of a multivariate distribution. In: Proc. Third Berkeley Symp. Math. Statist. Prob., vol. 1, pp. 197–206 (1956)

Chapter 15
On Explicit Estimation of the Growth Curve Model with a Block Circular Covariance Structure

Yuli Liang and Deliang Dai

Abstract Estimation of mean parameters in the growth curve model, when the covariance matrix has a block circular Toeplitz structure, is considered. The purpose of this article is to find the appropriate design matrices so that explicit mean estimators can be obtained.

15.1 Introduction

Nowadays it is very common to collect data hierarchically. Hierarchically structured data naturally arise in various applications including sociology, education, biology and life sciences. In particular, for each subject, there may be p variables measured at different sites or positions. The variables may have variations that differ within sites or positions and across subjects, which implies the presence of different block structures in the covariance matrices and the inference should take care concerning this. The area of block patterned covariance matrices has been intensively developed since the 1970s by Olkin [15, 16], Khatri [5], Anderson [1], Arnold [2], Krishnaiah and Lee [7], Arnold [3], among others. Olkin [16] considered a multivariate normal model in which the covariance matrix exhibits circularity in blocks, as an extension of the circularly symmetric model considered by Olkin and Press [17]. Although multivariate normal models with block covariance structures were studied extensively many decades ago, there is a variety of questions still to be addressed. Viana and Olkin [27] considered a statistical model with a block covariance matrix that can be used in medical stuides of paired organs. Roy and coauthors [22, 23] studied different block covariance structures which are present at two-level and three-level multivariate data. Liang et al. [11, 12] studied the problem of estimation

Y. Liang (✉)
Department of Statistics, Örebro Univeristy School of Business, Örebro, Sweden
e-mail: yuli.liang@oru.se

D. Dai
Department of Economics and Statistics, Linnaeus University, Växjö, Sweden
e-mail: deliang.dai@lnu.se

© Springer Nature Switzerland AG 2020

T. Holgersson, M. Singull (eds.), *Recent Developments in Multivariate and Random Matrix Analysis*, https://doi.org/10.1007/978-3-030-56773-6_15

in balanced multilevel models with a block circular symmetric covariance structure and showed that explicit maximum likelihood estimators (MLE) of variance-covariance components exist under certain restrictions on the parameter space and derived sufficient conditions for obtaining explicit and unique estimators.

The growth curve model (GCM), which is also known as generalized multivariate analysis of variance (GMANOVA) or bilinear regression model, was introduced in [18]. It has been studied for a long time and applied in many different areas [29]. As later defined in (15.1) the GCM has two design matrices: the within individual design matrix A takes care of the time dependency within the individuals, whereas the between individual design matrix C takes into account the group effects. For the GCM with an unstructured covariance matrix, von Rosen [28] derived the explicit MLEs. Lee [8] considered both estimation and prediction of GCM with the uniform covariance structure. Ohlson and von Rosen [14] proposed a residual based approach to obtain explicit estimators for a linearly structured covariance matrix in GCM. Rao and Reinsel [20, 21] have shown that under a certain covariance structure (called Rao's simple covariance structure in Sect. 15.3), the unweighted ordinary least estimator (OLS) is also the MLE. Explicit estimators are often meaningful because one can study their basic properties straightforwardly, in particular their sampling distributions, without worrying about convergence problems as in the case of numerical estimation methods. The GCM was later extended by Verbyla and Venables [26] in order to model different growth profiles and it is also called the "sum of profiles" model or extended bilinear regression model or extended growth curve model (EGCM). ECGM has been extensively studied by Kollo and von Rosen [6] in Chapter 4.

The aim of this paper is to extend the patterned covariance model in [12] to also include patterned mean structure, i.e., a bilinear mean structure, and study the explicit mean estimators of the growth curve model with a so-called block circular covariance structure. Our focus is to find the design matrices A and C so that an explicit estimator of B can be obtained. The structure of the paper is organized as follows. Section 15.2 introduces GCM and notation, as well as some results concerning explicit MLEs for mean parameters in multivariate normal models. In Sect. 15.3, the necessary and sufficient condition that the OLS is the BLUE (best linear unbiased estimator) for the classical GCM with a block circular covariance structure is presented. The results concerning EGCM with the sum of two profiles are derived in Sect. 15.4. A simulation study is given in Sect. 15.5 and conclusion and discussion are given in Sect. 15.6.

15.2 Preliminaries

In this section a GCM with block circular covariance structure is introduced and the existence of explicit MLEs for means in multivariate normal models is also presented.

15.2.1 GCM with Block Circular Covariance Structure

A GCM with block circular covariance matrix is defined by

$$Y_{p\times n} = A_{p\times q}B_{q\times k}C_{k\times n} + E, \tag{15.1}$$

where Y is the response matrix of n subjects measured at p time points. A and C are within- and between-subject design matrices with ranks q $(< p)$ and k $(< n)$, respectively. The regression coefficient matrix B is unknown, and the error matrix $E \sim N_{p,n}(0, \Sigma, I_n)$, i.e., $Cov(\text{vec}E) = I_n \otimes \Sigma$, where vec$(\bullet)$ denotes the vectorization of a matrix. In model (15.1) we suppose that the covariance matrix $\Sigma : p \times p$ has a block compound symmetric pattern with a symmetric circular Toeplitz (SC-Toeplitz) matrix in each block, i.e.

$$\Sigma = I_{n_2} \otimes \Sigma^{(1)} + (J_{n_2} - I_{n_2}) \otimes \Sigma^{(2)}, \tag{15.2}$$

where I_{n_2} is the $n_2 \times n_2$ identity matrix, $J_{n_2} = \mathbf{1}_{n_2}\mathbf{1}'_{n_2}$ and $\mathbf{1}_{n_2}$ is the n_2-dimensional vector of ones. The SC-Toeplitz matrix $\Sigma^{(h)} = (\sigma_{ij}^{(h)})$ depends on $[n_1/2] + 1$ parameters, the symbol $[\bullet]$ stands for the integer part, and for $i, j = 1, \ldots, n_1$, $h = 1, 2$,

$$\sigma_{ij}^{(h)} = \begin{cases} \tau_{|j-i|+(h-1)([n_1/2]+1)}, & \text{if } |j-i| \leq [\frac{n_1}{2}], \\ \tau_{n_1-|j-i|+(h-1)([n_1/2]+1)}, & \text{otherwise}, \end{cases} \tag{15.3}$$

and τ_q's are unknown parameters, $q = 0, \ldots, 2[n_1/2] + 1$. For example, when $n_1 = 4$,

$$\Sigma^{(1)} = \begin{pmatrix} \tau_0 & \tau_1 & \tau_2 & \tau_1 \\ \tau_1 & \tau_0 & \tau_1 & \tau_2 \\ \tau_2 & \tau_1 & \tau_0 & \tau_1 \\ \tau_1 & \tau_2 & \tau_1 & \tau_0 \end{pmatrix}, \quad \Sigma^{(2)} = \begin{pmatrix} \tau_3 & \tau_4 & \tau_5 & \tau_4 \\ \tau_4 & \tau_3 & \tau_4 & \tau_5 \\ \tau_5 & \tau_4 & \tau_3 & \tau_4 \\ \tau_4 & \tau_5 & \tau_4 & \tau_3 \end{pmatrix}.$$

15.2.2 Explicit MLEs for Mean Parameters: Zyskind's Condition

Consider the general linear model

$$\mathbf{y} = X\boldsymbol{\beta} + \boldsymbol{\epsilon}, \tag{15.4}$$

where $\mathbf{y} : n \times 1$ is a response vector; matrices $X : n \times m$ and $\boldsymbol{\beta} : m \times 1$ is a vector of unknown parameters; and $\boldsymbol{\epsilon} : n \times 1$ is a vector of random errors. Moreover, we assume that $E(\boldsymbol{\epsilon}) = \mathbf{0}$, $Var(\boldsymbol{\epsilon}) = \Sigma$. Under a normality assumption on $\boldsymbol{\epsilon}$, we have $\mathbf{y} \sim N_n(X\boldsymbol{\beta}, \Sigma)$, where Σ is assumed to be nonsingular.

Let $\widehat{X\boldsymbol{\beta}}$ denote the MLE of $X\boldsymbol{\beta}$. Using the normal equation $X'\Sigma^{-1}X\boldsymbol{\beta} = X'\Sigma^{-1}\mathbf{y}$, we have

$$\widehat{X\boldsymbol{\beta}} = X(X'\Sigma^{-1}X)^{-1}X'\Sigma^{-1}\mathbf{y}. \tag{15.5}$$

For (15.5), several authors have discussed the conditions of loosening dependence on Σ in $\widehat{X\boldsymbol{\beta}}$; for example, see [13, 19] and [31]. If $\widehat{X\boldsymbol{\beta}}$ does not depend on Σ, then $\widehat{X\boldsymbol{\beta}}$ results in an ordinary least square estimator in model (15.4). According to the result in [25], a necessary and sufficient condition for

$$(X'\Sigma^{-1}X)^{-1}X'\Sigma^{-1} = (X'X)^{-1}X'$$

is that there exists a subset of r orthogonal eigenvectors of Σ which form a basis of $\mathscr{C}(X)$, where $r = \text{rank}(X)$ and $\mathscr{C}(\bullet)$ denotes the column vector space. Alternatively, one can state that $\mathscr{C}(X)$ has to be Σ-invariant in order to obtain explicit estimators, i.e., $\boldsymbol{\beta}$ in (15.4) has explicit MLE if and only if $\mathscr{C}(\Sigma X) \subseteq \mathscr{C}(X)$. In the work of [19], an easy condition to check the invariance in practice is $P_X\Sigma = \Sigma P_X$, where $P_X = X(X'X)^{-1}X'$. This is to say the projection matrix P_X commutes with the covariance matrix Σ. We refer to this necessary and sufficient condition as Zyskind's condition.

15.3 Explicit Mean Parameters in the Growth Curve Model

We now present the corresponding Zyskind's condition for the explicit estimator of the matrix B in model (15.1).

Theorem 15.1 *Assuming the covariance structure given in* (15.2)*, the OLS estimator of the matrix B in model* (15.1)*, i.e.,*

$$\widehat{vecB} = (CC' \otimes A'A)^{-1}(C \otimes A')vecY, \tag{15.6}$$

is the BLUE if and only if the matrix A is Σ-invariant or equivalently $P_A\Sigma = \Sigma P_A$.

Proof After vectorizing the model in (15.1) we get

$$\text{vec}Y = (C' \otimes A)\text{vec}B + \text{vec}E,$$

where $\text{vec}E \sim N_{pn}(\mathbf{0}, I_n \otimes \Sigma)$. The condition for the existence of the explicit estimator of $\text{vec}B$ becomes

$$\mathscr{C}\left((I_n \otimes \Sigma)(C' \otimes A)\right) = \mathscr{C}(C' \otimes \Sigma A) \subseteq \mathscr{C}(C' \otimes A), \tag{15.7}$$

i.e., $C' \otimes A$ is $(I_n \otimes \Sigma)$-invariant. Using the result of [6] on page 55, it can be shown that (15.7) holds if and only if $\mathscr{C}(\Sigma A) \subseteq \mathscr{C}(A)$, which implies that vec$B$ has the explicit estimator given in (15.6) if and only if A is Σ-invariant. Equivalently, the Zynkind's condition is given by

$$P_{C' \otimes A}(I_n \otimes \Sigma) = (I_n \otimes \Sigma) P_{C' \otimes A}, \tag{15.8}$$

where $P_{C' \otimes A} = (C' \otimes A)\left[(C \otimes A')(C' \otimes A)\right]^{-1}(C \otimes A')$. Using the result of [6] on page 96, it can be shown that $P_{C' \otimes A} = P_{C'} \otimes P_A$. The condition in (15.8) becomes $P_{C'} \otimes P_A \Sigma = P_{C'} \otimes \Sigma P_A$, which implies the equality $P_A \Sigma = \Sigma P_A$. □

For the GCM, Rao [20] has showed that for certain covariance structures, the MLE of B is identical to the unweighted estimator (OLS) given in (15.6) if and only if Σ belongs to the class of so-called Rao's simple covariance (RSC) structure,

$$\Sigma = A \Gamma A' + Q \Theta Q', \tag{15.9}$$

where Γ and Θ are $q \times q$ and $(p-q) \times (p-q)$ unknown positive definite matrices, and Q is a $p \times (p-q)$ matrix orthogonal to the design matrix A, i.e., $Q'A = 0$. The covariance structure given in (15.9) contains many parsimonious structures. For example, [8, 9] showed that the compound symmetry structure (uniform covariance structure) $\Sigma = \sigma^2\left[(1-\rho)I_p + \rho J_p\right]$ is a special case of the RSC structure, where $\sigma^2 > 0$ and $-1/(p-1) < \rho < 1$, if and only if $A'\mathbf{1}_p\mathbf{1}_p'Q = 0$, which means that $\mathbf{1}_p$ is in the space generated by A or Q. In the next theorem we show that the block circular covariance structure given in (15.2) belongs to the class of the RSC structure.

Theorem 15.2 *The Rao's simple covariance structure* $\Sigma = A\Gamma A' + Q\Theta Q'$ *includes the covariance structure given in* (15.2) *as a special case with*

$$\Gamma = X\left[I_{n_2} \otimes \Sigma^{(1)} + (J_{n_2} - I_{n_2}) \otimes \Sigma^{(2)}\right]X', \tag{15.10}$$

$$\Theta = D\left[I_{n_2} \otimes \Sigma^{(1)} + (J_{n_2} - I_{n_2}) \otimes \Sigma^{(2)}\right]D', \tag{15.11}$$

if and only if $P_A\Sigma = \Sigma P_A$, *where* $X = (A'A)^{-1}A'$ *and* $D = (Q'Q)^{-1}Q'$.

Proof We use similar techniques as in [9] to prove this theorem. With Γ and Θ given in (15.10) and (15.11) respectively, we have

$$\begin{aligned}\Sigma = {} & AX\left[I_{n_2} \otimes \Sigma^{(1)} + (J_{n_2} - I_{n_2}) \otimes \Sigma^{(2)}\right]X'A' \\ & + QD\left[I_{n_2} \otimes \Sigma^{(1)} + (J_{n_2} - I_{n_2}) \otimes \Sigma^{(2)}\right]D'Q',\end{aligned}$$

$$= P_A \left[I_{n_2} \otimes \Sigma^{(1)} + (J_{n_2} - I_{n_2}) \otimes \Sigma^{(2)} \right] P_A$$
$$+ P_Q \left[I_{n_2} \otimes \Sigma^{(1)} + (J_{n_2} - I_{n_2}) \otimes \Sigma^{(2)} \right] P_Q.$$

Since $P_A \Sigma = \Sigma P_A$ and using the fact $P_A + P_Q = I_p$, we have

$$\Sigma = I_{n_2} \otimes \Sigma^{(1)} + (J_{n_2} - I_{n_2}) \otimes \Sigma^{(2)},$$

which implies that the "only if" part of the theorem is true. The "if" part can be shown due to the structure of Σ. Suppose that

$$\Sigma = A \Gamma A' + Q \Theta Q' = I_{n_2} \otimes \Sigma^{(1)} + (J_{n_2} - I_{n_2}) \otimes \Sigma^{(2)}.$$

Since $P_A + P_Q = I_p$, we have

$$\begin{pmatrix} A & Q \end{pmatrix}^{-1} = \begin{pmatrix} X \\ D \end{pmatrix}.$$

Moreover,

$$A \Gamma A' + Q \Theta Q' = \begin{pmatrix} A & Q \end{pmatrix} \begin{pmatrix} \Gamma & 0 \\ 0 & \Theta \end{pmatrix} \begin{pmatrix} A' \\ Q' \end{pmatrix},$$

therefore we have

$$\begin{pmatrix} \Gamma & 0 \\ 0 & \Theta \end{pmatrix} = \begin{pmatrix} X \\ D \end{pmatrix} \left[I_{n_2} \otimes \Sigma^{(1)} + (J_{n_2} - I_{n_2}) \otimes \Sigma^{(2)} \right] \begin{pmatrix} X' & D' \end{pmatrix},$$

which implies $P_A \Sigma D' = D \Sigma P_A = 0$, i.e., $P_A \Sigma = \Sigma P_A$, where Σ is given in (15.2). Hence, the proof is completed. □

From Theorem 15.2 it can be observed that the condition $P_A \Sigma = \Sigma P_A$ coincides with the Zyskind's condition given in Theorem 15.1.

15.4 Explicit Mean Parameters in the Extended Growth Curve Model

In this section we consider the following extended growth curve model (EGCM):

$$Y = A_1 B_1 C_1 + A_2 B_2 C_2 + E, \tag{15.12}$$

where B_1 and B_2 are unknown matrices. von Rosen [28] derived the MLEs of B_1 and B_2 by assuming the nested subspace condition $\mathscr{C}(C_2') \subseteq \mathscr{C}(C_1')$. In [4] the explicit MLEs are presented with the nested subspace condition on the within design matrices $\mathscr{C}(A_2) \subseteq \mathscr{C}(A_1)$. We first consider the model with the condition $\mathscr{C}(A_2) \subseteq \mathscr{C}(A_1)$ and it is called EBRM^2_W in [29]. All design matrices A_1, A_2, C_1 and C_2 are assumed to have full rank. Vectorizing the model (15.12) we get

$$\begin{aligned}\mathrm{vec}Y &= (C_1' \otimes A_1)\mathrm{vec}B_1 + (C_2' \otimes A_2)\mathrm{vec}B_2 + \mathrm{vec}E,\\ &= \underbrace{\left(C_1' \otimes A_1 \; C_2' \otimes A_2\right)}_{Z}\begin{pmatrix}\mathrm{vec}B_1\\ \mathrm{vec}B_2\end{pmatrix} + \mathrm{vec}E,\end{aligned}$$

where $\mathrm{vec}E \sim N_{pn}(\mathbf{0}, I_n \otimes \Sigma)$. To derive Zyskind's condition in model (15.12), we need the expression of P_Z. Using the similar techniques as in [30], we have

$$\begin{aligned}P_Z &= \left[(C_1' \otimes A_1)(a + bdc) - (C_2' \otimes A_2)dc\right](C_1 \otimes A_1')\\ &\quad + \left[(C_2' \otimes A_2)d - (C_1' \otimes A_1)bd\right](C_2 \otimes A_2'),\end{aligned}$$

where

$$\begin{aligned}a &= (C_1C_1')^{-1} \otimes (A_1'A_1)^{-1},\\ b &= (C_1C_1')^{-1}C_1C_2' \otimes (A_1'A_1)^{-1}A_1'A_2,\\ c &= C_2C_1'(C_1C_1')^{-1} \otimes A_2'A_1(A_1'A_1)^{-1},\\ d &= (C_2C_2' \otimes A_2'A_2 - C_2P_{C_1'}C_2' \otimes A_2'P_{A_1}A_2)^{-1}.\end{aligned}$$

Since $\mathscr{C}(A_2) \subseteq \mathscr{C}(A_1)$ implies that both $P_{A_1}A_2 = A_2$ and $P_{A_2} = P_{A_1}P_{A_2}$ hold, the expression of d becomes $(C_2(I - P_{C_1'})C_2')^{-1} \otimes (A_2'A_2)^{-1}$ and P_Z becomes

$$P_Z = P_{C_1'} \otimes P_{A_1} + P_{C_2',(I-P_{C_1'})^{-1}} \otimes P_{A_2},$$

where $P_{C_2',(I-P_{C_1'})^{-1}} = (I - P_{C_1'})C_2'(C_2(I - P_{C_1'})C_2')^{-1}C_2(I - P_{C_1'})$. Therefore, the Zynkind's condition in model (15.12) becomes

$$\begin{aligned}&P_{C_1'} \otimes P_{A_1}\Sigma + P_{C_2',(I-P_{C_1'})^{-1}} \otimes P_{A_2}\Sigma\\ &= P_{C_1'} \otimes \Sigma P_{A_1} + P_{C_2',(I-P_{C_1'})^{-1}} \otimes \Sigma P_{A_2},\end{aligned}$$

which implies that $P_{A_1}\Sigma = \Sigma P_{A_1}$ and $P_{A_2}\Sigma = \Sigma P_{A_2}$. These two conditions can be simplified to one condition of $P_{A_1}\Sigma = \Sigma P_{A_1}$ due to the facts $P_{A_1}A_2 = A_2$ and $P_{A_2} = P_{A_1}P_{A_2}$. The result is formulated in the next theorem.

Theorem 15.3 *Assuming the covariance structure given in* (15.2)*, the OLS estimator of the matrices B_1 and B_2 in model* (15.12)*, i.e.,*

$$\widehat{B}_1 = (A_1'A_1)^{-1}A_1'(Y - A_2\widehat{B}_2C_2)C_1'(C_1C_1')^{-1} \tag{15.13}$$

$$\widehat{B}_2 = (A_2'A_2)^{-1}A_2'Y(I - P_{C_1'})C_2'(C_2(I - P_{C_1'})C_2')^{-1} \tag{15.14}$$

is the BLUE if and only if the matrix A_1 is Σ-invariant or equivalently $P_{A_1}\Sigma = \Sigma P_{A_1}$.

It has been shown in Theorem 15.2 that the block circular Toeplitz covariance matrix belongs to the class of RSC structure and the results can also be applied to model (15.12) when Σ in (15.2) is assumed.

Corollary 15.1 *Suppose $\mathscr{C}(A_2) \subseteq \mathscr{C}(A_1)$ in model* (15.12) *with a covariance matrix Σ given in* (15.2)*. Then Σ belongs to the class of RSC structure if and only if $P_{A_1}\Sigma = \Sigma P_{A_1}$.*

Proof Model (15.12) can be rewritten as

$$Y = \underbrace{\begin{pmatrix} A_1 & A_2 \end{pmatrix}}_{A} \begin{pmatrix} B_1 & 0 \\ 0 & B_2 \end{pmatrix} \begin{pmatrix} C_1 \\ C_2 \end{pmatrix} + \text{vec}E.$$

Based on Theorem 15.2, the correspondingly necessary and sufficient condition for model (15.12) is $P_{(A_1\ A_2)}\Sigma = \Sigma P_{(A_1\ A_2)}$. The nested subspace condition $\mathscr{C}(A_2) \subseteq \mathscr{C}(A_1)$ implies that $P_{(A_1\ A_2)} = P_{A_1}$. The proof is complete. □

Model (15.12) with another nested subspace condition $\mathscr{C}(C_2') \subseteq \mathscr{C}(C_1')$ is further discussed. It is called EBRM$_B^2$ in [29]. At first, from $\mathscr{C}(C_2') \subseteq \mathscr{C}(C_1')$ indicates $C_2 = QC_1$ for some matrix Q, we have

$$Y = A_1B_1C_1 + A_1B_2Q'C_1 + E = (A_1\ A_2)\begin{pmatrix} B_1 \\ B_2Q \end{pmatrix} C_1 + E \text{ with } FBG = 0,$$

where $B = \begin{pmatrix} B_1 \\ B_2Q \end{pmatrix}$, $F = (0\ I)$, $G' = Q^o$, and Q^o denotes any matrix of full rank spanning the orthogonal complement to the space $\mathscr{C}(Q)$.

Solving the linear equation, $B = \theta_1 G^{o'} + (F')^o\theta_2G' = \theta_1 Q + \begin{pmatrix} I \\ 0 \end{pmatrix}\theta_2 Q^o$. Thus,

$$Y = \underbrace{(A_1\ A_2)}_{A_1^*}\underbrace{\theta_1}_{B_1^*}\underbrace{QC_1}_{C_1^*} + \underbrace{(A_1\ A_2)\begin{pmatrix} I \\ 0 \end{pmatrix}}_{A_2^*}\underbrace{\theta_2}_{B_2^*}\underbrace{Q^oC_1}_{C_2^*} + E, \tag{15.15}$$

and $\mathscr{C}(A_2^*) \subseteq \mathscr{C}(A_1^*)$. Model (15.12) under the condition $\mathscr{C}(C_2') \subseteq \mathscr{C}(C_1')$ is equivalent to the condition $\mathscr{C}(A_2) \subseteq \mathscr{C}(A_1)$ and Theorem 15.3 can be applied.

The equivalence between EBRM_B^m and EBRM_W^m, where m is the number of profile expressions, is presented by von Rosen [29] in Section 3.5.

15.5 Simulation Study

In this section a simulation study is conducted as a summary of the derived results. Three mean structures are studied and for each case, 500 replicates of the following procedure are performed.

1. Randomly generate a sample of size $n = 27$ from a growth curve model with a block circular Toepitz covariance structure given in (15.2) with $n_2 = 3$ and $n_1 = 4$.
2. Calculate the explicit mean estimates in each replicate.
3. Calculate the average values of the obtained estimates after 500 replicates.

For the mean structure $A_{12\times 2}B_{2\times 2}C_{2\times 27}$, data is generated with parameters

$$A = \mathbf{1}_3 \otimes \begin{pmatrix} 1 & 1 \\ 1 & 1 \\ 1 & -1 \\ 1 & -1 \end{pmatrix}, \quad B = \begin{pmatrix} 1 & 1 \\ 1 & 1 \end{pmatrix}, \quad C = \begin{pmatrix} \mathbf{1}'_{11} & \mathbf{0}'_{16} \\ \mathbf{0}'_{11} & \mathbf{1}'_{16} \end{pmatrix}.$$

The condition $P_A\Sigma = \Sigma P_A$ is satisfied here. The explicit OLS estimates of the matrix B is calculated according to (15.6) and based on 500 replicates the average estimates are given by

$$\widehat{B} = \begin{pmatrix} 0.9982 & 0.9991 \\ 1.0001 & 0.9964 \end{pmatrix}.$$

For the mean structure $A_1B_1C_1 + A_2B_2C_2$, where $\mathscr{C}(A_2) \subseteq \mathscr{C}(A_1)$, data is generated with parameters

$$A_1 = \mathbf{1}_3 \otimes \begin{pmatrix} 1 & 1 \\ 1 & 1 \\ 1 & -1 \\ 1 & -1 \end{pmatrix}, \quad B_1 = \begin{pmatrix} 1 \\ 1 \end{pmatrix}, \quad C_1 = \begin{pmatrix} \mathbf{1}'_{11} & \mathbf{0}'_{16} \end{pmatrix},$$

$$A_2 = \mathbf{1}_{12}, \quad B_2 = 1, \quad C_2 = \begin{pmatrix} \mathbf{0}'_{11} & \mathbf{1}'_{16} \end{pmatrix}.$$

Here $\mathscr{C}(A_2) \subseteq \mathscr{C}(A_1)$ is satisfied and $P_{A_1}\Sigma = \Sigma P_{A_1}$. The explicit OLS estimates of the matrices B_1 and B_2 are calculated according to (15.13) and based on 500 replicates the average estimates are given by

$$\widehat{B}_1 = \begin{pmatrix} 0.9982 \\ 1.0001 \end{pmatrix}, \quad \widehat{B}_2 = 0.9991.$$

For the mean structure $A_1B_1C_1 + A_2B_2C_2$, where $\mathscr{C}(C_2') \subseteq \mathscr{C}(C_1')$, data is generated with parameters

$$A_1 = \mathbf{1}_3 \otimes \begin{pmatrix} 1 & 1 \\ 1 & 1 \\ 1 & -1 \\ 1 & -1 \end{pmatrix}, \quad B_1 = \begin{pmatrix} 1 & 1 \\ 1 & 1 \end{pmatrix}, \quad C_1 = \begin{pmatrix} \mathbf{1}_{11}' & \mathbf{0}_{16}' \\ \mathbf{0}_{11}' & \mathbf{1}_{16}' \end{pmatrix},$$

$$A_2 = \mathbf{1}_3 \otimes \begin{pmatrix} 1 \\ -1 \\ 1 \\ -1 \end{pmatrix}, \quad B_2 = 1, \quad C_2 = \begin{pmatrix} \mathbf{0}_{11}' & \mathbf{1}_{16}' \end{pmatrix}.$$

Note that $\mathscr{C}(C_2') \subseteq \mathscr{C}(C_1')$ holds. After a reparametrization the model under $\mathscr{C}(C_2') \subseteq \mathscr{C}(C_1')$ can be converted to $\mathscr{C}(A_2^*) \subseteq \mathscr{C}(A_1^*)$, where $A_1^* = (\, A_1 \; A_2 \,)$ and $A_2^* = (\, A_1 \; A_2 \,) \begin{pmatrix} I_2 \\ \mathbf{0}_2' \end{pmatrix}$. Based on (15.15) we have $B = \begin{pmatrix} B_1 \\ B_2Q \end{pmatrix}$, where $Q = (0 \;\; 1)$. Moreover, $C_1^* = QC_1$ and $C_2^* = Q^oC_1$, where $Q^o = (-1 \;\; 0)$. The explicit OLS estimates of the matrices B_1^* and B_2^* were calculated according to (15.13) and based on 500 replicates the average estimates of B are given by

$$\widehat{B} = \begin{pmatrix} 0.9982 & 0.9991 \\ 1.0001 & 0.9964 \\ 0.0000 & 0.9980 \end{pmatrix}.$$

From the above simulations one conclusion is that with appropriate design matrices, the explicit estimates perform very well and are close to the true values.

15.6 Conclusion and Discussion

The explicit mean estimators of the patterned covariance GCM and EGCM are studied in this paper. It has been shown that the class of RSC structure contains the block circular covariance matrix in (15.2) as a special case and the explicit MLEs of the mean parameters are equal to the unweighted estimators and they are BLUE. It is worth mentioning that the eigenvectors of Σ in (15.2) are functionally independent of the Σ elements [11], which makes it is somehow easier to find the within design matrices that satisfy the necessary and sufficient conditions in Theorems 15.1 and 15.3. The methodology presented in Sects. 15.3 and 15.4 can be extended to the situation where the covariance matrix Σ has other block covariance structures, e.g., block compound symmetry (BCS) structure (see [24], for example) and circular blocks with compound symmetry matrices inside [10]. We hope to communicate the relevant results in a future correspondence.

Acknowledgments The authors thank Professor Dietrich von Rosen for the idea and inspiration of the method in the chapter during their Ph.D. studies. Yuli Liang's research is supported by the strategic funding of Örebro University. The authors are also grateful to both the two editors and the anonymous reviewer for improvements of the manuscript.

References

1. Anderson, T.W.: Asymptotically efficient estimation of covariance matrices with linear structure. Ann. Stat. **1**, 135–141 (1973)
2. Arnold, S.F.: Application of the theory of products of problems to certain patterned covariance matrices. Ann. Stat. **1**, 682–699 (1973)
3. Arnold, S.F.: Linear models with exchangeably distributed errors. J. Amer. Stat. Asso. **74**, 194–199 (1979)
4. Filipiak, K., von Rosen, D.: On MLEs in an extended multivariate linear growth curve model. Metrika **75**, 1069–1092 (2012)
5. Khatri, C.G.: Testing some covariance structures under a growth curve model. J. Multivar. Anal. **3**, 102–116 (1973)
6. Kollo, T., von Rosen, D.: Advanced Multivariate Statistics with Matrices. Springer, Dordrecht (2005)
7. Krishnaiah, P.R., Lee, J.C.: On covariance structures. Sankhyā: Indian J. Stat. Ser. A **38**, 357–371 (1974)
8. Lee, J.C.: Prediction and estimation of growth curves with special covariance structures. J. Amer. Stat. Asso. **83**, 432–440 (1988)
9. Lee, J.C., Geisser, S.: Growth curve prediction. Technical Report, University of Minnesota (1972)
10. Liang, Y., von Rosen, D., von Rosen, T.: On Properties of Toeplitz-type Covariance Matrices in Models with Nested Random Effects. Statistical Papers, Forthcoming (2020).
11. Liang, Y., von Rosen, D., von Rosen, T.: On estimation in multilevel models with block circular symmetric covariance structure. Acta et Comment. Univ. Tartu. de Math. **16**, 83–96 (2012)
12. Liang, Y., von Rosen, D., von Rosen, T.: On estimation in hierarchical models with block circular covariance structures. Ann. Instit. Stat. Math. **67**, 773–791 (2015)
13. Mitra, S.K., Moore, B.J.: Gauss-Markov estimation with an incorrect dispersion matrix. Sankhyā: Indian J. Stat. Ser. A **35**, 139–152 (1973)
14. Ohlson, M., von Rosen, D.: Explicit estimators of parameters in the growth curve model with linearly structured covariance matrices. J. Multivar. Anal. **101**, 1284–1295 (2010)
15. Olkin, I.: Testing and estimation for structures which are circularly symmetric in blocks. In: Kabe, D.G., Gupta, R.P. (eds.), Multivariate Statistical Inference, pp. 183–195. Amsterdam, North-Holland (1973)
16. Olkin, I.: Inference for a normal population when the parameters exhibit some structure. Technical Report No.77, Department of Statistics, Stanford University (1973)

17. Olkin, I., Press, S.: Testing and estimation for a circular stationary model. Ann. Math. Stat. **40**, 1358–1373 (1969)
18. Potthoff, R.F., Roy, S.N.: A generalized multivariate analysis of variance model useful especially for growth curve problems. Biometrika **51**, 313–326 (1964)
19. Puntanen, S., Styan, G.P.: The equality of the ordinary least squares estimator and the best linear unbiased estimator. Amer. Stat. **43**, 153–161 (1989)
20. Radhakrishna Rao, C.: Least squares theory using an estimated dispersion matrix and its application to measurement of signals. In: Proceedings of the Fifth Berkeley Symposium on Mathematical Statistics and Probability, vol. 1, pp. 355–372. University of California Press, Berkeley (1967)
21. Reinsel, G.: Multivariate repeated-measurement or growth curve models with multivariate random-effects covariance structure. J. Amer. Stat. Asso. **77**, 190–195 (1982)
22. Roy, A., Fonseca, M.: Linear models with doubly exchangeable distributed errors. Commun. Stat. Theory Method **41**, 2545–2569 (2012)
23. Roy, A., Leiva, R.: Likelihood ratio tests for triply multivariate data with structured correlation on spatial repeated measurements. Stat. Probab. Lett. **78**, 1971–1980 (2008)
24. Roy, A., Zmyślony, R., Fonseca, M., Leiva, R.: Optimal estimation for doubly multivariate data in blocked compound symmetric covariance structure. J. Multivar. Anal. **144**, 81–90 (2016)
25. Szatrowski, T.H.: Necessary and sufficient conditions for explicit solutions in the multivariate normal estimation problem for patterned means and covariances. Ann. Stat. **8**, 802–810 (1980)
26. Verbyla, A.P., Venables, W.N.: An extension of the growth curve model. Biometrika **75**, 129–138 (1988)
27. Viana, M., Olkin, I.: Symmetrically dependent models arising in visual assessment data. Biometrics **56**, 1188–1191 (2000)
28. von Rosen, D.: Maximum likelihood estimators in multivariate linear normal models. J. Multivar. Anal. **31**, 187–200 (1989)
29. von Rosen, D.: Bilinear Regression Analysis: An Introduction, vol. 220. Springer, Berlin (2018)
30. Žežula, I.: Remarks on unbiased estimation of the sum-of-profiles model parameters. Tatra Mt. Math. Publ. **39**, 45–52 (2008)
31. Zyskind, G.: On canonical forms, non-negative covariance matrices and best and simple least squares linear estimators in linear models. Ann. Math. Stat. **38**, 1092–1109 (1967)

Chapter 16
Space Decomposition and Estimation in Multivariate Linear Models

Joseph Nzabanita and Innocent Ngaruye

Abstract Linear models are important in statistical analysis. In many real situations, these models become more and more complex, as such the estimation of model parameters constitutes a big challenge. To overcome this challenge many approaches have been proposed and space decomposition has emerged as a powerful tool in handling these complex models. This work gives a comprehensive review of some of challenges related to complex multivariate models and how the space decomposition has been successfully used. In this review, we first present the space decomposition as a tool to decompose complex models into tractable models that are easy to handle for estimation and testing. On the other hand, we give another space decomposition approach used for obtaining explicit estimators in multivariate linear models. Some examples on how this decomposition is performed for specific multivariate linear models are presented for both approaches.

16.1 Introduction

The problems of estimation and testing, in linear models, were relatively straight forward until Pottoff and Roy [10] introduced the Generalized Multivariate Analysis of Variance (GMANOVA) model, which is also known as the Growth Curve Model.

Definition 16.1 Let $X : p \times n$, $A : p \times q, q \leq p$, $B : q \times k$, $C : k \times n$, $r(C) + p \leq n$, where $r(\cdot)$ represents the rank of a matrix. The growth curve model is given by

$$X = ABC + E, \tag{16.1}$$

J. Nzabanita (✉) · I. Ngaruye
Department of Mathematics, College of Science and Technology, University of Rwanda, Kigali, Rwanda
e-mail: j.nzabanita@ur.ac.rw; i.ngaruye@ur.ac.rw

© Springer Nature Switzerland AG 2020

T. Holgersson, M. Singull (eds.), *Recent Developments in Multivariate and Random Matrix Analysis*, https://doi.org/10.1007/978-3-030-56773-6_16

where columns of E are assumed to be independently distributed as a multivariate normal distribution with mean zero and a positive definite dispersion matrix Σ; i.e., $E \sim N_{p,n}(0, \Sigma, I_n)$.

The matrices A and C, often called, respectively, within-individuals and between-individuals design matrices, are known matrices whereas matrices B and Σ are unknown parameter matrices.

The aim of Pottoff and Roy [10] was to build a model that is particularly useful to study many growth curve problems. They considered the problem of estimating parameters in GMANOVA model and testing hypothesis of the form

$$GBH = 0, \tag{16.2}$$

where G and H are known matrices.

The strategy adopted to estimate and test the hypothesis in (16.2) was to reduce the model to the MANOVA model, which by the time had been already treated in literature. This transformation was achieved by pre-multiplying the model in (16.1) with $(AG^{-1}A')^{-1}AG^{-1}$, where $G : p \times p$ was allowed to be any symmetric positive definite matrix or any other non-singular matrix such that $AG^{-1}A'$ is invertible. Then, the model becomes the usual MANOVA with a new covariance matrix for each column of the form

$$\Omega = (AG^{-1}A')^{-1}AG^{-1}\Sigma G^{-1}A'(AG^{-1}A')^{-1}. \tag{16.3}$$

What they actually did was to decompose model (16.1) into two models one of which is a usual MANOVA.

Due to the complexity of data from various fields of applied research, it is inevitable that complex multivariate linear models are to be considered. For such kind of models, the estimation of model parameters and hypotheses testing often constitutes a big issue. For this class of multivariate linear models, it is essential to make a decomposition of subspaces in order to obtain simpler models that are easy to handle when estimating parameters or performing some hypothesis testing. This decomposition is based on some subspace relations about matrix algebra of vector spaces such as linear transformations and basic relations for column vector spaces. This consists of vector space decomposition into orthogonal direct sum of subspaces. Common applications induce the decomposition of the whole space according to design matrices and the estimators are obtained by projections of observations using induced space decomposition. The aim of this chapter is to illustrate, using examples, how the space decomposition is a powerful tool for estimating and testing in complex multivariate linear models.

Throughout this work the following notations will be used. $\mathscr{C}(A)$, $r(A)$ and $\mathrm{tr}(A)$ denote the column space, the rank and the trace of a matrix A, respectively. For a positive definite matrix S and any matrix A, $P_{A,S} = A(A'S^{-1}A)^{-}A'S^{-1}$ defines the projector onto the space $\mathscr{C}_S(A)$, where the subscript S in $\mathscr{C}_S(A)$ indicates that the inner products are defined via the positive definite matrix S^{-1}. Note that $P^{o}_{A,S} =$

$I - P'_{A,S} = P_{A^o,S^{-1}}$ is a projector onto the space $\mathscr{C}(A)^{\perp} = \mathscr{C}(A^o)$, where A^o denotes any matrix of full rank spanning the orthogonal complement to the space $\mathscr{C}(A)$. If $S = I$, we simply write P_A instead of $P_{A,I}$. The symbol $\otimes$ denotes the Kronecker product of matrices or the tensor product of linear spaces. The symbol $\oplus$ denotes the direct sum of linear spaces, while the symbol $\boxplus$ denotes the direct sum of tensor spaces.

16.2 Decomposition of Complex Models to Simpler Models

In this section a comprehensive review on how to decompose complex multivariate linear models to simpler ones is presented. The class of models considered are bilinear regression models that include random effects with and without latent variables [8, 13].

16.2.1 Multivariate Linear Model for Repeated Measures Data

Consider a multivariate linear regression model for repeated measurements over p time points for n units divided into k non-overlapping group units, which is given by

$$Y = ABC + B_1C_1 + UZ + E, \tag{16.4}$$

where $Y : p \times n$ is a data matrix, $B : q \times k$, $B_1 : p \times k_1$ and $U : p \times m$ are unknown parameter matrices, while $C_1 : k_1 \times n$ is known matrix of covariables. Moreover, $E \sim \mathscr{N}_{p,n}(0, \Sigma_e, I_n)$ and $U \sim \mathscr{N}_{p,m}(0, \Sigma_u, I_m)$ are independently distributed.

The matrices $A : p \times q$, $C : k \times n$ and $Z : m \times n$ are known design matrices of full row rank such that $\mathscr{C}(Z') \subseteq \mathscr{C}(C')$ and $ZZ' = I_m$, where $\mathscr{C}(A)$ denotes the column vector space generated by the columns of an arbitrary matrix A. This model has been considered in [8] for small area estimation of repeated measures data. For simplicity and without loss of generality, let us assume that $A = I$.

In order to perform the estimation of model parameters, we make an orthogonal transformation of model in (16.4) and partition it into three independent models. We make a one-one transformation by post-multiplying the model in (16.4) by a matrix $M : n \times n$

$$M = [M_1 : M_2] : \big(n \times r \quad n \times (n-r)\big)$$

such that

$$\mathscr{C}(M_1) = \mathscr{C}(C' : Z') \text{ and } \mathscr{C}(M_2) = \mathscr{C}(C' : Z')^{\perp},$$
$$M_1'M_1 = I_q, \; M_2'M_2 = I_{n-r} \text{ and } M_1'M_2 = 0,$$

where $r = rank(C' : Z')$ with $r > m$ and $\mathscr{C}(A)^{\perp}$ stands for the column space spanning the orthogonal complement of the matrix A.

Lemma 16.1 *Consider the orthogonal transformation obtained by post-multiplying the model in (16.4) by the matrix M.*

(i) The matrix $W = M_1'Z'ZM_1$ is idempotent.
(ii) Let W be diagonalized by $W = \Omega D \Omega'$ with $\Omega = \left[\Omega_1 : \Omega_2\right]$, where Ω_1 and Ω_2 are orthogonal matrices of eigenvectors for m and $n - m$ elements. Denote by

$$K_i = CR_i, \quad \text{and} \quad R_i = M_1\Omega_i, \quad i = 1, 2,$$

then with this orthogonal transformation, we obtain the three independently distributed models

$$YR_1 \sim \mathscr{N}_{p,m}\Big(BK_1 + B_1C_1R_1, \Sigma_u + \Sigma_e, I_m\Big),$$
$$YR_2 \sim \mathscr{N}_{p,r-m}\Big(BK_2 + B_1C_1R_2, \Sigma_e, I_{r-m}\Big),$$
$$YM_2 \sim \mathscr{N}_{p,n-r}\Big(B_1C_1M_2, \Sigma_e, I_{n-r}\Big).$$

Proof

(i) Straightforward calculations show that

$$WW = M_1'Z'ZM_1M_1'Z'ZM_1.$$

But $\mathscr{C}(Z') \subseteq \mathscr{C}(M_1)$ and $ZZ' = I$ implies that $ZM_1M_1Z' = I$ and hence $WW = W$. Moreover, since $\mathscr{C}(M_1) = \mathscr{C}(C' : Z')$ and $\mathscr{C}(M_2) = \mathscr{C}(C' : Z')^{\perp}$, it follows that $CM_2 = 0$ and $ZM_2 = 0$ and thus

$$YM_2 = B_1C_1M_2 + EM_2.$$

Furthermore,

$$YR_1 = BK_1 + B_1C_1R_1 + (UZ + E)R_1$$
$$YR_2 = BK_2 + B_1C_1R_2 + ER_2,$$

and

$$
\begin{aligned}
R_1' Z' Z R_1 &= \Omega_1' \Omega D \Omega' \Omega_1 = I_m, \\
R_2' Z' Z R_2 &= \Omega_2' \Omega D \Omega' \Omega_2 = I_{n-r} \\
M_2' M_2 &= I_{n-r}.
\end{aligned}
$$

Hence,

$$
\begin{aligned}
Y R_1 &\sim \mathcal{N}_{p,m}\Big(B K_1 + B_1 C_1 R_1, \Sigma_u + \Sigma_e, I_m\Big), \\
Y R_2 &\sim \mathcal{N}_{p,r-m}\Big(B K_2 + B_1 C_1 R_2, \Sigma_e, I_{r-m}\Big), \\
Y M_2 &\sim \mathcal{N}_{p,n-r}\Big(B_1 C_1 M_2, \Sigma_e, I_{n-r}\Big).
\end{aligned}
$$

The independence of YR_1, YR_2 and YM_2 follows by the fact that the matrices Ω_1, Ω_1 and M_2 are pairwise orthogonal. □

The following theorem summarize the estimation of mean and covariance which is performed using a likelihood based approach while the prediction of random effect is performed using Henderson's approach consisting of the maximization of the joint density $f(Y, U)$ with respect to U under the assumption of known Σ_e and Σ_u [3].

Theorem 16.1 *Consider the model (16.4). Assume that the matrices C_1 and $K_2^{o\prime} K_1$ are of full rank. Then the maximum likelihood estimators for B_1, B, Σ_u and the predictor of U are given by*

$$
\begin{aligned}
\widehat{B_1} =& Y P_1 C_1'(C_1 P_1 C_1')^{-1}, \\
\widehat{B} =& Y R_2 K_2'(K_2 K_2')^{-} - Y P_1 C_1'(C_1 P_1 C_1')^{-1} C_1 R_2 K_2'(K_2 K_2')^{-} + T K_2^{o\prime}, \\
\widehat{\Sigma}_u =& \frac{1}{m-1} Y P_2 R_1 Q_{K_1' K_2^o} R_1' P_2' Y' - \Sigma_e, \\
\widehat{U} =& (\Sigma_e + \widehat{\Sigma}_u)^{-1} \widehat{\Sigma}_u Y P_2 R_1 Q_{K_1' K_2^o} R_1' Z',
\end{aligned}
$$

where

$$
\begin{aligned}
P_1 =& M_2 M_2' + R_2 Q_{K_2'} R_2', \\
P_2 =& I_n - R_2 K_2'(K_2 K_2')^{-} C,
\end{aligned}
$$

and T is an arbitrary matrix.

Proof We consider the likelihood of joint density function of YR_2 and YM_2. Differentiating its log-likelihood with respect to B_1 and B leads to the likelihood equations

$$\begin{aligned} YM_2M_2'C_1' + YR_2R_2'C_1' - BK_2R_2'C_1' - B_1(C_1M_2M_2'C_1' + C_1R_2R_2'C_1') &= 0 \\ (YR_2 - BK_2 - B_1C_1R_2)K_2' &= 0. \end{aligned}$$

These equations lead to

$$B_1C_1(M_2M_2' + R_2R_2')C_1' = Y(M_2M_2' + R_2R_2')C_1' - BK_2R_2'C_1' \quad (16.5)$$

$$B = YR_2K_2'(K_2K_2')^- - B_1C_1R_2K_2'(K_2K_2')^- + TK_2^{o'}, \quad (16.6)$$

for arbitrary matrix T. Insertingx the expression of B from Eq. (16.6) into Eq. (16.5) yields

$$B_1C_1(M_2M_2' + R_2Q_{K_2'}R_2')C_1' = Y(M_2M_2' + R_2Q_{K_2'}R_2')C_1'.$$

Therefore, for matrix C_1 of full rank, we obtain

$$\begin{aligned} \widehat{B_1} &= YP_1C_1'(C_1P_1C_1')^{-1}, \\ \widehat{B} &= YR_2K_2'(K_2K_2')^- - YP_1C_1'(C_1P_1C_1')^{-1}C_1R_2K_2'(K_2K_2')^- + TK_2^{o'}, \end{aligned}$$

where

$$P_1 = M_2M_2' + R_2Q_{K_2'}R_2'.$$

Further, to estimate the covariance matrix Σ_u, we maximize the likelihood function of YR_1 with respect to Σ_u after inserting the expressions of B_1 and B. The prediction of random effect matrix U is derived by maximizing the joint density $f(Y, U)$ with respect to U assuming the other parameters B, B_1 and Σ_u to be known. Straightforward calculations show that

$$\begin{aligned} \widehat{\Sigma}_u &= \frac{1}{m-1}Y\Big(I_n - R_2K_2'(K_2K_2')^-C\Big)R_1Q_{K_1'K_2^o}R_1'\Big(I_n - R_2K_2'(K_2K_2')^-C\Big)'Y' - \Sigma_e, \\ \widehat{U} &= (\Sigma_e + \widehat{\Sigma}_u)^{-1}\widehat{\Sigma}_uY\Big(I_n - R_2K_2'(K_2K_2')^-C\Big)R_1Q_{K_1'K_2^o}R_1'Z', \end{aligned}$$

which completes the proof. □

16.2.2 Reduced Rank Bilinear Regression Model

We consider a bilinear regression model that includes random effects and latent variables. The vector space decomposition is then applied to remove random effects, in order to derive explicit estimators. This model is expressed as

$$Y = ABC + B_1C_1 + \Gamma X + UZ + E, \tag{16.7}$$

where $A : p \times q$, $C : k \times n$, $C_1 : k_1 \times n$, $X : k_2 \times n$ are all known matrices with $\mathscr{C}(X') \subseteq \mathscr{C}(C')$ and $\text{rank}(\Gamma) < \min(p, r)$. Moreover, $U \sim \mathscr{N}_{p,m}\Big(0, \Sigma_u, I_m\Big)$ and $E \sim \mathscr{N}_{p,m}\Big(0, \Sigma_e, \Psi\Big)$ are independently distributed. The matrix $Z : m \times n$ is standardized in such a way that $Z\Psi^{-1}Z' = I_m$. The covariance matrix Ψ is assumed to be known and the parameters of the model, which are to be estimated are $B, B_1, \Gamma, \Sigma_u, U$ and Σ_e. This model has been considered in [13] to produce small area estimates for repeated surveys with latent variables.

Similarly to the previous model, we can make a one-one transformation by post-multiplying the model (16.7) by a matrix $M : n \times n$

$$M = [M_1 : M_2] : \big(n \times r \quad n \times (n - r)\big)$$

such that

$$\mathscr{C}(M_1) = \mathscr{C}(\Psi^{-1/2}C' : \Psi^{-1/2}C_1' : Z') \text{ and } \mathscr{C}(M_2) = \mathscr{C}(C' : C_1' : Z')^{\perp},$$

assuming that $M_1'M_1 = I_r$, where $r = rank(C' : C_1' : Z')$ with $r > m$ and $\Psi^{-1/2}$ stands for a symmetric square root of Ψ.

Since the matrix $W = M_1'\Psi^{-1/2}Z'Z\Psi^{-1/2}$ is idempotent, it can be diagonalized by considering $W = \Omega D \Omega'$ with $\Omega = \big[\Omega_1 : \Omega_2\big]$, where Ω_1 and Ω_2 are orthogonal matrices of eigenvectors for m and $n - m$ elements.

Denote by

$$R_i = \Psi^{-1/2}M_i\Omega_i, \ i = 1, 2.$$

Then the one-one transformation of model (16.7) yields three independent models

$$YR_1 \sim \mathscr{N}_{p,m}\Big(ABCR_1 + B_1C_1R_1, \Sigma_u + \Sigma_e, I_m\Big), \tag{16.8}$$

$$YR_2 \sim \mathscr{N}_{p,r-m}\Big(ABCR_2 + B_1C_1R_2, \Sigma_e, I_{r-m}\Big), \tag{16.9}$$

$$YM_2 \sim \mathscr{N}_{p,n-r}\Big(0, \Sigma_e, I_{n-r}\Big). \tag{16.10}$$

The estimation is performed by utilizing model (16.9) and (16.10) to estimate B, B_1, Γ and Σ_e, and then inserting these estimators in model (16.8) to estimate Σ_u. The prediction of random effects matrix U can be done by the maximizing of the joint density $f(\text{vec}Y, \text{vec}U)$ with respect to vec$U$ assuming other parameters to be known, where vec$A$ denotes the columnwise vectorization of the matrix A. More details and further considerations can be found in [13].

16.3 Space Decomposition for Explicit Estimators in the Extended GMANOVA When the Covariance Matrix is Linearly Structured

This section gives another space decomposition approach used for obtaining explicit estimators in multivariate linear models. The example given on the extended GMANONVA when the covariance matrix is linearly structured has been treated in [9]. The GMANOVA model, also known as the classical growth curve model (GCM) by Pottoff and Roy [10] has emerged as a powerful tool to deal with the growth curve problems related to short time series. The GMANOVA model was extended later on by Verbyla and Venables [11] to handle different growth profiles. The resulted model was called sum of profiles model, also known as extended growth curve model (EGCM) or extended GMANOVA model in statistical literature. In [11] an iterative algorithm to obtain the maximum likelihood estimators (MLEs), which could not be obtained explicitly, was proposed. Later, von Rosen [12] studied the model and derived explicit MLEs under the additional nested subspaces condition on the between design matrices as in Definition 16.2.

Definition 16.2 Let $X : p \times n$, $A_i : p \times q_i$, $B_i : q_i \times k_i$, $C_i : k_i \times n$, $r_1 + p \leq n$, $i = 1, 2, \ldots, m$, $\mathscr{C}(C_i') \subseteq \mathscr{C}(C_{i-1}')$, $i = 2, 3, \ldots, m$, where $r_i = r(C_i)$. The Extended Growth Curve Model is given by

$$X = \sum_{i=1}^{m} A_i B_i C_i + E,$$

where columns of E are assumed to be independently distributed as a p-variate normal distribution with mean zero and a positive definite dispersion matrix Σ; i.e., $E \sim N_{p,n}(0, \Sigma, I_n)$. The matrices A_i and C_i, often called design matrices, are known matrices whereas matrices B_i and Σ are unknown parameter matrices.

The model in Definition 16.2 has been extensively studied by several authors and the book by [7] [Chapter 4] contains useful detailed information about uniqueness, estimability conditions, moments and approximative distributions of the maximum likelihood estimators. Some other authors considered the model with slightly different conditions. For example in [2] the explicit MLEs were presented with the nested subspace conditions on the within design matrices instead. In [4, 5] the extended GMANOVA model without nested subspace conditions but with orthogonal design matrices was considered and generalized least-squares estimators and their properties were studied. Nzabanita et al. [9] derived explicit estimators of parameters in the extended growth curve model with a linearly structured covariance matrix, which means that for $\Sigma = (\sigma_{ij})$ the only linear structure between the elements is given by $|\sigma_{ij}| = |\sigma_{kl}| \neq 0$ and there exist at least one $(i, j) \neq (k, l)$ so that $|\sigma_{ij}| = |\sigma_{kl}| \neq 0$. The proposed estimation procedure was based on the decomposition of the residual space into $m + 1$ subspaces, and on the study of residuals obtained from projecting observations onto those subspaces. In this section this idea is illustrated.

16.3.1 Main Idea on Space Decomposition

We start with recalling the result established in [12] about the estimation of the mean structure in the extended GMANOVA model.

Theorem 16.2 *Let $\widehat{B}_i$'s be the maximum likelihood estimators of B_i's in the model as in Definition 16.2. Then*

$$P_r \sum_{i=r}^{m} A_i \widehat{B}_i C_i = \sum_{i=r}^{m} (I - T_i) X C_i' (C_i C_i')^{-} C_i,$$

where

$$P_r = T_{r-1} T_{r-2} \times \cdots \times T_0, \; T_0 = I, \quad r = 1, 2, \ldots, m+1,$$

$$T_i = I - P_i A_i (A_i' P_i' S_i^{-1} P_i A_i)^{-} A_i' P_i' S_i^{-1}, \quad i = 1, 2, \ldots, m,$$

$$S_i = \sum_{j=1}^{i} K_j, \quad i = 1, 2, \ldots, m,$$

$$K_j = P_j X (C_{j-1}' (C_{j-1} C_{j-1}')^{-} C_{j-1} - C_j' (C_j C_j')^{-} C_j) X' P_j', \; C_0 = I.$$

A useful result is the corollary of this theorem when $r = 1$, which gives the estimated mean structure, i.e.,

$$\widehat{E[X]} = \sum_{i=1}^{m} A_i \widehat{B}_i C_i = \sum_{i=1}^{m} (I - T_i) X C_i' (C_i C_i')^{-} C_i. \qquad (16.11)$$

Replacing T_i in (16.11) by its expression given in Theorem 16.2 we get

$$\widehat{E[X]} = \sum_{i=1}^{m} P_i A_i (A_i' P_i' S_i^{-1} P_i A_i)^{-} A_i' P_i' S_i^{-1} X C_i' (C_i C_i')^{-} C_i,$$

or equivalently

$$\widehat{E[X]} = \sum_{i=1}^{m} P_{P_i A_i, S_i} X P_{C_i'}. \tag{16.12}$$

Noticing that the matrices $P_{P_i A_i, S_i}$ and $P_{C_i'}$ are projector matrices, we see that estimators of the mean structure are based on a projection of the observations on the space generated by the design matrices. Naturally, the estimators of the variance parameters are based on a projection of the observations on the residual space, that is the orthogonal complement to the design space.

If Σ was known, we would have obtained from the least squares theory the best linear estimator (BLUE) given by

$$\widetilde{E[X]} = \sum_{i=1}^{m} P_{\widetilde{P}_i A_i, \Sigma} X P_{C_i'}, \tag{16.13}$$

where S_i in P_i is replaced with Σ to get $\widetilde{P}_i$. Thus, we see that in the projections, if Σ is unknown, the parameter has been replaced with S_i's, which according to their expressions are not maximum likelihood estimators. However, S_i's define consistent estimators of Σ in the sense that $n^{-1} S_i \to \Sigma$ in probability.

Applying the vec-operator on both sides of (16.13) we get

$$\mathrm{vec}(\widetilde{E[X]}) = \sum_{i=1}^{m} (P_{C_i'} \otimes P_{\widetilde{P}_i A_i, \Sigma}) \mathrm{vec} X.$$

The next theorem is essential for the development of the sequel of this paper.

Theorem 16.3 *Let* $P = \sum_{i=1}^{m} P_{C_i'} \otimes P_{\widetilde{P}_i A_i, \Sigma}$ *and* $\mathcal{V}_i = \mathcal{C}_{\Sigma}(\widetilde{P}_i A_i)$, $i = 1, 2, \ldots, m$. *Then,*

(i) The subspaces $\mathcal{V}_i$*'s are mutually orthogonal and*

$$\mathcal{V}_1 \oplus \mathcal{V}_2 \oplus \cdots \oplus \mathcal{V}_i = \mathcal{C}_{\Sigma}(A_1 : A_2 : \cdots : A_i),\ i = 1, 2, \ldots, m;$$

(ii) The matrix P *is a projection matrix;*
(iii) $\mathcal{C}(P) = \sum_{i=1}^{m} \mathcal{C}(C_i') \otimes \mathcal{V}_i$.

Proof

(i) From the definition of $\widetilde{P}_i$'s and using Theorem 1.2.16. and Proposition 1.2.2. in [7], we have the following

$$\begin{aligned}\mathscr{V}_1 &= \mathscr{C}_\Sigma(\widetilde{P}_1 A_1) = \mathscr{C}_\Sigma(A_1),\\ \mathscr{V}_2 &= \mathscr{C}_\Sigma(\widetilde{P}_2 A_2)\\ &= \mathscr{C}_\Sigma((I - P_{A_1,\Sigma})A_2)\\ &= \mathscr{C}_\Sigma(A_1 : A_2) \cap \mathscr{C}_\Sigma(A_1)^\perp,\end{aligned}$$

$$\begin{aligned}\mathscr{V}_3 &= \mathscr{C}_\Sigma(\widetilde{P}_3 A_3)\\ &= \mathscr{C}_\Sigma((I - P_{\widetilde{P}_2 A_2,\Sigma})\widetilde{P}_2 A_3)\\ &= \mathscr{C}_\Sigma(\widetilde{P}_2(A_2 : A_3)) \cap \mathscr{C}_\Sigma(\widetilde{P}_2 A_2)^\perp,\\ &= \mathscr{C}_\Sigma((I - P_{A_1,\Sigma})(A_2 : A_3)) \cap \mathscr{C}_\Sigma((I - P_{A_1,\Sigma})A_2)^\perp,\\ &= (\mathscr{C}_\Sigma(A_1)^\perp \cap \mathscr{C}_\Sigma(A_1 : A_2 : A_3)) \cap (\mathscr{C}_\Sigma(A_1 : A_2) \cap \mathscr{C}_\Sigma(A_1)^\perp)^\perp,\\ &= (\mathscr{C}_\Sigma(A_1)^\perp \cap \mathscr{C}_\Sigma(A_1 : A_2 : A_3)) \cap (\mathscr{C}_\Sigma(A_1 : A_2)^\perp + \mathscr{C}_\Sigma(A_1)),\\ &= \mathscr{C}_\Sigma(A_1 : A_2 : A_3) \cap (\mathscr{C}_\Sigma(A_1)^\perp \cap \mathscr{C}_\Sigma(A_1 : A_2)^\perp + \mathscr{C}_\Sigma(A_1)^\perp \cap \mathscr{C}_\Sigma(A_1)),\\ &= \mathscr{C}_\Sigma(A_1 : A_2 : A_3) \cap \mathscr{C}_\Sigma(A_1 : A_2)^\perp,\end{aligned}$$

and in general

$$\mathscr{V}_i = \mathscr{C}_\Sigma(A_1 : A_2 : \cdots : A_i) \cap \mathscr{C}_\Sigma(A_1 : A_2 : \cdots : A_{i-1})^\perp,\ i = 1, 2, \ldots, m,\ A_0 = 0.$$

This shows that the subspaces $\mathscr{V}_i$'s are mutually orthogonal. Now we prove the second assertion. Clearly

$$\mathscr{V}_1 \oplus \mathscr{V}_2 \oplus \cdots \oplus \mathscr{V}_i \subseteq \mathscr{C}_\Sigma(A_1 : A_2 : \cdots : A_i).$$

Since $\mathscr{C}_\Sigma(A_1 : A_2 : \cdots : A_{i-1}) \subset \mathscr{C}_\Sigma(A_1 : A_2 : \cdots : A_i)$,

$$\dim(\mathscr{V}_i) = r(A_1 : \cdots : A_i) - r(A_1 : \cdots : A_{i-1}),$$

so that

$$\begin{aligned}\sum_{j=1}^{i} \dim(\mathscr{V}_j) &= r(A_1) + r(A_1 : A_2) - r(A_1) + \cdots + r(A_1 : \cdots : A_i) - r(A_1 : \cdots : A_{i-1})\\ &= r(A_1 : \cdots : A_i).\end{aligned}$$

This proves that the dimension of $\mathcal{V}_1 \oplus \mathcal{V}_2 \oplus \cdots \oplus \mathcal{V}_i$ equals the dimension of $\mathcal{C}_\Sigma(A_1 : A_2 : \cdots : A_i)$ and hence the result follows.

(ii) We need to show that P is an idempotent matrix. Let $H_i = P_{\widetilde{P}_i A_i, \Sigma}$ and $G_i = P_{C_i'} \otimes H_i$. Then, $P = \sum_{i=1}^{m} G_i$ and

$$PP = \sum_{i=1}^{m} G_i^2 + \sum_{i \neq j} G_i G_j.$$

On one hand, G_i is obviously an idempotent matrix as $P_{C_i'}$ and H_i are. On the other hand, from calculations in (i), H_i may be rewritten as

$$H_i = P_{\mathcal{A}_i, \Sigma} - P_{\mathcal{A}_{i-1}, \Sigma},$$

where $\mathcal{A}_i = (A_1 : \cdots : A_i)$ and thus,

$$\sum_{i=1}^{m} H_i = P_{\mathcal{A}_m, \Sigma} \text{ and } \sum_{i=1}^{m} r(H_i) = r(P_{\mathcal{A}_m, \Sigma}) = r(\mathcal{A}_m).$$

So, from Lemma 3 in [6] it follows that $H_i H_j = 0$ for $i \neq j$, which in turn implies that $G_i G_j = 0$ for $i \neq j$ since $G_i G_j = P_{C_i'} P_{C_j'} \otimes H_i H_j$. Hence $P^2 = P$ and P is a projector.

(iii) With notations and calculations introduced in the proof of (ii) it is clear that $\mathcal{C}(P) = \sum_{i=1}^{m} \mathcal{C}(G_i)$. But,

$$\mathcal{C}(G_i) = \mathcal{C}(P_{C_i'} \otimes H_i) = \mathcal{C}(P_{C_i'}) \otimes \mathcal{C}(H_i) = \mathcal{C}(P_{C_i'}) \otimes \mathcal{V}_i,$$

which completes the Proof of Theorem 16.3. □

We refer to the space $\mathcal{C}(P)$ as the mean space and it is used to estimate the mean parameters whereas $\mathcal{C}(P)^\perp$ is referred to as the residual space and it is used to create residuals.

When Σ is not known it should be estimated. The general idea is to use the variation in the residuals. For our purposes we decompose the residual space into $m + 1$ orthogonal subspaces and Theorem 16.4 shows how such a decomposition is made.

Theorem 16.4 *Let $\mathcal{C}(P)$ and $\mathcal{V}_i$ be given as in Theorem 16.3. Then*

$$\mathcal{C}(P)^\perp = I_1 \boxplus I_2 \boxplus \cdots \boxplus I_{m+1},$$

where

$$I_r = \mathscr{W}_{m-r+2} \otimes (\oplus_{i=1}^{r-1} \mathscr{V}_i)^{\perp},\ r = 1, 2, \ldots, m+1,$$

$$\mathscr{W}_r = \mathscr{C}(C'_{m-r+1}) \cap \mathscr{C}(C'_{m-r+2})^{\perp},\ r = 1, \ldots, m+1,$$

in which by convenience $(\oplus_{i=1}^{0} \mathscr{V}_i)^{\perp} = \emptyset^{\perp} = \mathscr{V}_0 = \mathbb{R}^p$, $C_0 = I$ *and* $C_{m+1} = 0$.

Proof On one hand, the conditions $\mathscr{C}(C'_i) \subseteq \mathscr{C}(C'_{i-1})$, $i = 2, 3, \ldots, m$, imply that $\mathscr{C}(C'_1)$ can be decomposed as a sum of orthogonal subspaces as follows:

$$\mathscr{C}(C'_1) = [\mathscr{C}(C'_1) \cap \mathscr{C}(C'_2)^{\perp}] \oplus [\mathscr{C}(C'_2) \cap \mathscr{C}(C'_3)^{\perp}] \oplus \cdots$$
$$\oplus [\mathscr{C}(C'_{m-1}) \cap \mathscr{C}(C'_m)^{\perp}] \oplus \mathscr{C}(C'_m).$$

On the other hand, by Theorem 16.3, the subspaces $\mathscr{V}_i$, $i = 1, 2, \ldots, m$; are orthogonal. Hence, the result follows by letting

$$\mathscr{W}_1 = \mathscr{C}(C'_m),\ \mathscr{W}_2 = \mathscr{C}(C'_{m-1}) \cap \mathscr{C}(C'_m)^{\perp},\ \ldots,$$
$$\mathscr{W}_m = \mathscr{C}(C'_1) \cap \mathscr{C}(C'_2)^{\perp},\ \mathscr{W}_{m+1} = \mathscr{C}(C'_1)^{\perp},$$
$$\mathscr{V}_{m+1} = (\oplus_{i=1}^{m} \mathscr{V}_i)^{\perp},\ \mathscr{V}_0 = \oplus_{i=1}^{m} \mathscr{V}_i \oplus \mathscr{V}_{m+1},$$

which completes the proof. □

The residuals obtained by projecting data to these subspaces are

$$H_r = (I - \sum_{i=1}^{r-1} P_{\widetilde{P}_i A_i, \Sigma}) X (P_{C'_{r-1}} - P_{C'_r}), \qquad (16.14)$$

$r = 1, 2, 3, \ldots, m+1$, and here we use for convenience $\sum_{i=1}^{0} P_{P_i A_i, S_i} = 0$.

In practice Σ is not known and should be estimated. A natural way to get an estimator of Σ is to use the sum of the squared estimated residuals. If Σ is not structured, we estimate the residuals in (16.14) with

$$R_r = (I - \sum_{i=1}^{r-1} P_{P_i A_i, S_i}) X (P_{C'_{r-1}} - P_{C'_r}),\ r = 1, 2, 3, \ldots, m+1,$$

where P_i and S_i are given as in Theorem 16.2. Thus, a natural estimator of Σ is obtained from the sum of the squared residuals, i.e.,

$$n\widehat{\Sigma} = R_1 R'_1 + R_2 R'_2 + \cdots + R_{m+1} R'_{m+1},$$

which is the maximum likelihood estimator.

16.3.2 Estimators When the Covariance Matrix is Linearly Structured

This section considers the extended growth curve model as in Definition 16.2, but with a linearly structured covariance matrix Σ. This Σ will be denoted $\Sigma^{(s)}$ so that $E \sim N_{p,n}(0, \Sigma^{(s)}, I_n)$.

The estimation procedure presented here was proposed in [9] and relies on the decomposition of the spaces done in Sect. 16.3.1. The inner products in the spaces $\mathscr{V}_i$ $(i = 1, 2, \ldots, m)$, the residuals in (16.14) are estimated sequentially, and finally the covariance matrix is estimated using all estimated residuals.

To start with, it is natural to use $Q_1 = H_1 H_1'$, $H_1 = X(I - P_{C_1'})$, to estimate the inner product in the space $\mathscr{V}_1$. We apply the general least squares approach and minimize $\mathrm{tr}\left\{(Q_1 - (n - r_1)\Sigma^{(s)})()'\right\}$ with respect to $\Sigma^{(s)}$, where the notation $(Y)()'$ stands for $(Y)(Y)'$. This is done in Lemma 16.2.

By $\mathrm{vec}\Sigma(K)$ we mean the patterned vectorization of the linearly structured matrix $\Sigma^{(s)}$, that is the columnwise vectorization of $\Sigma^{(s)}$ where all 0's and repeated elements (by modulus) have been disregarded. Then there exists a transformation matrix T such that

$$\mathrm{vec}\Sigma(K) = T\mathrm{vec}\Sigma^{(s)} \quad \text{or} \quad \mathrm{vec}\Sigma^{(s)} = T^{+}\mathrm{vec}\Sigma(K), \tag{16.15}$$

where T^{+} denotes the Moore-Penrose generalized inverse of T. Furthermore, from [7], we have

$$\frac{\mathrm{d}\Sigma^{(s)}}{\mathrm{d}\Sigma(K)} = (T^{+})'. \tag{16.16}$$

Lemma 16.2 *Let* $Q_1 = H_1 H_1' = X(I - P_{C_1'})X'$, $\widehat{\Upsilon}_1 = (n - r_1)I$. *Then, the minimizer of*

$$f_1(\Sigma^{(s)}) = \mathrm{tr}\left\{(Q_1 - (n - r_1)\Sigma^{(s)})()'\right\}$$

is given by

$$\mathrm{vec}\widehat{\Sigma}_1^{(s)} = T^{+}\left((T^{+})'\widehat{\Upsilon}_1'\widehat{\Upsilon}_1 T^{+}\right)^{-}(T^{+})'\widehat{\Upsilon}_1'\mathrm{vec}Q_1.$$

Proof We may write

$$\begin{aligned}\mathrm{tr}\left\{\left(\widehat{Q}_1 - (n - r_1)\Sigma^{(s)}\right)()'\right\} &= \left(\mathrm{vec}\left(\widehat{Q}_1 - (n - r_1)\Sigma^{(s)}\right)\right)'()\\ &= \left(\mathrm{vec}\widehat{Q}_1 - \widehat{\Upsilon}_1\mathrm{vec}\Sigma^{(s)}\right)'(),\end{aligned}$$

and thus

$$f_1(\Sigma^{(s)}) = \left(\mathrm{vec}\widehat{Q}_1 - \widehat{\Upsilon}_1 \mathrm{vec}\Sigma^{(s)}\right)' \left(\right). \tag{16.17}$$

Differentiating the expression in the right hand side of (16.17) with respect to $\mathrm{vec}\Sigma(K)$ and equalizing to 0, we get

$$-2\frac{\mathrm{d}\Sigma^{(s)}}{\mathrm{d}\Sigma(K)}\widehat{\Upsilon}_1'(\mathrm{vec}\widehat{Q}_1 - \widehat{\Upsilon}_1 \mathrm{vec}\Sigma^{(s)}) = 0. \tag{16.18}$$

From (16.15), (16.16) and (16.18) we obtain the linear equation

$$(T^+)'\widehat{\Upsilon}_1'\widehat{\Upsilon}_1 T^+ \mathrm{vec}\Sigma(K) = (T^+)'\widehat{\Upsilon}_1' \mathrm{vec}\widehat{Q}_1,$$

which is consistent and a general solution is given by

$$\mathrm{vec}\Sigma(K) = \left((T^+)'\widehat{\Upsilon}_1'\widehat{\Upsilon}_1 T^+\right)^- (T^+)'\widehat{\Upsilon}_1'\mathrm{vec}\widehat{Q}_1 + ((T^+)'\widehat{\Upsilon}_1'\widehat{\Upsilon}_1 T^+)^o z,$$

where z is an arbitrary vector. Hence, using (16.15) we obtain a unique minimizer of (16.17) given by

$$\mathrm{vec}\widehat{\Sigma}_1^{(s)} = T^+ \left((T^+)'\widehat{\Upsilon}_1'\widehat{\Upsilon}_1 T^+\right)^- (T^+)'\widehat{\Upsilon}_1'\mathrm{vec}\widehat{Q}_1, \tag{16.19}$$

and the proof is complete. □

Lemma 16.2 gives the first estimator of $\Sigma^{(s)}$. Assuming that $\widehat{\Sigma}_1^{(s)}$ is positive definite (which always holds for large n), we can use $\widehat{\Sigma}_1^{(s)}$ to define the inner product in the space $\mathcal{V}_1$. Therefore we consider $\mathcal{C}_{\widehat{\Sigma}_1^{(s)}}(A_1)$ instead of $\mathcal{C}_{\Sigma^{(s)}}(A_1)$. At the same time, an estimator of M_1 and H_2 are obtained by projecting the observations on $\mathcal{C}(C_1') \otimes \mathcal{V}_1$ and $\left(\mathcal{C}(C_1') \cap \mathcal{C}(C_2')^\perp\right) \otimes \mathcal{V}_1^\perp$ respectively, i.e.,

$$\begin{aligned} \widehat{M}_1 &= P_{A_1, \widehat{\Sigma}_1^{(s)}} X P_{C_1'}, \\ \widehat{H}_2 &= (I - P_{A_1, \widehat{\Sigma}_1^{(s)}}) X (P_{C_1'} - P_{C_2'}). \end{aligned} \tag{16.20}$$

At the second step, the estimator of $\Sigma^{(s)}$ is obtained using the sum of Q_1 and $\widehat{H}_2\widehat{H}_2'$. Note that

$$\widehat{H}_2\widehat{H}_2' = (I - P_{A_1, \widehat{\Sigma}_1^{(s)}}) X (P_{C_1'} - P_{C_2'}) X' (I - P_{A_1, \widehat{\Sigma}_1^{(s)}})'.$$

Put $\widehat{\mathscr{T}}_1 = I - P_{A_1,\widehat{\Sigma}_1^{(s)}}$ and $W_1 = X(P_{C_1'} - P_{C_2'})X'$. Then $\widehat{H}_2\widehat{H}_2' = \widehat{\mathscr{T}}_1 W_1 \widehat{\mathscr{T}}_1'$ and Q_1 is independent of W_1. Therefore, the conditional distribution of $\widehat{H}_2\widehat{H}_2'$ with respect to Q_1 is

$$\widehat{H}_2\widehat{H}_2'|Q_1 \sim W_p\left(\widehat{\mathscr{T}}_1\Sigma^{(s)}\widehat{\mathscr{T}}_1', r_1 - r_2\right),$$

where $W_p(\cdot,\cdot)$ stands for the Wishart matrix.

The following lemma gives a second estimator of $\Sigma^{(s)}$ where again the general least squares approach is employed.

Lemma 16.3 *Let* $\widehat{Q}_2 = Q_1 + \widehat{H}_2\widehat{H}_2'$, $\widehat{\Upsilon}_2 = (n - r_1)I + (r_1 - r_2)\widehat{\mathscr{T}}_1 \otimes \widehat{\mathscr{T}}_1$. *Then, the minimizer of*

$$f_2(\Sigma^{(s)}) = \mathrm{tr}\left\{(\widehat{Q}_2 - [(n - r_1)\Sigma^{(s)} + (r_1 - r_2)\widehat{\mathscr{T}}_1\Sigma^{(s)}\widehat{\mathscr{T}}_1')()'\right\}$$

is given by

$$\mathrm{vec}\widehat{\Sigma}_2^{(s)} = T^+\left((T^+)'\widehat{\Upsilon}_2'\widehat{\Upsilon}_2T^+\right)^-(T^+)'\widehat{\Upsilon}_2'\mathrm{vec}\widehat{Q}_2.$$

Proof Similar as in Lemma 16.2. □

Now assume that $\widehat{\Sigma}_2^{(s)}$ is positive definite and use it to define the inner product in $\mathscr{V}_2$, i.e., consider $\mathscr{C}_{\widehat{\Sigma}_2^{(s)}}(\widehat{\mathscr{T}}_1A_2)$ instead of $\mathscr{C}_{\Sigma^{(s)}}(\mathscr{T}_1A_2)$. This gives us an estimator of M_2 and an estimator of H_3 by projecting the observations on $\mathscr{C}(C_2')\otimes\mathscr{C}_{\widehat{\Sigma}_2^{(s)}}(\widehat{\mathscr{T}}_1A_2)$ and $(\mathscr{C}(C_2')\cap\mathscr{C}(C_3')^\perp)\otimes\left(\mathscr{C}_{\widehat{\Sigma}_1^{(s)}}(A_1) + \mathscr{C}_{\widehat{\Sigma}_2^{(s)}}(\widehat{\mathscr{T}}_1A_2)\right)^\perp$ respectively, i.e.,

$$\begin{aligned}
\widehat{M}_2 &= P_{\widehat{\mathscr{T}}_1A_2,\widehat{\Sigma}_2^{(s)}}XP_{C_2'},\\
\widehat{H}_3 &= \widehat{\mathscr{T}}_2X(P_{C_2'} - P_{C_3'}),\\
\widehat{\mathscr{T}}_2 &= I - P_{A_1,\widehat{\Sigma}_1^{(s)}} - P_{\widehat{\mathscr{T}}_1A_2,\widehat{\Sigma}_2^{(s)}}.
\end{aligned}$$

To derive a third estimator of $\Sigma^{(s)}$, the idea is to use the sum $\widehat{Q}_3 = \widehat{Q}_2 + \widehat{H}_3\widehat{H}_3'$ in a similar way as in Lemma 16.2 or Lemma 16.3. We continue the same process until all residuals are estimated and then use the sum of the estimated residuals to estimate the covariance matrix that can be understood as a dispersion matrix. After the $(r-1)$th stage of the process, we have already $r-1$ estimates of $\Sigma^{(s)}$. In Lemma 16.4, we show how to obtain the rth estimate of $\Sigma^{(s)}$. Before we state it, note that after the $(r-1)$th stage, we have also the following quantities:

$$\begin{aligned}
W_i &= X(P_{C_i'} - P_{C_{i+1}'})X' \sim W_p(\Sigma^{(s)}, r_i - r_{i+1}), \qquad (16.21)\\
&\quad i = 0, 1, 2, \ldots, m,
\end{aligned}$$

$$\widehat{P}_j = \widehat{T}_{j-1}\widehat{T}_{j-2} \times \cdots \times \widehat{T}_0,\ \widehat{T}_0 = I,\ j = 1, 2, \ldots, r-1, \tag{16.22}$$

$$\widehat{T}_i = I - P_{\widehat{P}_i A_i, \widehat{\Sigma}_i^{(s)}},\ i = 1, 2, \ldots, r-1, \tag{16.23}$$

$$\widehat{\mathscr{T}}_i = I - \sum_{j=1}^{i} P_{\widehat{P}_j A_j, \widehat{\Sigma}_j^{(s)}},\ i = 1, 2, \ldots, r-1, \tag{16.24}$$

$$\widehat{H}_i\widehat{H}_i' = \widehat{\mathscr{T}}_{i-1} W_{i-1} \widehat{\mathscr{T}}_{i-1}',\ \widehat{\mathscr{T}}_0 = I,\ i = 1, 2, \ldots, r, \tag{16.25}$$

$$\widehat{Q}_i = \sum_{j=1}^{i} \widehat{H}_j\widehat{H}_j' = \sum_{j=1}^{i} \widehat{\mathscr{T}}_{j-1} W_{j-1} \widehat{\mathscr{T}}_{j-1}', \tag{16.26}$$

$$i = 1, 2, \ldots, r,$$

$$\widehat{\Upsilon}_i = \sum_{j=1}^{i} (r_{j-1} - r_j)\widehat{\mathscr{T}}_{j-1} \otimes \widehat{\mathscr{T}}_{j-1},\ i = 1, 2, \ldots, r. \tag{16.27}$$

Lemma 16.4 *Let $\widehat{\mathscr{T}}_i$, $\widehat{Q}_r$ and $\widehat{\Upsilon}_r$ be defined as in (16.24), (16.26) and (16.27), respectively. Then, the minimizers of*

$$f_r(\Sigma^{(s)}) = \mathrm{tr}\left\{\left(\widehat{Q}_r - \sum_{i=1}^{r}(r_{i-1} - r_i)\widehat{\mathscr{T}}_{i-1}\Sigma^{(s)}\widehat{\mathscr{T}}_{i-1}'\right)()'\right\},\ r = 1, 2, \ldots, m+1,$$

are given by

$$\mathrm{vec}\widehat{\Sigma}_r^{(s)} = T^+\left((T^+)'\widehat{\Upsilon}_r'\widehat{\Upsilon}_r T^+\right)^-(T^+)'\widehat{\Upsilon}_r'\mathrm{vec}\widehat{Q}_r.$$

Proof Similar as in Lemma 16.2. □

Lemma 16.4 gives the rth estimate of $\Sigma^{(s)}$ and the $(m+1)$th estimate of $\Sigma^{(s)}$ can be understood as the estimator for the dispersion matrix as it uses all information that we have in the residuals. Now the results at our disposal permit us to propose estimators of parameters in the extended growth curve model with a linearly structured covariance matrix.

Theorem 16.5 *Consider the extended growth curve model given by (16.2). Then*

(i) A consistent estimator of the structured covariance matrix $\Sigma^{(s)}$ is given by

$$\mathrm{vec}\widehat{\Sigma}_{m+1}^{(s)} = T^+\left((T^+)'\widehat{\Upsilon}_{m+1}'\widehat{\Upsilon}_{m+1}T^+\right)^-(T^+)'\widehat{\Upsilon}_{m+1}'\mathrm{vec}\widehat{Q}_{m+1}. \tag{16.28}$$

(ii) An unbiased estimator of the mean is given by

$$\widehat{E[X]} = \sum_{i=1}^{m}(I - \widehat{T}_i)XC_i'(C_iC_i')^{-}C_i, \tag{16.29}$$

where $\widehat{T}_i$, $\widehat{Q}_r$ and $\widehat{\Upsilon}_r$ are defined as in (16.23), (16.26) and (16.27), respectively.

Proof

(i) The consistency of the estimator in (16.28) is established through the following implications, where the notation '$\xrightarrow{p}$' means convergence in probability.

$$\begin{array}{c}
\frac{1}{n-r_1}Q_1 \xrightarrow{p} \Sigma^{(s)} \\
\Downarrow \\
\widehat{\Sigma}_1^{(s)} \xrightarrow{p} \Sigma^{(s)} \\
\Downarrow \\
\widehat{\Sigma}_2^{(s)} \xrightarrow{p} \Sigma^{(s)} \\
\Downarrow \\
\vdots \\
\Downarrow \\
\widehat{\Sigma}_{m+1}^{(s)} \xrightarrow{p} \Sigma^{(s)}.
\end{array}$$

The above implications can be easily checked. The first line follows from the well known fact that the sample covariance matrix is a consistent estimator of the true covariance matrix. The rest is established using Cramér-Slutsky's theorem [1].

(ii) Next we establish the unbiasedness of the estimator given in (16.29). From (i) Theorem 16.3, and using uniqueness property of projectors, it is possible to rewrite the estimated mean as

$$\widehat{E[X]} = \sum_{i=1}^{m} P_{\mathcal{A}_i, \widehat{\Sigma}_i^{(s)}} X(P_{C_i'} - P_{C_{i+1}'}).$$

Using the linearity of the expectation operator and independence of $P_{\mathscr{A}_i, \widehat{\Sigma}_i^{(s)}}$ and $X(P_{C_i'} - P_{C_{i+1}'})$, we have

$$
\begin{aligned}
E[\widehat{E[X]}] &= \sum_{i=1}^{m} E\left[P_{\mathscr{A}_i, \widehat{\Sigma}_i^{(s)}}\right] E\left[X(P_{C_i'} - P_{C_{i+1}'})\right] \\
&= E\left[P_{\mathscr{A}_m, \widehat{\Sigma}_m^{(s)}}\right] E\left[X P_{C_m'}\right] \\
&\quad + E\left[P_{\mathscr{A}_{m-1}, \widehat{\Sigma}_{m-1}^{(s)}}\right] E\left[X(P_{C_{m-1}'} - P_{C_m'})\right] \\
&\quad + E\left[P_{\mathscr{A}_{m-2}, \widehat{\Sigma}_{m-2}^{(s)}}\right] E\left[X(P_{C_{m-2}'} - P_{C_{m-1}'})\right] + \cdots \\
&\quad + E\left[P_{\mathscr{A}_1, \widehat{\Sigma}_1^{(s)}}\right] E\left[X(P_{C_1'} - P_{C_2'})\right].
\end{aligned}
$$

Since $E[X] = \sum_{i=1}^{m} A_i B_i C_i$, using the facts that $\mathscr{C}(C_i') \subseteq \mathscr{C}(C_{i-1}')$ and

$$
P_{\mathscr{A}_n, \widehat{\Sigma}_n^{(s)}} \left(\sum_{i=1}^{n} A_i B_i C_i \right) = \sum_{i=1}^{n} A_i B_i C_i,
$$

we get

$$
\begin{aligned}
E[\widehat{E[X]}] &= \left(\sum_{i=1}^{m} A_i B_i C_i \right) P_{C_m'} + \left(\sum_{i=1}^{m-1} A_i B_i C_i \right) (P_{C_{m-1}'} - P_{C_m'}) + \cdots \\
&\quad + \left(\sum_{i=1}^{2} A_i B_i C_i \right) (P_{C_2'} - P_{C_3'}) + A_1 B_1 C_1 (P_{C_1'} - P_{C_2'}) \\
&= A_m B_m C_m + \left(\sum_{i=1}^{m-1} A_i B_i C_i \right) P_{C_m'} + A_{m-1} B_{m-1} C_{m-1} \\
&\quad + \left(\sum_{i=1}^{m-2} A_i B_i C_i \right) P_{C_{m-1}'} - \left(\sum_{i=1}^{m-1} A_i B_i C_i \right) P_{C_m'} + \cdots \\
&\quad + A_2 B_2 C_2 + A_1 B_1 C_1 P_{C_2'} - \left(\sum_{i=1}^{2} A_i B_i C_i \right) P_{C_3'} \\
&\quad + A_1 B_1 C_1 - A_1 B_1 C_1 P_{C_2'}.
\end{aligned}
$$

Canceling the opposite terms in the last expression, we get the desired result.

□

16.4 Concluding Remarks

We considered a class of multivariate models that present some challenges about estimation of model parameters and hypothesis testing. We systematically reviewed how space decomposition has been successful to address these challenges. The multivariate models presented is this work are some extensions of Generalized Multivariate Analysis of Variance (GMANOVA) namely the bilinear regression that include random effects and latent variables and the extended GMANOVA with linearly structured covariance matrix. In this review, we pointed out two approaches used by several authors. The first consists on the decomposition of complex models into tractable ones easy to handle for estimation and testing. The main idea used in this approach is to make an orthogonal decomposition of the working model and partition it into independent models. The second approach is about obtaining explicit estimators where the estimation of the mean structure is based on the projection of the observations on the space generated by the design matrices while the estimation of the covariance parameters is based on the projection of the observations on the residual space that is orthogonal complement to the design matrices.

References

1. Cramér, H.: *Mathematical Methods in Statistics*. Princeton University Press, Princeton (1946)
2. Filipiak, K., von Rosen, D.: On MLEs in an extended multivariate linear growth curve model. Metrika **75**(8), 1069–1092 (2012)
3. Henderson, C.R.: Sire evaluation and genetic trends. J. Anim. Sci. **1973**, 10–41 (1973). (Symposium)
4. Hu, J.: Properties of the explicit estimators in the extended growth curve model. Statistics **44**(5), 477–492 (2010)
5. Hu, J., Yan, G., You, J.: Estimation for an additive growth curve model with orthogonal design matrices. Bernoulli **17**(4), 1400–1419 (2011)
6. Khatri, C.G.: Some results for the singular normal multivariate regression models. Sankhyā: Indian J. Stat. Ser. A **30**, 267–280 (1968)
7. Kollo, T., von Rosen, D.: Advanced Multivariate Statistics with Matrices. Springer, Dordrecht (2005)
8. Ngaruye, I., Nzabanita, J., von Rosen, D.: Small area estimation under a multivariate linear model for repeated measures data. Commun. Stat. Theory Method **46**(21), 10835–10850 (2017)
9. Nzabanita, J., von Rosen, D., Singull, M.: Extended GMANOVA model with a linearly structured covariance matrix. Math. Methods Stat. **24**(4), 280–291 (2015)
10. Pothoff, R.F., Roy, S.N.: A generalized multivariate analysis of variance model useful especially for growth curve problems. Biometrika **51**, 313–326 (1964)
11. Verbyla, A.P., Venables, W.N.: An extension of the growth curve model. Biometrika **75**(1), 129–138 (1988)
12. von Rosen, D.: Maximum likelihood estimators in multivariate linear normal models. J. Multivar. Anal. **31**(2), 187–200 (1989)
13. von Rosen, T., von Rosen, D.: Small area estimation using reduced rank regression models. Commun. Stat. Theory Method **49**, 3286–3297 (2019). https://doi.org/10.1080/03610926.2019.1586946

Chapter 17
Detection of Sparse and Weak Effects in High-Dimensional Feature Space, with an Application to Microbiome Data Analysis

Tatjana Pavlenko, Annika Tillander, Justine Debelius, and Fredrik Boulund

Abstract We present a family of goodness-of-fit (GOF) test statistics specifically designed for detection of *sparse-weak* mixtures, where only a small fraction of the observational units are contaminated arising from a different distribution. The test statistics are constructed as sup-functionals of weighted empirical processes where the weight functions employed are the Chibisov-O'Reilly functions of a Brownian bridge. The study recovers and extends a number of previously known results on sparse detection using a weighted GOF (wGOF) approach. In particular, the results obtained demonstrate the advantage of our approach over a common approach that utilizes a family of regularly varying weight functions. We show that the Chibisov-O'Reilly family has important advantages over better known approaches as it allows for optimally adaptive, fully data-driven test procedures. The theory is further developed to demonstrate that the entire family is a flexible device that adapts to many interesting situations of modern scientific practice where the number of observations stays fixed or grows very slowly while the number of automatically measured features grows dramatically and only a small fraction of these features are useful. Numerical studies are performed to investigate the finite sample properties of the theoretical results. We shown that the Chibisov-O'Reilly family compares favorably to related test statistics over a broad range of sparsity and weakness regimes for the Gaussian and high-dimensional Dirichlet types of sparse mixture. Finally, an example of human gut microbiome data set is presented to illustrate

---

T. Pavlenko (✉)
Department of Mathematics, KTH Royal Institute of Technology, Stockholm, Sweden
e-mail: pavlenko@math.kth.se

A. Tillander
Department of Statistics and Machine Learning, Linköping University, Linköping, Sweden
e-mail: annika.tillander@liu.se

J. Debelius · F. Boulund
The Centre for Translational Microbiome Research (CTMR), Department of Microbiology, Tumor, and Cell Biology, Karolinska Institutet, Stockholm, Sweden
e-mail: justine.debelius@ki.se; fredrik.boulund@ki.se

© Springer Nature Switzerland AG 2020

T. Holgersson, M. Singull (eds.), *Recent Developments in Multivariate and Random Matrix Analysis*, https://doi.org/10.1007/978-3-030-56773-6_17

that the family of tests has found applications in real-life sparse signal detection problems where the sample size is small in relation to the features dimension.

17.1 Introduction

One of the problems of sparse signal detection is to detect a small fraction of relatively weak signals hidden among the large number of those which are pure noise. In its generally, this is a problem of sparse-weak mixture detection which is stated as follows. Let $\mathbb{X}$ be a set of independent observational units, $\mathbb{X} = (X_1, X_2 \dots, X_n)$. The goal is to assess the validity of the null hypothesis that $\mathbb{X}$ follows a known and fully specified continuous distribution F, against an alternative where a *small* proportion ε_n of the variables are contaminated and have a different, usually unknown distribution $Q_n \neq F$. More precisely, given a random sample $\mathbb{X}$ the corresponding hypothesis to test is given as

$$H_0 : X_i \overset{iid}{\sim} F, \tag{17.1}$$

$$\text{versus} \quad H_{1,n} : X_i \overset{iid}{\sim} (1 - \varepsilon_n)F + \varepsilon_n Q_n,$$

where Q_n can be thought as a distribution modeling the statistical variation of the non-null effects, i.e., sparse contamination phenomena. The contamination effect is calibrated according to $\varepsilon_n = n^{-\beta}$, with $1/2 < \beta < 1$.

In what follows, we will call a *relevant* feature a *signal* and *irrelevant* feature *noise*.

The problem of sparse signal recovery, or what is known in the literature as *variable selection*, is of fundamental importance for the theory of statistical learning in infinite or high dimensions. The development of learning techniques that can accommodate sparsity is driven by the a general phenomenon arising in many areas of modern applied science and engineering: the number of feature variables is much higher than the number of observations, but the degree of freedom of the underlying model is relatively low.

A theoretical analysis of this problem can be found in, e.g., Arias-Castro [2, 4] and references therein, and related applications in signal detection and processing are extensive, see, e.g., [3] and [5].

17.1.1 Hypothesis Testing Based Sparse Signal Detection

Let us briefly review some specific examples where the sparse signal detection problem is formulated as a hypothesis testing problem.

Example 17.1 Theory of radar, sparse signal detection in the presence of noise in n frequency diapasons. A very early study on sparse mixture detection is the paper by Dobrushin, [13], where the hypothesis testing based approach was exploited to solve the problem of multi-channel signal detection in radio-location. Specifically, the data observed were assumed to be a sample of independent random variables $X_1, \ldots, X_n$, representing the voltages measured over n frequency diapasons and following the Rayleigh distribution with the density

$$f_{X_i}(x|\gamma_i) = \frac{2x}{\gamma_i}\exp(-x^2/\gamma_i), \quad x \geq 0,$$

where $\gamma_i = E(X_i^2)$, $i \in [\![n]\!]$ and $[\![n]\!]$ denotes the set $\{1, \ldots, n\}$. In communication theory, Rayleigh distributions are used to model scattered signals that reach a receiver by multiple paths. In the absence of noise, γ_is are all equal to 1, whereas in the presence of signal, exactly one of γ_is becomes a known value $\gamma > 1$. Hence, there are two competing hypothesis concerning γ_i:

$$\begin{aligned} &H_0 : \gamma_i = 1,\ i \in [\![n]\!] \\ \text{versus} \quad &H_{1,n} : \gamma_i = 1 + (\gamma - 1)\mathbb{1}_{i=J}, \quad J \sim \mathcal{U}_n, \end{aligned} \tag{17.2}$$

where $\mathcal{U}_n$ denotes the uniform distribution defined on $[\![n]\!]$. In words, H_0 states the absence of signals, meaning that the source of voltage observed in all n channels is noise. $H_{1,n}$ states that all but one γ_i is equal a known value γ, i.e., the signal appears only in channel j, where index j follows the uniform distribution $\mathcal{U}_n$ defined on $[\![n]\!]$. To have the signal distinguishable from noise given that the signal appears only in one channel, it is necessary that the strength of the signal, γ, grows with the number of channels observed. In [13], Dobrushin derived a sharp, growing n asymptotic boundary for the minimum signal strength to be reliably detected.

Example 17.2 Sparse-weak, two-components Gaussian mixture. Theoretical developments in sparse signal detection have largely been concentrated on Gaussian signals. Specifically, in the context of general testing problem (17.1), the sparse non-Gaussianity detection problem is formulated as follows. By setting $F = \mathcal{N}(0, 1)$ and $Q_n = \mathcal{N}(\mu_n, \sigma^2)$ in (17.1), the goal is to test

$$\begin{aligned} &H_0 : X_i \overset{iid}{\sim} \mathcal{N}(0, 1), \\ \text{versus} \quad &H_{1,n} : X_i \overset{iid}{\sim} (1 - \varepsilon_n)\mathcal{N}(0, 1) + \varepsilon_n\mathcal{N}(\mu_n, \sigma^2). \end{aligned} \tag{17.3}$$

Interesting cases correspond to specification of (ε_n, μ_n) that are calibrated with a pair of parameters (β, r) as follows

$$\varepsilon_n = n^{-\beta}, \quad \mu_n = \sqrt{2r\log(n)}, \quad 1/2 < \beta < 1, \quad 0 < r < 1, \tag{17.4}$$

where β is the *sparsity index* and r may be thought as the signal strength or *weakness* parameter. This parametrization can be explained as follows: recall first that for large n, the maximum of n iid standard Gaussian random variables $Z_1, Z_2, \ldots$, is sharply concentrated around $\sqrt{2\log n}$ and

$$P\left(\max_{1\leq i\leq n}|Z_i| \geq \sqrt{2\log n}\right) \to 0, \quad \text{as} \quad n \to \infty.$$

Hence, for any fixed ε_n, as $n \to \infty$, the contamination strength of the level of $\mu_n > \sqrt{2\log n}(1+\delta)$ would be perfectly detectable by for example, the maximum-type statistics. In the same way, for any fixed $\mu_n = \mu$, sparsity levels of $\varepsilon_n \gg n^{-1/2}$ are perfectly detectable by the sum-type statistics. These properties explain the choice of $\varepsilon_n = n^{-\beta}$ and the contamination strength in the sparse Gaussian model, which makes the problem very hard but yet solvable. Since any mixture parametrized with such (r, β) where either $r > 1$ or $\beta < 0.5$ are easily detectable, the interesting regime is in the region of *sparsity and weakness*, i.e., where both $0 < r < 1$ and $0.5 < \beta < 1$. The testing problem (17.3) can be extended to the case where μ_n varies under the sparse-mixture alternative with some distribution.

The asymptotic optimality theory developed for these types of detection problems provides a benchmark for the development of data-driven procedures. It turns out that there is a general *phase transition phenomena*, meaning that in the two-dimensional parametric phase space $\{(\beta, r) : 1/2 < \beta < 1, r > 0\}$ there is a demarcation line which splits the whole space into two separate regions where detection problem is distinctly different. Ingster [18] and later, Donoho and Jin [14] derived the class of sparse and weak settings in (17.4) where the true signals are so sparse and so weak that reliable detection is impossible.

Recent developments in detecting sparse and weak signals are also motivated by modern experiments in astronomy and cosmology where the data produced can be interpreted as images and where there is a well-defined null hypothesis. Examples given by, e.g., Vielva [37] and Bennet et al. [6] deal with studies of the Cosmic Microwave Background (CMB) which is a relic of radiation emitted when the Universe was about 370,000 years old. Using the easiest inflation model CMB temperature fluctuations should behave as a realization of a zero-mean Gaussian random variable in each pixel of an images. The resulting Gaussian field is completely determined by its power spectrum and the goal is to detect sparsely scattered non-Gaussian signatures (e.g., hot and cold spots) in the CMB.

Example 17.3 High-dimensional classification when relevant features are sparse and weak. In applied classification problems, examples include but are not restricted to genomic and metabolomics, (see e.g., [23, 24] and [25]) the number of feature variables p is large compared to the number of training samples, n. This so-called *high-dimensional* scenario when $p \gg n$ (for example, $p = 10{,}000$, $n = 100$) usually means that the informative features are *sparse*, and a small n relates to *weak* separation power. In this setting, the problem of detection and identification of a set of informative features is closely related to the testing problem in Example 17.2.

For illustration, assume that we have the training samples (X_i, c_i), $i \in [\![n]\!]$ from two different classes and the goal is to predict the class label $c \in \{-1, +1\}$ for a fresh feature vector X. To model the high-dimensional setting with $p \gg n$, consider a sequence of classification problems indexed by (n, p) where feature vectors $X_i \sim \mathcal{N}_p(c_i \cdot \mu, I_p)$ with $Y_i = \pm 1$ denoting the class label and vector μ denoting the constant mean. Using a vector of feature scores $Z = (\sqrt{n})^{-1} \sum_{i=1}^{n} (X_i \cdot c_i)$ which contains the evidence in favor of each feature informativeness, we note that $Z \in \mathcal{N}_p(\sqrt{n}\mu, I_p)$. Furthermore, let the number of features p be the driving index, the sparse and weak model for classification is then

$$\sqrt{n}\mu_j \overset{iid}{\sim} (1 - \varepsilon_p)\delta_0 + \varepsilon_p \delta_{\mu_\tau}, \quad j \in [\![p]\!],$$

where sparsity and weakness parameters are calibrated as

$$\varepsilon_p = p^{-\beta}, \quad \mu_p = \sqrt{2r \log p}, \quad \text{and} \quad n = n_p = p^{\tau},$$

for $\tau \in (0, 1)$ (regular growth regime), $r > 0$ and $0 < \beta < 1 - \tau$. This asymptotics is called *regular growth* regime: when $p \to \infty$, n_p grows with p but is still much smaller than p. In this regime, only a small fraction of entries of μ is non-zero, and the non-zero entries are individually low informative, contributing weakly to the classification decision. The goal is to survey many features simultaneously and identify a few useful ones allowing successful classification.

Calibrating the classification model with four parameters, p, n, β and r, Ingster et al. [19] and later Fan et al. [17] and Ji et al. [22], show that the detection boundary phenomenon which has been proved for the sparse Gaussian mixture of Example 17.2, extends to high-dimensional classification. Specifically, the sparsity-weakness domain $(\beta, r) \in (0, 1)$ splits by a so-called *classification boundary*, which is a curve such that certain classification procedures succeed asymptotically with n and p tending to infinity, when (β, r) lies above the curve and fail if (β, r) is below the curve. These asymptotic results are further extended by Tillander [36] and Stepanova and Pavlenko [35] to the sparse χ^2- and F-types mixtures which are exploited for detection group-sparse signals in classification settings.

Despite very different nature of the signal detection examples above, all of them can be viewed as a particular case of the general problem of testing *goodness-of-fit* (GOF) to a presumptive model. Using a random sample or a set of input statistics which is the data summary, the goal is to determine whether there is a significant statistical evidence for the presence of signals. A common approach in such problems is based on the deviations of the empirical distribution functions (EDF) resulting in the classical Kolmogorov-Smirnov (KS) test statistic, which has a number of attractive theoretical properties such as consistency, high power in detecting the a shift in the median of the distribution and existing of efficient procedures to compute p-values. In what follows, we give analytical arguments why the KS does not have power enough to detect the sparse-weak contamination, i.e., when only a few of n signals are contaminated and the contamination strength is small. In words, this flaw stems from poor sensitivity of the tests to the quality of the fit in the tails of the tested distribution, when these tail events are primarily concerned with in the detection of sparse-weak mixture.

Hypothesis testing based approach to sparse signal detection is extensively studied in the literature, see, e.g., [2, 3] and [28] for theoretical results, and [5, 27] and [40] and references therein, for related applied problems of sparse signal detection and processing. We contribute to this line of research by taking the weighted empirical process view point on the GOF tests, we show that in the context of sparse alternatives, certain classical tests, including KS and Anderson-Darling tests [1], may benefit essentially from using properly selected weight functions. Specifically, we study a class of GOF test statistics based on the supremum of weighted empirical processes.

17.1.2 Content

On the theoretical front, in Sect. 17.2, we describe our analytical framework, relate the KS family of sup-norm statistics to the weighted empirical processes and provide a general consistency result for the weighted KS. We further define a family of wGOF statistics where the weighting procedure is governed by the Chibisov-O'Reilly type of functions and investigate theoretical properties of the family of tests. In Sect. 17.3, we show that the proposed family may be considered as a competitor to the so-called *Higher Criticism* (HC) statistic, presented in [14] and widely used in the literature. Section 17.4 provides results of empirical study. We point out that the whole Chibisov-O'Reilly family of statistics is optimally adaptive, i.e., its use does not require knowledge of sparsity and weakness parameters. Using this property, we perform systematic power comparisons among the related optimally adaptive tests and develop a general framework for applying the wGOF tests under high-dimensional, sparse-weak Dirichlet mixture. For the latter we specify the empirical detection boundary and showcase scenarios when Chibisov-O'Reilly's test has improved power compared to other wGOF tests commonly used in the literature.

Recent advances in biotechnological analysis methods, including high-throughput sequencing, proteomics, and metabolomics, provide new opportunities for sparse signal detection. To illustrate the utility of our approach, we apply it to human microbiome data from the American Gut Project. A common goal of the data analysis is to detect differentially abundant features.

Sparse feature detection in microbiome data has proven to be a challenging statistical problem [30, 38]. Because of several non-quantitative steps in the sample processing, microbiome data becomes inherently compositional; it is constrained to the unit p-dimensional simplex [23] where all feature observations are interdependent. Analysis is further hampered by the large number of zero-counts in the data [25]. These zeros are likely structural, sampling, and outlier zeros, present due to the biological realities of free-living organisms and technological under-sampling (under-sampled because of the current technical inability to exhaustively capture all DNA fragments in microbiome sample). Finally, the data is high-dimensional, where a study of even a few hundred samples may contain ten times that many features, adding to challenges for power and detection. As a result, microbiome studies present a potentially interesting application of sparsity signal detection.

17.1.3 Notational Convention

Some notations used throughout the paper are as follows. The symbols $\stackrel{\mathscr{D}}{=}$ and $\stackrel{\text{a.s.}}{=}$ are used for equality in distribution and almost surely, respectively. The symbols $\stackrel{\mathscr{D}}{\longrightarrow}$ and $\stackrel{P}{\longrightarrow}$ denote convergence in distribution and convergence in probability, respectively. We denote by $a_n = o(b_n)$ if $\limsup_{n\to\infty} \frac{|a_n|}{|b_n|} = 0$ and $a_n = O(b_n)$ if $\limsup_{n\to\infty} \frac{|a_n|}{|b_n|} < \infty$, these asymptotic notations extend naturally to probabilistic setups, denoted by o_P and O_P where limits are in the sense of convergence in probability.

17.2 The Weighted Kolmogorov-Smirnov Statistics

For the following, we need some further notations. For a given sequence $\mathbb{X} = (X_1, X_2, \ldots, X_n, \ldots)$ of i.i.d random variables with the continuous cumulative distribution function (CDF) F on $\mathbb{R}$, we let $\mathbb{F}_n(x) = n^{-1} \sum_{i=1}^{n} \mathbf{1}(X_i \le x)$ be the EDF based on a given sample $X_1, \ldots, X_n$. The goal is to test the hypothesis of GOF $H_0 : F = F_0$, some specific CDF, versus either a two-sided alternative $H_1 : F \neq F_0$ or a one-sided alternative $H_1^{'} : F > F_0$. The KS statistics the two-sided version is defined by an L_∞-distance (a sup-norm metrics) measuring discrepancy of the observed data from the postulated CDF F_0) over a continuous variable x as

$$\mathrm{KS}_n = \sqrt{n}\|\mathbb{F}_n - F_0\|_\infty = \sup_{x\in\mathbb{R}} \sqrt{n}\, |\mathbb{F}_n(x) - F_0(x)| \tag{17.5}$$

To explain the problem with poor tail sensitivity to deviations from a null distribution, observe first that the difference $|\mathbb{F}_n(x) - F_0(x)|$ is considered *without* taking into account the value of $F_0(x)$ at x. Indeed, since under the joint null hypothesis that all X_i follow the same distribution F_0, we have $n\mathbb{F}_n(x) \sim Bin(n, F_0(x))$ for any fixed $x \in \mathbb{R}$ and $\mathrm{Var}(\mathbb{F}_n(x)) = \frac{1}{n} F_0(x)(1 - F_0(x))$. The variance of $\mathbb{F}_n(x)$ varies considerably throughout the range of x, having a maximum at the median of the distribution and smaller values around the tails. This implies that the EDF $\mathbb{F}_n$ is a moderate distance away from the hypothesized $F_0(x)$ over large parts of the $\mathbb{R}$. This phenomena is of special importance for reliable detection of sparse and weak effects where all the actions takes place at the tails and where the KS statistic may fail to detect the falsity of the postulated null hypothesis. Rényi [33] proposed that, instead of the absolute deviations $|\mathbb{F}_n(x) - F_0(x)|$, the relative values $|\mathbb{F}_n(x) - F_0(x)|/F_0(x)$ should be considered. In the same spirit, Anderson and Darling [1] analyses asymptotic properties of the statistics like

$$\sup_{a \le x \le b} \frac{\sqrt{n}\sum_{i=1}^{n}(\mathbf{1}(F_0(X_i) \le x) - x)}{\sqrt{x(1-x)}}, \qquad 0 < a < b < 1,$$

and their two-sided versions, measuring distances in terms of their standard deviation.

A conceivable idea of the more general nature is to consider a variance stabilizing or a kind of *weighted* versions of the classical KS in order to accentuate its behavior at the tails. The problem of specification of the weighting function such that the resulting test statistic provides a desirable tail sensitivity can be solved by exploiting *weighted empirical processes* techniques along with the related weak convergence theory. However it turns out that the limiting behavior of the weighted EDF-based test statistics in sparse and weak regimes is a delicate area so that adapting of conventional asymptotic theory to newly proposed test statistics is not straightforward. On the other hand, there are a number of results in weak convergence of the weighted empirical processes whose potential use in deriving methods for detecting sparse-weak effects has not been fully explored. In this study, we show that a reasonable compromise between the statistics focusing on the center of the distribution and statistics demonstrating sensitivity on the tails can be achieved by using the *Chibisov-O'Reilly* weight function of the sup-functional metrics in (17.5) and the characterization of the weighted weak convergence provided first by Chibisov (see Theorem 3 in [8]) and later by O'Reilly (see Theorem 2 in [32]) .

Before presenting the main object of our study, the class of test statistics based on the Chibisov-O'Reilly weight function, we need to introduce some auxiliary concepts, definitions and results.

17.2.1 Weighted Empirical Processes and Properties of Weight Functions

By the CLT for iid random variables with a finite variance, for any fixed x, and with $n \to \infty$

$$\sqrt{n}(\mathbb{F}_n(x) - F_0(x)) \xrightarrow{\mathscr{D}} N\left(0, F_0(x)(1 - F_0(x)\right).$$

Thinking now of $x \in \mathbb{R}$ as a running time parameter, we can interpret $\sqrt{n}(\mathbb{F}_n(x) - F_0(x))$ as the one-dimensional (normalized) *empirical process*. Further, let U be a

$\mathscr{U}[0, 1]$ random variable. The function of U defined as $X = F_0^{-1}(U)$ is called the *quantile transformation of* U and has exactly the CDF F_0; see, e.g., [12, Chapter 1]. Now, if H_0 is true, i.e., all X_i follow F_0 on $\mathbb{R}$ the quantile transformation shows that for any n, the distribution of KS_n is the same for all F_0.

A useful reduction which will help us in pinning down finer properties of $\|\mathbb{F}_n - F_0\|_\infty$ is achieved by the probability integral transformation: let $U_i = 1 - F_0(X_i)$. If all X_i's have the same continuous distribution F_0 then U_i's are iid uniform $U(0, 1)$ random variables and for each n, we have

$$\mathrm{KS}_n \stackrel{\mathscr{D}}{=} \sup_{0<u<1} \sqrt{n}|\mathbb{U}_n(u) - u|,$$

where $\mathbb{U}_n(u) = n^{-1} \sum_{i=1}^n \mathbf{1}(U_i \leq u)$. The construction under the supremum, $\xi_n(u) = \sqrt{n}(\mathbb{U}_n(u) - u), 0 < u < 1$, is the classical *uniform* empirical process whose weak convergence to the Brownian bridge was shown by Donsker in [15]. We thus have that

$$\mathrm{KS}_n \stackrel{\mathscr{D}}{\to} \sup_{0<u<1} |B(u)| \quad \text{as} \quad n \to \infty,$$

where $\{B(u), 0 \leq u \leq 1\}$ denotes a Brownian bridge process on $[0, 1]$.

Let w be strictly positive function defined on $(0, 1)$ with the property $w(u) = w(1 - u)$ for $u \in (0, 1/2)$, nondecreasing in a neighborhood of zero and non-increasing in a neighborhood of one. We consider the *weighted* empirical process and define the corresponding family of weighted KS statistics as

$$\mathrm{KS}_n(w) := \sup_{x\in\mathbb{R}} \frac{\sqrt{n}|\mathbb{F}_n(x) - F_0(x)|}{w(F_0(x))}, \quad \mathrm{KS}_n^+(w) := \sup_{x\in\mathbb{R}} \frac{\sqrt{n}(\mathbb{F}_n(x) - F_0(x))}{w(F_0(x))}. \tag{17.6}$$

If H_0 is true, then by the probability integral transformation, for each n,

$$\begin{aligned} \mathrm{KS}_n(w) &\stackrel{\mathscr{D}}{=} \sup_{0<u<1} \frac{\sqrt{n}|\mathbb{U}_n(u) - u|}{w(u)} = \sup_{0<u<1} |\xi_{w,n}(u)|, \\ \mathrm{KS}_n^+(w) &\stackrel{\mathscr{D}}{=} \sup_{0<u<1} \frac{\sqrt{n}(\mathbb{U}_n(u) - u)}{w(u)} = \sup_{0<u<1} \xi_{w,n}(u), \end{aligned} \tag{17.7}$$

where $\xi_{w,n}(u) := \xi_n(u)/w(u)$ is a weighted uniform empirical process.

In what follows, for functions Q, G on $[0, 1]$ and the weight function w, we define the *weighted sup-norm metric* $\| \cdot /w\|$ by

$$\|(Q - G)/w\| := \sup_{0<u<1} |(Q(u) - G(u))/w(u)|$$

whenever it is well defined, i.e., when $\limsup_{u\to 0} |(Q(u) - G(u))/w(u)|$ is finite. The weighted sup-norm metric appeared in, e.g., [9]. We note that $\|(Q - G)/w\|$ with $w = 1$ coincides with the sup-norm metric we have used so far for functions Q, G on $[0, 1]$.

The convergence results (17.7) suggest to use $\mathrm{KS}_n(w)$ and $\mathrm{KS}_n^+(w)$ to conduct the tests of significance for a prescribed level $0 < \alpha < 1$ as follows. One would reject H_0 in favor of H_1 when $\mathrm{KS}_n(w) > h(n, \alpha)$, where the critical point $h(n, \alpha)$ is chosen to have $\Pr(\mathrm{KS}_n(w) > h(n, \alpha)) = \alpha$; and one would reject H_0 in favor of $H_1^{'}$ when $\mathrm{KS}_n^+(w) > h^+(n, \alpha)$, where $h^+(n, \alpha)$ is determined by $\Pr(\mathrm{KS}_n^+(w) > h^+(n, \alpha)) = \alpha$.

The test procedures are consistent against both alternatives $H_1 : F \neq F_0$ and $H_1' : F > F_0$, respectively. Indeed, testing $H_0 : F = F_0$ versus two-sided alternative $H_1 : F \neq F_0$ we observe that for any fixed alternative $F \neq F_0$ the statistic $\|(\mathbb{F}_n - F_0)/w(F_0)\|_\infty := \sup_{0<F_0(t)<1} |\mathbb{F}_n(t) - F_0(t)|/w(F_0(t))$ satisfies

$$\left\| \frac{\mathbb{F}_n - F_0}{w(F_0)} \right\|_\infty \geq \left\| \frac{F - F_0}{w(F_0)} \right\|_\infty + o_P(1),$$

implying $\mathrm{KS}_n(w) \xrightarrow{P} \infty$ whenever $F \neq F_0$. From this,

$$\mathrm{KS}_n(w) = \sqrt{n}\|(\mathbb{F}_n - F_0)/w(F_0)\|_\infty = O_P(\sqrt{n}),$$

and we get for any $\alpha \in (0, 1)$

$$\Pr_{H_1}(\mathrm{KS}_n(w) > h(n, \alpha)) \to 1 \quad \text{as} \quad n \to \infty,$$

where $h(n, \alpha)$ is the top α-percentile of the null distribution of $\mathrm{KS}_n(w)$ which in turn is given by the distribution of $\sup_{0<u<1} |\xi_{w,n}(u)|$ and $h(n, \alpha) = O(1)$. The case of the one-sided alternative is treated similarly.

17.2.2 Theoretical Properties of ChOR Statistic

Statistical properties of the family of test statistics defined in (17.6) are essentially determined by the behavior of the used weighting functions. Specifically, variance stabilizing weighting poses the problem of weak convergence of the empirical processes in the $\| \cdot /w\|$ metric, i.e., in the weighted sup-norm metric to a Brownian bridge process B. The idea of considering a weighted version of KS in order to shift actions to the middle of the distribution while properly regulated on the tails is useful but the limiting behavior of weighted sup-functionals of GOF statistics is not always what one can expect intuitively. For example, $\mathrm{KS}_n(w)$ does not necessarily converge in distribution to the supremum of $|B(x)|/w(x)$. In words, in order to achieve a non-degenerate asymptotic null-distribution of $\mathrm{KS}_n(w)$, i.e., to be able to control the probability of Type I Error when using (17.6), the tails of the weighting function w in (17.6) must behave in such a way that $|B(x)|/w(x)$ does not blow up in the vicinity of zero and one. More precisely, if we denote $\mathscr{W}$ the class of positive functions on (0, 1) that are bounded away from zero on $(\eta, 1-\eta)$ for all $0 \leq \eta < 1/2$), the goal is to characterize those functions $w \in \mathscr{W}$ for which we have a weak convergence of the weighted uniform empirical processes $\xi_n(x)/w(x)$ to $|B(x)|/w(x)$, i.e., for which the tests statistics (17.6) admit nontrivial asymptotic distribution. This characterization is available from the following facts:

Fact 1 *Let $\mathscr{W}$ be the class of strictly positive functions defined on* (0, 1) *with the property $w(u) = w(1-u)$ for $u \in (0, 1/2)$, which are nondecreasing in a neighborhood of* 0 *and non-increasing in the a neighborhood of one. Such class of functions is called Chibisov-O'Reilly weight functions of a Brownian bridge $\{B(u), 0 \leq u \leq 1\}$ if*

$$\limsup_{u\to 0} \frac{|B(u)|}{w(u)} \overset{\text{a.s.}}{=} 0.$$

Fact 2 (see [8], [32]) *Let $w \in \mathscr{W}$. The sequence of random variables*

$$\sup_{0<u<1} \sqrt{n}\frac{|\mathbb{U}_n(u) - u|}{w(u)}$$

converges in distribution to a non-degenerate random variable if and only if

$$I(w, \epsilon) = \int_0^1 \frac{\exp\left(-\epsilon\frac{w^2(t)}{t(1-t)}\right)}{t(1-t)} dt < \infty \quad \textit{for some} \quad \epsilon > 0,$$

in which case

$$\sup_{0<u<1} \frac{\sqrt{n}|\mathbb{U}_n(u) - u|}{w(u)} \xrightarrow{\mathscr{D}} \sup_{0<u<1} \frac{|B(u)|}{w(u)}.$$

The integral $I(w, \epsilon)$ appeared for the first time in Chibisov [8], and later in [32]. A special case of Chibisov-O'Reilly theorem is given below.

Fact 3 (See [12, Theorem 16.4, p. 535.]) *For each* $0 < \nu < 1/2$,

(a) $\sup_{0<u<1} \frac{|B(u)|}{[u(1-u)]^\nu} < \infty$, *a.s.*
(b) *and, the following result holds true*

$$\sup_{x\in\mathbb{R}} \frac{\sqrt{n}|(\mathbb{F}(x) - F_0(x))|}{[F_0(x)(1-F_0(x))]^\nu} \xrightarrow{\mathscr{D}} \sup_{0<u<1} \frac{|B(u)|}{[u(1-u)]^\nu} \quad \textit{as} \quad n \to \infty.$$

The above convergence results suggest the function $w = w_\nu$ in the weighted KS statistics (17.6) to be selected as

$$w_\nu(u) = [u(1-u)]^{1/2-\nu}, \quad \text{where} \quad 0 < \nu < 1/2. \tag{17.8}$$

Using this family of functions we define the *two-sided* and *one-sided Chibisov-O'Reilly* (ChOR) statistics as

$$\mathrm{ChOR}_n(w_\nu) := \sup_{x\in\mathbb{R}} \frac{\sqrt{n}|\mathbb{F}_n(x) - F_0(x)|}{w_\nu(F_0(x))}, \quad \mathrm{ChOR}_n^+(w_\nu) := \sup_{x\in\mathbb{R}} \frac{\sqrt{n}\,(\mathbb{F}_n(x) - F_0(x))}{w_\nu(F_0(x))}. \tag{17.9}$$

The parameter ν in (17.8) will be used as a tuning parameter to guarantee a balanced behavior of $B(u)$ near zero and one. In practical applications, ν must be chosen from the data. In Sect. 17.3, we show that this structure of the weight function makes it possible to construct fully adapted procedure which optimally fits the data at hand.

17.3 Connection of Anderson-Darling and HC-like Constructions

Some of the earliest works on modifications of the classical KS statistic date back to Anderson and Darling [1] and Rényi [33] who suggested a variance stabilizing transformation and a weighted version of $\mathrm{KS}_n(w)$ measuring distances in terms of their standard deviations. Specifically, Anderson and Darling studied the asymptotics of the weighted sup-norm statistics as

$$\mathrm{AD}_n(w) := \sup_{x\in\mathbb{R}} \frac{\sqrt{n}\,(\mathbb{F}_n(x) - F(x))}{w(F(x))} \tag{17.10}$$

and their two-sided versions, where $w(u) = \sqrt{u(1-u)}$ for $u \in (0,1)$ by its construction, is called *standard deviation proportional* (SDP) weight function. The weighted KS test statistic (17.10) standardized by the SDP have further been extensively studied in the literature (see for example [10, 14, 16, 20, 21]).

Closely related to (17.10) is the popular family of Tukey's *Higher Criticism* (HC) statistic, which in its one-sided version is defined as

$$\mathrm{HC}_n := \sup_{0<u<\alpha_0} \frac{\sqrt{n}(\mathbb{U}_n(u) - u)}{\sqrt{u(1-u)}}, \quad 0 < \alpha_0 < 1, \tag{17.11}$$

where $0 < \alpha_0 < 1$ defines the range of significance levels in the multiple-comparisons testing and therefore is a number like 0.1 or 0.2. HC_n was introduced by Donoho and Jin [14] for multiple testing settings where dominating part of the component problems correspond to the null hypothesis and a small proportion correspond to non-null hypotheses. Indeed, by setting $\alpha_0 = 1$ in (17.11) we obtain a one-sided version of (17.10) with the weighting function $w(u) = \sqrt{u(1-u)}$.

By Fact 1, we immediately observe that HC-statistics does not have a non-degenerated asymptotic null distribution. Precisely, under H_0, the statistic HC_n (as well as its two-sided version) tend to infinity in probability (see [20] and [16]), and even almost surely, (see Chapter 16 in [34]). Theoretically, the problem stems from the normalizing factor $1/\sqrt{u(1-u)}$ of the SDP weight function which grows too quickly in the vicinity of $u = 0$ and 1 for the supremum of the weighted empirical process to settle down. The almost sure rate at which HC_n blows up is studied by Shorak and Wellner, see [34, Chapter 16].

As a computationally simple and non-expensive procedure, the HC_n-type statistics attracts a lot of attention in the recent literature on the sparsity detection and variable selection in high dimensions. A number of recent studies (see, e.g., [31] and [28]) exploited the sup-norm construction with SDP weight functions and admitted the problems with the finite sample behavior of (17.11), providing in fact, a numerical justification to the theoretical properties of HC_n. As noticed in Remark on p. 611 of [11], "*It is not clear for what n the asymptotics start to give reasonably accurate description of the actual finite sample performance and actual finite sample comparison...*", which complicates the use of HC_n in practice. But the fact that the limiting null distribution is degenerated has never been recognized as the actual cause of the problem.

A number of strategies to alleviate the problems with weak convergence of AD and HC has been discussed in the literature. A seemingly better modification of HC_n suggested in [21] has the form

$$\mathrm{HC}_n^* := \sup_{u_{(1)} < u < u_{(\lfloor \alpha_0 \cdot n \rfloor)}} \frac{\sqrt{n}(\mathbb{U}_n(u) - u)}{\sqrt{u(1-u)}}, \quad 0 < \alpha_0 < 1, \tag{17.12}$$

where $u_{(i)}$ are observed values of the corresponding order statistics $U_{(1)} < \cdots < U_{(n)}$. A general idea of truncating the domain over which supremum is taken in (17.11) is also addressed in [40], where the focus was placed on the computational efficiency rather than the degenerated limiting null distribution.

Viewing the problem of finding the distribution of HC as a computationally intensive numerical problem, Wu et al. [39] device a system of two Volterra-type integral equations for $\Pr(\mathrm{KS}_n(w) < x)$, $x > 0$ with w from the class of SDP weight functions.

Unfortunately, neither truncating the range as in (17.12) nor integral equations-type approach eliminate the problem of weak convergence of (17.11). But this phenomenon of normalizing the process $\sqrt{n}(\mathbb{U}_n(u) - u)$ by SDP family of weight functions, seems to be well known for experts in the theory of empirical processes. See for example results by Eicker [16, Theorem 2] and Jaeshke [20] which show that the finite limit distribution can be obtained by an appropriate normalization of (17.11). The precise statement of the asymptotic result is the following: for any

$0 < \alpha_0 < 1$ and any $x \in \mathbb{R}$

$$\lim_{n\to\infty} \Pr\left(a_n \sup_{0<u<\alpha_0} \frac{\sqrt{n}(\mathbb{U}_n(u) - u)}{\sqrt{u(1-u)}} - b_n \le x\right) = E(x), \tag{17.13}$$

where

$$a_n = \sqrt{2\log\log n}, \quad b_n = 2\log\log n + \frac{1}{2}\log\log\log n - \frac{1}{2}\log\pi,$$

and $E(x) = \exp(-\exp(-x))$ is the CDF of Gumbel or extreme value distribution. In words, result (17.13) implies that an appropriately normalized version of HC-statistic, under the null hypothesis, will always have the same extreme value distribution *regardless* of a particular value of $0 < \alpha_0 < 1$. Jager and Wellner [21] also study an appropriately normalized version of $(1/2)(\mathrm{HC}_n^*)^2$ with HC_n^* as in (17.13) in terms of an extreme value distribution, see Theorem 3.1 of [21].

The main problem of using these limiting results in practical applications is that the convergence in distribution in (17.13) is very slow. Jaeschke [20] showed that the convergence rate is $O((\log n)^{-1/2})$, so that the asymptotic approximation (17.13) can be inaccurate for n as large as 10^6, which even with this correct normalization makes the asymptotic theory-based p-values of HC-statistics unreliable.

All this leads us to study a different type of weight functions in the sup-norm scenario, namely the class of Chibisov-O'Railly functions which provide a more modest weighting and result in a non-trivial limit for the wGOF statistics.

17.4 Empirical Study

In this section, we present simulation results, where we study the finite sample performance of three wGOF tests: the $\mathrm{ChOR}_n^+(w_\nu)$, the HC^* and the CsCsHM_n^+ (which stands for Csörgő et al. [9] family of statistics; to be defined below). The $\mathrm{ChOR}_n^+(w_\nu)$ is defined in (17.9), where the tuning parameter ν is taken to be optimal the value in the interval 0.05–0.45 with increment 0.05, that results in the smallest sum of Type I and Type II errors; the HC^* is defined in (17.12), where α_0 is set to 0.1 in the truncation domain.

17.4.1 Gaussian Mixture

For the specific model (17.3) in Example 17.2, in the idealized case where ε_n, μ_n and σ are known, both H_0 and $H_{1,n}$ are simple hypothesis, and the Neyman-Pearson lemma asserts that the optimal test is obtained with the likelihood ratio approach. Its performance was studied first by Ingster [18] for the homoscedastic case, and later by Cai et al. [7], who determine the fundamental detection limits in testing as follows: when testing the hypothesis (17.3), the strict region of detectability in the (β, r) plane is specified by the function curve

$$\beta^*(r, \sigma^2) = \begin{cases} \dfrac{1}{2} + \dfrac{r}{2-\sigma^2} & \text{if } 2\sqrt{r} + \sigma^2 \geq 2 \\ 1 - \dfrac{(1-\sqrt{r})_+^2}{\sigma^2} & \text{if } 2\sqrt{r} + \sigma^2 < 2, \end{cases} \tag{17.14}$$

which gives the smallest possible β such that hypothesis in (17.3) can be distinguished in the sense that the sum of Type-I and II error probabilities goes to zero as $n \to \infty$ and x_+ denotes $\max(x, 0)$. Namely, the detectable region is given by the set $\{(\beta, r) : \beta < \beta^*(r, \sigma^2)\}$.

While being optimal, the likelihood ratio-based test is inapplicable in practice since it requires explicit knowledge of the model parameters, ε_n, μ_n and σ. It is desirable therefore to design *adaptive* test procedures that are simultaneously optimal for a class of alternatives over the whole region of detectability. A general asymptotic theory of optimal adaptivity for the GOF tests based on the sup-functionals of the weighted empirical processes was developed by Stepanova and Pavlenko [35] for a broad class of sparse mixture problems. Specifically, the family of weight functions $\mathscr{W}$ was defined by the *Erdős Feller Kolmogorov Petrovski* (EFKP) *upper-class function* of a Brownian bridge $\{B(u), 0 \leq u \leq 1\}$, assuming that there exists a constant $0 \leq b < \infty$ such that

$$\limsup_{u \to 0} \frac{|B(u)|}{w(u)} \overset{\text{a.s.}}{=} b. \tag{17.15}$$

Two specific examples of this family are defined by

$$\text{CsCsHM}_n(w) := \sup_{x \in \mathbb{R}} \frac{\sqrt{n}|\mathbb{F}_n(x) - F_0(x)|}{w(F_0(x))}, \quad \text{CsCsHM}_n^+(w) := \sup_{x \in \mathbb{R}} \frac{\sqrt{n}\,(\mathbb{F}_n(t) - F_0(x))}{w(F_0(x))}, \tag{17.16}$$

where w belongs to the family of the EFKP upper-class functions of a Brownian bridge $\{B(u), 0 \leq u \leq 1\}$. Since $\text{CsCsHM}_n(w)$, in this generality, appeared for the first time in the paper of Csörgő, Csörgő, Horváth, and Mason [9], the statistics $\text{CsCsHM}_n(w)$ and $\text{CsCsHM}_n^+(w)$ are called the *two-sided* and *one-sided Csörgő Csörgő Horváth Mason* (CsCsHM) *statistics*, respectively.

It was shown that the test statistics of type (17.16) constructed by w from the EFKP class of weight functions distinguish between H_0 and $H_{1,n}$ *without* knowledge of ε_n and μ_n and are therefore optimally adaptive over the whole region of detectability; (see Theorem 4.1 of [35]). One can see that the Chibisov-O'Reilly weight function is a special case of the EFKP weight for which $b = 0$, and therefore the optimal adaptivity holds for the whole family of ChOR-type statistics.

Now, using optimal adaptivity of $\mathrm{ChOR}_n(w_\nu)$ we can investigate its finite sample performance in comparison with other adaptive techniques. Theoretical studies have proposed a number of test statistics that are asymptotically optimal in the sense that they can asymptotically reach the detection boundary (17.14), see, e.g., [35] for CsCsHM and ChOR families, and [21] for the ϕ-divergence family of statistics which includes both the square of the supremum form of the Anderson-Darling statistic defined by (17.10) as well as its classical integral form introduced in [1]. However, while sharing the same optimality property with $n \to \infty$, they could have different detection power in practical data analysis, where n is always finite and can be small. To compare the finite sample performance, we consider the problem of detecting the change of the mean stated in Example 17.2 assuming homoscedasticity.

The practical, fully data-driven, mixture detection procedures for the optimally adaptive tests can be readily obtained using the following simulation steps. First, fixing parameters (n, β, r) in the region of detectability according to (17.14) with $\sigma = 1$, we let $\mu_n = \sqrt{2r \log n}$, $\varepsilon_n = p^{-\beta}$ for $\beta > 1/2$. Second, we set $\sigma = 1$ in (17.3) and draw $n = 10{,}000$ samples from $N(0, 1)$ for the null hypothesis; for the alternative hypothesis, we first draw $n(1 - \varepsilon_n)$ samples from $N(0, 1)$, and then draw $n\varepsilon_n$ samples from $N(\mu_n, 1)$. Third, we implement all three tests for the two levels of sparsity, $\varepsilon_n = 0.008$ and $\varepsilon_n = 0.01$, and for μ ranging from 0 to 2.5 with an increment of 0.05.

All the considered test statistics involve a supremum over $x \in \mathbb{R}$. In what follows we use an equivalent discrete formulation, whereby $\sup_{x\in\mathbb{R}}$ will be replaced by $\max_{1\leq i\leq n}$, and $\sqrt{n}(\mathbb{F}_n(x) - F_0(x))$ will be replaced by $\sqrt{n}(i/n - u_{(i)})$. Recall that u_i's denote realizations of n iid samples $U_i \in \mathscr{U}[0, 1]$ and $U_{(i)}$ are the corresponding order statistics. By the probability integral transform, $U_i = 1 - F_0(X_i)$ are equivalent to p-values for each of the individual hypothesis, which are iid $\mathscr{U}[0, 1]$ under H_0. This collection is the input p-values to be used to form a summary of the test statistics HC_n^*, $\mathrm{ChOR}_n^+(w_\nu)$ and $\mathrm{CsCsHM}_n^+(w)$.

To capture the evidence against the joint null of (17.3) with each test, we need to find the critical value so that the Type I error is controlled at a specified level α for any fixed $\alpha \in (0, 1)$ and n. This is done in the following way. For each of the three tests, we simulate the corresponding test statistics under the null for a large number M such that $M\alpha \gg 1$, (e.g., $M\alpha = 50$) and denote by $h(n, \alpha)$ the top α percentile of empirical distribution of the simulated scores representing the critical value. The null hypotheses is then rejected if the test score exceeds $h(n, \alpha)$. For each test score, the empirical distribution is obtained by generating data under H_0 for $n = 10{,}000$ with $M = 1000$ replicates, averaged over 100 iterations.

The empirical way of determining the critical value for a α-level test seems to be most suitable for performing a fair comparison of the test procedures discussed above: the precise form of limit distribution of HC_n is not specified even for its truncated version (17.12) and the distribution of CsCsHM_n^+ statistic is tabulated only for the specific choice of $w(u)$ from the class of EFKP upper-class functions, namely $w(u) = \sqrt{(u(1-u)\log\log(1/u(1-u))}$. See [35] for more details.

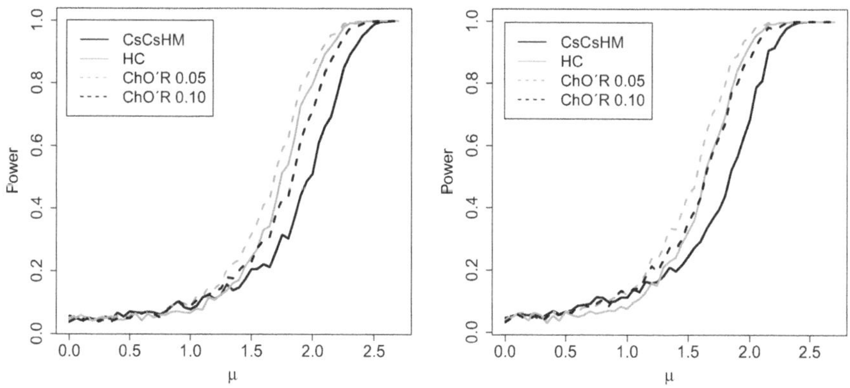

Fig. 17.1 Power comparison among one-sided wGOF tests for detecting sparse-weak Gaussian mixture $(1-\varepsilon_n)\mathcal{N}(0,1)+\varepsilon_n\mathcal{N}(\mu_n,1)$ versus $\mathcal{N}(0,1)$ at significance level $\alpha = 0.05$. Left panel: $\varepsilon_n = 0.008$. Right panel: $\varepsilon_n = 0.01$

With the significance level $\alpha = 0.05$, the left and right panels of Fig. 17.1 provide the empirical power of $\text{ChOR}_n(w_\nu)$, CsCsHM and HC_n^* (all one-sided) under the two sparse alternatives. For detecting a change in the mean, the $\text{ChOR}_n(w_\nu)$ test with the optimal value ν has high detection power and outperforms both CsCsHM and the improved version of HC, HC_n^* of (17.12) for higher contamination levels μ_n. As seen in the figure, the gap in detection power is not very large. This is due to the specific normalization of the deviations at the extreme indices is HC_n^* applied to avoid the situations where the test power is very close to zero. We also observe an accurate size control for all the tests for both sparsity levels. In all the simulations, the estimated standard errors of the results are generally small, this is because each point on the power curves is the mean of 100 iterations.

Observe that for some low (μ, ε_n) values CsCsHM_n^+ achieves better performance, whereas for higher ν, $\text{ChOR}_n^+(w_\nu)$ with the optimally selected value of ν is performing better. The natural questions is how to compare the finite sample performance of the tests simultaneously over both sparsity and weakness parameters? Borrowing ideas from [31], we study this question by conducting an extensive simulation which exploits the mis-detection evaluation technique. Specifically, for a variety of different values of (ε_n, μ), we empirically computed the detection power of these tests at a significance level $\alpha = 0.05$ for sample sizes $n = 1000$ and $n = 10{,}000$. For each value of μ and ε_n, we state that a test T_1 is a *clear winner* if $\min_{k=2,3,4} \Pr_{H_1}(T_k = H_0)/\Pr_{H_1}(T_1 = H_0) > 1.1$. In words, T_1 is a clear winner if its mis-detection rate is significantly lower.

Figure 17.2 shows the regions of the phase space (μ, ε_n) where the different colors represent the tests which are asserted as clear winners. As seen in the left panel of the figure, at the upper left part of the phase space (μ, ε), CsCsHM is the best statistic. This is due to relatively high ε_n which corresponds to a more *dense* contamination regime, where CsCsHM is shown to better capture denser signals;

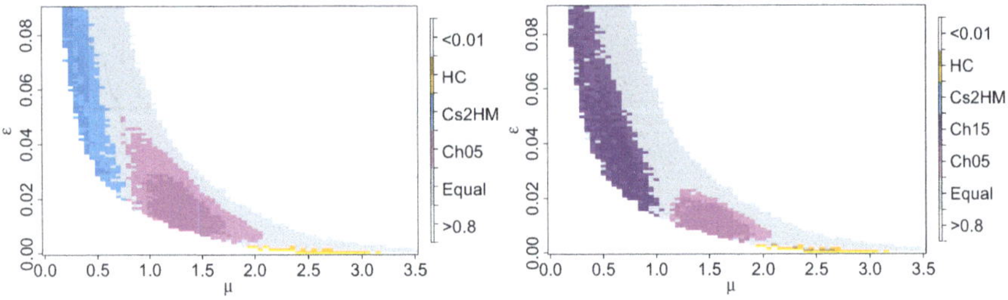

Fig. 17.2 Comparison of tests for detecting sparse and weak Gaussian mixture $(1-\varepsilon_n)\mathcal{N}(0,1)+\varepsilon_n\mathcal{N}(\mu_n,1)$ versus $\mathcal{N}(0,1)$. Colored patterns show regions where the mis-detection level with the second-best testing procedure divided by that of the best one exceeds 1.1. Color intensity signifies the regions where this ratio exceeds 1.5. The grey color represent the regions where the mis-detection level is in the range 0.1–80%. Left panel: ChOR with parameter $\nu = 0.05$. Right panel: ChOR with two values of $\nu = 0.05$ and $\nu = 0.15$

see, e.g., [35]. At the other extreme, in the lower right parts of both left and right panels, where ϵ_n is small but μ is high (sparse and strong signals regime), the HC_n^* works best with the $\text{ChOR}_n^+(w_\nu)$ statistic with $\nu = 0.05$ being a close second. In the intermediate regions of both panes, which represent the sparse and weak regime, it is $\text{ChOR}_n^+(w_\nu)$ with both $\nu = 0.15$ and $\nu = 0.05$ outperforms the other tests. Finally, in this simulation study the HC_n statistic defined by (17.11) (i.e., without truncation of the range of p-values) performed worse than at least one the other tests over the whole phase space.

17.4.2 *Sparse and Weak Effects in High Dimensions: Dirichlet Mixture*

Motivated by the problem of detecting sparse microbiome features (e.g., microbial taxa, certain genes, etc.) exhibiting differences in abundance between samples, we exploit the hypothesis testing signal detection technique based on wGOF statistics constructed as sup-functionals of weighted empirical processes. Spectroscopy-based metabolomic sequencing, the method used to quantitatively characterize microbiomes, tends to screen as many features as possible, while truly highly informative features often have a relatively small number, and their effects (signal strength) are often low. Therefore, it is appealing to apply a sparse-weak setting when modeling microbiome data. A specific example where detecting of a sparse-weak mixture is combined with high-dimensional, low-sample regime is the problem of detection and identification of the covariates highly affecting the microbiome composition.

We develop the modeling framework to mimic a real microbiome data set which we further analyze in Sect. 17.4.3. Suppose that we have n microbiome samples measured on p distinct bacterial taxa/species whose counts for each sample are modeled by a vector of random variables $\mathbb{Y} = (Y_1, \ldots, Y_p)$ where $Y^+ = \sum_{i=1}^p Y_i$ and $p \gg n$.

The total taxon count, Y^+ is in part determined by the sequencing depth and will be treated as an ancillary statistic whose distribution does not depend on the parameters in the model. The standard statistical model for count data is the multinomial distribution but in the microbiome composition samples the variation in the data is usually larger than that what would be predicted by the multinomial model with the fixed underlying proportions. This effect of increased variation stems from the combination of heterogeneity of microbiome samples and high sparsity of the underlying proportions which vary among n samples. To account for this combined effect, a sparse-weak mixture model for compositional data analysis is designed.

Specifically, we let $\mathbb{X} = (X_1, \ldots, X_p)$ to denote the composition vector of species proportions obtained by normalizing $\mathbb{Y}$ with Y^+. Then the elements of the composition X_i's are themselves positive random variables subject to the constraint that $\sum_{j=1}^p X_j = 1$ where $X_j \in (0, 1)$, $j \in [\![p]\!]$ and the support of $\mathbb{X}$ is the p-dimensional simplex. The most suitable model for the data on bounded domain is the *p-dimensional Dirichlet distribution* whose density is given by

$$f_{\mathbb{X}}(x_1, \ldots, x_p; \boldsymbol{\theta}) = \frac{\Gamma(\sum_{j=0}^p \theta_j)}{\prod_{j=1}^p \Gamma(\theta_j)} \left[1 - \sum_{j=1}^p\right]^{\theta_0 - 1} \prod_{j=1}^p x_j^{\theta_j - 1}, \quad x_j \geq 0, \quad \sum_{j=1}^p x_j \leq 1,$$

where $\boldsymbol{\theta} = (\theta_0, \theta_1, \ldots, \theta_p)$, are positive shape parameters, $\theta^+ = \sum_{j=1}^p \theta_j$ and $\Gamma(\cdot)$ denotes the Gamma function.

In what follows, we will write $\mathbb{X} \sim Dir(\theta_0, \ldots, \theta_p)$ and let $\mathbb{Y}_n = (Y_{ij})_{n \times p}$ denote the observed count matrix for n samples and $\mathbb{X}_n = (X_{ij})_{n \times p}$ be the corresponding matrix of estimated proportions.

The straightforward way to model sparse-weak effects in Dirichlet distribution would be to calibrate parameters θ_j in a way similar to Example 17.3, where the signal strength, i.e., a proportion level of a specific bacterial taxon and a fraction of informative taxons must be linked to both p and n assuming a specific regime of growth of p relative to n.

Due to its structure, the Dirichlet distribution with the parameter vector having $\theta_j = 1$ for $j \in [\![p]\!]$ would represent the null hypothesis that all p bacterial taxa/species are entirely non-informative in n given samples. The alternative hypothesis would be represented by a mixture of *large* amount $1 - \varepsilon_p$ of non-informative covariates and a *small* fraction ε_p of those which have significantly elevated values of θ_j.

The main problem with this approach to sparse mixture modeling is that covariates X_j in $\mathbb{X}$ are negatively correlated (see, e.g., Kotz et al. [26, Chapter 49]) and to avoid the highly non-trivial problem of incorporating covariance into the mixture parametrization, we suggest to use the following characterization of the Dirichlet distribution: the random variables $Z_j = X_j / \sum_{\ell=1}^p X_\ell$, are mutually independent standard *Beta*-distributed random variable with parameters θ_j and $\sum_{\ell=j+1}^{p} \theta_\ell$, respectively, i.e., if $\mathbb{X} \sim Dir(\theta_0, \ldots, \theta_p)$ then the vector of transformed variables $\mathbb{Z}$ consists of $Z_j \sim Beta(\theta_j, \sum_{\ell=j+1}^{p} \theta_\ell)$ which are mutually independent for $j \in [\![p]\!]$; see Kotz et al. [26, Chapter 25]. Hence, it is possible to simulate a p-dimensional Dirichlet random vector by simulating $(p+1)$ independent Beta variables.

An alternative, a computationally efficient way to simulate from a Dirichlet distribution is by using the *systematic scan Gibbs sampler*, which uses the fact that the full conditionals of $Dir(\theta_0, \ldots, \theta_p)$ are simply Beta distributions, see, e.g., [12].

Now, to detect the presence of informative signals (relevant taxa), we test the null hypothesis that $\mathbb{X} \sim Dir(\boldsymbol{\theta})$ where $\theta_j = 1, j \in [\![p]\!]$, that is taxa proportions are uniformly distributed in the p-dimensional simplex. We calibrate each (taxa) proportion θ with $n\theta_j \sim (1-\varepsilon_p)\delta_1 + \varepsilon_p \delta_{\theta_\tau}$, where δ_1 denotes the point mass at 1 and $\theta_\tau > 1$, i.e., the distribution δ_{θ_τ} has no mass at 1. To link the sample size n to the number of proportions in the high-dimensional asymptotics discussed in Example 17.3, we use p as the driving index, let $n_p = p^\tau$ for $\tau \in (0, 1)$ and allow θ_τ to depend on p. Recall that ε_p is parametrized as $p^{-\beta}$ for $\beta \in (1/2, 1)$, hence as $p \to \infty$ the informative signals become increasingly rare being asymptotically negligible fraction of the coordinates in the vector $\mathbb{X}$.

To represent a finite sample scenario of the high-dimensional asymptotics, we set $p = 5000$ and $n = 50$, which corresponds the case of regular growth of p with $\tau = 0.4593$. We further generate n Dirichlet samples using the systematic scan Gibbs sampler and obtain the data matrix $\mathbb{Z}_n = (Z_{ij})_{n \times p}$. For each $j \in [\![p]\!]$, we test whether $Z_{1j}, \ldots, Z_{nj} \stackrel{iid}{\sim} Beta(1, \cdot)$ and obtain a collection of the KS-based p-values to be used as the input statistics for construction of HC_n^*, $\text{ChOR}_n^+(w_\nu)$ and CsCsHM_n^+ scores.

The significance thresholds $h(p, \alpha)$ are obtained using the simulation scheme of Sect. 17.4.1 as the top α percentile of the null empirical distribution of each of the three test scores. The empirical distributions are obtained with $M = 500$ replicates averaged over 200 iterations for each test procedure. The values of the test scores exceeding the empirical threshold $h(p, \alpha)$ are interpreted as evidence for the presence of informative signals sparsely scattered in the $\mathbb{X}$ vector.

For the power computation, we generate a sparse-weak mixture Dirichlet data under $H_{1,n}$ as a function of θ_τ varying from 1 to 1.5 with the two levels of sparsity $\varepsilon_p = 0.02$ and 0.01, and using $M = 50$ replicates repeated over 500 iterations. From both panels of Fig. 17.3, we observe power of $\text{ChOR}_n(w_\nu)$ with optimal value of ν slightly higher than its competitors where the curves comes very close for lower values of θ and have more pronounced gaps for higher signal levels. This phenomena is very similar to the Gaussian mixture case as well as the accurate size control for all the tests, as may be observed in Fig. 17.3.

To study the power as a function of both sparsity and weakness parameters, we consider θ ranging from 1 to 1.5 and ε_p ranging from 0 to 0.1, both with the increment of 0.05. Figure 17.4 show the regions of the phase space (θ, ε_p) where the different colors represent the tests which are asserted as clear winners

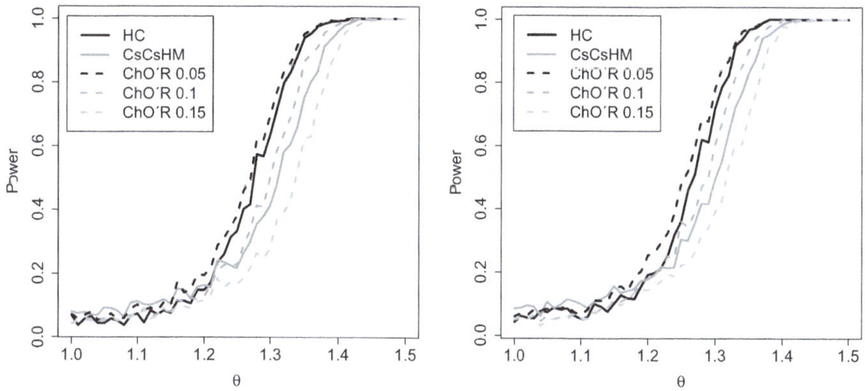

Fig. 17.3 Power comparison of one-sided tests for detecting sparse-weak Dirichlet mixture at significance level $\alpha = 0.05$. Left panel: $\varepsilon_p = 0.01$. Right panel: $\varepsilon_p = 0.02$

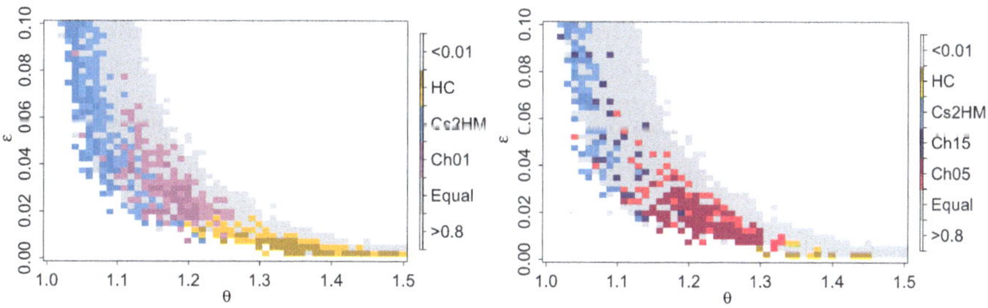

Fig. 17.4 Comparison of tests for detecting sparse-weak Dirichlet mixture. Colored patterns show regions where the mis-detection level with the second-best testing procedure divided by that of the best one exceeds 1.1. Color intensity with the dark centers signify the regions where this ratio exceeds 1.5. The grey color represent the regions where the mis-detection level is in the range 0.1–80%. Left panel: $\text{ChOR}_n^+(w_\nu)$ with parameter $\nu = 0.1$. Right panel: $\text{ChOR}_n^+(w_\nu)$ with two values of $\nu = 0.15$ and $\nu = 0.05$

according to the evaluation technique described in Sect. 17.4.1. It is seen that the pattern of comparative performance of the tests is similar to the Gaussian mixture, but the choice of parameter ν effects the detection accuracy. With increasing ν, the performance of $\text{ChOR}_n(w_\nu)$ is par with CsCsHM which is known to perform well for very wreak but denser signals, whereas with ν decreasing the performance of $\text{ChOR}_n(w_\nu)$ approaches HC_n^*.

17.4.3 Human Gut Microbiome Composition

For illustration, we employ the sparse-weak signal detection methodologies to analyze differentially abundant features in the gut microbiome. The data came from the American Gut Project, which is a crowd sourced characterization of the human gut microbiome with samples from more than 10,000 participants in total [29]. In the project, fecal samples are collected at home and sent to a central lab for sequencing. Sequencing data is deposited in the publicly available Qiita database [24].

In the present study, we conduct the analysis focusing the 150 nucleotide denoised sequences available under Qiita study ID 10317, from which we used the first fecal sample from individuals who had completed a food frequency questionnaire. The *amplicon sequence variant* (ASV) table represents the observed count matrix $\mathbb{Y}_n = (Y_{ij})_{n\times p}$ comprising $n = 2821$ observation over $p = 11{,}496$ features. We are interested in testing hypothesis that there are systematic changes in microbial features which are not due to random noise originating from inter-individual differences.

In the light of shortcomings of count data modeling in microbiome study discussed in Sect. 17.4.2, we transform $\mathbb{Y}_n$ to get a composition data $\mathbb{X}_n$ where each $\mathbb{X}$ is modeled as p-dimensional Dirichlet distribution $Dir(\boldsymbol{\theta})$ with the vector of shape parameters $\boldsymbol{\theta}$. After this transformation, the null hypothesis of pure noise is expressed in terms of the shape parameter: $\theta_j = 1$ for $j \in [\![p]\!]$ in $\boldsymbol{\theta}$, meaning that the taxa proportions are uniformly distributed in the unit p-dimensional simplex.

We further transform $\mathbb{X}_{n\times p}$ to $\mathbb{Z}_{n\times p}$ and obtain a collection of p-values by testing for each $j \in [\![p]\!]$ the GOF of $(Z_{1j}, \ldots, Z_{nj})$ with $Beta(1, \cdot)$ as explained in Sect. 17.4.2.

The obtained p-values are used as the input statistics to calculate the three wGOF scores, resulting in $\mathrm{HC}^* = 488.8456$, $\mathrm{CsCsHM}^+ = 1269.217$, and ChOR^+ varying between 1249.1 for $\nu = 0.05$ and 97.9 for $\nu = 0.45$. To make the tests run smoothly and avoid division with 0, p-values of 0 or 1 were recoded to 0.0001 and 0.999, respectively.

The test significance thresholds are obtained using the simulation scheme of Sect. (17.4.1) as the top $\alpha = 0.05$ percentile of the null empirical distribution of each of the three test scores obtained with $M = 500$ replicates averaged over 200 iterations for each test procedure. The resulting values are $h_{\mathrm{HC}^*} = 3.6$, $h_{\mathrm{CsCsHM}} = 5.2$, and h_{ChOR_ν} varying between 3.6 for $\nu = 0.05$ and and 1.6 for $\nu = 0.45$. By the results above, for all three tests of significance, the null hypothesis is rejected providing statistical evidence for the presence of informative signals sparsely scattered over the composition vector $\mathbb{X}$.

17.5 Discussion and Scope for the Future

By analyzing wGOF-based approach to detection of sparse-weak effects, we have shown that the Chibisov-O'Reilly is the class of weight functions for which the weak convergence of the empirical processes in the $\|\cdot/w\|$-metrics to a Brownian bridge process holds. This is an important result, which in turn guarantees that the corresponding ChOR family of tests statistics admits a non-degenerated asymptotic null distribution, unlike the class of wGOF tests which exploits SDP weight functions, such as for example statistics with HC-like constructions.

Our study provided a thorough framework for applying and comparing existing wGOF test statistics for detection of sparse-weak signals in number of high-dimensional scenarios far beyond the standard zero-inflated, Gaussian distribution mixture. In particular, we have shown that our methodology is suitable for detection of sparse-weak effects in the high-dimensional models with compositional constrains, such as Dirichlet mixture which is used to model microbiome data residing on in a unite p-dimensional simplex, where $p \gg n$. Unlike the log-ratio transformation which is conventionally used in the literature to remove the simplex constraint in the compositional data, we exploited a theoretical characterization of the Dirichlet distribution and propose a transformation which alleviates the interdependence among the microbiome features. This make it possible to operate with the sparse-weak detection techniques using a set independent p-values as the input statistics for any of the wGOF tests.

The attractive feature of the ChOR family is that is forms a broad family of supremum-based, one- and two-sided tests statistics which only requires a group of input statistics (or, which is equivalent, the input p-values) to device an adaptive, fully data-driven procedure for the sparse-weak signal detection.

The signal detection problem studied in this paper is closely connect to other important problems in large-scale inference under sparsity, including estimation of non-null effects and signal *identification*. This latter problem, while being more complex than the signal detection is in fact more interesting from the practical point of view. Indeed, after detecting the presence of signals, a natural next step is to identify the location of informative signals. While the scope of this paper is limited to the detection problem, we are currently working on extension of ChOR strategy to the problem of signal identification as well as to another closely related problem of feature selection in high-dimensional classification models discussed in Example 17.3.

References

1. Anderson, B.J., Darling, D.A.: Asymptotic theory of certain "goodness of fit" criteria based on stochastic processes. Ann. Math. Statist. **23**, 193–212 (1952)
2. Arias-Castro, E., Wang, M.: Distribution-free tests for sparse heterogeneous mixtures. Test **26**, 1–94.R (2017)
3. Arias-Castro, E., Donoho, D., Huo, X.: Near-optimal detection of geometric objects by fast multiscale methods. IEEE Trans. Inf. Theory **51**(7), 2402–2425 (2005)
4. Arias-Castro, E., Candés, E. Durand, A.: Detection of an anomalous cluster in a network. Ann. Statist. **39**(1), 278–304 (2011)
5. Bacher, R., Meillier, C., Chatelain, F., Michel, O.: Robust control of varying weak hyper-spectral target detection with sparse nonnegative representation. IEEE Trans. Signal Process. **65**(13), 3538–3550 (2017)
6. Bennett, M., Melatos, A., Delaigle, A., Hall, P.: Reanalysis of F-statistics gravitational-wave search with the higher criticism statistics. Astrophys. J. **766**, 1–10 (2012)

7. Cai, T.T., Jeng, X.J., Jin, J.: Optimal detection of heterogeneous and heteroscedastic mixtures. J. R. Statist. Soc. B **73**, 629–662 (2011)
8. Chibisov, D.M.: Some theorems on the limiting behaviour of empirical distribution functions. Selected Transl. Math. Statist. Probab. **6**, 147–156 (1964)
9. Csörgő, M., Csörgő, S., Horváth, L., Mason, D.: Weighted empirical and quantile processes. Ann. Probab. **14**, 31–85 (1986)
10. Darling, D.A., Erdős, P.: A limit theorem for the maximum of normalized sums of independent random variables. Duke Math. J. **23**, 143–145 (1956)
11. DasGupta, A.: Asymptotic Theory of Statistics and Probability. Springer Science+Business Media, LLC, Berlin (2008)
12. DasGupta, A.: Probability for Statistics and Machine Learning: Fundamentals and Advanced Topics. Springer Texts in Statistics. Springer, New-York (2011)
13. Dobrushin, R.: A statistical problem arising in the theory of detection of signals in the presence of noise in a multi-channel system and leading to stable distribution laws. Theory Probab. Appl. **3**(2), 161–173 (1958)
14. Donoho, D., Jin, J.: Higher criticism for detecting sparse heterogeneous mixtures. Ann. Statist. **32**, 962–994 (2004)
15. Donsker, M.: Justification and extension of Doob's heuristic approach to the Kolmogorov-Smirnov theorem. Ann. Math. Statist. **23**, 277–283 (1958)
16. Eicker, F.: The asymptotic distribution of the suprema of the standardized empirical processes. Ann. Statist. **7**, 116–138 (1979)
17. Fan, Y., Jin, J., Yao, Z.: Optimal classification in sparse Gaussian graphic model. Ann. Statist. **5**(41), 2537–2571 (2013)
18. Ingster, Yu.I.: Some problems of hypothesis testing leading to infinitely divisible distribution. Math. Meth. Statist. **6**, 47–69 (1997)
19. Ingster, Yu.I., Pouet, C., Tsybakov, A.B.: Classification of sparse high-dimensional vectors. R. Soc. Philos. Trans. A **367**, 4427–4448 (2009)
20. Jaeschke, D.: The asymptotic distribution of the supremum of the standardized empirical distribution function on subintervals. Ann. Statist. **7**, 108–115 (1979)
21. Jager, L., Wellner, J.: Goodness-of-fit test via phi-divergences. Ann. Statist. **35**, 2018–2035 (2007)
22. Ji, P., Jin, J.: UPS delivers optimal phase diagram in high-dimensional variable selection. Ann. Statist. **1**(40), 73–103 (2012)
23. Gloor, G.B., Macklaim, J.M., Pawlowsky-Glahn, V., Egozcue, J.J.: Microbiome datasets are compositional: and this is not optional. Front. Microbiol. **8**, 22–24 (2017)
24. Gonzalez, A., et al.: Qiita: rapid, web-enabled microbiome meta-analysis. Nat. Method **15**(10), 796–798 (2018)
25. Kaul, A., Mandal, S., Davidov, O., Peddada, S.D.: Analysis of microbiome data in the presence of excess zeros. Front. Microbiol. **8**, 2114 (2017)
26. Kotz, S., Balakrishnan, N., Johnson, N.: Continuous Multivariate Distributions: Models and Applications, vol. 1. Wiley, New-York (2000)
27. Kundargi, N., Liu, Y., Tewfik, A.: A framework for inference using goodness of fit tests based on ensemble of Phi-divergences. IEEE Trans. Signal Process. **61**(4), 945–955 (2013)
28. Li, J., Siegmund, D.: Higher criticism: p -values and criticism. Ann. Statist. **43**(3), 1323–1350 (2015)
29. McDonald, D., et al.: American gut: an open platform for citizen science microbiome research. MSystems **3**(3), e00031–18 (2018)
30. McMurdie, P.J., Holmes, S.: Waste not, want not: why rarefying microbiome data is inadmissible. PLoS Comput. Biol. **10**(4), e1003531 (2014)
31. Moscovich, A., Nadler, B., Spiegelman, C.: On the exact Berk-Jones statistics and their p-value calculation. Electr. J. Statist. **10**(2), 2329–2354 (2016)
32. O'Reilly, N.: On the weak convergence of empirical processes in sup-norm metrics. Ann. Probab. **2**, 642–651 (1974)
33. Rényi, A.: On the theory of order statistics. Acta Math. Acad. Sci. Hungar. **4**, 191–231 (1953)

34. Shorack, G.R., Wellner, J.A.: Empirical Processes with Applications to Statistics. Wiley, New York (1986)
35. Stepanova, N., Pavlenko, T.: Goodness-of-fit tests based on sup-functionals of weighted empirical processes. SIAM Theory Probab. Appl. **63**(2), 292–317 (2018)
36. Tillander, A.: Classification models for high-dimensional data with sparsity patterns. PhD Dissertation. Stockholm University, Stockholm, 2013
37. Vielva, P.: A comprehensive overview of the cold spot. Adv. Astron. **2010**, Article ID 592094, 20 (2010)
38. Weiss, S., et al.: Normalization and microbial differential abundance strategies depend upon data characteristics. Microbiome **5**(1), 27 (2017)
39. Wu, D., Hin, L., Kordzakhia, N., Novikov, A.: Numerical approximations to distributions of weighted Kolmogorov-Smirnov statistics via integral equations. In: Wood, D., de Gier, J., Praeger, C., Tao, T. (eds.) 2017 MATRIX Annals. MATRIX Book Series, vol. 2. Springer, Berlin (2019)
40. Zhang, H, Jin, J., Wu, Z.: Distributions and power of optimal signal-detection statistics in finite case. IEEE Trans. Signal Proc. **68**, 1021–1033 (2020)

Chapter 18
Exploring Consistencies of Information Criterion and Test-Based Criterion for High-Dimensional Multivariate Regression Models Under Three Covariance Structures

Tetsuro Sakurai and Yasunori Fujikoshi

Abstract In this paper, we consider the high-dimensional consistency properties of an information criterion and a test-based criterion (KOO method) for the selection of variables in multivariate regression models with covariance structures. The covariance structures considered are (1) an independent covariance structure, (2) a uniform covariance structure and (3) an autoregressive covariance structure. In our model the sample size is not necessarily larger than the dimensionality (number) of response variables. Sufficient conditions for these criteria to be consistent are derived under a high-dimensional asymptotic framework such that the sample size and the dimensionality proceed to infinity together, with their ratio converging to a finite nonzero constant. Our results, and tendencies therein, are explored numerically through a Monte Carlo simulation.

18.1 Introduction

We consider a multivariate linear regression of p response variables $y_1, \ldots, y_p$ on a subset of k explanatory variables $x_1, \ldots, x_k$. Suppose that there are n observations on $\boldsymbol{y} = (y_1, \ldots, y_p)'$ and $\boldsymbol{x} = (x_1, \ldots, x_k)'$, and let $\mathbf{Y} : n \times p$ and $\mathbf{X} : n \times k$ be the observation matrices of $\boldsymbol{y}$ and $\boldsymbol{x}$, respectively, for a sample size n. The multivariate linear regression model including all explanatory variables is written as

$$\mathbf{Y} = \mathbf{X}\boldsymbol{\Theta} + \mathbf{E}, \tag{18.1}$$

T. Sakurai
School of General and Management Studies, Suwa University of Science, Nagano, Japan
e-mail: sakurai@rs.sus.ac.jp

Y. Fujikoshi (✉)
Department of Mathematics, Graduate School of Science, Hiroshima University, Hiroshima, Japan

© Springer Nature Switzerland AG 2020

T. Holgersson, M. Singull (eds.), *Recent Developments in Multivariate and Random Matrix Analysis*, https://doi.org/10.1007/978-3-030-56773-6_18

where $\mathbf{\Theta}$ is a $k \times p$ unknown matrix of regression coefficients and $\mathbf{E} = (\boldsymbol{e}_1, \ldots, \boldsymbol{e}_n)'$ is an error matrix. It is assumed that $\boldsymbol{e}_1, \ldots, \boldsymbol{e}_n$ are independently and identically distributed as a p-dimensional normal distribution $\mathrm{N}_p(\mathbf{0}, \mathbf{\Sigma})$. The multivariate normal linear regression model is also written as

$$\mathbf{Y} \sim \mathrm{N}_{n \times p}(\mathbf{X}\mathbf{\Theta}, \mathbf{\Sigma} \otimes \mathbf{I}_n). \tag{18.2}$$

The notation $\mathrm{N}_{n \times p}(\cdot, \cdot)$ denotes the matrix normal distribution such that the mean of $\mathbf{Y}$ is $\mathbf{X}\mathbf{\Theta}$ and the covariance matrix of vec $(\mathbf{Y})$ is $\mathbf{\Sigma} \otimes \mathbf{I}_n$. Here, vec$(\mathbf{Y})$ is the $np \times 1$ column vector obtained by vertically stacking the columns of $\mathbf{Y}$.

An important issue in theoretically and empirical contexts concerns selecting the best or true model from a suite of candidate models all specified by a linear regression of $\boldsymbol{y}$ on subvectors of $\boldsymbol{x}$. When $\mathbf{\Sigma}$ is an unknown positive definite matrix, a well-known and often applied method involves using one or more information information criteria, principally AIC or BIC. It has been shown that AIC has consistency properties when $p/n \to c \in (0, 1)$, under some conditions, but BIC is not necessarily consistent, [8, 17]. It may be noted that such a property does not transfer to the large sample case since BIC under a large-sample framework is consistent, but AIC is not consistent, (see, e.g., [11]).

In the context of high-dimensional data, it is important to consider the selection of regression variables for the case where p is larger than n, and k is also large. When p is large, it is natural to consider the covariance structure, since the covariance matrix with no structures involves many unknown parameters. One approach to this involves employing a sparse method or joint regularization of the regression parameters and the inverse covariance matrix (see, e.g., [12, 13]). Another approach, considered herein, would be to select the regression variables, assuming a simple covariance structure.

In this paper, we first consider the variable selection problem by a general information criterion including AIC and BIC, under three simple covariance structures; (i) an independent covariance structure, (ii) a uniform covariance structure, and (iii) an autoregressive covariance structure. An information criterion including AIC and BIC as special cases is proposed. A sufficient condition for the general information to be consistent is provided. However, this criterion is computationally onerous when k is large. To circumvent this issue, we consider a test-based criterion drawing on the significance of each variable, which was used in [5]. The idea of this method is based on [18] and [11], in which they consider a selection method based on comparing information criterion for the subset removing each variable and the full set of variables. The method is called a KOO (Kick One Out) method by [2]. In this paper it is shown that the variable selection methods are also consistent in a high-dimensional context. Our results are validated numerically by conducting a Monte Carlo simulation.

Our high-dimensional consistency property is derived when $p/n \to c \in (0, 1)$ and k is fixed, though it is hoped that the condition on k is extended to $k/n \to \tilde{c} \in (0, 1)$. From a practical point, there are many high-dimensional data with large k, but there are also high-dimensional data with small or moderate k. As examples of the latter, see (1) DNA finger print data ([7, 16]) with $p = 84, n = 89$ and $k = 4$, and (2) Arabidopsis thaliana data [10] with $p = 795, n = 118$ and $k = 39$.

The remainder of the present paper is organized as follows. In Sect. 18.2, we present the relevant notation and preliminaries for the test-based method. In Sects. 18.3, 18.4 and 18.5, we focus on the high-dimensional consistencies of information criteria under covariance structures (i), (ii) and (iii), respectively. These are also numerically studied. In Sect. 18.6, their test-based criteria are proposed and shown to be consistent under the same condition as in the case of information criteria. In Sect. 18.7, conclusions are offered. All proofs of our results are provided in the Appendix.

18.2 Notation and Preliminaries

This paper is concerned with the selection of explanatory variables in multivariate regression model (18.1). Suppose that j denotes a subset of $\omega = \{1, \ldots, k\}$ containing k_j elements, and $\mathbf{X}_j$ denotes the $n \times k_j$ matrix consisting of columns of $\mathbf{X}$ indexed by elements of j. Then, $\mathbf{X}_\omega = \mathbf{X}$. Further, we assume that the covariance matrix $\boldsymbol{\Sigma}$ have a covariance structure $\boldsymbol{\Sigma}_s$. A generic candidate model can then be expressed as

$$M_{s,j} : \mathbf{Y} \sim \mathrm{N}_{n\times p}(\mathbf{X}_j\boldsymbol{\Theta}_j, \boldsymbol{\Sigma}_s \otimes \mathbf{I}_n), \tag{18.3}$$

where $\boldsymbol{\Theta}_j$ is a $k_j \times p$ unknown matrix of regression coefficients. We assume that rank$(\mathbf{X}) = k(< n)$. The selection of the i-th explanatory variable x_i is corresponding to whether the i-th row of $\boldsymbol{\Theta}$ is the null vector or not. Note that the row vectors of $\boldsymbol{\Theta}$ are p dimension, and asymptotic properties of our selection methods closely depend on p.

As a model selection method, let us consider an information criterion based on AIC [1]. When $\boldsymbol{\Sigma}_s$ is a $p \times p$ unknown covariance matrix, the AIC (see, e.g., [3, 4] for $M_{s,j}$ is given by

$$\mathrm{AIC}_{g,j} = n \log|\hat{\boldsymbol{\Sigma}}_{s,j}| + np(\log 2\pi + 1) + 2\left\{k_j p + \frac{1}{2}p(p+1)\right\}, \tag{18.4}$$

where $n\hat{\boldsymbol{\Sigma}}_{s,j} = \mathbf{Y}'(\mathbf{I}_n - \mathbf{P}_j)\mathbf{Y}$ and $\mathbf{P}_j = \mathbf{X}_j(\mathbf{X}_j'\mathbf{X}_j)^{-1}\mathbf{X}_j'$. The sum of the first two terms, "$n \log|\hat{\boldsymbol{\Sigma}}_{s,j}| + np(\log 2\pi + 1)$" is "$-2\log\max_{M_{s,j}} f(\mathbf{Y}; \boldsymbol{\Theta}_j, \boldsymbol{\Sigma}_{s,j})$", where $f(\mathbf{Y}; \boldsymbol{\Theta}_j, \boldsymbol{\Sigma}_{s,j})$ is the density function of $\mathbf{Y}$ under $M_{s,j}$. The AIC was introduced as an asymptotic unbiased estimator for the risk function defined as the expected log-predictive-likelihood or equivalently the Kullback-Leibler information, for a candidate model $M_{s,j}$ (see, e.g., [4]. When $j = \omega$, model $M_{s,\omega}$ is the full model. Note that $\hat{\boldsymbol{\Sigma}}_{s,\omega}$ and $\mathbf{P}_\omega$ are defined from $\hat{\boldsymbol{\Sigma}}_{s,j}$ and $\mathbf{P}_j$ as $j = \omega$, $k_\omega = k$ and $\mathbf{X}_\omega = \mathbf{X}$.

We consider the case where the covariance matrix $\boldsymbol{\Sigma}$ is described by each of the following three classes:

(i) Independent covariance structure (IND); $\boldsymbol{\Sigma}_v = \sigma_v^2 \mathbf{I}_p$,

(ii) Uniform covariance structure (UNIF); $\boldsymbol{\Sigma}_u = \sigma_u^2 (\rho_u^{1-\delta_{ii'}})_{1 \le i,\, i' \le p}$,

(iii) Autoregressive covariance structure (AUTO) ; $\boldsymbol{\Sigma}_a = \sigma_a^2 (\rho_a^{|i-i'|})_{1 \le i,\, i' \le p}$.

The candidate model can be expressed as (18.3) with Σ_v, Σ_u or Σ_a for Σ_s. For deriving the maximum likelihood under $M_{s,j}$, we use the fact that for any positive definite $\boldsymbol{\Sigma}_s$,

$$\begin{aligned} \max_{\boldsymbol{\Theta}_j} f(\mathbf{Y}; \boldsymbol{\Theta}_j, \boldsymbol{\Sigma}_s) &= np \log |\boldsymbol{\Sigma}_s| + np(\log 2\pi + 1) \\ &\quad + \min_{\boldsymbol{\Theta}_j} \mathrm{tr} \boldsymbol{\Sigma}_s^{-1} (\mathbf{Y} - X_j \boldsymbol{\Theta}_j)' (\mathbf{Y} - X_j \boldsymbol{\Theta}_j) \qquad (18.5) \\ &= np \log |\boldsymbol{\Sigma}_{g,j}| + np \log 2\pi + \mathrm{tr} \boldsymbol{\Sigma}_s^{-1} \mathbf{Y} (\mathbf{I}_n - \mathbf{P}_j) \mathbf{Y}. \end{aligned}$$

Let $\hat{\boldsymbol{\Sigma}}_{s,j}$ be the quantity minimizing the right side of (18.5). Then, in our problem, it satisfies

$$\mathrm{tr} \hat{\boldsymbol{\Sigma}}_{s,j}^{-1} \mathbf{Y} (\mathbf{I}_n - \mathbf{P}_j) \mathbf{Y} = np.$$

We consider a general information criterion defined by

$$\begin{aligned} \mathrm{IC}_{d,s,j} &= -2 \log f(\mathbf{Y}; \hat{\boldsymbol{\Theta}}_j, \hat{\boldsymbol{\Sigma}}_{s,j}) + d m_{s,j} \\ &= np \log |\boldsymbol{\Sigma}_{s,j}| + np(\log 2\pi + 1) + d m_{s,j}, \qquad (18.6) \end{aligned}$$

where $m_{s,j}$ is the number of independent unknown parameters under $M_{s,j}$, and d is a positive constant which may depend on n. When $d = 2$ and $d = \log n$,

$$\mathrm{IC}_{2,s,j} = \mathrm{AIC}_{s,j}, \quad \mathrm{IC}_{\log n,s,j} = \mathrm{BIC}_{s,j}.$$

For each of the three covariance structures, consider that the objective is to select the best model from all candidate models or a subset from the candidate models. Let $\mathscr{F}_\omega$ be the entire suite of candidate models, denoted by

$$\mathscr{F}_\omega = \{\{1\}, \ldots, \{k\}, \{1, 2\}, \ldots, \{1, \ldots, k\}\},$$

or its subfamily $\mathscr{F}$. Then, our model selection criterion is to select the model M_j or subset j minimizing $\mathrm{IC}_{s,d,j}$, which is written as

$$\hat{j}_{\mathrm{IC}d,s} = \arg \min_{j \in \mathscr{F}} \mathrm{IC}_{d,s,j}. \qquad (18.7)$$

To explore the consistency properties of $\hat{j}_{\mathrm{IC}d,s}$, it is assumed that the true model $M_{s,*}$ is included in the full model, i.e.,

$$M_{s,*} : \mathbf{Y} \sim \mathrm{N}_{n\times p}(\mathbf{X}_*\boldsymbol{\Theta}_*, \boldsymbol{\Sigma}_{s,*} \otimes \mathbf{I}_n). \tag{18.8}$$

Let us denote the minimum model including the true model by M_{s,j_*}. The true mean of $\mathbf{Y}$ is expressed as

$$\mathbf{X}_*\boldsymbol{\Theta}_* = \mathbf{X}_{j_*}\boldsymbol{\Theta}_{j_*} \tag{18.9}$$

for some $k_{j_*} \times p$ matrix $\boldsymbol{\Theta}_{j_*}$, so the notation $\mathbf{X}_{j_*}\boldsymbol{\Theta}_{j_*}$ is also used for the true mean of $\mathbf{Y}$. Let $\mathscr{F}$ separate into two mutually exclusive sets: a set of overspecified models, $\mathscr{F}_+ = \{j \in \mathscr{F} \mid j_* \subseteq j\}$ and a set of underspecified models, $\mathscr{F}_- = \mathscr{F}_+^c \cap \mathscr{F}$.

Here we list our main assumptions:

A1 (The true model): $M_{s,*} \in \mathscr{F}$.
A2 (The asymptotic framework): $p \to \infty,\ n \to \infty,\ p/n \to c \in (0, \infty)$.

A general model selection criterion $\hat{j}_{\mathrm{IC}d,s}$ is called consistent if

$$\lim_{p/n \to c \in (0,\infty)} \Pr(\hat{j}_{\mathrm{IC}d,s} = j_*) = 1.$$

To obtain $\hat{j}_{\mathrm{IC}d,s}$, $\mathrm{IC}_{s,d}$ needs to be calculated for all subsets of $\mathscr{F}$. When $\mathscr{F} = \mathscr{F}_\omega$, there are $2^k - 1$ subsets. This is computationally onerous as k becomes large. To overcome this challenge, we consider a test-based criterion (TC) based on a significance test of each $\boldsymbol{\theta}_i$, where $\boldsymbol{\Theta} = (\boldsymbol{\theta}_1, \ldots, \boldsymbol{\theta}_k)'$. Note that a critical region for testing "$\boldsymbol{\theta}_i = \mathbf{0}$" based on the likelihood ratio principle and AIC is expressed as

$$\mathrm{IC}_{2,s,(-i)} - \mathrm{IC}_{2,s,\omega} > 0, \tag{18.10}$$

where $(-i)$ is the subset of $\omega = \{1, \ldots, k\}$ obtained by omitting the i $(1 \le i \le k)$ from ω. According to [5], we consider a $\mathrm{TC}_{d,s}$ defined to select

$$\hat{j}_{\mathrm{TC}_{d,s}} = \{i \in \omega \mid \mathrm{IC}_{d,s,(-i)} > \mathrm{IC}_{d,s,\omega},\ i = 1, \ldots, k\}. \tag{18.11}$$

The idea has hitherto been considered by [18]. [11] have studied a consistency property in a large-sample asymptotic framework. In Sect. 18.6, we identify the consistency property of $\hat{j}_{\mathrm{TC}_{d,s}}$.

18.3 IC Under an Independent Covariance Structure

In this section, we consider the problem of selecting variables in a multivariate regression model under the assumption that the covariance matrix has an independent covariance structure. A generic candidate model can be expressed as

$$M_{v,j} : \mathbf{Y} \sim \mathrm{N}_{n\times p}(\mathbf{X}_j \boldsymbol{\Theta}_j, \boldsymbol{\Sigma}_v \otimes \mathbf{I}_n), \tag{18.12}$$

where $\boldsymbol{\Sigma}_v = \sigma_v^2 \mathbf{I}_p$ and $\sigma_v > 0$. Then, we have

$$\begin{aligned} -2\log f(\mathbf{Y}; \boldsymbol{\Theta}_j, \sigma_v^2) &= np\log(2\pi) + np\log\sigma_v^2 \\ &\quad + \frac{1}{\sigma_v^2}\mathrm{tr}(\mathbf{Y} - \mathbf{X}_j\boldsymbol{\Theta}_j)'(\mathbf{Y} - \mathbf{X}_j\boldsymbol{\Theta}_j). \end{aligned}$$

Therefore, it is straighforward to identify that the maximum likelihood estimators of $\boldsymbol{\Theta}_j$ and σ_v^2 under $M_{v,j}$ are given as

$$\hat{\boldsymbol{\Theta}}_j = (\mathbf{X}_j'\mathbf{X}_j)^{-1}\mathbf{X}_j'\mathbf{Y}, \quad \hat{\sigma}_{v,j}^2 = \frac{1}{np}\mathrm{tr}\mathbf{Y}'(\mathbf{I}_n - \mathbf{P}_j)\mathbf{Y}. \tag{18.13}$$

The information criterion (18.6) is given by

$$\mathrm{IC}_{v,d,j} = np\log\hat{\sigma}_{v,j}^2 + np(\log 2\pi + 1) + dm_{v,j}, \tag{18.14}$$

where d is a positive constant, which may depend on n, and $m_{v,j} = k_j p + 1$. Assume that the true model is expressed as

$$M_{v,*} : \mathbf{Y} \sim \mathrm{N}_{n\times p}(\mathbf{X}_*\boldsymbol{\Theta}_*, \sigma_{v,*}^2\mathbf{I}_p \otimes \mathbf{I}_n), \tag{18.15}$$

with (18.9), denoting the minimum model including the true model $M_{v,*}$ by M_{v,j_*}. In general, $\mathbf{Y}'(\mathbf{I}_n - \mathbf{P}_j)\mathbf{Y}$ is distributed as a nocentral Wishart distribution, more precisely

$$\mathbf{Y}'(\mathbf{I}_n - \mathbf{P}_j)\mathbf{Y} \sim \mathrm{W}_p(n - k_j, \boldsymbol{\Sigma}_{v,*}; (\mathbf{X}_{j_*}\boldsymbol{\Theta}_{j_*})'(\mathbf{I}_n - \mathbf{P}_j)\mathbf{X}_{j_*}\boldsymbol{\Theta}_{j_*}),$$

which implies the following lemma.

Lemma 18.1 *Under* (18.15), *$np\hat{\sigma}_{v,j}^2/\sigma_{v,*}^2$ is distributed as a noncentral chi-square distribution $\chi^2_{(n-k_j)p}(\delta^2_{v,j})$, where*

$$\delta_{v,j}^2 = \frac{1}{\sigma_{v,*}^2}\mathrm{tr}(\mathbf{X}_{j_*}\boldsymbol{\Theta}_{j_*})'(\mathbf{P}_\omega - \mathbf{P}_j)\mathbf{X}_{j_*}\boldsymbol{\Theta}_{j_*}. \tag{18.16}$$

Here we write tr(**ABC**) *as* tr**ABC** *simply. If* $j \in \mathscr{F}_+$, *then* $\delta^2_{v,j} = 0$.

For a sufficient condition for the consistency of $\mathrm{IC}_{d,v}$, we assume

$$\text{A3v : For any } j \in \mathscr{F}_-,\ \delta^2_{v,j} = \mathrm{O}(np), \text{ and } \lim_{p/n \to c} \frac{1}{np}\delta^2_{v,j} = \eta^2_{v,j} > 0. \qquad (18.17)$$

Theorem 18.1 *Suppose that assumptions* A1, A2, *and* A3v *are satisfied. Then, the information criteria* $\mathrm{IC}_{d,v}$ *defined by* (18.14) *is consistent if* $d > 1$ *and* $d/n \to 0$.

AIC and BIC satisfy conditions "$d > 1$ and $d/n \to 0$", and we have the following result.

Corollary 18.1 *Under assumptions* A1, A2, *and* A3v, AIC *and* BIC *are consistent.*

Next we numerically explore the validity of our claims. The true model was assumed as

$$M_{v,*} : \mathbf{Y} \sim \mathrm{N}_{n\times p}(\mathbf{X}_*\mathbf{\Theta}_*, \sigma^2_{v,*}\mathbf{I}_p \otimes \mathbf{I}_n).$$

Here $\mathbf{\Theta}_* : 3 \times p$ was determined by random sampling from a uniform distribution on the interval $(1, 2)$, i.e., i.i.d. from $\mathrm{U}(1, 2)$. The first column of $\mathbf{X}_\omega : n \times 10$ is $\mathbf{1}_n$, and the other elements are i.i.d. from $\mathrm{U}(-1, 1)$. The true variance was set as $\sigma^2_{v,*} = 2$. In this paper we consider to select the true model from the set of all candidate models or from its subset. Here, we considered a selection from five candidate models $M_{j_\alpha}, \alpha = 1, 2, \ldots, 5$, where $j_\alpha = \{1, \ldots, \alpha\}$. We focused on selection percentages of the true model for 10^4 replications under AIC and BIC for

$$(n, p) = (50, 15), (100, 30), (200, 60), (50, 100), (100, 200), (200, 400).$$

Results are shown in Table 18.1. The true model is selected in all cases except for $(n, p) = (50, 15)$ of AIC_v. In the case $(n, p) = (50, 15)$ of AIC_v, the selection percentage is not 100, but it is still very high.

We do not present the selection percentages of AIC_v and BIC_v for $p/n = 2$, $n = 50, 100, 200$, since the selection percentage was 100.

Table 18.1 Selection percentages of AIC_v and BIC_v for $p/n = 0.3$

	AIC_v			BIC_v		
j	(50, 15)	(100, 30)	(200, 60)	(50, 15)	(100, 30)	(200, 60)
1	0.0	0.0	0.0	0.0	0.0	0.0
2	0.0	0.0	0.0	0.0	0.0	0.0
3	97.3	99.9	100.0	100.0	100.0	100.0
4	2.4	0.1	0.0	0.0	0.0	0.0
5	0.3	0.0	0.0	0.0	0.0	0.0
≥ 6	0.0	0.0	0.0	0.0	0.0	0.0

18.4 IC Under a Uniform Covariance Structure

In this section we consider the model selection criterion when the covariance matrix has a uniform covariance structure

$$\boldsymbol{\Sigma}_u = \sigma_u^2(\rho_u^{1-\delta_{ii'}}) = \sigma_u^2\{(1-\rho_u)\mathbf{I}_p + \rho_u \mathbf{1}_p\mathbf{1}_p'\}. \tag{18.18}$$

The covariance structure is expressed as

$$\boldsymbol{\Sigma}_u = \tau_1\left(\mathbf{I}_p - \frac{1}{p}\mathbf{G}_p\right) + \tau_2\frac{1}{p}\mathbf{G}_p,$$

where

$$\tau_1 = \sigma_u^2(1-\rho_u), \quad \tau_2 = \sigma_u^2\{1+(p-1)\rho_u\}, \quad \mathbf{G}_p = \mathbf{1}_p\mathbf{1}_p',$$

and $\mathbf{1}_p = (1,\ldots,1)'$. Noting that $\mathbf{I}_p - \frac{1}{p}\mathbf{G}_p$ and $\frac{1}{p}\mathbf{G}_p$ are orthogonal idempotent matrices, we have

$$|\boldsymbol{\Sigma}_u| = \tau_2\tau_1^{p-1}, \quad \boldsymbol{\Sigma}_u^{-1} = \frac{1}{\tau_1}\left(\mathbf{I}_p - \frac{1}{p}\mathbf{G}_p\right) + \frac{1}{\tau_2}\frac{1}{p}\mathbf{G}_p.$$

Now, consider the model $M_{u,j}$ given by

$$M_{u,j} : \mathbf{Y} \sim \mathrm{N}_{n\times p}(\mathbf{X}_j\boldsymbol{\Theta}_j, \boldsymbol{\Sigma}_u \otimes \mathbf{I}_n), \tag{18.19}$$

where $\boldsymbol{\Sigma}_u = \tau_1\left(\mathbf{I}_p - p^{-1}\mathbf{G}_p\right) + \tau_2 p^{-1}\mathbf{G}_p$. Let $\mathbf{H} = (\boldsymbol{h}_1, \mathbf{H}_2)$ be an orthogonal matrix where $\boldsymbol{h}_1 = p^{-1/2}\mathbf{1}_p$, and let

$$\mathbf{U}_j = \mathbf{H}'\mathbf{W}_j\mathbf{H}, \quad \mathbf{W}_j = \mathbf{Y}'(\mathbf{I}_n - \mathbf{P}_j)\mathbf{Y}.$$

The density function of $\mathbf{Y}$ under $M_{u,j}$ is denoted by $f(\mathbf{Y}; \boldsymbol{\Theta}, \tau_1, \tau_2)$. From (18.5) we have

$$\begin{aligned} g(\tau_1,\tau_2) &= -2\log\max_{\boldsymbol{\Theta}_j} f(\mathbf{Y}; \boldsymbol{\Theta}_j, \tau_1, \tau_2) \\ &= np\log(2\pi) + n\log\tau_2 + n(p-1)\log\tau_1 + \mathrm{tr}\boldsymbol{\Psi}^{-1}\mathbf{U}_j, \end{aligned}$$

where $\boldsymbol{\Psi} = \mathrm{diag}(\tau_2, \tau_1, \ldots, \tau_1)$. Then, it can be shown that the maximum likelihood estimators of τ_1 and τ_2 under $M_{u,j}$ are given by

$$\begin{aligned} \hat{\tau}_{1j} &= \frac{1}{n(p-1)}\mathrm{tr}\,\mathbf{D}_1\mathbf{U}_j = \frac{1}{n(p-1)}\mathrm{tr}\mathbf{H}_2'\mathbf{Y}'(\mathbf{I}_n - \mathbf{P}_j)\mathbf{Y}\mathbf{H}_2, \\ \hat{\tau}_{2j} &= \frac{1}{n}\mathrm{tr}\,\mathbf{D}_2\mathbf{U}_j = \frac{1}{n}\boldsymbol{h}_1'\mathbf{Y}'(\mathbf{I}_n - \mathbf{P}_j)\mathbf{Y}\boldsymbol{h}_1, \end{aligned}$$

where $\mathbf{D}_1 = \text{diag}(0, 1, \ldots, 1)$ and $\mathbf{D}_2 = \text{diag}(1, 0, \ldots, 0)$. The number of independent parameters under $M_{u,j}$ is $m_j = k_j p + 2$. Noting that $\boldsymbol{\Psi}$ is diagonal, we can derive the information criterion (18.6) as

$$\text{IC}_{d,u,j} = n(p-1)\log\hat{\tau}_{1j} + n\log\hat{\tau}_{2j} + np(\log 2\pi + 1) + d(k_j p + 2). \tag{18.20}$$

Assume that the true model is expressed as

$$M_{u,*} : \mathbf{Y} \sim \text{N}_{n\times p}(\mathbf{X}_*\boldsymbol{\Theta}_*, \boldsymbol{\Sigma}_{u,*} \otimes \mathbf{I}_n), \tag{18.21}$$

with (18.9), denoting the minimum model including the true model $M_{u,*}$ by M_{u,j_*}, where $\boldsymbol{\Sigma}_{u,*} = \tau_{1*}\left(\mathbf{I}_p - p^{-1}\mathbf{G}_p\right) + \tau_{2*}p^{-1}\mathbf{G}_p$. In general, it holds that $\mathbf{U}_j \sim \text{W}_p(n-k_j, \boldsymbol{\Psi}_*; \boldsymbol{\Delta}_j)$, where

$$\boldsymbol{\Delta}_j = (\mathbf{X}_{j_*}\boldsymbol{\Theta}_{j_*}\mathbf{H})'(\mathbf{I}_n - \mathbf{P}_j)\mathbf{X}_{j_*}\boldsymbol{\Theta}_{j_*}\mathbf{H},$$

and $\boldsymbol{\Psi}_* = \text{diag}(\tau_{2*}, \tau_{1*}, \ldots, \tau_{1*})$. Therefore, we have the following Lemma (see, e.g., [9]).

Lemma 18.2 *Under the true model* (18.21), *it holds that*

(1) $n(p-1)\tau_{1*}^{-1}\hat{\tau}_{1j}$ *is noncetrally distributed as* $\chi^2_{(p-1)(n-k_j)}(\delta^2_{1j})$, *where*

$$\delta^2_{1j} = \frac{1}{\tau_{1*}}\text{tr}\mathbf{H}_2'(\mathbf{X}_{j_*}\boldsymbol{\Theta}_{j_*})'(\mathbf{I}_n - \mathbf{P}_j)(\mathbf{X}_{j_*}\boldsymbol{\Theta}_{j_*})\mathbf{H}_2.$$

(2) $n\tau_{2*}^{-1}\hat{\tau}_{2j}$ *is noncentrally distributed as* $\chi^2_{n-k_j}(\delta^2_{2j})$, *where*

$$\delta^2_{2j} = \frac{1}{\tau_{2*}}\text{tr}\boldsymbol{h}_1'(\mathbf{X}_{j_*}\boldsymbol{\Theta}_{j_*})'(\mathbf{I}_n - \mathbf{P}_j)(\mathbf{X}_{j_*}\boldsymbol{\Theta}_{j_*})\boldsymbol{h}_1.$$

(3) *If* $j \in \mathscr{F}_+$, *then* $\delta_{1j} = 0$ *and* $\delta_{2j} = 0$.

To identify a sufficient condition for the consistency of $\text{IC}_{d,u}$, we assume
A3u: For any $j \in \mathscr{F}_-$, $\delta^2_{1j} = \text{O}(np)$, $\delta^2_{2j} = \text{O}(n)$ and

$$\lim_{p/n\to c}\frac{1}{np}\delta^2_{1j} = \eta^2_{1j} > 0, \quad \lim_{p/n\to c}\frac{1}{n}\delta^2_{2j} = \eta^2_{2j} > 0. \tag{18.22}$$

Theorem 18.2 *Suppose that assumptions* A1, A2 *and* A3u *are satisfied. Then, the information criteria* $\text{IC}_{d,u}$ *defined by* (18.20) *is consistent if* $d > 1$ *and* $d/n \to 0$.

Corollary 18.2 *Under assumptions* A1, A2 *and* A3u, AIC *and* BIC *are consistent.*

We ran a numerical experiment under the same settings as in the independence covariance structure case except for the covariance structure itself. The true-uniform

Table 18.2 Selection percentages of AIC and BIC for $p/n = 0.3$

	AIC			BIC		
j	(50, 15)	(100, 30)	(200, 60)	(50, 15)	(100, 30)	(200, 60)
1	0.2	0.0	0.0	72.8	1.3	0.0
2	0.6	0.0	0.0	7.6	3.0	0.0
3	96.5	99.9	100.0	19.6	95.7	100.0
4	2.4	0.1	0.0	0.0	0.0	0.0
5	0.3	0.0	0.0	0.0	0.0	0.0
≥ 6	0.0	0.0	0.0	0.0	0.0	0.0

Table 18.3 Selection percentages of AIC and BIC for $p/n = 2$

	AIC			BIC		
j	(50, 100)	(100, 200)	(200, 400)	(50, 100)	(100, 200)	(200, 400)
1	0.1	0.0	0.0	100.0	99.6	0.0
2	3.1	0.0	0.0	0.0	0.4	0.0
3	96.9	100.0	100.0	0.0	0.0	100.0
4	0.0	0.0	0.0	0.0	0.0	0.0
5	0.0	0.0	0.0	0.0	0.0	0.0
≥ 6	0.0	0.0	0.0	0.0	0.0	0.0

covariance structure was set as $\sigma_{u,*}^2 = 2$, $\rho_{u,*} = 0.2$. Results are given in Tables 18.2 and 18.3. In general, AIC selects the true model with a high probability even in a finite setting. probability. However, BIC does not always select the true model when (n, p) is small, though it exhibits a consistency property.

18.5 IC Under an Autoregressive Covariance Structure

In this section, we consider the model selection criterion when the covariance matrix $\mathbf{\Sigma}$ has an autoregressive covariance structure

$$\mathbf{\Sigma}_a = \sigma_a^2(\rho_a^{|i-i'|})_{1\leq i,\ j\leq p}. \tag{18.23}$$

Then, it is well-known that (see, e.g., [6])

$$|\mathbf{\Sigma}_a| = (\sigma_a^2)^p(1-\rho_a^2)^{p-1}, \quad \mathbf{\Sigma}_a^{-1} = \frac{1}{\sigma_a^2(1-\rho_a^2)}(\rho_a^2\mathbf{C}_1 - 2\rho_a\mathbf{C}_2 + \mathbf{C}_0),$$

where $\mathbf{C}_0 = \mathbf{I}_p$,

$$\mathbf{C}_1 = \begin{pmatrix} 0 & 0 & \cdots & 0 & 0 \\ 0 & 1 & \cdots & 0 & 0 \\ \vdots & \vdots & \ddots & \vdots & \vdots \\ 0 & 0 & \cdots & 1 & 0 \\ 0 & 0 & \cdots & 0 & 0 \end{pmatrix}, \quad \mathbf{C}_2 = \frac{1}{2} \begin{pmatrix} 0 & 1 & \cdots & 0 & 0 \\ 1 & 0 & \cdots & 0 & 0 \\ \vdots & \vdots & \ddots & \vdots & \vdots \\ 0 & 0 & \cdots & 0 & 1 \\ 0 & 0 & \cdots & 1 & 0 \end{pmatrix}.$$

Now, consider the model $M_{a,j}$ given by

$$M_{a,j} : \mathbf{Y} \sim \mathrm{N}_{n\times p}(X_j \boldsymbol{\Theta}_j, \boldsymbol{\Sigma}_a \otimes \mathbf{I}_n), \tag{18.24}$$

where $\boldsymbol{\Sigma}_a = \sigma_a^2(\rho_a^{|i-i'|})$. Then, from (18.5) the maximum likelihood estimator of $\boldsymbol{\Theta}_j$ is given by

$$\hat{\boldsymbol{\Theta}}_j = (\mathbf{X}'\mathbf{X})^{-1}\mathbf{X}'\mathbf{Y},$$

and the maximum likelihood estimators of ρ_a and σ_a^2 under $M_{a,j}$ can be obtained through the minimization of

$$\begin{aligned} -2\log f(\mathbf{Y}; \hat{\boldsymbol{\Theta}}_{a,j}, \sigma_a^2, \rho) &= np\log(2\pi) + np\log\sigma_a^2 + n(p-1)\log(1-\rho_a^2) \\ &+ \frac{1}{\sigma_a^2(1-\rho_a^2)} + \mathrm{tr}(\rho_a^2\mathbf{C}_1 - 2\rho_a\mathbf{C}_2 + \mathbf{C}_0)\mathbf{Y}'(\mathbf{I}_n - \mathbf{P}_j)\mathbf{Y} \end{aligned}$$

with respect to σ_a and ρ_a. Therefore, the maximum likelihood estimators $\hat{\sigma}_{a,j}$ and $\hat{\rho}_{a,j}$ of σ_a^2 and ρ_a under $M_{a,j}$ are given as solutions to the following two equations (see [6]):

$$\hat{\sigma}_{a,j}^2 = \frac{n-k_j}{n}\frac{1}{p(1-\hat{\rho}_{a,j}^2)}(a_{1j}\hat{\rho}_{a,j}^2 - 2a_{2j}\hat{\rho}_{a,j} + a_{0j}), \tag{18.25}$$

$$(p-1)a_{1j}\hat{\rho}_{a,j}^3 - (p-2)a_{2j}\hat{\rho}_{a,j}^2 - (pa_{1j} + a_{0j})\hat{\rho}_{a,j} + pa_{2j} = 0, \tag{18.26}$$

where $a_{ij} = \mathrm{tr}\mathbf{C}_i\mathbf{S}_j$, $i = 0, 1, 2$, and $\mathbf{S}_j = (n-k_j)^{-1}\mathbf{Y}'(\mathbf{I}_n - \mathbf{P}_j)\mathbf{Y}$. The information criterion $\mathrm{IC}_{d,a,j}$ can then be written as

$$\mathrm{IC}_{d,a,j} = np\log\hat{\sigma}_{a,j}^2 + n(p-1)\log(1-\hat{\rho}_{a,j}^2) + np(\log 2\pi + 1) + d(k_jp + 2). \tag{18.27}$$

Note that the maximum likelihood estimators of ρ and σ^2 are expressed in terms of a_{0j}, a_{1j} and a_{2j} or

$$b_{0j} = \mathrm{tr}\mathbf{C}_0\mathbf{W}_j, \quad b_{1j} = \mathrm{tr}\mathbf{C}_1\mathbf{W}_j, \quad b_{2j} = \mathrm{tr}\mathbf{C}_2\mathbf{W}_j, \tag{18.28}$$

where

$$\mathbf{W}_j = (n - k_j)\mathbf{S}_j = \mathbf{Y}'(\mathbf{I}_n - \mathbf{P}_j)\mathbf{Y}. \tag{18.29}$$

Assume that the true model is expressed as

$$M_{a,*} : \mathbf{Y} \sim \mathrm{N}_{n\times p}(\mathbf{X}_*\boldsymbol{\Theta}_*, \boldsymbol{\Sigma}_{a,*} \otimes \mathbf{I}_n), \tag{18.30}$$

with (18.9) denoting the minimum model including the true model M_* by M_{j_*}, where $\boldsymbol{\Sigma}_{a,*} = \sigma^2_{a,*}(\rho_{a,*}^{|i-i'|})$. Then,

$$\mathbf{W}_j = \mathbf{Y}'(\mathbf{I}_n - \mathbf{P}_j)\mathbf{Y} \sim \mathrm{W}_p(n - k_j, \boldsymbol{\Sigma}_{a,*}; \boldsymbol{\Omega}_j),$$

where $\boldsymbol{\Omega}_j = (\mathbf{X}_{j_*}\boldsymbol{\Theta}_{j_*})'(\mathbf{I}_n - \mathbf{P}_j)\mathbf{X}_{j_*}\boldsymbol{\Theta}_{j_*}$. To relate to the noncentrality matrix $\boldsymbol{\Omega}_j$, we use the following three quantities:

$$\delta_{ij} = \mathrm{tr}\mathbf{C}_i\boldsymbol{\Omega}_j, \quad i = 0, 1, 2. \tag{18.31}$$

As a sufficient condition for consistency, we assume A3a: For any $j \in \mathscr{F}_-$, the order of each element of $\boldsymbol{\Omega}_j$ is $\mathrm{O}(n)$, $\delta^2_{ij} = \mathrm{O}(np)$, and

$$\lim_{p/n\to c} \frac{1}{np}\delta^2_{ij} = \eta^2_{ij} > 0, \; i = 0, 1, 2. \tag{18.32}$$

Theorem 18.3 *Suppose that the assumptions* A1, A2 *and* A3a *are satisfied. Then, the information criteria* $\mathrm{IC}_{d,a}$ *defined by* (18.27) *are consistent if* $d > 1$ *and* $d/n \to 0$.

Corollary 18.3 *Under assumptions* A1, A2 *and* A3a, AIC *and* BIC *are consistent.*

We ran a numerical experiment under the same settings as in the independence covariance structure and the uniform covariance structure cases except for the covariance structure itself. Here the true covariance structure was set as the autoregressive covariance structure $\boldsymbol{\Sigma}_{a,*} = \sigma^2_{a,*}(\rho_{a,*}^{|i-i'|})$ with $\sigma^2_{a,*} = 2$, $\rho_{a,*} = 0.2$. Results are shown in given Tables 18.4 and 18.5 and indicate that AIC and BIC select the true model in all cases.

Table 18.4 Selection percentages of AIC and BIC for $p/n = 0.3$

	AIC			BIC		
j	(50, 15)	(100, 30)	(200, 60)	(50, 15)	(100, 30)	(200, 60)
1	0.0	0.0	0.0	0.5	0.0	0.0
2	0.0	0.0	0.0	0.1	0.0	0.0
3	97.2	99.9	100.0	99.4	100.0	100.0
4	2.4	0.1	0.0	0.0	0.0	0.0
5	0.3	0.0	0.0	0.0	0.0	0.0
≥ 6	0.0	0.0	0.0	0.0	0.0	0.0

Table 18.5 Selection percentages of AIC and BIC for $p/n = 2$

	AIC			BIC		
j	(50, 100)	(100, 200)	(200, 400)	(50, 100)	(100, 200)	(200, 400)
1	0.0	0.0	0.0	0.0	0.0	0.0
2	0.0	0.0	0.0	0.0	0.0	0.0
3	100.0	100.0	100.0	100.0	100.0	100.0
4	0.0	0.0	0.0	0.0	0.0	0.0
5	0.0	0.0	0.0	0.0	0.0	0.0
≥ 6	0.0	0.0	0.0	0.0	0.0	0.0

18.6 Consistency Properties of TC

In this section, we explore the consistency properties of three TC: $\mathrm{TC}_{d,v}$, $\mathrm{TC}_{d,u}$, and $\mathrm{TC}_{d,a}$ in (18.11), based on $\mathrm{IC}_{d,v}$, $\mathrm{IC}_{d,u}$, and $\mathrm{IC}_{d,a}$, respectively. Note that consistency properties of $\mathrm{IC}_{d,v}$, $\mathrm{IC}_{d,u}$, and $\mathrm{IC}_{d,a}$ are given in Theorems 18.1–18.3. Using a part of conditions used in the proofs of these consistencies, it is expected that $\mathrm{TC}_{d,v}$, $\mathrm{TC}_{d,u}$ and $\mathrm{TC}_{d,a}$ have similar consistency properties. In fact, we use the following assumptions $\widetilde{\mathrm{A}}$3v, $\widetilde{\mathrm{A}}$3u and $\widetilde{\mathrm{A}}$3a instead of A3v, A3u and A3a, respectively.

$\widetilde{\mathrm{A}}3\mathrm{v}$: For any $i \in j_*$, $\delta^2_{v,j_{(-i)}} = \mathrm{O}(np)$, and

$$\lim_{p/n \to c} \frac{1}{np} \delta^2_{v,j_{(-i)}} = \eta^2_{v,j_{(-i)}} > 0. \tag{18.33}$$

$\widetilde{\mathrm{A}}3\mathrm{u}$: For any $i \in j_*$, $\delta^2_{1j_{(-i)}} = \mathrm{O}(np)$, $\delta^2_{2j_{(-i)}} = \mathrm{O}(n)$, and

$$\lim_{p/n \to c} \frac{1}{np} \delta^2_{1j_{(-i)}} = \eta^2_{1j_{(-i)}} > 0, \quad \lim_{p/n \to c} \frac{1}{n} \delta^2_{2j_{(-i)}} = \eta^2_{2j_{(-i)}} > 0. \tag{18.34}$$

$\widetilde{\mathrm{A}}3\mathrm{a}$: For any$i \in j_*$, $\boldsymbol{\Omega}_j = \mathrm{O}(n)$, $\delta^2_{ij_{(-i)}} = \mathrm{O}(np)$, and

$$\lim_{p/n \to c} \frac{1}{np} \delta^2_{ij} = \eta^2_{ij} > 0, \; i = 0, 1, 2. \tag{18.35}$$

The assumption $\widetilde{\text{A}}$3s is more weaker than the assumption A3s for s = v, u, a.

Theorem 18.4 *Suppose that assumptions* A1 *and* A2 *are satisfied. Then it holds that*

(1) the test-based criterion $\mathrm{TC}_{d,v}$ *is consistent under assumption* $\widetilde{\text{A}}$3v *if* $d > 1$ *and* $d/n \to 0$.
(2) the test-based criterion $\mathrm{TC}_{d,u}$ *is consistent under assumption* $\widetilde{\text{A}}$3u *if* $d > 1$ *and* $d/n \to 0$.
(3) the test-based criterion $\mathrm{TC}_{d,a}$ *is consistent under assumption* $\widetilde{\text{A}}$3a *if* $d > 1$ *and* $d/n \to 0$.

To explore both the validity of the results and the convergence speeds, we ran a numerical experiment. The simulation settings are similar to the cases of $\mathrm{IC}_{d,v}$, $\mathrm{IC}_{d,u}$, and $\mathrm{IC}_{d,a}$ except in two respects. First, the following points: The total number of explanatory variables to be selected is 5 rather than 10. Second, the true covariance structures were set as follows:

$$\text{(i)}: \quad \boldsymbol{\Sigma} = \sigma_{v,*}^2 \mathbf{I}_p, \quad \sigma_{v,*}^2 = 2.$$

$$\text{(ii)}: \quad \boldsymbol{\Sigma} = \sigma_{u,*}^2 (\rho_{u,*}^{1-\delta_{ii'}}), \quad \sigma_{u,*}^2 = 2, \quad \rho_{u,*} = 0.9.$$

$$\text{(iii)}: \quad \boldsymbol{\Sigma} = \sigma_{a,*}^2 (\rho_{a,*}^{|i-i'|}), \quad \sigma_{a,*}^2 = 2, \quad \rho_{a,*} = 0.9.$$

Let TC_{A} and TC_{B} be the test-based criteria based on AIC and BIC, respectively. The selection rates associated with these criteria are given in Tables 18.6, 18.7 and 18.8 for each of the three covariance structures. Therein, column x_i denotes the

Table 18.6 Selection percentages of TC_{A} and TC_{B} for $(n, p) = (20, 10)$

$n = 20$, $p = 10$		Under	True	Over	x_1	x_2	x_3	x_4	x_5
TC_{A}	IND	0.03	0.74	0.23	1.00	1.00	0.97	0.13	0.12
	UNIF	0.36	0.47	0.17	1.00	0.83	0.73	0.14	0.12
	AUTO	0.40	0.44	0.17	1.00	0.84	0.68	0.14	0.13
TC_{B}	IND	0.21	0.77	0.02	1.00	1.00	0.79	0.01	0.01
	UNIF	0.78	0.22	0.01	1.00	0.50	0.37	0.01	0.01
	AUTO	0.81	0.18	0.01	1.00	0.49	0.30	0.01	0.01

Table 18.7 Selection percentages of TC_{A} and TC_{B} for $(n, p) = (200, 100)$

$n = 200$, $p = 100$		Under	True	Over	x_1	x_2	x_3	x_4	x_5
TC_{A}	IND	0.00	1.00	0.00	1.00	1.00	1.00	0.00	0.00
	UNIF	0.00	1.00	0.00	1.00	1.00	1.00	0.00	0.00
	AUTO	0.00	1.00	0.00	1.00	1.00	1.00	0.00	0.00
TC_{B}	IND	0.00	1.00	0.00	1.00	1.00	1.00	0.00	0.00
	UNIF	0.00	1.00	0.00	1.00	1.00	1.00	0.00	0.00
	AUTO	0.00	1.00	0.00	1.00	1.00	1.00	0.00	0.00

Table 18.8 Selection percentages of TC_A and TC_B for $(n, p) = (10, 20)$

$n = 10, p = 20$		Under	True	Over	x_1	x_2	x_3	x_4	x_5
TC_A	IND	0.07	0.41	0.53	1.00	1.00	0.94	0.37	0.34
	UNIF	0.47	0.15	0.38	0.97	0.85	0.62	0.39	0.35
	AUTO	0.47	0.15	0.38	0.95	0.81	0.65	0.37	0.35
TC_B	IND	0.15	0.54	0.31	1.00	0.99	0.86	0.20	0.19
	UNIF	0.68	0.15	0.17	0.93	0.70	0.43	0.20	0.20
	AUTO	0.71	0.13	0.15	0.87	0.63	0.46	0.20	0.20

selection rate for the i-th explanatory variable x_i. "Under", "True", and "Over" denote the underspecified models, the true model, and the over-specified models, respectively. When $(n, p) = (200, 100)$ or $(100, 200)$, the probabilities of selecting the true model are equal almost one. However, when $(n, p) = (20, 10)$ or $(10, 20)$, the probabilities are not so large and depend on the covariance structures. It is seen that the probabilities in the case of independence covariance structure are larger than the ones in the case of uniform and autoregressive covariance structures.

We do not present the selection percentages of TC_A and TC_B for $(n, p) = (100, 200)$, since it was the same as Table 18.7.

18.7 Concluding Remarks

This paper is concerned with the problem of variable selecting in p variate regression models with one of three covariance structures: (i) an independent covariance structure (IND); (ii) a uniform covariance structure (UNIF); (iii) an autoregressive covariance structure (AUTO). As a selection method, an information $IC_{d,s}$ was considered for each of the three covariance structures, where d is a positive constant and may depend on the sample size n. When $d = 2$ and $\log n$, $IC_{g,d}$ becomes equal to the AIC and BIC. Under a high-dimensional asymptotic framework $p/n \to c \in (0, \infty)$, it was shown that $IC_{d,s}$ with $s = v, u$, or a is consistent under assumption A3s if $d/n \to 0$ and $d > 1$. It may be noted that the sufficient condition does not depend on the covariance structures. Numerical experiments were done for a restrictive case of $\mathscr{F} = \{\{1\}, \{1, 2\}, \ldots, \{1, \ldots, k\}\}$. However, we do not examine for all the subsets. To circumvent computational complexities associated with $IC_{d,s}$, we examined $TC_{d,s}$. It was identified that $TC_{d,s}$ has a consistency property similar to $IC_{d,s}$. The result was first obtained by assuming that the number k of explanatory variables is fixed and the distribution of errors is normal. It is expected that the results extend to the cases when k is large and the distribution of errors is non-normal. It is also important to weaken condition A3.

In addition to the three covariance structures (i), (ii) and (iii) focused on in this research, there are additional such structures including [6]; general independent covariance structure and [7]; no covariance structure. These cases also warrants exploration but are left for future research.

Acknowledgments We thank a referee for careful reading of our manuscript and many helpful comments which improved the presentation of this paper.

Appendix

Outline of Proofs and Preliminary Lemma

First, we provide an outline of our proofs of Theorems 1, 2, and 3. In general, let $\mathscr{F}$ be a finite set of candidate models j(or M_j). Assume that j_* is the minimum model including the true model and $j_* \in \mathscr{F}$. Let $\mathrm{T}_j(n)$ be a general criterion for model j, which depends on parameters p and n. The best model chosen by minimizing $\mathrm{T}_j(p,n)$ is written as $\hat{j}_{\mathrm{T}}(p,n) = \arg\min_{j\in\mathscr{F}} \mathrm{T}_j(p,n)$. Suppose that we are interested in the asymptotic behavior of $\hat{j}_{\mathrm{T}}(p,n)$ when p/n tends to $c > 0$. To show the consistency of $\mathrm{T}_j(p,n)$, we may check a sufficient condition such that for any $j \neq j_* \in \mathscr{F}$, there exists a sequence $\{a_{p,n}\}$ with $a_{p,n} > 0$,

$$a_{p,n}\left\{T_j(p,n) - T_{j_*}(p,n)\right\} \xrightarrow{p} b_j > 0.$$

In fact, the condition implies that for any $j \neq j_* \in \mathscr{F}$,

$$P(\hat{j}_{\mathrm{T}}(p,n) = j) \leq P(T_j(p,n) < T_{j_*}(p,n)) \to 0,$$

and

$$P(\hat{j}_{\mathrm{T}}(p,n) = j_*) = 1 - \sum_{j\neq j_*\in\mathscr{F}} P(\hat{j}_{\mathrm{T}}(p,n) = j) \to 1.$$

For the Proofs of Theorems 18.1–18.3, we use the following lemma frequently.

Lemma 18.3 *Suppose that a $p \times p$ symmetric random matrix $\mathbf{W}$ follows a noncentral Wishart distribution $W_p(n-k, \boldsymbol{\Sigma}; \boldsymbol{\Omega})$. Let $\mathbf{A}$ be a given $p \times p$ positive semidefinite matrix. We consider the asymptotic behavior of* $\mathrm{tr}\mathbf{AW}$ *when p and n are large such that $p/n \to c \in (0, \infty)$, where k is fixed. Suppose that*

$$\lim \frac{1}{p}\mathrm{tr}\mathbf{A}\boldsymbol{\Sigma} = a^2 > 0, \quad \lim \frac{1}{np}\mathrm{tr}\mathbf{A}\boldsymbol{\Omega} = \eta^2 > 0. \tag{18.36}$$

Then, it holds that

$$T_{p,n} = \frac{1}{np}\text{tr}\mathbf{AW} \xrightarrow{p} a^2 + \eta^2. \tag{18.37}$$

Proof Let $m = n - k$. We write $\mathbf{W}$ as $\mathbf{W} = \boldsymbol{\Sigma}^{1/2}(z_1 z_1' + \cdots + z_m z_m')\boldsymbol{\Sigma}^{1/2}$, where $z_i \sim \text{N}_p(\boldsymbol{\zeta}_i, \mathbf{I}_p)$, $i = 1, \ldots, m$ and z_i's are independent. Here, $\boldsymbol{\Omega} = \boldsymbol{\mu}_1\boldsymbol{\mu}_1' + \cdots + \boldsymbol{\mu}_m\boldsymbol{\mu}_m'$ and $\boldsymbol{\zeta}_i = \boldsymbol{\Sigma}^{-1/2}\boldsymbol{\mu}_i$, $i = 1, \ldots, m$. Note that $\text{tr}\mathbf{AW}$ is expressed as a quadratic form of $z = (z_1', \ldots, z_m')'$ as follows: $\text{tr}\mathbf{AW} = z'\mathbf{B}z$, where $\mathbf{B} = \mathbf{I}_m \otimes \boldsymbol{\Sigma}^{1/2}\mathbf{A}\boldsymbol{\Sigma}^{1/2}$. Note also that $z \sim \text{N}_{mp}(\boldsymbol{\zeta}, \mathbf{I}_{mp})$, where $\boldsymbol{\zeta} = (\boldsymbol{\zeta}_1, \ldots, \boldsymbol{\zeta}_m')'$. Then, it is known (see, e.g., [15]) that for any symmetric matrix $\mathbf{B}$,

$$\text{E}[z'\mathbf{B}z] = \text{tr}\mathbf{B} + \boldsymbol{\zeta}'\mathbf{B}\boldsymbol{\zeta}, \quad \text{Var}(z'\mathbf{B}z) = 2\text{tr}\mathbf{B}^2 + 4\boldsymbol{\zeta}'\mathbf{B}^2\boldsymbol{\zeta}.$$

Especially, when $\mathbf{B} = \mathbf{I}_m \otimes \boldsymbol{\Sigma}^{1/2}\mathbf{A}\boldsymbol{\Sigma}^{1/2}$, we have

$$\text{E}[z'\mathbf{B}z] = m\text{tr}\mathbf{A}\boldsymbol{\Sigma} + \text{tr}\mathbf{A}\boldsymbol{\Omega}, \quad \text{Var}(z'\mathbf{B}z) = 2m\text{tr}\,(\mathbf{A}\boldsymbol{\Sigma})^2 + 4\text{tr}\mathbf{A}\boldsymbol{\Sigma}\mathbf{A}\boldsymbol{\Omega}.$$

Under assumption (18.36), we have

$$\text{E}(T_{p,n}) = \frac{m}{np}\text{tr}\mathbf{A}\boldsymbol{\Sigma} + \frac{1}{np}\,\text{tr}\mathbf{A}\boldsymbol{\Omega} \to a^2 + \eta^2.$$

Further, using $\text{tr}\mathbf{A}\boldsymbol{\Sigma}\mathbf{A}\boldsymbol{\Omega} \leq \text{tr}\mathbf{A}\boldsymbol{\Sigma}\text{tr}\mathbf{A}\boldsymbol{\Omega}$ (see, e.g., [15]),

$$\begin{aligned}\text{Var}(T_{p,n}) &= \frac{2m}{(np)^2}\text{tr}(\mathbf{A}\boldsymbol{\Sigma})^2 + \frac{4}{(np)^2}\,\text{tr}\mathbf{A}\boldsymbol{\Sigma}\mathbf{A}\boldsymbol{\Omega} \\ &\leq \frac{2m}{(np)^2}(\text{tr}\mathbf{A}\boldsymbol{\Sigma})^2 + \frac{4}{(np)^2}\,\text{tr}\mathbf{A}\boldsymbol{\Sigma}\text{tr}\mathbf{A}\boldsymbol{\Omega} \to 0.\end{aligned}$$

These imply our conclusion. □

In the special case $\mathbf{A} = \mathbf{I}_p$ and $\boldsymbol{\Sigma} = \sigma^2\mathbf{I}_p$, the assumptions in (18.36) become

$$\lim \frac{1}{p}\text{tr}\mathbf{A}\boldsymbol{\Sigma} = \lim \frac{1}{p}p\sigma^2 = \sigma^2, \quad \lim \frac{1}{np}\text{tr}\mathbf{A}\boldsymbol{\Omega} = \lim \frac{1}{np}\text{tr}\boldsymbol{\Omega} \xrightarrow{p} \eta^2.$$

The conclusion is that

$$T_{p,n} = \frac{1}{np}\text{tr}\mathbf{W} \to \sigma^2(1 + \delta^2), \tag{18.38}$$

where $\delta^2 = (1/\sigma^2)\eta^2$. Note that $(1/\sigma^2)\mathrm{tr}\mathbf{W} \sim \chi^2_{(n-k)p}(\delta^2)$. This implies that if $(np)^{-1}\mathrm{tr}\boldsymbol{\Omega} \to \eta^2$, $\frac{1}{\sigma^2}\mathrm{tr}\mathbf{W} \sim \chi^2_{(n-k)p}(\delta^2)$. Therefore, under a high-dimensional asymptotic framework $p/n \to c \in (0, \infty)$,

$$\frac{1}{np}\chi^2_{(n-k)p}(\delta^2) \xrightarrow{p} 1 + \eta^2, \tag{18.39}$$

if $(np)^{-1}\delta^2 \to \eta^2$. More generally, we use the following result.

$$\frac{1}{np}\chi^2_{(n-k)(p-h)}(\delta^2) \xrightarrow{p} 1 + \eta^2, \tag{18.40}$$

if $(np)^{-1}\delta^2 \to \eta^2$, where k and h are constants or more generally, they may be constants satisfying $k/n \to 0$ and $h/n \to 0$. Further, we use the following property of a noncentral χ^2-square distribution.

$$\frac{1}{n}\chi^2_{n-k}(\delta^2) \xrightarrow{p} 1 + \eta^2, \tag{18.41}$$

if $n^{-1}\delta^2 \to \eta^2$.

Proof of Theorem 18.1

Using (18.14) we can write

$$\mathrm{IC}_{v,d,j} - \mathrm{IC}_{v,d,j_*} = np \log \hat{\sigma}^2_{v,j} - np \log \hat{\sigma}^2_{v,j_*} + d(k_j - k_{j_*})p.$$

Lemma 18.1 shows that

$$\frac{np}{\sigma^2}\hat{\sigma}^2_j \sim \chi^2_{(n-k_j)p}(\delta^2_j), \quad \delta^2_j = \frac{1}{\sigma^2}\mathrm{tr}(\mathbf{X}_{j_*}\boldsymbol{\Theta}_{j_*})'(\mathbf{I}_n - \mathbf{P}_j)(\mathbf{X}_{j_*}\boldsymbol{\Theta}_{j_*}).$$

In particular,

$$\frac{np}{\sigma^2}\hat{\sigma}^2_{j_*} \sim \chi^2_{(n-k_{j*})p}.$$

First, consider the case $j \not\supset j_*$. Under assumption A3v,

$$\frac{1}{np}\frac{np}{\sigma^2}\hat{\sigma}^2_j = \frac{(n-k_j)p}{np}\frac{1}{(n-k_j)p}\frac{1}{\sigma^2}\hat{\sigma}^2_j \xrightarrow{p} 1 + \eta^2_j,$$

$$\frac{1}{np}\frac{np}{\sigma^2}\hat{\sigma}^2_{j_*} = \frac{(n-k_{j_*})p}{np}\frac{1}{(n-k_{j_*})p}\frac{1}{\sigma^2}\hat{\sigma}^2_{j_*} \xrightarrow{p} 1.$$

Therefore,

$$\begin{aligned}\frac{1}{np}\left(\mathrm{IC}_{v,g,j}-\mathrm{IC}_{v,g,j_*}\right) &= \log\hat{\sigma}_j^2-\log\hat{\sigma}_{j_*}^2+\frac{d}{n}(k_j-k_{j_*})\\ &= \log\left(\frac{1}{np}\frac{np}{\sigma^2}\hat{\sigma}_j^2\right)-\log\left(\frac{1}{np}\frac{np}{\sigma^2}\hat{\sigma}_{j_*}^2\right)+\frac{d}{n}(k_j-k_{j_*})\\ &\xrightarrow{p}\log(1+\eta_j^2)+\log 1+0=\log(1+\eta_j^2)>0,\end{aligned}$$

when $d/n\to 0$.

Next, consider the case $j\supset j_*$. Then

$$\mathrm{IC}_{v,g,j}-\mathrm{IC}_{v,g,j_*}=np\log\frac{\hat{\sigma}_{v,j}^2}{\hat{\sigma}_{v,j_*}^2}+d(k_j-k_{j_*})p.$$

Further, $\hat{\sigma}_j^2/\hat{\sigma}_{j_*}^2=\mathrm{tr}\mathbf{Y}'(\mathbf{I}_n-\mathbf{P}_j)\mathbf{Y}/\mathrm{tr}\mathbf{Y}'(\mathbf{I}_n-\mathbf{P}_{j_*})\mathbf{Y}$, and we have

$$\log\frac{\hat{\sigma}_j^2}{\hat{\sigma}_{j_*}^2}=-\log\left(1+\frac{\chi^2_{(k_j-k_{j*})p}}{\chi^2_{(n-k_j)p}}\right).$$

Given $\chi_m^2/m\to 1$ as $m\to\infty$, we have

$$n\log\frac{\hat{\sigma}_j^2}{\hat{\sigma}_{j_*}^2}=-n\log\left\{1+\frac{(k_j-k_{j_*})}{(n-k_j)}\frac{\chi^2_{(k_j-k_{j_*})p}/((k_j-k_{j_*})p)}{\chi^2_{(n-k_j)p}/((n-k_j)p)}\right\}\to-(k_j-k_{j_*}),$$

and hence

$$\begin{aligned}\frac{1}{p}\{\mathrm{IC}_{v,d,j}-\mathrm{IC}_{v,d,j_*}\} &= n\log\frac{\hat{\sigma}_j^2}{\hat{\sigma}_{j_*}}+d(k_j-k_{j_*})\\ &\xrightarrow{p}-(k_j-k_{j_*})+d(k_j-k_{j_*})=(d-1)(k_j-k_{j_*})>0,\end{aligned}$$

if $d>1$.

Proofs of Theorems 18.2 and 18.3

The results can be proved by the same line as in the Proof of Theorem 18.1. For details, see [14].

Proof of Theorem 18.4

For a notational simplicity, let $\mathrm{IC}_{d,s,j}$ and $\mathrm{TC}_{d,s}$ be denoted by $\mathrm{IC}_{(j)}$ and TC, respectively. Further, $\omega = \{1, 2, \ldots, k\}$ and $(-i)$ are also denoted by j_ω and $j_{(-i)}$, respectively. Note that k is finite. Without loss of generality, we may assume that the true model is $j_* = \{1, \ldots, b\}$ and $b = k_{j_*}$. Under the assumption A1, A2 and A3(v,u,a), it was shown that our information criterion $\hat{j}_{\mathrm{IC}}$ has a high-dimensional consistency property under some condition. The consistency property was shown by proving that

(1) for $j \in \mathscr{F}_-$,

$$\frac{1}{m_1}\{\mathrm{IC}(j) - \mathrm{IC}(j_*)\} \xrightarrow{p} \gamma_j > 0, \tag{18.42}$$

and

(2) for $j \in \mathscr{F}_+$ and $j \neq j_*$,

$$\frac{1}{m_2}\{\mathrm{IC}(j) - \mathrm{IC}(j_*)\} \xrightarrow{p} (d-1)(k_j - k_{j_*}) > 0, \tag{18.43}$$

where $d > 1$, $m_1 = np$ and $m_2 = p$. Further, we use the following inequality (see, e.g., [5]):

$$\begin{aligned} P(\hat{j}_{\mathrm{TC}} = j_*) \geq & 1 - \sum_{i \in j_*}\{1 - P(\mathrm{IC}(j_{(-i)}) - \mathrm{IC}(j_\omega) \geq 0)\} \\ & - \sum_{i \in j_\omega / j_*} \{1 - P(\mathrm{IC}(j_{(-i)}) - \mathrm{IC}(j_\omega) \leq 0)\}. \end{aligned} \tag{18.44}$$

Now, consider to evaluate the following probabilities:

$$\begin{aligned} i \in j_*, \quad & P(\mathrm{IC}(j_{(-i)}) - \mathrm{IC}(j_\omega) \geq 0), \\ i \in j_\omega / j_*, \quad & P(\mathrm{IC}(j_{(-i)}) - \mathrm{IC}(j_\omega) \leq 0). \end{aligned}$$

When $i \in j_*$, $j_{(-i)} \in \mathscr{F}_-$, and hence, using (18.42), it holds that

$$\begin{aligned} \frac{1}{m_1}(\mathrm{IC}(j_{(-i)}) - \mathrm{IC}(j_\omega)) &= \frac{1}{m_1}(\mathrm{IC}(j_{(-i)}) - \mathrm{IC}(j_*)) - \frac{1}{m_1}(\mathrm{IC}(j_\omega) - \mathrm{IC}(j_*)) \\ &\xrightarrow{p} \gamma_i + 0 = \gamma_i > 0. \end{aligned}$$

When $i \in j_\omega/j_*$, $j_{(-i)} \in \mathscr{F}_+$ and hence, using (18.43), it holds that

$$\frac{1}{m_2}(\mathrm{IC}(j_{(-i)}) - \mathrm{IC}(j_\omega)) = \frac{1}{m_2}(\mathrm{IC}(j_{(-i)}) - \mathrm{IC}(j_*)) - \frac{1}{m_2}(\mathrm{IC}(j_\omega) - \mathrm{IC}(j_*))$$
$$\xrightarrow{p} (d-1)(k-1-b) - (d-1)(k-b) = -(d-1) < 0.$$

These results imply that

$$i \in j_*, \quad \lim P(\mathrm{IC}(j_{(-i)}) - \mathrm{IC}(j_\omega) > 0) = 1,$$
$$i \in j_\omega/j_*, \quad \lim P(\mathrm{IC}(j_{(-i)}) - \mathrm{IC}(j_\omega) < 0) = 1.$$

Using the above results, we can see that the right-hand side of (18.44) tends to

$$1 - \left[\sum_{j \in j_*}\{1-1\} + \sum_{j \in j_k/j_*}\{1-1\}\right] = 1,$$

and $P(\hat{j}_{\mathrm{TC}} = j_*) \to 1$. This completes the Proof of Theorem 18.4.

References

1. Akaike, H.: Information theory and an extension of the maximum likelihood principle. In: Petrov, B.N., Csáki, F. (eds.) 2nd. International Symposium on Information Theory, pp. 267–281. Akadémiai Kiadó, Budapest (1973)
2. Bai, Z., Fujikoshi, Y., Hu, J.: Strong consistency of the AIC, BIC, C_p and KOO methods in high-dimensional multivariate linear regression. Hiroshima Statistical Research Group, TR; 18–09 (2018). arXiv:1810.12609
3. Bedrick, E.J., Tsai, C.-L.: Model selection for multivariate regression in small samples. Biometrics **50**, 226–231 (1994)
4. Fujikoshi, Y., Satoh, K.: Modified AIC and C_p in multivariate linear regression. Biometrika **84**, 707–716 (1997)
5. Fujikoshi, Y., Sakurai, T.: Consistency of test-based method for selection of variables in high-dimensional two group-discriminant analysis. Jpn. J. Stat. Data Sci. **2**, 155–171 (2019)
6. Fujikoshi, Y., Kanda, T., Tanimura, N.: The growth curve model with an autoregressive covariance structure. Ann. Inst. Statist. Math. **42**, 533–542 (1990)
7. Fujikoshi, Y., Sakurai, T., Abe, Y., Yakushinji, Y., Sugiyama, T.: Variable selection in discriminant analysis with high-dimensional binary data and its application to DNA fingerprint data. J. Inst. Sci. Eng. Chuo Univer., **12**, 1–18 (2006) (in Japanese)
8. Fujikoshi, Y., Sakurai, T., Yanagihara, H.: Consistency of high-dimensional AIC-type and C_p-type criteria in multivariate linear regression. J. Multivariate Anal. **149**, 199–212 (2014)
9. Fujikoshi, Y., Ulyanov, V.V., Shimizu, R.: *Multivariate Statistics: High-Dimensional and Large-Sample Approximations*. Wiley, Hobeken (2010)
10. Lutz, R.W., Bühlmann, P.: Boosting for high-multivariate regressions in high-dimensional linear regression. Stati. Sinica **16**, 471–494 (2006)
11. Nishii, R., Bai, Z.D., Krishnaia, P.R.: Strong consistency of the information criterion for model selection in multivariate analysis. Hiroshima Math. J. **18**, 451–462 (1988)

12. Pourahmadi, M.: *High-Dimensional Covariance Estimation*. Wiley, Hobeken (2013)
13. Rothman, A., Levina, E., Zhu, J.: Sparse multivariate regression with covariance estimation. J. Comput. Graph.. Stat. **19**, 947–962 (2010)
14. Sakurai, T., Fujikoshi, Y.: High-dimensional properties of information criteria and their efficient criteria for multivariate linear regression models with covariance structures. Hiroshima Statistical Research Group, TR; 17–13 (2017)
15. Seber, G.A.F.: *A Matrix Handbook for Statisticians*. Wiley, Hobeken (2008)
16. Wilbur, J.D., Ghosh, J.K., Nakatsu, C.H., Brouder, S.M., Doerge, R.W.: Variable selection in high-dimensional multivariate binary data with application to the analysis of microbial community DNA fingerprints. Biometrics **58**, 378–386 (2002)
17. Yanagihara, H., Wakaki, H., Fujikoshi, Y.: A consistency property of the AIC for multivariate linear models when the dimension and the sample size are large. Electron. J. Stat. **9**, 869–897 (2015)
18. Zhao, L.C., Krishnaia, P.R., Bai, Z.D.: On detection of the number of signals in presence of white noise. J. Multivariate Anal., **20**, 1–25 (1986)

Chapter 19
Mean Value Test for Three-Level Multivariate Observations with Doubly Exchangeable Covariance Structure

Ivan Žežula, Daniel Klein, and Anuradha Roy

Abstract We consider matrix-valued multivariate observation model with three-level doubly-exchangeable covariance structure. We derive estimators of unknown parameters and their distributions under multivariate normality assumption. Test statistic for testing a mean value is proposed, and its exact distribution is derived. Several methods of computing p-values and critical values of the distribution are compared in real data example.

19.1 Introduction

Multivariate data are very common in current scientific research. If the dimension of the data is large, we often cannot use classical methods of multivariate analysis due to insufficient sample size. On the other hand, the design of the experiment and/or properties of the investigated individuals may induce a special covariance structure in the data. We have to respect such special covariance structure in order to get valid results. Also, such a structure reduces (sometimes substantially) the number of parameters to be estimated, so that the requirement on the sample size decreases.

One of the special covariance structures we can meet in experiments, is the doubly exchangeable covariance structure. Let us consider a pqr-dimensional

I. Žežula (✉) · D. Klein
Institute of Mathematics, Faculty of Science, P. J. Šafárik University, Košice, Slovakia
e-mail: ivan.zezula@upjs.sk; daniel.klein@upjs.sk

A. Roy
Department of Management Science and Statistics, The University of Texas at San Antonio, San Antonio, TX, USA
e-mail: Anuradha.Roy@utsa.edu

© Springer Nature Switzerland AG 2020
T. Holgersson, M. Singull (eds.), *Recent Developments in Multivariate and Random Matrix Analysis*, https://doi.org/10.1007/978-3-030-56773-6_19

random vector $\boldsymbol{x}$ consisting of pq r-dimensional subvectors $\boldsymbol{x}_{ij}$ such that

$$\operatorname{cov}\left[\boldsymbol{x}_{ij};\boldsymbol{x}_{kl}\right]=\begin{cases}\boldsymbol{U}_0 & \text{if } i=k \text{ and } j=l,\\ \boldsymbol{U}_1 & \text{if } i=k \text{ and } j\neq l,\\ \boldsymbol{U}_2 & \text{if } i\neq k,\end{cases}$$

where $\boldsymbol{U}_0$, $\boldsymbol{U}_1$ and $\boldsymbol{U}_2$ are symmetric $r \times r$ matrices. The mean value of $\boldsymbol{x}$ will be denoted by $\boldsymbol{\mu}$. Should the structure bear the name "doubly exchangeable" legitimately, we must suppose that $p, q \geq 2$.

This kind of dependence structure can be met in some repeated measurements designs. E.g., r-dimensional measurements are done in p times at q locations in such a way that times and locations are exchangeable. It implies the overall covariance matrix of $\boldsymbol{x}$ to be

$$\boldsymbol{\Gamma}=\begin{pmatrix}\boldsymbol{\Gamma}_0 & \boldsymbol{\Gamma}_1 & \dots & \boldsymbol{\Gamma}_1\\ \boldsymbol{\Gamma}_1 & \boldsymbol{\Gamma}_0 & \dots & \boldsymbol{\Gamma}_1\\ \vdots & \vdots & \ddots & \vdots\\ \boldsymbol{\Gamma}_1 & \boldsymbol{\Gamma}_1 & \dots & \boldsymbol{\Gamma}_0\end{pmatrix},\quad \boldsymbol{\Gamma}_0=\begin{pmatrix}\boldsymbol{U}_0 & \boldsymbol{U}_1 & \dots & \boldsymbol{U}_1\\ \boldsymbol{U}_1 & \boldsymbol{U}_0 & \dots & \boldsymbol{U}_1\\ \vdots & \vdots & \ddots & \vdots\\ \boldsymbol{U}_1 & \boldsymbol{U}_1 & \dots & \boldsymbol{U}_0\end{pmatrix},\quad \boldsymbol{\Gamma}_1=\begin{pmatrix}\boldsymbol{U}_2 & \dots & \boldsymbol{U}_2\\ \vdots & \ddots & \vdots\\ \boldsymbol{U}_2 & \dots & \boldsymbol{U}_2\end{pmatrix},$$

so that

$$\boldsymbol{\Gamma}=\boldsymbol{I}_{pq}\otimes(\boldsymbol{U}_0-\boldsymbol{U}_1)+\boldsymbol{I}_p\otimes\boldsymbol{J}_q\otimes(\boldsymbol{U}_1-\boldsymbol{U}_2)+\boldsymbol{J}_{pq}\otimes\boldsymbol{U}_2, \tag{19.1}$$

where $\boldsymbol{I}_a$ is the $a \times a$ identity matrix, $\boldsymbol{J}_a=\mathbf{1}_a\mathbf{1}_a'$, $\mathbf{1}_a$ is the $a-$vector of ones, and $\otimes$ denotes the Zehfuss (Kronecker) product.

The problem we are interested in is testing whether the mean value of a normal distribution with doubly exchangeable covariance structure assumes a specific given value, i.e., a generalization of the one-sample t-test for this situation. Useful algebraic tools are derived in Sect. 19.2. Precise formulation of the hypothesis is given at the beginning of Sect. 19.3. Estimators of parameters of interest, their distributions, the proposed test statistic and its distribution are also derived in Sect. 19.3. Application of the method to real data is the content of Sect. 19.4.

It is important to note that the matrix $\boldsymbol{\Gamma}$ can be decomposed into three mutually orthogonal parts:

$$\boldsymbol{\Gamma}=\boldsymbol{I}_p\otimes\boldsymbol{Q}_q\otimes\boldsymbol{\Delta}_1+\boldsymbol{Q}_p\otimes\boldsymbol{P}_q\otimes\boldsymbol{\Delta}_2+\boldsymbol{P}_p\otimes\boldsymbol{P}_q\otimes\boldsymbol{\Delta}_3, \tag{19.2}$$

where $\boldsymbol{P}_a=\frac{1}{a}\boldsymbol{J}_a$ and $\boldsymbol{Q}_a=\boldsymbol{I}_a-\boldsymbol{P}_a$ are mutually orthogonal projectors on the linear subspace generated by the vector $\mathbf{1}_a$ and its orthogonal complement, and

$$\boldsymbol{\Delta}_1=\boldsymbol{U}_0-\boldsymbol{U}_1, \tag{19.3}$$

$$\boldsymbol{\Delta}_2=\boldsymbol{U}_0+(q-1)\boldsymbol{U}_1-q\boldsymbol{U}_2=(\boldsymbol{U}_0-\boldsymbol{U}_1)+q(\boldsymbol{U}_1-\boldsymbol{U}_2), \tag{19.4}$$

$$\boldsymbol{\Delta}_3=\boldsymbol{U}_0+(q-1)\boldsymbol{U}_1+(p-1)q\boldsymbol{U}_2=(\boldsymbol{U}_0-\boldsymbol{U}_1)+q(\boldsymbol{U}_1-\boldsymbol{U}_2)+pq\boldsymbol{U}_2. \tag{19.5}$$

Moreover, it is clear that $\boldsymbol{\Gamma}$ is a positive definite matrix iff $\boldsymbol{\Delta}_1$, $\boldsymbol{\Delta}_2$, and $\boldsymbol{\Delta}_3$ are all positive definite matrices. One can also note that these matrices are the result of a canonical transformation using Helmert matrices (denoted by $\boldsymbol{H}_a$): it holds that $\left(\boldsymbol{H}_p \otimes \boldsymbol{H}_q \otimes \boldsymbol{I}_r\right) \boldsymbol{\Gamma} \left(\boldsymbol{H}'_p \otimes \boldsymbol{H}'_q \otimes \boldsymbol{I}_r\right)$ is a block diagonal matrix with blocks

$$\begin{bmatrix} \boldsymbol{\Delta}_3 & \mathbf{0} \\ \mathbf{0} & \boldsymbol{I}_{q-1} \otimes \boldsymbol{\Delta}_1 \end{bmatrix}, \quad \text{and } p-1 \text{ times} \quad \begin{bmatrix} \boldsymbol{\Delta}_2 & \mathbf{0} \\ \mathbf{0} & \boldsymbol{I}_{q-1} \otimes \boldsymbol{\Delta}_1 \end{bmatrix}. \tag{19.6}$$

Due to the orthogonality of the components and positive definiteness of the $\boldsymbol{\Delta}$'s, it is easy to verify that

$$\boldsymbol{\Gamma}^{-1} = \boldsymbol{I}_p \otimes \boldsymbol{Q}_q \otimes \boldsymbol{\Delta}_1^{-1} + \boldsymbol{Q}_p \otimes \boldsymbol{P}_q \otimes \boldsymbol{\Delta}_2^{-1} + \boldsymbol{P}_p \otimes \boldsymbol{P}_q \otimes \boldsymbol{\Delta}_3^{-1}. \tag{19.7}$$

19.2 Algebraic Tools

Definition 19.1 Let A be a square matrix of order bc, consisting of $b \times b$ square blocks of order c:

$$\boldsymbol{A} = \begin{pmatrix} \boldsymbol{A}_{11} & \boldsymbol{A}_{12} & \dots & \boldsymbol{A}_{1b} \\ \vdots & \ddots & & \vdots \\ \vdots & & \ddots & \vdots \\ \boldsymbol{A}_{b1} & \boldsymbol{A}_{b2} & \dots & \boldsymbol{A}_{bb} \end{pmatrix}.$$

The block-trace and block-sum operators are defined by the relations:

$$\mathrm{BTr}_c[\boldsymbol{A}] = \sum_{i=1}^{b} \boldsymbol{A}_{ii} \quad \text{and} \quad \mathrm{BSum}_c[\boldsymbol{A}] = \sum_{i=1}^{b} \sum_{j=1}^{b} \boldsymbol{A}_{ij}\,.$$

We will use these operators for the estimation of blocks of covariance matrices. First, we derive some basic properties.

Lemma 19.1 *Let* $\boldsymbol{A} = \left(\boldsymbol{A}_{ij}\right)_{ij}$ *for* $i, j = 1, 2, \dots, q$ *be a block matrix with* $mp \times mp$ *square blocks. Let further each block* $\boldsymbol{A}_{ij} = \left(\boldsymbol{A}_{ijkl}\right)_{kl}$ *for* $k, l = 1, 2, \dots, p$ *be a block matrix with* $m \times m$ *square blocks. Then*

(i) $\mathrm{BTr}_m\left[\mathrm{BTr}_{mp}\left[\boldsymbol{A}\right]\right] = \mathrm{BTr}_m\left[\boldsymbol{A}\right]$,
(ii) $\mathrm{BSum}_m[\mathrm{BTr}_{mp}[\boldsymbol{A}]] = \mathrm{BTr}_m\left[(\boldsymbol{I}_q \otimes \boldsymbol{J}_p \otimes \boldsymbol{I}_m)\boldsymbol{A}\right]$,
(iii) $\mathrm{BSum}_m[\mathrm{BSum}_{mp}[\boldsymbol{A}]] = \mathrm{BTr}_m[(\boldsymbol{J}_q \otimes \boldsymbol{J}_p \otimes \boldsymbol{I}_m)\boldsymbol{A}]$.

Proof

(i) According to the definition we have

$$\mathrm{BTr}_m[\mathrm{BTr}_{mp}[\boldsymbol{A}]] = \mathrm{BTr}_m\left[\sum_{i=1}^{q}\boldsymbol{A}_{ii}\right] = \sum_{i=1}^{q}\mathrm{BTr}_m[\boldsymbol{A}_{ii}] = \sum_{i=1}^{q}\sum_{j=1}^{p}\boldsymbol{A}_{ii_{jj}} = \mathrm{BTr}_m[\boldsymbol{A}].$$

(ii) For the left-hand side we have

$$\mathrm{BSum}_m[\mathrm{BTr}_{mp}[\boldsymbol{A}]] = \mathrm{BSum}_m\left[\sum_{i=1}^{q}\boldsymbol{A}_{ii}\right] = \sum_{i=1}^{q}\mathrm{BSum}_m[\boldsymbol{A}_{ii}] = \sum_{i=1}^{q}\sum_{k=1}^{p}\sum_{l=1}^{p}\boldsymbol{A}_{ii_{kl}}.$$

Now, observe that on the main diagonal of the matrix $(\boldsymbol{I}_q \otimes \boldsymbol{J}_p \otimes \boldsymbol{I}_m)\boldsymbol{A}$ the $m \times m$ blocks are $\sum_{l=1}^{p}\boldsymbol{A}_{ii_{kl}}$ for $i = 1, \ldots, q$ and $k = 1, \ldots, p$. Therefore

$$\mathrm{BTr}_m[(\boldsymbol{I}_q \otimes \boldsymbol{J}_p \otimes \boldsymbol{I}_m)\boldsymbol{A}] = \sum_{i=1}^{q}\sum_{k=1}^{p}\sum_{l=1}^{p}\boldsymbol{A}_{ii_{kl}}.$$

(iii) Observe that $\mathrm{BSum}_m[\boldsymbol{A}] = \mathrm{BTr}_m[(\boldsymbol{J}_n \otimes \boldsymbol{I}_m)\boldsymbol{A}]$ for any $mn \times mn$ block matrix $\boldsymbol{A}$. Then

$$\mathrm{BSum}_m[\mathrm{BSum}_{mp}[\boldsymbol{A}]] = \mathrm{BSum}_m[\mathrm{BTr}_{mp}[(\boldsymbol{J}_q \otimes \boldsymbol{I}_p \otimes \boldsymbol{I}_m)\boldsymbol{A}]] =$$
$$= \mathrm{BTr}_m[(\boldsymbol{I}_q \otimes \boldsymbol{J}_p \otimes \boldsymbol{I}_m)(\boldsymbol{J}_q \otimes \boldsymbol{I}_p \otimes \boldsymbol{I}_m)\boldsymbol{A}] = \mathrm{BTr}_m[(\boldsymbol{J}_q \otimes \boldsymbol{J}_p \otimes \boldsymbol{I}_m)\boldsymbol{A}].$$

□

More properties of block-trace operator can be found in [1]. Now, we can derive the following useful relations:

Lemma 19.2 *For a doubly exchangeable covariance matrix*

$$\boldsymbol{\Gamma} = \boldsymbol{I}_{pq} \otimes (\boldsymbol{U}_0 - \boldsymbol{U}_1) + \boldsymbol{I}_p \otimes \boldsymbol{J}_q \otimes (\boldsymbol{U}_1 - \boldsymbol{U}_2) + \boldsymbol{J}_{pq} \otimes \boldsymbol{U}_2$$

it holds

$$\boldsymbol{U}_0 = \frac{1}{pq}\,\mathrm{BTr}_r[\boldsymbol{\Gamma}],$$
$$\boldsymbol{U}_1 = \frac{1}{pq(q-1)}\,\mathrm{BTr}_r[(\boldsymbol{I}_p \otimes (\boldsymbol{J}_q - \boldsymbol{I}_q) \otimes \boldsymbol{I}_r)\boldsymbol{\Gamma}], \quad \textit{and}$$
$$\boldsymbol{U}_2 = \frac{1}{p(p-1)q^2}\,\mathrm{BTr}_r[((\boldsymbol{J}_p - \boldsymbol{I}_p) \otimes \boldsymbol{J}_q \otimes \boldsymbol{I}_r)\boldsymbol{\Gamma}].$$

Proof It is easy to see that

$$\boldsymbol{\Gamma}_0 = \frac{1}{p}\,\mathrm{BTr}_{rq}[\boldsymbol{\Gamma}], \quad \text{and} \quad \boldsymbol{\Gamma}_1 = \frac{1}{p(p-1)}\Big(\mathrm{BSum}_{rq}[\boldsymbol{\Gamma}] - \mathrm{BTr}_{rq}[\boldsymbol{\Gamma}]\Big).$$

Then using Lemma 19.1 we have

$$\begin{aligned}
\boldsymbol{U}_0 &= \frac{1}{q}\,\mathrm{BTr}_r[\boldsymbol{\Gamma}_0] = \frac{1}{pq}\,\mathrm{BTr}_r[\mathrm{BTr}_{rq}[\boldsymbol{\Gamma}]] = \frac{1}{pq}\,\mathrm{BTr}_r[\boldsymbol{\Gamma}],\\
\boldsymbol{U}_1 &= \frac{1}{q(q-1)}\Big(\mathrm{BSum}_r[\boldsymbol{\Gamma}_0] - \mathrm{BTr}_r[\boldsymbol{\Gamma}_0]\Big)\\
&= \frac{1}{pq(q-1)}\Big(\mathrm{BSum}_r[\mathrm{BTr}_{rq}[\boldsymbol{\Gamma}]] - \mathrm{BTr}_r[\mathrm{BTr}_{rq}[\boldsymbol{\Gamma}]]\Big)\\
&= \frac{1}{pq(q-1)}\,\mathrm{BTr}_r[(\boldsymbol{I}_p \otimes (\boldsymbol{J}_q - \boldsymbol{I}_q) \otimes \boldsymbol{I}_r)\boldsymbol{\Gamma}],
\end{aligned}$$

and

$$\begin{aligned}
\boldsymbol{U}_2 &= \frac{1}{q^2}\,\mathrm{BSum}_r[\boldsymbol{\Gamma}_1]\\
&= \frac{1}{p(p-1)q^2}\Big(\mathrm{BSum}_r[\mathrm{BSum}_{rq}[\boldsymbol{\Gamma}]] - \mathrm{BSum}_r[\mathrm{BTr}_{rq}[\boldsymbol{\Gamma}]]\Big)\\
&= \frac{1}{p(p-1)q^2}\,\mathrm{BTr}_r[((\boldsymbol{J}_p - \boldsymbol{I}_p) \otimes \boldsymbol{J}_q \otimes \boldsymbol{I}_r)\boldsymbol{\Gamma}].
\end{aligned}$$

□

Lemma 19.3 *It holds:*

$$\boldsymbol{\Delta}_1 = \frac{1}{p(q-1)}\,\mathrm{BTr}_r[(\boldsymbol{I}_p \otimes \boldsymbol{Q}_q \otimes \boldsymbol{I}_r)\boldsymbol{\Gamma}], \tag{19.8}$$

$$\boldsymbol{\Delta}_2 = \frac{1}{p-1}\,\mathrm{BTr}_r\Big[\Big(\boldsymbol{Q}_p \otimes \boldsymbol{P}_q \otimes \boldsymbol{I}_r\Big)\boldsymbol{\Gamma}\Big], \tag{19.9}$$

$$\boldsymbol{\Delta}_3 = \mathrm{BTr}_r[(\boldsymbol{P}_p \otimes \boldsymbol{P}_q \otimes \boldsymbol{I}_r)\boldsymbol{\Gamma}]. \tag{19.10}$$

Proof

$$\begin{aligned}
\boldsymbol{\Delta}_1 = \boldsymbol{U}_0 - \boldsymbol{U}_1 &= \frac{1}{pq}\,\mathrm{BTr}_r[\boldsymbol{\Gamma}] - \frac{1}{pq(q-1)}\,\mathrm{BTr}_r[(\boldsymbol{I}_p \otimes (\boldsymbol{J}_q - \boldsymbol{I}_q) \otimes \boldsymbol{I}_r)\boldsymbol{\Gamma}]\\
&= \frac{1}{pq}\,\mathrm{BTr}_r\left[\left(\boldsymbol{I}_p \otimes \left(\boldsymbol{I}_q - \frac{1}{q-1}\boldsymbol{J}_q + \frac{1}{q-1}\boldsymbol{I}_q\right) \otimes \boldsymbol{I}_r\right)\boldsymbol{\Gamma}\right]\\
&= \frac{1}{p(q-1)}\,\mathrm{BTr}_r[(\boldsymbol{I}_p \otimes \boldsymbol{Q}_q \otimes \boldsymbol{I}_r)\boldsymbol{\Gamma}],
\end{aligned}$$

$$
\begin{aligned}
\boldsymbol{\Delta}_2 &= (\boldsymbol{U}_0 - \boldsymbol{U}_1) + q\,(\boldsymbol{U}_1 - \boldsymbol{U}_2) = \boldsymbol{\Delta}_1 + q\,(\boldsymbol{U}_1 - \boldsymbol{U}_2) \\
&= \frac{1}{p(q-1)} \operatorname{BTr}_r[(\boldsymbol{I}_p \otimes \boldsymbol{Q}_q \otimes \boldsymbol{I}_r)\boldsymbol{\Gamma}] + \frac{1}{p(q-1)} \operatorname{BTr}_r[(\boldsymbol{I}_p \otimes (\boldsymbol{J}_q - \boldsymbol{I}_q) \otimes \boldsymbol{I}_r)\boldsymbol{\Gamma}] \\
&\quad - \frac{1}{p(p-1)q} \operatorname{BTr}_r[((\boldsymbol{J}_p - \boldsymbol{I}_p) \otimes \boldsymbol{J}_q \otimes \boldsymbol{I}_r)\boldsymbol{\Gamma}] \\
&= \frac{1}{pq} \operatorname{BTr}_r[(\boldsymbol{I}_p \otimes \boldsymbol{J}_q \otimes \boldsymbol{I}_r)\boldsymbol{\Gamma}] - \frac{1}{p(p-1)q} \operatorname{BTr}_r[((\boldsymbol{J}_p - \boldsymbol{I}_p) \otimes \boldsymbol{J}_q \otimes \boldsymbol{I}_r)\boldsymbol{\Gamma}] \\
&= \frac{1}{p} \operatorname{BTr}_r[(\boldsymbol{I}_p \otimes \boldsymbol{P}_q \otimes \boldsymbol{I}_r)\boldsymbol{\Gamma}] - \frac{1}{p(p-1)} \operatorname{BTr}_r[((\boldsymbol{J}_p - \boldsymbol{I}_p) \otimes \boldsymbol{P}_q \otimes \boldsymbol{I}_r)\boldsymbol{\Gamma}] \\
&= \frac{1}{p} \operatorname{BTr}_r\left[\left(\left(\boldsymbol{I}_p - \frac{1}{p-1}\boldsymbol{J}_p + \frac{1}{p-1}\boldsymbol{I}_p\right) \otimes \boldsymbol{P}_q \otimes \boldsymbol{I}_r\right)\boldsymbol{\Gamma}\right] \\
&= \frac{1}{p-1} \operatorname{BTr}_r\left[\left(\boldsymbol{Q}_p \otimes \boldsymbol{P}_q \otimes \boldsymbol{I}_r\right)\boldsymbol{\Gamma}\right],
\end{aligned}
$$

$$
\begin{aligned}
\boldsymbol{\Delta}_3 &= (\boldsymbol{U}_0 - \boldsymbol{U}_1) + q\,(\boldsymbol{U}_1 - \boldsymbol{U}_2) + pq\boldsymbol{U}_2 = \boldsymbol{\Delta}_2 + pq\boldsymbol{U}_2 \\
&= \frac{1}{p-1} \operatorname{BTr}_r\left[\left(\boldsymbol{Q}_p \otimes \boldsymbol{P}_q \otimes \boldsymbol{I}_r\right)\boldsymbol{\Gamma}\right] + \frac{1}{(p-1)q} \operatorname{BTr}_r[((\boldsymbol{J}_p - \boldsymbol{I}_p) \otimes \boldsymbol{J}_q \otimes \boldsymbol{I}_r)\boldsymbol{\Gamma}] \\
&= \frac{1}{p-1} \operatorname{BTr}_r\left[\left(\boldsymbol{Q}_p \otimes \boldsymbol{P}_q \otimes \boldsymbol{I}_r\right)\boldsymbol{\Gamma}\right] + \frac{1}{p-1} \operatorname{BTr}_r[((\boldsymbol{J}_p - \boldsymbol{I}_p) \otimes \boldsymbol{P}_q \otimes \boldsymbol{I}_r)\boldsymbol{\Gamma}] \\
&= \operatorname{BTr}_r[(\boldsymbol{P}_p \otimes \boldsymbol{P}_q \otimes \boldsymbol{I}_r)\boldsymbol{\Gamma}].
\end{aligned}
$$

□

19.3 Mean Value Test

The very basic test needed for almost any kind of data is a test for a given mean value. Our data are pqr-dimensional, and—as the first step—we have to choose a reasonable mathematical representation of them. The most natural representation would be a $p \times q \times r$ tensor, but general 3D tensor structures do not have as many elegant and powerful mathematical tools as matrices and vectors do. That is why we will use for the representation of such data either $r \times pq$ matrices (denoted by $\boldsymbol{X}$), or pqr-vectors (denoted by $\boldsymbol{x} = \operatorname{vec} \boldsymbol{X}$).

Let $\boldsymbol{X}_1, \ldots, \boldsymbol{X}_n$ be a random sample from $N_{r\times pq}(\boldsymbol{M}, \boldsymbol{\Gamma})$, where $\boldsymbol{\Gamma}$ has the doubly exchangeable structure (19.1), and $\overline{\boldsymbol{X}}$ be the sample mean. Equivalently, we will write a vectorized form of the sample as $\boldsymbol{x}_1, \ldots, \boldsymbol{x}_n$ i.i.d. $N_{pqr}(\boldsymbol{\mu}, \boldsymbol{\Sigma})$ and $\overline{\boldsymbol{x}}$ for the sample mean. We want to test the hypothesis

$$H_0 : \boldsymbol{M} = \boldsymbol{M}_0 \quad \text{against} \quad H_1 : \boldsymbol{M} \neq \boldsymbol{M}_0. \tag{19.11}$$

Let $\mathscr{X} = (\boldsymbol{x}_1, \ldots, \boldsymbol{x}_n)$ be the $(pqr \times n)$-dimensional data matrix (see [7], Section 3.3, p. 64).

It is easy to see that the $(pqr \times pqr)$-dimensional sample variance-covariance matrix $\boldsymbol{S}$ can be expressed in the form

$$\boldsymbol{S} = \frac{1}{n-1}\mathscr{X}\boldsymbol{Q}_n\mathscr{X}' = \begin{pmatrix} \boldsymbol{S}_{11} & \boldsymbol{S}_{12} & \ldots & \boldsymbol{S}_{1p} \\ \boldsymbol{S}_{21} & \boldsymbol{S}_{22} & \ldots & \boldsymbol{S}_{2p} \\ \vdots & \vdots & \ddots & \vdots \\ \boldsymbol{S}_{p1} & \boldsymbol{S}_{p2} & \ldots & \boldsymbol{S}_{pp} \end{pmatrix} = \begin{pmatrix} \mathfrak{S}_{1,1} & \mathfrak{S}_{1,2} & \ldots & \mathfrak{S}_{1,pq} \\ \mathfrak{S}_{2,1} & \mathfrak{S}_{2,2} & \ldots & \mathfrak{S}_{2,pq} \\ \vdots & \vdots & \ddots & \vdots \\ \mathfrak{S}_{pq,1} & \mathfrak{S}_{pq,2} & \ldots & \mathfrak{S}_{pq,pq} \end{pmatrix},$$

where $\boldsymbol{S}_{ij}$ are blocks of order qr and $\mathfrak{S}_{i,j}$ are blocks of order r.

Using Corollary 3.3.3.2 of Theorem 3.3.3 from [7] Section 3.3, p. 66, we conclude that $\overline{\boldsymbol{x}} = \frac{1}{n}\mathscr{X}\boldsymbol{1}_n$ is independent of $\boldsymbol{S} = \frac{1}{n-1}\mathscr{X}\boldsymbol{Q}_n\mathscr{X}'$. From 8b.2(ii) in [12] we know that $\boldsymbol{S} \sim W_{pqr}\left(n-1, \frac{1}{n-1}\boldsymbol{\Gamma}\right)$ (even if possibly singular), therefore $\boldsymbol{S}$ is an unbiased estimator of $\boldsymbol{\Gamma}$. However, it has not the structure of $\boldsymbol{\Gamma}$. For finding an unbiased estimator of $\boldsymbol{\Gamma}$ with the desired structure we need find unbiased estimators of $\boldsymbol{U}_0$, $\boldsymbol{U}_1$, and $\boldsymbol{U}_2$, or $\boldsymbol{\Delta}_1$, $\boldsymbol{\Delta}_2$, and $\boldsymbol{\Delta}_3$.

19.3.1 Unbiased Estimators of Covariance Matrix Blocks

Since $\boldsymbol{S}$ is an unbiased estimator of $\boldsymbol{\Gamma}$ and both matrix multiplication and block-trace are linear operators, we can use Lemma 19.2 to immediately conclude that unbiased estimators of $\boldsymbol{U}_0$, $\boldsymbol{U}_1$, and $\boldsymbol{U}_2$ are

$$\widehat{\boldsymbol{U}}_0 = \frac{1}{pq}\,\mathrm{BTr}_r[\boldsymbol{S}],$$

$$\widehat{\boldsymbol{U}}_1 = \frac{1}{pq(q-1)}\,\mathrm{BTr}_r[(\boldsymbol{I}_p \otimes (\boldsymbol{J}_q - \boldsymbol{I}_q) \otimes \boldsymbol{I}_r)\boldsymbol{S}],$$

and

$$\widehat{\boldsymbol{U}}_2 = \frac{1}{p(p-1)q^2}\,\mathrm{BTr}_r[((\boldsymbol{J}_p - \boldsymbol{I}_p) \otimes \boldsymbol{J}_q \otimes \boldsymbol{I}_r)\boldsymbol{S}].$$

By the same logic, using Lemma 19.3, unbiased estimators of $\boldsymbol{\Delta}_1$, $\boldsymbol{\Delta}_2$ and $\boldsymbol{\Delta}_3$ are

$$\widehat{\boldsymbol{\Delta}}_1 = \frac{1}{p(q-1)}\,\mathrm{BTr}_r[(\boldsymbol{I}_p \otimes \boldsymbol{Q}_q \otimes \boldsymbol{I}_r)\boldsymbol{S}], \tag{19.12}$$

$$\widehat{\boldsymbol{\Delta}}_2 = \frac{1}{p-1}\,\mathrm{BTr}_r\left[\left(\boldsymbol{Q}_p \otimes \boldsymbol{P}_q \otimes \boldsymbol{I}_r\right)\boldsymbol{S}\right], \tag{19.13}$$

$$\widehat{\boldsymbol{\Delta}}_3 = \mathrm{BTr}_r[(\boldsymbol{P}_p \otimes \boldsymbol{P}_q \otimes \boldsymbol{I}_r)\boldsymbol{S}]. \tag{19.14}$$

One can note that all three estimators are non-singular with probability 1 if and only if $n-1 \geq r$.

19.3.2 Distributions of Estimators $\widehat{\boldsymbol{\Delta}}_1$, $\widehat{\boldsymbol{\Delta}}_2$ and $\widehat{\boldsymbol{\Delta}}_3$

The Blocks $\boldsymbol{U}_0$, $\boldsymbol{U}_1$, and $\boldsymbol{U}_2$ are easily interpretable, and so are their estimators, but a canonical transformation makes things much better tractable. It is shown in the following theorem.

Theorem 19.1 *The estimators $\widehat{\boldsymbol{\Delta}}_1$, $\widehat{\boldsymbol{\Delta}}_2$ and $\widehat{\boldsymbol{\Delta}}_3$ are mutually independent and*

$$(n-1)p(q-1)\widehat{\boldsymbol{\Delta}}_1 \sim W_r\left((n-1)p(q-1), \boldsymbol{\Delta}_1\right),$$
$$(n-1)(p-1)\widehat{\boldsymbol{\Delta}}_2 \sim W_r\left((n-1)(p-1), \boldsymbol{\Delta}_2\right),$$
$$(n-1)\widehat{\boldsymbol{\Delta}}_3 \sim W_r\left(n-1, \boldsymbol{\Delta}_3\right).$$

Proof Denoting $\boldsymbol{R}_1 = (\boldsymbol{I}_p \otimes \boldsymbol{Q}_q \otimes \boldsymbol{I}_r)\mathscr{X}$, $\boldsymbol{R}_2 = (\boldsymbol{Q}_p \otimes \boldsymbol{P}_q \otimes \boldsymbol{I}_r)\mathscr{X}$ and $\boldsymbol{R}_3 = (\boldsymbol{P}_p \otimes \boldsymbol{P}_q \otimes \boldsymbol{I}_r)\mathscr{X}$ the statistics $\widehat{\boldsymbol{\Delta}}_1$, $\widehat{\boldsymbol{\Delta}}_2$ and $\widehat{\boldsymbol{\Delta}}_3$ are mutually independent iff $\boldsymbol{R}_1\boldsymbol{Q}_n\boldsymbol{R}_1'$, $\boldsymbol{R}_2\boldsymbol{Q}_n\boldsymbol{R}_2'$ and $\boldsymbol{R}_3\boldsymbol{Q}_n\boldsymbol{R}_3'$ are mutually independent, and this is true iff $\boldsymbol{R}_1$, $\boldsymbol{R}_2$ and $\boldsymbol{R}_3$ are mutually independent. However, the independence of $\boldsymbol{R}_1$, $\boldsymbol{R}_2$ and $\boldsymbol{R}_3$ is implied by the following facts:

- vec $\boldsymbol{R}_1$, vec $\boldsymbol{R}_2$ and vec $\boldsymbol{R}_3$ are normally distributed random vectors,
- $\operatorname{cov}[\operatorname{vec}\boldsymbol{R}_1, \operatorname{vec}\boldsymbol{R}_2] = \boldsymbol{I}_n \otimes (\boldsymbol{I}_p \otimes \boldsymbol{Q}_q \otimes \boldsymbol{I}_r)\boldsymbol{\Gamma}(\boldsymbol{Q}_p \otimes \boldsymbol{P}_q \otimes \boldsymbol{I}_r) = 0$,
- $\operatorname{cov}[\operatorname{vec}\boldsymbol{R}_1, \operatorname{vec}\boldsymbol{R}_3] = \boldsymbol{I}_n \otimes (\boldsymbol{I}_p \otimes \boldsymbol{Q}_q \otimes \boldsymbol{I}_r)\boldsymbol{\Gamma}(\boldsymbol{P}_p \otimes \boldsymbol{P}_q \otimes \boldsymbol{I}_r) = 0$,
- $\operatorname{cov}[\operatorname{vec}\boldsymbol{R}_1, \operatorname{vec}\boldsymbol{R}_2] = \boldsymbol{I}_n \otimes (\boldsymbol{Q}_p \otimes \boldsymbol{P}_q \otimes \boldsymbol{I}_r)\boldsymbol{\Gamma}(\boldsymbol{P}_p \otimes \boldsymbol{P}_q \otimes \boldsymbol{I}_r) = 0$.

Let $\boldsymbol{e}_{i,q}$ be the i-th column of $\boldsymbol{I}_q$. Then for any r-dimensional vector $\boldsymbol{t}$ we have

$$(n-1)\boldsymbol{t}'\operatorname{BTr}_r[(\boldsymbol{I}_p \otimes \boldsymbol{Q}_q \otimes \boldsymbol{I}_r)\boldsymbol{S}]\boldsymbol{t} = (n-1)\boldsymbol{t}'\operatorname{BTr}_r[(\boldsymbol{I}_p \otimes \boldsymbol{Q}_q \otimes \boldsymbol{I}_r)\boldsymbol{S}(\boldsymbol{I}_p \otimes \boldsymbol{Q}_q \otimes \boldsymbol{I}_r)]\boldsymbol{t}$$
$$= (n-1)\boldsymbol{t}'\sum_{i=1}^{p}\sum_{j=1}^{q}(\boldsymbol{e}_{i,p}' \otimes \boldsymbol{e}_{j,q}'\boldsymbol{Q}_q \otimes \boldsymbol{I}_r)\boldsymbol{S}(\boldsymbol{e}_{i,p} \otimes \boldsymbol{Q}_q\boldsymbol{e}_{j,q} \otimes \boldsymbol{I}_r)\boldsymbol{t}$$
$$= (n-1)\sum_{i=1}^{p}\sum_{j=1}^{q}(\boldsymbol{e}_{i,p}' \otimes \boldsymbol{e}_{j,q}'\boldsymbol{Q}_q \otimes \boldsymbol{t}')\boldsymbol{S}(\boldsymbol{e}_{i,p} \otimes \boldsymbol{Q}_q\boldsymbol{e}_{j,q} \otimes \boldsymbol{t})$$
$$= (n-1)\operatorname{Tr}\left[(\boldsymbol{I}_p \otimes \boldsymbol{Q}_q \otimes \boldsymbol{t}')\boldsymbol{S}(\boldsymbol{I}_p \otimes \boldsymbol{Q}_q \otimes \boldsymbol{t})\right].$$

Since $(\boldsymbol{I}_p \otimes \boldsymbol{Q}_q \otimes \boldsymbol{t}')\boldsymbol{\Gamma}(\boldsymbol{I}_p \otimes \boldsymbol{Q}_q \otimes \boldsymbol{t}) = \boldsymbol{t}'\boldsymbol{\Delta}_1\boldsymbol{t} \cdot (\boldsymbol{I}_p \otimes \boldsymbol{Q}_q)$ is a multiple of an idempotent matrix, its only positive eigenvalue is $\boldsymbol{t}'\boldsymbol{\Delta}_1\boldsymbol{t}$ with multiplicity $r(\boldsymbol{I}_p \otimes \boldsymbol{Q}_q) = p(q-1)$. Therefore from Lemma 2 from [3] it follows that

$$(n-1)\operatorname{Tr}\left[(\boldsymbol{I}_p \otimes \boldsymbol{Q}_q \otimes \boldsymbol{t}')\boldsymbol{S}(\boldsymbol{I}_p \otimes \boldsymbol{Q}_q \otimes \boldsymbol{t})\right] \sim \boldsymbol{t}'\boldsymbol{\Delta}_1\boldsymbol{t} \cdot \chi^2_{(n-1)p(q-1)}.$$

Hence, we have that

$$(n-1)p(q-1)\widehat{\boldsymbol{\Delta}}_1 \sim W_r\left((n-1)p(q-1), \boldsymbol{\Delta}_1\right).$$

Similarly, for any r-dimensional vector $\boldsymbol{t}$ we have

$$(n-1)\boldsymbol{t}'\,\mathrm{BTr}_r[(\boldsymbol{Q}_p \otimes \boldsymbol{P}_q \otimes \boldsymbol{I}_r)\boldsymbol{S}]\boldsymbol{t} = (n-1)\,\mathrm{Tr}\Big[(\boldsymbol{Q}_p \otimes \boldsymbol{P}_q \otimes \boldsymbol{t}')\boldsymbol{S}(\boldsymbol{Q}_p \otimes \boldsymbol{P}_q \otimes \boldsymbol{t})\Big].$$

Since $(\boldsymbol{Q}_p \otimes \boldsymbol{P}_q \otimes \boldsymbol{t}')\boldsymbol{\Gamma}(\boldsymbol{Q}_p \otimes \boldsymbol{P}_q \otimes \boldsymbol{t}) = \boldsymbol{t}'\boldsymbol{\Delta}_2\boldsymbol{t}\cdot(\boldsymbol{Q}_p \otimes \boldsymbol{P}_q)$ is a multiple of an idempotent matrix, its only positive eigenvalue is $\boldsymbol{t}'\boldsymbol{\Delta}_2\boldsymbol{t}$ with multiplicity $r(\boldsymbol{Q}_p \otimes \boldsymbol{P}_q) = p-1$. Therefore $(n-1)\,\mathrm{Tr}\Big[(\boldsymbol{Q}_p \otimes \boldsymbol{P}_q \otimes \boldsymbol{t}')\boldsymbol{S}(\boldsymbol{Q}_p \otimes \boldsymbol{P}_q \otimes \boldsymbol{t})\Big] \sim \boldsymbol{t}'\boldsymbol{\Delta}_2\boldsymbol{t}\cdot\chi^2_{(n-1)(p-1)}$. Then it follows that

$$(n-1)(p-1)\widehat{\boldsymbol{\Delta}}_2 \sim W_r\left((n-1)(p-1), \boldsymbol{\Delta}_2\right).$$

Finally, for any r-dimensional vector $\boldsymbol{t}$ we have

$$(n-1)\boldsymbol{t}'\,\mathrm{BTr}_r[(\boldsymbol{P}_p \otimes \boldsymbol{P}_q \otimes \boldsymbol{I}_r)\boldsymbol{S}]\boldsymbol{t} = (n-1)\,\mathrm{Tr}\big[(\boldsymbol{P}_p \otimes \boldsymbol{P}_q \otimes \boldsymbol{t}')\boldsymbol{S}(\boldsymbol{P}_p \otimes \boldsymbol{P}_q \otimes \boldsymbol{t})\big].$$

Since $(\boldsymbol{P}_p \otimes \boldsymbol{P}_q \otimes \boldsymbol{t}')\boldsymbol{\Gamma}(\boldsymbol{P}_p \otimes \boldsymbol{P}_q \otimes \boldsymbol{t}) = \boldsymbol{t}'\boldsymbol{\Delta}_3\boldsymbol{t}\cdot(\boldsymbol{P}_p \otimes \boldsymbol{P}_q)$ is a multiple of an idempotent matrix, its only positive eigenvalue is $\boldsymbol{t}'\boldsymbol{\Delta}_3\boldsymbol{t}$ with multiplicity $r(\boldsymbol{P}_p \otimes \boldsymbol{P}_q) = 1$. Therefore $(n-1)\,\mathrm{Tr}\big[(\boldsymbol{P}_p \otimes \boldsymbol{P}_q \otimes \boldsymbol{t}')\boldsymbol{S}(\boldsymbol{P}_p \otimes \boldsymbol{P}_q \otimes \boldsymbol{t})\big] \sim \boldsymbol{t}'\boldsymbol{\Delta}_3\boldsymbol{t}\cdot\chi^2_{(n-1)}$. Then it follows that

$$(n-1)\widehat{\boldsymbol{\Delta}}_3 \sim W_r\left(n-1, \boldsymbol{\Delta}_3\right). \qquad \square$$

19.3.3 The Test Statistic

To test the hypothesis (19.11), we can now use a Mahalanobis-type test statistic since the covariance matrix has explicit-form inversion and we know the distribution of its estimator components. To distinguish it from Hotelling T^2, we denote it D^2:

$$\begin{aligned}
D^2 &= n\left(\overline{\boldsymbol{x}} - \boldsymbol{\mu}_0\right)'\widehat{\boldsymbol{\Gamma}}^{-1}\left(\overline{\boldsymbol{x}} - \boldsymbol{\mu}_0\right) \\
&= n\left(\overline{\boldsymbol{x}} - \boldsymbol{\mu}_0\right)'\left[\boldsymbol{I}_p \otimes \boldsymbol{Q}_q \otimes \widehat{\boldsymbol{\Delta}}_1^{-1} + \boldsymbol{Q}_p \otimes \boldsymbol{P}_q \otimes \widehat{\boldsymbol{\Delta}}_2^{-1} + \boldsymbol{P}_p \otimes \boldsymbol{P}_q \otimes \widehat{\boldsymbol{\Delta}}_3^{-1}\right]\left(\overline{\boldsymbol{x}} - \boldsymbol{\mu}_0\right) \\
&= n\,\mathrm{Tr}\left[\widehat{\boldsymbol{\Delta}}_1^{-1}\left(\overline{\boldsymbol{x}} - \boldsymbol{M}_0\right)\left(\boldsymbol{I}_p \otimes \boldsymbol{Q}_q\right)\left(\overline{\boldsymbol{x}} - \boldsymbol{M}_0\right)'\right] \\
&\quad + n\,\mathrm{Tr}\left[\widehat{\boldsymbol{\Delta}}_2^{-1}\left(\overline{\boldsymbol{x}} - \boldsymbol{M}_0\right)\left(\boldsymbol{Q}_p \otimes \boldsymbol{P}_q\right)\left(\overline{\boldsymbol{x}} - \boldsymbol{M}_0\right)'\right] \\
&\quad + n\,\mathrm{Tr}\left[\widehat{\boldsymbol{\Delta}}_3^{-1}\left(\overline{\boldsymbol{x}} - \boldsymbol{M}_0\right)\left(\boldsymbol{P}_p \otimes \boldsymbol{P}_q\right)\left(\overline{\boldsymbol{x}} - \boldsymbol{M}_0\right)'\right],
\end{aligned} \tag{19.15}$$

where $\boldsymbol{\mu}_0 = \mathrm{vec}\,\boldsymbol{M}_0$.

Theorem 19.2 *Let* $\boldsymbol{X}_1, \ldots, \boldsymbol{X}_n$ *be a random sample from* $N_{r\times pq}(\boldsymbol{M}_0, \boldsymbol{\Gamma})$, *where* $\boldsymbol{\Gamma}$ *has the structure (19.2), and let estimators* $\widehat{\boldsymbol{\Delta}}_1$, $\widehat{\boldsymbol{\Delta}}_2$, *and* $\widehat{\boldsymbol{\Delta}}_3$ *are given by (19.12), (19.13), and (19.14), respectively. Let* D^2 *be given by (19.15). Then it holds*

$$D^2 \sim T_0^2(r;\, p(q-1), (n-1)p(q-1)) \oplus T_0^2(r;\, p-1, (n-1)(p-1)) \oplus T_0^2(r;\, 1, n-1),$$

where $\oplus$ *denotes convolution and* $T_0^2(r;\, a, b)$ r*-dimensional Lawley-Hotelling trace distribution with* a *and* b *degrees of freedom.*

Proof Let us denote $\boldsymbol{b} = \left(\boldsymbol{H}_p \otimes \boldsymbol{H}_q \otimes \boldsymbol{I}_q\right)\left(\overline{\boldsymbol{x}} - \boldsymbol{\mu}_0\right)$. We consider the vector $\boldsymbol{b}$ to be partitioned in pq subvectors as $\boldsymbol{b} = (\boldsymbol{b}'_{11}, \boldsymbol{b}'_{12} \ldots, \boldsymbol{b}'_{pq})'$, where $\boldsymbol{b}_{ij}$, $i = 1, \ldots, p$, $j = 1, \ldots, q$, is r-dimensional subvector. Then, the relation (19.6) implies that $\boldsymbol{b}_{11}, \ldots, \boldsymbol{b}_{pq}$ are independently normally distributed with

$$\boldsymbol{b}_{11} \sim N_r\left(\boldsymbol{0};\, \frac{1}{n}\boldsymbol{\Delta}_3\right),$$

$$\boldsymbol{b}_{i1} \sim N_r\left(\boldsymbol{0};\, \frac{1}{n}\boldsymbol{\Delta}_2\right) \quad \text{for } i = 2, \ldots, p,$$

$$\boldsymbol{b}_{ij} \sim N_r\left(\boldsymbol{0};\, \frac{1}{n}\boldsymbol{\Delta}_1\right) \quad \text{for } i = 1, \ldots, p,\ j = 2, \ldots, q,$$

under the null hypothesis H_0. Since

$$\boldsymbol{H}_a \boldsymbol{P}_a \boldsymbol{H}'_a = \begin{pmatrix} 1 & 0 & \cdots & 0 \\ 0 & 0 & \cdots & 0 \\ \vdots & \vdots & \ddots & \vdots \\ 0 & 0 & \cdots & 0 \end{pmatrix} = \boldsymbol{e}_{1,a}\boldsymbol{e}'_{1,a},$$

and

$$\boldsymbol{H}_a \boldsymbol{Q}_a \boldsymbol{H}'_a = \begin{pmatrix} 0 & 0 & \cdots & 0 \\ 0 & 1 & \cdots & 0 \\ \vdots & \vdots & \ddots & \vdots \\ 0 & 0 & \cdots & 1 \end{pmatrix} = \sum_{i=2}^{a} \boldsymbol{e}_{i,a}\boldsymbol{e}'_{i,a},$$

where $\boldsymbol{e}_{i,a}$ is the i-th column of $\boldsymbol{I}_a$, the statistic D^2 can be written in the form

$$\begin{aligned} D^2 &= n\left(\overline{\boldsymbol{x}} - \boldsymbol{\mu}_0\right)'\left[\boldsymbol{I}_p \otimes \boldsymbol{Q}_q \otimes \widehat{\boldsymbol{\Delta}}_1^{-1} + \boldsymbol{Q}_p \otimes \boldsymbol{P}_q \otimes \widehat{\boldsymbol{\Delta}}_2^{-1} + \boldsymbol{P}_p \otimes \boldsymbol{P}_q \otimes \widehat{\boldsymbol{\Delta}}_3^{-1}\right]\left(\overline{\boldsymbol{x}} - \boldsymbol{\mu}_0\right) \\ &= n\boldsymbol{b}'\left(\boldsymbol{H}_p \otimes \boldsymbol{H}_q \otimes \boldsymbol{I}_q\right)\left[\boldsymbol{I}_p \otimes \boldsymbol{Q}_q \otimes \widehat{\boldsymbol{\Delta}}_1^{-1} + \boldsymbol{Q}_p \otimes \boldsymbol{P}_q \otimes \widehat{\boldsymbol{\Delta}}_2^{-1} + \boldsymbol{P}_p \otimes \boldsymbol{P}_q \otimes \widehat{\boldsymbol{\Delta}}_3^{-1}\right] \\ &\qquad \times \left(\boldsymbol{H}_p \otimes \boldsymbol{H}_q \otimes \boldsymbol{I}_q\right)'\boldsymbol{b} \\ &= n\boldsymbol{b}'\left[\boldsymbol{I}_p \otimes \boldsymbol{H}_q\boldsymbol{Q}_q\boldsymbol{H}'_q \otimes \widehat{\boldsymbol{\Delta}}_1^{-1} + \boldsymbol{H}_p\boldsymbol{Q}_p\boldsymbol{H}'_p \otimes \boldsymbol{H}_q\boldsymbol{P}_q\boldsymbol{H}'_q \otimes \widehat{\boldsymbol{\Delta}}_2^{-1}\right. \\ &\qquad \left. + \boldsymbol{H}_p\boldsymbol{P}_p\boldsymbol{H}'_p \otimes \boldsymbol{H}_q\boldsymbol{P}_q\boldsymbol{H}'_q \otimes \widehat{\boldsymbol{\Delta}}_3^{-1}\right]\boldsymbol{b} \end{aligned}$$

$$= n \sum_{i=1}^{p} \sum_{j=2}^{q} \boldsymbol{b}_{ij}' \widehat{\boldsymbol{\Delta}}_1^{-1} \boldsymbol{b}_{ij} + n \sum_{i=2}^{p} \boldsymbol{b}_{i1}' \widehat{\boldsymbol{\Delta}}_2^{-2} \boldsymbol{b}_{i1} + n \boldsymbol{b}_{11}' \widehat{\boldsymbol{\Delta}}_3^{-1} \boldsymbol{b}_{11}$$

$$= \mathrm{Tr}\left[\sum_{i-1}^{p} \sum_{j-2}^{q} n \boldsymbol{b}_{ij} \boldsymbol{b}_{ij}' \widehat{\boldsymbol{\Delta}}_1^{-1} \right] + \mathrm{Tr}\left[\sum_{i-2}^{p} n \boldsymbol{b}_{i1} \boldsymbol{b}_{i1}' \widehat{\boldsymbol{\Delta}}_2^{-1} \right] + \mathrm{Tr}\left[n \boldsymbol{b}_{11} \boldsymbol{b}_{11}' \widehat{\boldsymbol{\Delta}}_3^{-1} \right] \quad (19.16)$$

$$\stackrel{df}{=} T_{01}^2 + T_{02}^2 + T_{03}^2.$$

Clearly T_{01}^2, T_{02}^2, and T_{03}^2 are independent, so that the resulting distribution is the convolution of the three ones. Taking into account independence of all $\boldsymbol{b}_{ij}$'s and parameters of their distributions, it is clear that $n\boldsymbol{b}_{11}\boldsymbol{b}_{11}' \sim W_r(1, \boldsymbol{\Delta}_3)$, $\sum_{i=2}^{p} n\boldsymbol{b}_{i1}\boldsymbol{b}_{i1}' \sim W_r(p-1, \boldsymbol{\Delta}_2)$, and $\sum_{i=1}^{p} \sum_{j=2}^{q} n\boldsymbol{b}_{ij}\boldsymbol{b}_{ij}' \sim W_r(p(q-1), \boldsymbol{\Delta}_1)$. Using Theorem 19.1, we can immediately conclude that T_{01}^2 has the Lawley-Hotelling trace (LH-trace) distribution $T_0^2(r; p(q-1), (n-1)p(q-1))$ if $(n-1)p(q-1) \geq r$, T_{02}^2 has the LH-trace distribution $T_0^2(r; p-1, (n-1)(p-1))$ if $(n-1)(p-1) \geq r$, and T_{03}^2 has the LH-trace distribution $T_0^2(r; 1, n-1))$ if $n-1 \geq r$. □

Thus, the distribution of D^2 is the convolution of three LH-trace distributions. Let us note that since $p, q \geq 2$, all three distributions exist if $n-1 \geq r$, which is a very mild condition. To compare, the standard Hotelling test for the same dimension but with general covariance matrix would require $n-1 > pqr$.

Unfortunately, there is no simple way of obtaining critical values of this convolution. One possibility is to use simulations. The other possibility is to approximate the LH-trace distribution by an F-distribution (see [8]) and then use the characteristic function inversion method (see [18]). Witkovský [18] also developed a freely available software package enabling such computations. Using this package is the most easy way of carrying out the test. That is why we have investigated the precision of the McKeon's approximation in the section on a real data example.

The approximations are as follows:

- $T_0^2(r; p(q-1), (n-1)p(q-1))$ can be approximated by

$$\frac{(n-1)p^2(q-1)^2 r}{((n-1)p(q-1)-r-1)} \frac{b_1-2}{b_1} F\left(p(q-1)r, b_1\right),$$

where $b_1 = 4 + \dfrac{(p(q-1)r+2)((n-1)p(q-1)-r-3)((n-1)p(q-1)-r)}{(n-1)p^2(q-1)^2 + p(q-1)(nr+n-r-2)-(r-1)^2+2}$;

- $T_0^2(r; p-1, (n-1)(p-1))$ can be approximated by

$$\frac{(n-1)(p-1)^2 r}{np-n-p-r} \frac{b_2-2}{b_2} F\left((p-1)r, b_2\right),$$

where $b_2 = 4 + \dfrac{(pr-r+2)(np-n-p-r-2)(np-n-p-r+1)}{(np-n-p)(p+r)-(r-1)(r+2)}$;

- $T_0^2(r; 1, n-1)$ is equal to the usual Hotelling $T^2_{r,n-1}$, which is equivalent to $\frac{(n-1)r}{n-r} F(r, n-r)$.

Note that all multiplicative constants and degrees of freedom in these approximations are positive for $n - 1 \geq r \geq 1$ and $p, q \geq 2$.

19.4 A Real Data Example

To illustrate our proposed testing method, we test the hypothesis (19.11) on a real data set. The data set is from the textbook [2]. An investigator measured the mineral content of bones (radius, humerus, and ulna) by photon absorptiometry to examine whether dietary supplements would slow bone loss in 25 older women. Measurements were recorded for the three bones on the dominant and non-dominant sides ([2], p. 43). Thus, the data is two-level multivariate and clearly $q = 2$ and $r = 3$.

The bone mineral contents for the first 24 women 1 year after their participation in an experimental program is also given in [2], p. 353. Thus, for our analysis we take only the first 24 women in the first data set, and combine these two data sets side by side into a new one. Adding a time factor, this new data set has a three-level multivariate structure with $p = 2, q = 2$ and $r = 3$.

We rearrange the variables in the new data set by grouping together the mineral content of the dominant sides of radius, humerus, and ulna as the first three variables, that is, the variables in the first location ($s = 1$) and then the mineral contents for the non-dominant side of the same bones ($s = 2$) in the first year ($t = 1$) of the experiment, and then the same thing in the second year ($t = 2$) of the experiment.

We test the hypothesis (19.11) for one sample. To do this we take the three-level data of 24 women as the population and we randomly sample 12 women from this population and call this set of 12 women our sample. We test whether the mean of 12 women from our sample is equal

- to the mean of 24 women from our population, which is

$$\boldsymbol{M}_0^I = \begin{pmatrix} 0.841 & 0.813 & 0.841 & 0.810 \\ 1.785 & 1.729 & 1.778 & 1.717 \\ 0.698 & 0.687 & 0.713 & 0.687 \end{pmatrix}, \tag{19.17}$$

  where the rows corresponds to three bones (radius, humerus, and ulna); the first two columns correspond to measurements on dominant and non-dominant side, respectively, at the start of the experiment and similarly last two columns to measurements on both sides after 1 year;

- to the value

$$\boldsymbol{M}_0^{II} = \begin{pmatrix} 0.85 & 0.85 & 0.85 & 0.85 \\ 1.75 & 1.75 & 1.75 & 1.75 \\ 0.70 & 0.70 & 0.70 & 0.70 \end{pmatrix}, \qquad (19.18)$$

  which corresponds to the hypothesis of the same values of dominant and non-dominant sides and also the same values before and after the experimental treatment, while we assume that the values are given by the experimenter.

Our sample consists of women with subject number 2, 3, 4, 7, 9, 10, 11, 15, 17, 19, 21 and 23, and we want to test whether these 12 women have the mean equal to $\boldsymbol{M}_0^{I}$ and $\boldsymbol{M}_0^{II}$ given in (19.17) and (19.18), respectively. The calculated mean and covariance components are

$$\overline{\boldsymbol{X}} = \begin{pmatrix} 0.824 & 0.813 & 0.833 & 0.790 \\ 1.702 & 1.635 & 1.690 & 1.623 \\ 0.685 & 0.690 & 0.694 & 0.693 \end{pmatrix},$$

$$\widehat{\boldsymbol{\Delta}}_1 = \begin{pmatrix} 0.00324 & 0.00475 & 0.00145 \\ 0.00475 & 0.01206 & 0.00195 \\ 0.00145 & 0.00195 & 0.00305 \end{pmatrix},$$

$$\widehat{\boldsymbol{\Delta}}_2 = \begin{pmatrix} 0.00047 & 0.00096 & -0.00024 \\ 0.00096 & 0.00495 & -0.00081 \\ -0.00024 & -0.00081 & 0.00090 \end{pmatrix},$$

$$\widehat{\boldsymbol{\Delta}}_3 = \begin{pmatrix} 0.05861 & 0.08768 & 0.03514 \\ 0.08768 & 0.24712 & 0.06886 \\ 0.03514 & 0.06886 & 0.03040 \end{pmatrix}.$$

The values of the calculated test statistics from (19.15) are $D_I^2 = 8.66048$ and $D_{II}^2 = 14.4572$. We have compared three different methods to calculate the critical values and p-values (in simulations we have used 100 000 trials):

(A) simulating directly the test statistic D^2;
(B) simulating the convolution of the approximating F-distributions of the independent components of D^2;
(C) calculating the critical value and p-value by the method of [18], i.e., calculating numerically the values of cumulative distribution function of linear combination of three independent approximating F-distributions from its characteristic function.

The results are given in Table 19.1. As it can be seen, the critical values and p-values are very close to each other, so that all three methods can be viewed as

Table 19.1 Critical values of the test statistic D^2 for $n = 12, r = 3, p = 2$ and $q = 2$ and corresponding p-values

		p-values	
	Critical values	D_I^2	D_{II}^2
(A)	33.2492	0.8521	0.5225
(B)	33.4572	0.8510	0.5229
(C)	33.4991	0.8524	0.5235

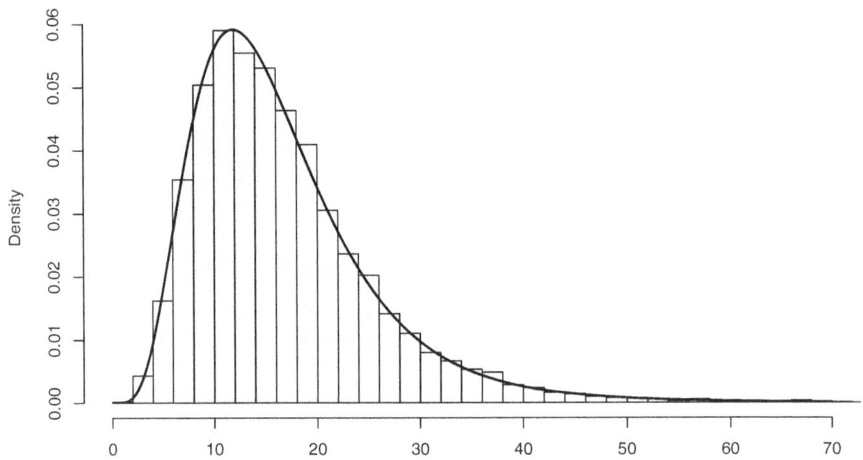

Fig. 19.1 Approximating and simulated distribution of the D^2-statistic

equivalent. Neither hypothesis has been rejected. Figure 19.1 shows very good fit of the approximating and simulated distribution of the D^2-statistic.

19.5 Conclusion

The doubly exchangeable covariance structure is a realistic assumption in many cases, and substantially reduces the number of parameters that have to be estimated. Hence, in our case $pqr = 12$, so that $n = 12$ would be an insufficient number of observations for estimating the unstructured covariance matrix. Moreover, the proposed method increases the power of the test by using information about the covariance structure. The test is readily applicable with current software means.

Acknowledgments Žežula's and Klein's research was supported by the Slovak Research and Development Agency under the Contract No. APVV-17-0568 and by grant VEGA MŠ SR 1/0311/18.

References

1. Filipiak, K., Klein, D., Vojtková, E.: The properties of partial trace and block trace operators of partitioned matrices. Electr. J. Linear Algebra **33**, 3–15 (2018)
2. Johnson, R.A., Wichern, D.W.: Applied Multivariate Statistical Analysis, 6th edn. Pearson Prentice Hall, New Jersey (2007)
3. Klein, D., Žežula, I.: Orthogonal decompositions in growth curve models. Acta et Comment. Univ. Tartu. de Math. **14**, 35–44 (2010)
4. Leiva, R.: Linear discrimination with equicorrelated training vectors. J. Multivar. Anal. **98**, 384–409 (2007)
5. Leiva, R., Roy, A.: Linear Discrimination for Multi-level Multivariate Data with Separable Means and Jointly Equicorrelated Covariance Structure. J. Stat. Plan. Inference **141**(5), 1910–1924 (2011)
6. Leiva, R., Roy, A.: Linear discrimination for three-level multivariate data with separable additive mean vector and doubly exchangeable covariance structure. Comput. Stat. Data Anal. **56**(6), 1644–1661 (2012)
7. Mardia, K.V., Kent, J.T., Bibby, J.M.: Multivariate Analysis. Academic, New York (1979)
8. McKeon, J.J.: F-approximations to the distribution of Hotelling's T_0^2. Biometrika **61**(2), 381–383 (1974)
9. Muirhead, R.J.: Aspects of Multivariate Statistical Theory, Second Ed. New York: John Wiley & Sons (2005)
10. Rao, C.R.: Familial correlations or the multivariate generalizations of the intraclass correlation. Curr. Sci. **14**, 66–67 (1945)
11. Rao, C.R.: Discriminant functions for genetic differentiation and selection. Sankhyā **12**, 229–246 (1953)
12. Rao, C.R.: Linear Statistical Inference and Its Applications, 2nd edn. Wiley, New York (1973)
13. Regalia, P., Mitra, S.: Kronecker products, unitary matrices and signal processing applications, SIAM review **31**(4), 586–613 (1989)
14. Roy, A., Leiva, R.: Likelihood Ratio Tests for Triply Multivariate Data with Structured Correlation on Spatial Repeated Measurements. Statist. Probab. Lett. **78**(13), 1971 1980 (2008)
15. Roy, A., Leiva, R.: Estimating and Testing a Structured Covariance Matrix for Three-level Multivariate Data. Commun. Stat. - Theory Methods **40**(10), 1945–1963 (2011)
16. Roy, A., Fonseca, M.: Linear Models with Doubly Exchangeable Distributed Errors. Commun. Stat. - Theory Methods **41**(13), 2545–2569 (2012)
17. Roy, A., Leiva, R., Žežula, I., Klein, D.: Testing the equality of mean vectors for paired doubly multivariate observations in blocked compound symmetric covariance matrix setup. J. Multivar. Anal. **137**, 50–60 (2015)
18. Witkovský, V.: Computing the exact distribution of the Bartlett's test statistic by numerical inversion of its characteristic function. J. Appl. Stat. 1–16 (2019). https://doi.org/10.1080/02664763.2019.1675608
19. Žežula, I., Klein, D., Roy, A.: Testing of multivariate repeated measures data with block exchangeable covariance structure. Test **27**(2), 360–378 (2018)

Chapter 20
Estimation of the Common Mean of Two Multivariate Normal Distributions Under Symmetrical and Asymmetrical Loss Functions

Dan Zhuang, S. Ejaz Ahmed, Shuangzhe Liu, and Tiefeng Ma

Abstract In this paper, the estimation of the common mean vector of two multivariate normal populations is considered and a new class of unbiased estimators is proposed. Several dominance results under the quadratic loss and LINEX loss functions are established. To illustrate the usefulness of these estimators, a simulation study with finite samples is conducted to compare them with four existing estimators, including the sample mean and the Graybill-Deal estimator. Based on the comparison studies, we found that the numerical performance of the proposed estimators is almost as good as $\tilde{\mu}_{CC}$ proposed by Chiou and Cohen (Ann Inst Stat Math 37:499–506, 1985) in terms of the risks. Its theoretical dominance over the sample mean of a single population under the sufficient conditions given is also established.

20.1 Introduction

Issues in the estimation and inference of the common mean of several distributions are very important. They are involved, for example, in balanced incomplete block design (BIBD) with uncorrelated random block effects and fixed treatment effects,

D. Zhuang
Fujian Normal University, Fuzhou, China
e-mail: zhgdan@fjnu.edu.cn

S. E. Ahmed (✉)
Brock University, St. Catharines, ON, Canada
e-mail: sahmed5@brocku.ca

S. Liu
University of Canberra, Canberra, ACT, Australia
e-mail: shuangzhe.liu@canberra.edu.au

T. Ma
Southwestern University of Finance and Economics, Chengdu, China
e-mail: matiefeng@swufe.edu.cn

© Springer Nature Switzerland AG 2020
T. Holgersson, M. Singull (eds.), *Recent Developments in Multivariate and Random Matrix Analysis*, https://doi.org/10.1007/978-3-030-56773-6_20

as well as in meta-analysis with applications. They are studied in normal and other distributions; see, for example, [1, 6, 18, 22], and the references therein.

Let X_i for $i = 1, 2, \ldots, n$ be a random sample of size n from a p-dimensional multivariate normal distribution with mean vector μ and covariance matrix Σ_x, denoted as $X \sim N(\mu, \Sigma_x)$, and let Y_i for $i = 1, 2, \ldots, n$ be a random sample of size n from a p-dimensional multivariate normal distribution with the same mean vector μ and covariance matrix Σ_y. Assuming the X-sample and Y-sample are independent, we consider the problem of estimating the common mean vector μ when Σ_x and Σ_y are unknown. This is because the multivariate normal distributions with unknown covariance matrices lay the essential ground work for multivariate statistics, see [21].

For simplicity, we introduce

$$S_x = \frac{1}{n-1}\sum_{i=1}^{n}(X_i - \bar{X})(X_i - \bar{X})', \quad S_y = \frac{1}{n-1}\sum_{i=1}^{n}(Y_i - \bar{Y})(Y_i - \bar{Y})', \tag{20.1}$$

where $\bar{X} = \frac{1}{n}\sum_{i=1}^{n} X_i$ and $\bar{Y} = \frac{1}{n}\sum_{i=1}^{n} Y_i$ are $p \times 1$ vectors, and S_x and S_y are $p \times p$ matrices.

When $p = 1$, this estimation problem has been extensively studied and a number of estimators have been proposed. It is well known that the estimator proposed by Graybill and Deal [4] is given by

$$\hat{\mu}_{\mathrm{GD}} = (S_y\bar{X} + S_x\bar{Y})/(S_x + S_y),$$

which is unbiased and has a smaller variance than either $\bar{X}$ or $\bar{Y}$ as long as $n > 10$. Brown and Cohen [2] also presented unbiased estimators with a smaller variance than $\bar{X}$. Misra and Van der Meulen [12] proposed a simple estimator which is better than $\hat{\mu}_{\mathrm{GD}}$ in terms of stochastic dominance and the Pitman measure of closeness with order-restricted unknown variances. Pal et al. [17] considered the maximum likelihood estimator of the common mean and its properties, compared to those of $\hat{\mu}_{\mathrm{GD}}$.

For the discussion of the finite sample properties of $\hat{\mu}_{\mathrm{GD}}$, the variance is very important. However the expression for the variance of $\hat{\mu}_{\mathrm{GD}}$ is very complex (see [13], for a detailed discussion of its complexities), so some simple approximations of it were derived by Nanayakkara and Cressie [14] and Pal and Lim [16]. But, it is impossible to obtain accurate finite sample properties based on approximate variance expression. In order to discuss the properties of finite samples, another popular technique is to construct unbiased estimation of variance of estimators. Some unbiased estimators of the variance of $\hat{\mu}_{\mathrm{GD}}$ were given by Sinha [19] and Hartung [5]. For more than two populations, [15] investigated the properties of $\hat{\mu}_{\mathrm{GD}}$, and [11] extended their studies and presented several results of $\hat{\mu}_{\mathrm{GD}}$ under the LINEX loss function.

For the multivariate case ($p \geq 2$), the Graybill-Deal estimator is given by

$$\tilde{\mu}_{\text{MGD}} = S_y(S_x + S_y)^{-1}\bar{X} + S_x(S_x + S_y)^{-1}\bar{Y}. \tag{20.2}$$

It is difficult to find a uniformly dominated result for (20.2). Chiou and Cohen [3] made progress by showing that for every n there is some (Σ_x, Σ_y) for which neither $\bar{X}$ nor $\bar{Y}$ is dominated by (20.2) under a covariance criterion. They considered an alternative class of estimators $\tilde{\mu}_{\text{CC}}$ which do not dominate $\bar{X}$ or $\bar{Y}$. It is presented in the following three steps:

(i) Let $U = (u_{ij})$ be an $n \times n$ known orthogonal matrix with n-th row $u_{(n)} = (1/\sqrt{n}, \ldots, 1/\sqrt{n})$. Define $U_i = \sum_{k=1}^{n} u_{ik}X_k$ so that $U_i \sim N(0, \Sigma_x)$ for $i = 1, \ldots, n-1$ and $U_n \sim N(\sqrt{n}\mu, \Sigma_x)$. Similarly, let $V = (v_{ik})$ be an $n \times n$ orthogonal matrix with random vectors $V_i = \sum_{k=1}^{n} v_{ik}Y_k$, $i = 1, \ldots, n$. So U_i, U_j, V_k and V_l are independent for $i \neq j$ and $k \neq l$.

(ii) Let r be an integer satisfying $1 \leq r \leq n-7$ and define $\bar{S} = \sum_{i=r+1}^{n-1}(U_i + V_i)(U_i + V_i)'$ and $\bar{S}_x = \sum_{i=1}^{r} U_iU_i'$. Note that $\bar{S} \sim W_p(n-r-1, \Sigma_x + \Sigma_y)$, $\bar{S}_x \sim W_p(r, \Sigma_x)$, and $\bar{S}_x$ is independent of $\bar{S}$, where $W_p(n, \Sigma)$ denotes a Wishart distribution with n degrees of freedom and the mean matrix $n\Sigma$.

(iii) Then let Chiou and Cohen's [3] estimator be defined by

$$\tilde{\mu}_{\text{CC}} = \bar{X} + b\bar{S}^{-1}\bar{S}_x(\bar{Y} - \bar{X}), \tag{20.3}$$

where b is a positive constant.

However, [3] only gave some theoretical results in the case of $p = 2$ based on some limit assumptions.

Different from what [3] did, [8] also gave a better estimator than $\bar{X}$ based on a special construction. Consider the transformation $G_i = X_i - Y_i$, $i = 1, 2, \ldots, n$. Define $\bar{G} = \frac{1}{n}\sum_{i=1}^{n} G_i$ and $S_G = \sum_{i=1}^{n}(G_i - \bar{G})(G_i - \bar{G})'/(n-1)$. Based on these, a simple unbiased estimator proposed by Krishnamoorthy [8] is given as follows

$$\tilde{\mu}_{\text{K}} = \bar{X} - dS_xS_G^{-1}\bar{G}, \tag{20.4}$$

where d is a positive constant. It has risk lower than $\bar{X}$, for any $p \geq 1$ and $n \geq p+5$, only under the quadratic loss function (20.6) given below.

The two approaches mentioned above have no symmetry, so we cannot give the dominance result under which they uniformly dominate $\bar{Y}$. Apart from these, to our knowledge, no dominance results have been reported. In order to establish such results in this paper, we propose the following class of unbiased estimators in the multivariate case:

$$\tilde{\mu}_x = \bar{X} + c\bar{S}_x\bar{S}^{-1}(\bar{Y} - \bar{X}), \tag{20.5}$$

where c is a positive constant, $\bar{X}$, $\bar{Y}$, $\bar{S}_x$ and $\bar{S}$ are defined above (20.3). Note that $\tilde{\mu}_x$ is a transformation of $\tilde{\mu}_{CC}$, with the locations of $\bar{S}_x$ and $\bar{S}^{-1}$ swapped. As we can see below, this swap makes a significant improvement in terms of theoretical derivation and risk comparison, although being a very simple alteration.

It is well known that the quadratic loss is a useful comparison criterion to evaluate point estimators. However, the use of asymmetric loss functions is quite common in many practical cases where overestimation is considered more serious than underestimation, or vice-versa. For instance, central banks, while lending money, are likely to have asymmetric preferences. The LINEX loss function proposed by Varian [20] is a widely used asymmetric loss function in this kind of scenario. Therefore, in this paper, we consider the following four loss functions:

(a) Quadratic loss

$$L_1(\hat{\mu}, \mu) = (\hat{\mu} - \mu)' \Sigma_x^{-1} (\hat{\mu} - \mu), \tag{20.6}$$

$$L_2(\hat{\mu}, \mu) = (\hat{\mu} - \mu)' \Sigma_y^{-1} (\hat{\mu} - \mu), \tag{20.7}$$

$$L_3(\hat{\mu}, \mu) = (\hat{\mu} - \mu)' (\Sigma_x + \Sigma_y)^{-1} (\hat{\mu} - \mu), \tag{20.8}$$

(b) LINEX loss

$$L_4(\hat{\mu}, \mu) = \exp\{a'(\hat{\mu} - \mu)\} - a'(\hat{\mu} - \mu) - 1, \tag{20.9}$$

where a is a non-zero constant vector. For this loss function, see [10] and [23, 24].

We calculate the risk of $\hat{\mu}$ by

$$R_i(\hat{\mu}, \mu) = E[L_i(\hat{\mu}, \mu)] \qquad (i = 1, 2, 3, 4). \tag{20.10}$$

The structure of the rest of this paper will be as follows: In Sect. 20.2, we study finite sample properties of our advocated class of estimators $\tilde{\mu}_x$ under the quadratic and LINEX loss functions. In Sect. 20.3, we conduct a simulation study to illustrate the performance of $\tilde{\mu}_x$, compared with the existing estimators mentioned above. Finally, we make concluding remarks in Sect. 20.4. The Appendix contains all the proofs.

20.2 Main Results

In this section, we present our results giving sufficient conditions for $\tilde{\mu}_x$ to dominate $\bar{X}$ and $\bar{Y}$ under the quadratic and LINEX loss functions respectively.

20.2.1 Quadratic Loss

Before giving out the main theorems, the following lemmas are useful; see e.g., [7, 10] and [21].

Lemma 20.1 *Let $S \sim W_p(n, \Sigma)$ and D be a known positive definite matrix, then*

$$\begin{aligned} E(S) &= n\Sigma, \\ E(SDS) &= n\, tr\,(D\Sigma)\Sigma + n(n+1)\Sigma D\Sigma, \\ E(S^{-1}) &= \frac{1}{n-p-1}\Sigma^{-1}, \\ E(S^{-1}DS^{-1}) &= c_1\Sigma^{-1}D\Sigma^{-1} + c_2\, tr\,(D\Sigma^{-1})\Sigma^{-1}, \end{aligned} \tag{20.11}$$

where $c_1 = 1/((n-p)(n-p-3))$ and $c_2 = 1/((n-p)(n-p-1)(n-p-3))$.

Lemma 20.2 *Let $a_1 \geq a_2 \geq \cdots \geq a_n$, $b_1 \geq b_2 \geq \cdots \geq b_n$, then*

$$n\text{tr}(AB) > \text{tr}(A)\text{tr}(B) \tag{20.12}$$

for the special matrices $A = diag(a_1, a_2, \ldots, a_n)$ and $B = diag(b_1, b_2, \ldots, b_n)$.

Lemma 20.3 *For $X \sim N(\mu, \sigma^2)$, a normal distribution with mean μ and variance σ^2, we have $E[\, exp\,(bX)] = \ exp\,(\mu b + \frac{1}{2}b^2\sigma^2)$.*

Lemma 20.4 *(Jensen Inequality) Let X be a random variable, and let the expectations $E(X)$ and $E(g(X))$ exist. If g is a convex function, then $E(g(X)) \geq g(E(X))$. If g is a concave function, then $E(g(X)) \leq g(E(X))$.*

Here, we compare $\tilde{\mu}_x$ with $\bar{X}$ and then with $\bar{Y}$. First we present sufficient conditions for $\tilde{\mu}_x$ to be better than $\bar{X}$ in terms of the risks under three different quadratic loss functions. By defining $T = \Sigma_x(\Sigma_x + \Sigma_y)^{-1}$ and $c_0 = \frac{(n-r-p-1)(n-r-p-4)}{(n-r-2)(r+p+1)}$, the theorems are given as follows.

Theorem 20.1 *If $0 \leq c \leq 2c_0$, then the risk of $\tilde{\mu}_x$ is less than $\bar{X}$ under the loss function $L_1(\hat{\mu}, \mu)$ in (20.6).*

Proof Using the unbiased property of (20.5), we get

$$\begin{aligned} &R_1(\bar{X}, \mu) - R_1(\tilde{\mu}_x, \mu) \\ &\quad = -2cE[\bar{X}'\Sigma_x^{-1}\bar{S}_x\bar{S}^{-1}(\bar{Y} - \bar{X})] - c^2E[(\bar{Y} - \bar{X})'\bar{S}^{-1}\bar{S}_x\Sigma_x^{-1}\bar{S}_x\bar{S}^{-1}(\bar{Y} - \bar{X})] \\ &\quad = -2c\,\text{tr}\,\{E[\Sigma_x^{-1}\bar{S}_x\bar{S}^{-1}(\bar{Y} - \bar{X})\bar{X}']\} \\ &\qquad - c^2\,\text{tr}\,\{E[\bar{S}^{-1}\bar{S}_x\Sigma_x^{-1}\bar{S}_x\bar{S}^{-1}(\bar{Y} - \bar{X})(\bar{Y} - \bar{X})']\}. \end{aligned} \tag{20.13}$$

As $\bar{X}$, $\bar{Y}$, $\bar{S}_x$ and $\bar{S}$ are mutually independent, it follows from (20.13) that

$$\begin{aligned}
&R_1(\bar{X}, \mu) - R_1(\tilde{\mu}_x, \mu) \\
&= -2c \operatorname{tr}[\Sigma_x^{-1} E(\bar{S}_x) E(\bar{S}^{-1}) E((\bar{Y} - \bar{X})\bar{X}')] \\
&\quad - c^2 \operatorname{tr}[E(\bar{S}_x \Sigma_x^{-1} \bar{S}_x) E(\bar{S}^{-1} E((\bar{Y} - \bar{X})(\bar{Y} - \bar{X})')\bar{S}^{-1})] \\
&= \frac{2c}{n} \operatorname{tr}[\Sigma_x^{-1} E(\bar{S}_x) E(\bar{S}^{-1}) \Sigma_x] \\
&\quad - \frac{c^2}{n} \operatorname{tr}[E(\bar{S}_x \Sigma_x^{-1} \bar{S}_x) E(\bar{S}^{-1}(\Sigma_x + \Sigma_y)\bar{S}^{-1})].
\end{aligned} \tag{20.14}$$

By Lemma 20.1, after some manipulation, we get

$$\begin{aligned}
R_1(\bar{X}, \mu) - R_1(\tilde{\mu}_x, \mu) &= \operatorname{tr}(\Sigma_x(\Sigma_x + \Sigma_y)^{-1}) \\
\times &\left[\frac{2cr}{n(n-r-p-2)} - \frac{c^2 r(n-r-2)(r+p+1)}{n(n-r-p-1)(n-r-p-2)(n-r-p-4)} \right].
\end{aligned} \tag{20.15}$$

One can then see that $R_1(\bar{X}, \mu) - R_1(\tilde{\mu}_x, \mu) \geq 0$ for $0 \leq c \leq \frac{2(n-r-p-1)(n-r-p-4)}{(n-r-2)(r+p+1)}$. This completes the proof. □

Theorem 20.2 *If* $0 \leq c \leq 2c_0$, *then the risk of* $\tilde{\mu}_x$ *is less than* $\bar{X}$ *under the loss function* $L_2(\hat{\mu}, \mu)$ *in* (20.7).

Theorem 20.3 *If* $0 \leq c \leq 2c_0$, *then the risk of* $\tilde{\mu}_x$ *is less than* $\bar{X}$ *under the loss function* $L_3(\hat{\mu}, \mu)$ *in* (20.8).

Remark 20.1 The intervals of c in the above three theorems are all the same $0 \leq c \leq \frac{2(n-r-p-1)(n-r-p-4)}{(n-r-2)(r+p+1)}$. In Theorem 20.1, it can be shown that $c = c_0 = \frac{(n-r-p-1)(n-r-p-4)}{(n-r-2)(r+p+1)}$ is the optimal selection and the degree of risk reduction is the largest and is given as follows

$$\begin{aligned}
&R_1(\bar{X}, \mu) - R_1(\tilde{\mu}_x, \mu) \\
&= \frac{r(n-r-p-1)(n-r-4)}{n(n-r-p-2)(n-r-2)(r+p+1)} \operatorname{tr}(\Sigma_x(\Sigma_x + \Sigma_y)^{-1}).
\end{aligned} \tag{20.16}$$

Although c_0 is not the optimal selection for c under loss functions $L_2(\hat{\mu}, \mu)$ and $L_3(\hat{\mu}, \mu)$, it is still a good selection in practice.

Next we present sufficient conditions for $\tilde{\mu}_x$ to dominate $\bar{Y}$ in terms of the risks under the three quadratic loss functions. Note that (20.5) can also be written as

$$\tilde{\mu}_x = \bar{Y} + (c\bar{S}_x\bar{S}^{-1} - I_p)(\bar{Y} - \bar{X}). \tag{20.17}$$

Unfortunately, there are no uniform dominance results, as indicated by our simulation in Sect. 20.3 where we can see that $\tilde{\mu}_x$ is always better than $\bar{X}$ but not always better than $\bar{Y}$. However, we find the following theorems with those conditions for $\tilde{\mu}_x$ to dominate $\bar{Y}$.

Theorem 20.4 *Let* $c_m = \frac{r(n-r-p-1)(n-r-p-4)}{(n-r-p-2)(n-r-2)(r+p+1)}$. *If* $c = c_0$ *and*

$$tr\,(T) \leq \left(\frac{1}{c_m} - \sqrt{\frac{1}{c_m} - c_m}\right) p,$$

then the risk of $\tilde{\mu}_x$ *is less than* $\bar{Y}$ *under the loss function* $L_1(\hat{\mu}, \mu)$ *in (20.6).*

Theorem 20.5 *Let* $c_t = \frac{(n-r-p-1)(n-r-p-4)}{2(r+1)(n-r-2)}$ *and* $c_s = \frac{r(n-r-p-1)(n-r-p-4)}{2(r+1)(n-r-2)(n-r-p-2)}$. *If* $c = c_t$ *and* $tr\,(\Sigma_x \Sigma_y^{-1}) \leq \frac{p}{\sqrt{1-c_s}}$, *then the risk of* $\tilde{\mu}_x$ *is less than* $\bar{Y}$ *under the loss function* $L_2(\hat{\mu}, \mu)$ *in (20.7).*

Theorem 20.6 *If* $c = c_0$ *and* $tr\,(T) \leq \frac{1}{1+\sqrt{1-c_m}} p$, *then the risk of* $\tilde{\mu}_x$ *is less than* $\bar{Y}$ *under the loss function* $L_3(\hat{\mu}, \mu)$ *in (20.8).*

Remark 20.2 Unlike the Graybill-Deal estimator (20.2), our estimator (20.5) is asymmetrical. It has only uniform dominance properties against $\bar{X}$, but needs extra conditions to dominate $\bar{Y}$. However, similar to (20.5), another unbiased estimator

$$\tilde{\mu}_y = \bar{Y} + c\bar{S}_y\bar{S}^{-1}(\bar{X} - \bar{Y}), \tag{20.18}$$

can be proposed and given similar dominance results against $\bar{Y}$ corresponding to Theorems 20.1–20.6.

Remark 20.3 Consider the unbiased estimator

$$\tilde{\mu}^* = c(\bar{S}_y\bar{S}^{-1}\bar{X} + \bar{S}_x\bar{S}^{-1}\bar{Y}). \tag{20.19}$$

It is unbiased when $c = \frac{n-r-p-2}{r}$. Unfortunately, it cannot be proven to uniformly dominate $\bar{X}$ or $\bar{Y}$ under the loss functions (20.6) through (20.8), as the risk functions are all dependent on μ.

20.2.2 LINEX Loss

A dominance result for $\tilde{\mu}_x$ under the LINEX loss function is given in the following theorem.

Theorem 20.7 *If* $0 \le c \le 2c_0$, *then the risk of* $\tilde{\mu}_x$ *is less than* $\bar{X}$ *under the loss function* $L_4(\hat{\mu}, \mu)$ *in* (20.9).

Proof Note that

$$\begin{aligned}
&R_4(\tilde{\mu}_x, \mu) - R_4(\bar{X}, \mu)\\
&= E[\exp\{a'(\hat{\mu}_x - \mu)\} - a'(\hat{\mu}_x - \mu) - 1] - E[\exp\{a'(\bar{X} - \mu)\} - a'(\bar{X} - \mu) - 1]\\
&= E[\exp\{a'(\hat{\mu}_x - \mu)\} - \exp\{a'(\bar{X} - \mu)\}]\\
&= E[\exp\{a'(\bar{X} - \mu + c\bar{S}_x\bar{S}^{-1}(\bar{Y} - \bar{X}))\} - \exp\{a'(\bar{X} - \mu)\}]\\
&= E[E[\exp\{a'(\bar{X} - \mu + c\bar{S}_x\bar{S}^{-1}(\bar{Y} - \bar{X}))\}|\bar{S}_x, \bar{S}]] - E[\exp\{a'(\bar{X} - \mu)\}].
\end{aligned} \tag{20.20}$$

After some manipulation, we have

$$a'(\hat{\mu}_x - \mu)|\bar{S}_x, \bar{S} \sim N(0, \sigma_1^2), a'(\bar{X} - \mu) \sim N(0, \sigma_2^2),$$

where $\sigma_1^2 = \frac{1}{n}a'\left[\Sigma_x - c\bar{S}_x\bar{S}^{-1}\Sigma_x - c\Sigma_x\bar{S}^{-1}\bar{S}_x + c^2\bar{S}_x\bar{S}^{-1}(\Sigma_x + \Sigma_y)\bar{S}^{-1}\bar{S}_x\right]a$ and $\sigma_2^2 = \frac{1}{n}a'\Sigma_x a$.

By using Lemma 20.3, it follows from (20.20) that

$$\begin{aligned}
R_4(\tilde{\mu}_x, \mu) - R_4(\bar{X}, \mu) &= E(\exp\{\frac{1}{2}\sigma_1^2\}) - \exp\{\frac{1}{2}\sigma_2^2\}\\
&= \exp\{\frac{1}{2}\sigma_2^2\}\left(E\left(\frac{\exp\{\frac{c^2}{n}a'\bar{S}_x\bar{S}^{-1}(\Sigma_x + \Sigma_y)\bar{S}^{-1}\bar{S}_x a\}}{\exp\{\frac{c}{n}a'(\bar{S}_x\bar{S}^{-1}\Sigma_x + \Sigma_x\bar{S}^{-1}\bar{S}_x)a\}}\right) - 1\right).
\end{aligned} \tag{20.21}$$

To get $R_4(\tilde{\mu}_x, \mu) - R_4(\bar{X}, \mu) \le 0$, we only need to prove

$$E\left(\frac{\exp\{\frac{c^2}{n}a'\bar{S}_x\bar{S}^{-1}(\Sigma_x + \Sigma_y)\bar{S}^{-1}\bar{S}_x a\}}{\exp\{\frac{c}{n}a'(\bar{S}_x\bar{S}^{-1}\Sigma_x + \Sigma_x\bar{S}^{-1}\bar{S}_x)a\}}\right) \le 1,$$

which is equivalent to

$$E\left(\frac{\exp\{\frac{c}{n}a'(\bar{S}_x\bar{S}^{-1}\Sigma_x + \Sigma_x\bar{S}^{-1}\bar{S}_x)a\}}{\exp\{\frac{c^2}{n}a'\bar{S}_x\bar{S}^{-1}(\Sigma_x + \Sigma_y)\bar{S}^{-1}\bar{S}_x a\}}\right) \ge 1. \tag{20.22}$$

As the exponential function is convex, a sufficient condition for (20.22) is given by using Lemma 20.4:

$$E\left(\frac{c}{n}a'(\bar{S}_x\bar{S}^{-1}\Sigma_x + \Sigma_x\bar{S}^{-1}\bar{S}_x)a - \frac{c^2}{n}a'\bar{S}_x\bar{S}^{-1}(\Sigma_x + \Sigma_y)\bar{S}^{-1}\bar{S}_x a\right) \geq 0. \tag{20.23}$$

Similar to the proof of Theorem 20.1, it can easily be shown that $R_4(\tilde{\mu}_x, \mu) - R_4(\bar{X}, \mu) \leq 0$ for any non-zero vector a when $0 \leq c \leq \frac{2(n-r-p-1)(n-r-p-4)}{(n-r-2)(r+p+1)}$.

The proof is completed. □

Remark 20.4 From Theorem 20.7, we can see that the dominance result does not depend on non-zero vector a and a good selection of c is also $c = c_0$.

Remark 20.5 Under the LINEX loss function, $\tilde{\mu}_x$ is still not uniformly better than $\bar{Y}$.

20.3 Numerical Comparison

In this section we simulate and use three settings of (finite) samples to numerically compare the performances of $\bar{X}$ (sample mean), $\bar{Y}$ (sample mean), $\tilde{\mu}_{\mathrm{MGD}}$ (Graybill-Deal estimator), $\tilde{\mu}_{\mathrm{K}}$ (unbiased estimator proposed by Krishnamoorthy [8] unbiased estimator), $\tilde{\mu}_{\mathrm{CC}}$ (unbiased estimator proposed by Chiou and Cohen [3] and $\tilde{\mu}_x$ (proposed in this paper).

For d in (20.4), we use $d = \frac{(n-p)(n-p-3)}{(n-1)(n+p+1)}$ given by Krishnamoorthy [8], where n is the sample size and p is the dimension of X. As [3] did not give a value of b in (20.5), we choose b to be $c_0 = \frac{(n-r-p-1)(n-r-p-4)}{(n-r-2)(r+p+1)}$ in our simulations. For $\tilde{\mu}_{\mathrm{CC}}$ in (20.3) and $\tilde{\mu}_x$ in (20.5), we choose r to be the maximum odd number smaller than $n/2$. For these six estimators, we report the risk values R_i $(i = 1, 2, 3, 4)$ in (20.10) in three tables; since the dominance result of $\tilde{\mu}_x$ does not depend on a under LINEX loss, a is chosen to be a unit vector for R_4. We use $n = 20$ and $p = 2$ in Table 20.1, $n = 30$ and $p = 5$ in Table 20.2, and $n = 60$ and $p = 20$ in Table 20.3.

For all these calculations, we first set $\Sigma_x = \Sigma_1$ and $\Sigma_y = \lambda\Sigma_2$, where λ is chosen to be 0.05, 0.1, 0.2, 0.5, 1, 2, 5, 10, 20 and 30. We then simulate 10,000 p-dimensional X-samples and Y-samples of size n using Σ_x and Σ_y, respectively, to calculate the risks. For the simulations in Tables 20.1, we consider a simple case with $\Sigma_1 = I_p$ and $\Sigma_2 = I_p$, where I_p is an $p \times p$ identity matrix. In Table 20.2, we set $\Sigma_1 = (\sigma_{1(ij)}) = (0.6^{|i-j|})$ and $\Sigma_2 = (\sigma_{2(ij)}) = ((-0.6)^{|i-j|})$. In Table 20.3, we use two $p \times p$ positive definite matrices Σ_1 and Σ_2 randomly generated for the computations.

Table 20.1 The risks of six estimators

$n=20, p=2$		R_1	R_2	R_3	R_4		R_1	R_2	R_3	R_4
$\lambda=0.05$	$\bar{X}$	0.1010	2.0207	0.0962	0.0525	$\lambda=2$	0.0983	0.0492	0.0328	0.0515
	$\bar{Y}$	0.0051	0.1012	0.0048	0.0025		0.1964	0.0982	0.0655	0.1035
	$\tilde{\mu}_{MGD}$	0.0049	0.0984	0.0047	0.0024		0.0706	0.0353	0.0235	0.0360
	$\tilde{\mu}_K$	0.0202	0.4044	0.0193	0.0101		0.0731	0.0365	0.0244	0.0374
	$\tilde{\mu}_{CC}$	0.0577	1.1548	0.0550	0.1479		0.0787	0.0393	0.0262	0.0393
	$\tilde{\mu}_x$	0.0544	1.0886	0.0518	0.0408		0.0784	0.0392	0.0261	0.0392
$\lambda=0.1$	$\bar{X}$	0.0984	0.9844	0.0895	0.0504	$\lambda=5$	0.1002	0.0200	0.0167	0.0506
	$\bar{Y}$	0.0101	0.1013	0.0092	0.0050		0.5051	0.1010	0.0842	0.2958
	$\tilde{\mu}_{MGD}$	0.0096	0.0960	0.0087	0.0047		0.0887	0.0177	0.0148	0.0454
	$\tilde{\mu}_K$	0.0240	0.2399	0.0218	0.0121		0.0873	0.0175	0.0146	0.0445
	$\tilde{\mu}_{CC}$	0.0547	0.5468	0.0497	0.0302		0.0731	0.0365	0.0244	0.0374
	$\tilde{\mu}_x$	0.0506	0.5065	0.0460	0.0273		0.0932	0.0186	0.0155	0.0473
$\lambda=0.2$	$\bar{X}$	0.0992	0.4958	0.0826	0.0503	$\lambda=10$	0.0992	0.0099	0.0090	0.0512
	$\bar{Y}$	0.0198	0.0992	0.0165	0.0101		1.0111	0.1011	0.0919	0.0664
	$\tilde{\mu}_{MGD}$	0.0173	0.0865	0.0144	0.0088		0.0945	0.0094	0.0086	0.0484
	$\tilde{\mu}_K$	0.0301	0.1503	0.0251	0.0152		0.0930	0.0093	0.0085	0.0477
	$\tilde{\mu}_{CC}$	0.0586	0.2931	0.0488	0.0302		0.0943	0.0094	0.0086	0.0489
	$\tilde{\mu}_x$	0.0560	0.2799	0.0466	0.0304		0.0943	0.0094	0.0086	0.0488
$\lambda=0.5$	$\bar{X}$	0.0999	0.1998	0.0666	0.0512	$\lambda=20$	0.1002	0.0050	0.0048	0.0515
	$\bar{Y}$	0.0502	0.1003	0.0334	0.0254		1.9991	0.1000	0.0952	1.6968
	$\tilde{\mu}_{MGD}$	0.0358	0.0717	0.0239	0.0181		0.0979	0.0049	0.0047	0.0505
	$\tilde{\mu}_K$	0.0457	0.0915	0.0305	0.0231		0.0969	0.0048	0.0046	0.0499
	$\tilde{\mu}_{CC}$	0.0629	0.1258	0.0419	0.0416		0.0973	0.0049	0.0046	0.0494
	$\tilde{\mu}_x$	0.0616	0.1233	0.0411	0.0440		0.0974	0.0049	0.0046	0.0494
$\lambda=1$	$\bar{X}$	0.0989	0.0989	0.0494	0.0503	$\lambda=30$	0.1015	0.0034	0.0033	0.0527
	$\bar{Y}$	0.1011	0.1011	0.0506	0.0512		2.9615	0.0987	0.0955	3.3075
	$\tilde{\mu}_{MGD}$	0.0541	0.0541	0.0271	0.0271		0.1002	0.0033	0.0032	0.0523
	$\tilde{\mu}_K$	0.0602	0.0602	0.0301	0.0300		0.0985	0.0033	0.0032	0.0505
	$\tilde{\mu}_{CC}$	0.0713	0.0713	0.0357	0.0374		0.0984	0.0033	0.0032	0.0505
	$\tilde{\mu}_x$	0.0700	0.0700	0.0350	0.0375		0.0983	0.0033	0.0032	0.0504

It can be seen from the above tables that $\tilde{\mu}_{\text{MGD}}$ is far better than $\tilde{\mu}_{\text{K}}$, $\tilde{\mu}_{\text{CC}}$ and $\tilde{\mu}_x$ when λ is small. However, the gaps between $\tilde{\mu}_{\text{MGD}}$ and $\tilde{\mu}_{\text{K}}$, $\tilde{\mu}_{\text{CC}}$ or $\tilde{\mu}_x$ become very small when λ is big. Although $\tilde{\mu}_{\text{MGD}}$ has no known uniform dominance results against $\bar{X}$ or $\bar{Y}$, it is the most robust and performs the best overall. After we swap the positions of $\bar{S}_x$ and $\bar{S}^{-1}$, the performance of $\tilde{\mu}_x$ becomes significantly better than $\tilde{\mu}_{\text{CC}}$. Unfortunately, when λ is small, the performances of $\tilde{\mu}_{\text{CC}}$ and $\tilde{\mu}_x$ are not as good as $\tilde{\mu}_{\text{K}}$. We think that this is the price we have to pay for $\bar{S}_x$ and $\bar{S}^{-1}$ to be independent. When λ is big, all the estimators perform almost the same.

Table 20.2 The risks of six estimators

$n = 30, p = 5$		R_1	R_2	R_3	R_4		R_1	R_2	R_3	R_4
$\lambda = 0.05$	$\bar{X}$	0.1678	8.5381	0.1488	0.0338	$\lambda = 2$	0.1656	0.2331	0.0650	0.0336
	$\bar{Y}$	0.0256	0.1670	0.0188	0.0006		0.9402	0.1673	0.1018	0.0230
	$\tilde{\mu}_{MGD}$	0.0205	0.1561	0.0153	0.0006		0.1110	0.0729	0.0286	0.0086
	$\tilde{\mu}_K$	0.0545	2.1106	0.0462	0.0082		0.1194	0.1075	0.0359	0.0141
	$\tilde{\mu}_{CC}$	0.3564	5.8702	0.2866	0.0181		0.1531	0.1431	0.0478	0.0188
	$\tilde{\mu}_x$	0.0908	4.3684	0.0798	0.0175		0.1271	0.1446	0.0437	0.0199
$\lambda = 0.1$	$\bar{X}$	0.1653	4.5366	0.1369	0.0330	$\lambda = 5$	0.1658	0.0931	0.0427	0.0338
	$\bar{Y}$	0.0465	0.1661	0.0285	0.0011		2.3208	0.1659	0.1231	0.0580
	$\tilde{\mu}_{MGD}$	0.0316	0.1467	0.0203	0.0009		0.1335	0.0473	0.0248	0.0151
	$\tilde{\mu}_K$	0.0619	1.1782	0.0471	0.0084		0.1359	0.0554	0.0277	0.0186
	$\tilde{\mu}_{CC}$	0.2739	2.9410	0.2005	0.0168		0.1535	0.0657	0.0325	0.0224
	$\tilde{\mu}_x$	0.0927	2.3253	0.0748	0.0171		0.1398	0.0663	0.0317	0.0232
$\lambda = 0.2$	$\bar{X}$	0.1656	2.3377	0.1231	0.0343	$\lambda = 10$	0.1675	0.0469	0.0287	0.0341
	$\bar{Y}$	0.0933	0.1671	0.0430	0.0022		4.7380	0.1670	0.1387	0.1239
	$\tilde{\mu}_{MGD}$	0.0472	0.1335	0.0248	0.0016		0.1490	0.0316	0.0204	0.0207
	$\tilde{\mu}_K$	0.0733	0.6513	0.0472	0.0092		0.1479	0.0333	0.0212	0.0225
	$\tilde{\mu}_{CC}$	0.2424	1.4826	0.1540	0.0186		0.1572	0.0367	0.0232	0.0253
	$\tilde{\mu}_x$	0.1011	1.2594	0.0718	0.0201		0.1505	0.0377	0.0235	0.0264
$\lambda = 0.5$	$\bar{X}$	0.1678	0.9428	0.1022	0.0342	$\lambda = 20$	0.1672	0.0232	0.0174	0.0338
	$\bar{Y}$	0.2305	0.1658	0.0647	0.0057		9.3910	0.1672	0.1496	0.2566
	$\tilde{\mu}_{MGD}$	0.0725	0.1113	0.0284	0.0033		0.1578	0.0191	0.0145	0.0267
	$\tilde{\mu}_K$	0.0921	0.3032	0.0446	0.0104		0.1554	0.0190	0.0144	0.0267
	$\tilde{\mu}_{CC}$	0.1913	0.5600	0.0941	0.0176		0.1590	0.0198	0.0150	0.0281
	$\tilde{\mu}_x$	0.1100	0.5099	0.0612	0.0181		0.1558	0.0200	0.0151	0.0284
$\lambda = 1$	$\bar{X}$	0.1662	0.4671	0.0834	0.0335	$\lambda = 30$	0.1676	0.0156	0.0127	0.0339
	$\bar{Y}$	0.4509	0.1649	0.0816	0.0112		13.7716	0.1661	0.1534	0.3925
	$\tilde{\mu}_{MGD}$	0.0917	0.0928	0.0292	0.0055		0.1620	0.0137	0.0113	0.0289
	$\tilde{\mu}_K$	0.1055	0.1758	0.0406	0.0117		0.1592	0.0135	0.0111	0.0285
	$\tilde{\mu}_{CC}$	0.1727	0.2806	0.0682	0.0178		0.1625	0.0139	0.0114	0.0297
	$\tilde{\mu}_x$	0.1191	0.2747	0.0536	0.0203		0.1596	0.0140	0.0115	0.0299

20.4 Concluding Remarks

In this paper we have proposed a class of estimators $\tilde{\mu}_x$ and studied its finite sample properties. Numerically its performance is almost as good as $\tilde{\mu}_{\mathrm{CC}}$ in terms of the risks. Theoretically its dominance over the sample mean of a single population under the sufficient conditions given is established. There are no such results for $\tilde{\mu}_{\mathrm{CC}}$ and $\tilde{\mu}_{\mathrm{MGD}}$. We have designed the structure of $\tilde{\mu}_x$ in order to make these results possible.

Table 20.3 The risks of six estimators

$n=60,\ p=20$		R_1	R_2	R_3	R_4		R_1	R_2	R_3	R_4
$\lambda=0.05$	$\bar{X}$	0.3324	4.5597	0.2890	3.4237×10^{-5}	$\lambda=2$	0.3347	0.1261	0.0782	4.8280×10^{-5}
	$\bar{Y}$	0.0557	0.3353	0.0437	7.5897×10^{-6}		2.2609	0.3338	0.2562	6.6540×10^{-4}
	$\tilde{\mu}_{MGD}$	0.0522	0.3376	0.0415	6.9063×10^{-6}		0.3049	0.0950	0.0624	4.7911×10^{-5}
	$\tilde{\mu}_K$	0.1825	2.3235	0.1568	1.9460×10^{-5}		0.2957	0.1000	0.0640	4.5458×10^{-5}
	$\tilde{\mu}_{CC}$	0.5607	5.3658	0.4585	5.5809×10^{-5}		0.3430	0.1187	0.0752	5.2428×10^{-5}
	$\tilde{\mu}_x$	0.2455	3.3074	0.2129	2.5781×10^{-5}		0.3085	0.1124	0.0704	4.5280×10^{-5}
$\lambda=0.1$	$\bar{X}$	0.3327	2.6199	0.2663	4.7788×10^{-5}	$\lambda=5$	0.3327	0.0593	0.0448	3.5784×10^{-5}
	$\bar{Y}$	0.0989	0.3333	0.0667	3.0016×10^{-5}		4.5935	0.3328	0.2880	7.5296×10^{-4}
	$\tilde{\mu}_{MGD}$	0.0806	0.3148	0.0566	2.0335×10^{-5}		0.3349	0.0543	0.0419	3.7759×10^{-5}
	$\tilde{\mu}_K$	0.1935	1.3780	0.1513	3.1803×10^{-5}		0.3108	0.0516	0.0396	3.4757×10^{-5}
	$\tilde{\mu}_{CC}$	0.4917	2.7789	0.3642	7.0173×10^{-5}		0.3393	0.0573	0.0437	3.6872×10^{-5}
	$\tilde{\mu}_x$	0.2525	1.9406	0.2010	3.7529×10^{-5}		0.3175	0.0552	0.0420	3.4599×10^{-5}
$\lambda=0.2$	$\bar{X}$	0.3337	1.2320	0.2244	4.0044×10^{-5}	$\lambda=10$	0.3328	0.0243	0.0215	2.9266×10^{-5}
	$\bar{Y}$	0.2197	0.3331	0.1092	4.7863×10^{-5}		10.0238	0.3319	0.3106	8.3571×10^{-4}
	$\tilde{\mu}_{MGD}$	0.1339	0.2708	0.0753	2.3775×10^{-5}		0.3477	0.0249	0.0220	3.0776×10^{-5}
	$\tilde{\mu}_K$	0.2169	0.6941	0.1390	2.9643×10^{-5}		0.3228	0.0228	0.0202	2.8458×10^{-5}
	$\tilde{\mu}_{CC}$	0.4706	1.2532	0.2747	6.1886×10^{-5}		0.3345	0.0239	0.02114	2.9177×10^{-5}
	$\tilde{\mu}_x$	0.2644	0.9434	0.1758	3.3023×10^{-5}		0.3257	0.0235	0.0208	2.8647×10^{-5}

$\lambda = 0.5$	$\bar{X}$	0.3337	0.5472	0.1680	3.2773×10^{-5}	$\lambda = 20$	0.3334	0.0139	0.0128	3.6711×10^{-5}
	$\bar{Y}$	0.5317	0.3317	0.1646	7.5092×10^{-5}		19.6976	0.3335	0.3208	2.6335×10^{-3}
	$\tilde{\mu}_{MGD}$	0.1992	0.2022	0.0810	2.4215×10^{-5}		0.3467	0.0144	0.0133	3.8095×10^{-5}
	$\tilde{\mu}_K$	0.2461	0.3420	0.1143	2.6376×10^{-5}		0.3273	0.0133	0.0123	3.6126×10^{-5}
	$\tilde{\mu}_{CC}$	0.3731	0.5077	0.1692	3.3891×10^{-5}		0.3344	0.0137	0.0127	3.6875×10^{-5}
	$\tilde{\mu}_x$	0.2816	0.4441	0.1389	2.8297×10^{-5}		0.3291	0.0136	0.0125	3.6257×10^{-5}
$\lambda = 1$	$\bar{X}$	0.3336	0.2482	0.1193	4.5049×10^{-5}	$\lambda = 30$	0.3329	0.0088	0.0084	3.8374×10^{-5}
	$\bar{Y}$	1.0742	0.3306	0.2118	2.7035×10^{-4}		31.4259	0.3343	0.3259	5.9803×10^{-3}
	$\tilde{\mu}_{MGD}$	0.2572	0.1456	0.0783	4.0146×10^{-5}		0.3438	0.0091	0.0087	3.9006×10^{-5}
	$\tilde{\mu}_K$	0.2729	0.1791	0.0904	3.9586×10^{-5}		0.3288	0.0086	0.0082	3.8047×10^{-5}
	$\tilde{\mu}_{CC}$	0.3679	0.2361	0.1191	6.8904×10^{-5}		0.3346	0.0087	0.0083	3.9510×10^{-5}
	$\tilde{\mu}_x$	0.2947	0.2112	0.1030	4.0906×10^{-5}		0.3299	0.0087	0.0082	3.8155×10^{-5}

Acknowledgments The research of S. Ejaz Ahmed is supported by the Natural Sciences and the Engineering Research Council of Canada (NSERC). The work by Tiefeng Ma is supported by the Foundation of Zhejiang Provincial Natural Science Foundation of China (No. Y6100053).

Appendix

Proof of Lemma 20.2

Proof We prove it by mathematical induction principle. Here we deal with the open statement

$$S(n) : n\mathrm{tr}(AB) \geq \mathrm{tr}(A)\mathrm{tr}(B),$$

that is

$$S(n) : n\sum_{i=1}^{n} a_i b_i \geq \sum_{i=1}^{n} a_i \sum_{i=1}^{n} b_i.$$

Basic Step It is easy to get that $S(1)$ is true.

Inductive Step Now we assume the validity of $S(k)$ for some $k \in \mathbb{N}^+$. That is we assume that

$$k\sum_{i=1}^{k} a_i b_i \geq \sum_{i=1}^{k} a_i \sum_{i=1}^{k} b_i, \tag{20.24}$$

holds (n is replaced by k). Based on this assumption, next we need to verify the truth of

$$S(k+1) : (k+1)\sum_{i=1}^{k+1} a_i b_i \geq \sum_{i=1}^{k+1} a_i \sum_{i=1}^{k+1} b_i. \tag{20.25}$$

For the left of inequality (20.25), we have

$$\text{Left} = (k+1)\sum_{i=1}^{k+1} a_i b_i = (k+1)\left(\sum_{i=1}^{k} a_i b_i + a_{n+1}b_{n+1}\right).$$

According to inequality (20.24), the formula above have

$$\text{Left} \geq \frac{k+1}{k}\sum_{i=1}^{k} a_i \sum_{i=1}^{k} b_i + (k+1)a_{k+1}b_{k+1}. \tag{20.26}$$

For the right of inequality (20.25), we have

$$\begin{aligned}
\text{Right} &= \sum_{i=1}^{k+1} a_i \sum_{i=1}^{k+1} b_i = \left(\sum_{i=1}^{k} a_i + a_{k+1}\right)\left(\sum_{i=1}^{k} b_i + b_{k+1}\right)\\
&= \sum_{i=1}^{k} a_i \sum_{i=1}^{k} b_i + \sum_{i=1}^{k} a_i b_{k+1} + \sum_{i=1}^{k} b_i a_{k+1} + a_{k+1} b_{k+1}.
\end{aligned}$$

Since $a_1 \geq a_2 \geq \cdots \geq a_n$, $b_1 \geq b_2 \geq \cdots \geq b_n$, it can easily be seen that $(\frac{1}{k}\sum_{i=1}^{k} a_i - a_{n+1})(\frac{1}{k}\sum_{i=1}^{k} b_i - b_{n+1}) \geq 0$, and then it obviously has Left $\geq$ Right. The proof is completed. □

Proof of Theorem 20.2

Proof Consider the difference between $R_2(\bar{X}, \mu)$ and $R_2(\tilde{\mu}_x, \mu)$:

$$\begin{aligned}
&R_2(\bar{X}, \mu) - R_2(\tilde{\mu}_x, \mu)\\
&\quad = -2cE[\bar{X}'\Sigma_y^{-1}\bar{S}_x\bar{S}^{-1}(\bar{Y} - \bar{X})] - c^2E[(\bar{Y} - \bar{X})'\bar{S}^{-1}\bar{S}_x\Sigma_y^{-1}\bar{S}_x\bar{S}^{-1}(\bar{Y} - \bar{X})]\\
&\quad = 2c \operatorname{tr}[E(\Sigma_y^{-1}\bar{S}_x\bar{S}^{-1}(\bar{Y} - \bar{X})\bar{X}')]\\
&\qquad - c^2 \operatorname{tr}[E(\bar{S}^{-1}\bar{S}_x\Sigma_y^{-1}\bar{S}_x\bar{S}^{-1}(\bar{Y} - \bar{X})(\bar{Y} - \bar{X})')].
\end{aligned} \tag{20.27}$$

Since $\bar{X}$, $\bar{Y}$, $\bar{S}_x$ and $\bar{S}$ are mutually independent, it follows from (20.27) that

$$\begin{aligned}
R_2(\bar{X}, \mu) - R_2(\tilde{\mu}_x, \mu) &= \frac{2c}{n} \operatorname{tr}[\Sigma_y^{-1}E(\bar{S}_x)E(\bar{S}^{-1})\Sigma_x]\\
&\quad - \frac{c^2}{n} \operatorname{tr}[E(\bar{S}_x\Sigma_y^{-1}\bar{S}_x)E(\bar{S}^{-1}(\Sigma_x + \Sigma_y)\bar{S}^{-1})].
\end{aligned} \tag{20.28}$$

Using Lemma 20.1, after some manipulation, we get

$$\begin{aligned}
&R_2(\bar{X}, \mu) - R_2(\tilde{\mu}_x, \mu)\\
&= \frac{2cr}{n(n-r-p-2)} \operatorname{tr}(\Sigma_y^{-1}\Sigma_x(\Sigma_x + \Sigma_y)^{-1}\Sigma_x)\\
&\quad - \frac{c^2(n-r-2)}{n(n-r-p-1)(n-r-p-2)(n-r-p-4)}\\
&\quad \times \operatorname{tr}\{[(r-1)\operatorname{tr}(\Sigma_x\Sigma_y^{-1})\Sigma_x + r(r-1)\Sigma_x(\Sigma_x + \Sigma_y)^{-1}\Sigma_x](\Sigma_x + \Sigma_y)^{-1}\}\\
&= \left[\frac{2cr}{n(n-r-p-2)} - \frac{c^2(n-r-2)r(r+1)}{n(n-r-p-1)(n-r-p-2)(n-r-p-4)}\right]
\end{aligned}$$

$$
\begin{aligned}
&\times\ \mathrm{tr}\,(\Sigma_y^{-1}\Sigma_x(\Sigma_x+\Sigma_y)^{-1}\Sigma_x)\\
&-\frac{c^2(n-r-2)r}{n(n-r-p-1)(n-r-p-2)(n-r-p-4)}\\
&\times\ \mathrm{tr}\,(\Sigma_x\Sigma_y^{-1})\ \mathrm{tr}\,(\Sigma_x(\Sigma_x+\Sigma_y)^{-1})\\
=&\left[\frac{2cr}{n(n-r-p-2)}-\frac{c^2(n-r-2)r(r+1)}{n(n-r-p-1)(n-r-p-2)(n-r-p-4)}\right]\\
&\times\ \mathrm{tr}\,(W^{-1}(I_p+W)^{-1})\\
&-\frac{c^2(n-r-2)r}{n(n-r-p-1)(n-r-p-2)(n-r-p-4)}\ \mathrm{tr}\,(W^{-1})\ \mathrm{tr}\,((I_p+W)^{-1}),
\end{aligned}
\tag{20.29}
$$

where $W=\Sigma_x^{-1}\Sigma_y$. Since

$$
\begin{aligned}
\mathrm{tr}\,(W^{-1})\ \mathrm{tr}\,((I_p+W)^{-1}) =&\ \mathrm{tr}\,(A_2^{-1}A_1^{-1})\ \mathrm{tr}\,(I_p+A_1A_2)^{-1}\\
=&\ \mathrm{tr}\,(A_2^{-1/2}A_1^{-1}A_2^{-1/2})\ \mathrm{tr}\,(A_2^{-1/2}A_2^{1/2}(I_p+A_1A_2)^{-1})\\
=&\ \mathrm{tr}\,(D^{-1})\ \mathrm{tr}\,(A_2^{1/2}(I_p+A_1A_2)^{-1}A_2^{-1/2})\\
=&\ \mathrm{tr}\,(D^{-1})\ \mathrm{tr}\,((I_p+D)^{-1}),
\end{aligned}
$$

where $A_1=\Sigma_x^{-1}$, $A_2=\Sigma_y$, $D=A_2^{1/2}A_1A_2^{1/2}$, Furthermore, we have

$$
\begin{aligned}
\mathrm{tr}\,(W^{-1}(I_p+W)^{-1}) =&\ \mathrm{tr}\,(A_2^{-1}A_1^{-1}(I_p+A_1A_2)^{-1})\\
=&\ \mathrm{tr}\,(A_2^{-1/2}A_1^{-1}A_2^{-1/2}A_2^{1/2}(I_p+A_1A_2)^{-1}A_2^{-1/2})\\
=&\ \mathrm{tr}\,((A_2^{1/2}A_1A_2^{1/2})^{-1}(I_p+A_2^{1/2}A_1A_2^{1/2})^{-1})\\
=&\ \mathrm{tr}\,(D^{-1}(I_p+D)^{-1})
\end{aligned}
$$

Obviously D is a positive definite matrix. By the spectral decomposition, we have $D=U\Lambda U'$, so Λ is a diagonal matrix. Now it is easy to verify that

$$
\begin{aligned}
\mathrm{tr}\,(D^{-1}) =&\ \mathrm{tr}\,(\Lambda^{-1}),\\
\mathrm{tr}\,((I_p+D)^{-1}) =&\ \mathrm{tr}\,((I_p+\Lambda)^{-1}),\\
\mathrm{tr}\,(D^{-1}(I_p+D)^{-1}) =&\ \mathrm{tr}\,(U\Lambda^{-1}U'(I_p+U\Lambda U')^{-1}) =\ \mathrm{tr}\,(\Lambda^{-1}(I_p+\Lambda)^{-1})
\end{aligned}
$$

and by Lemma 20.2, we can obtain that $p\ \mathrm{tr}\,(W^{-1}(I_p+W)^{-1}) \geq\ \mathrm{tr}\,(W^{-1})\ \mathrm{tr}\,((I_p+W)^{-1})$.

It follows from (20.29) that

$$R_2(\bar{X}, \mu) - R_2(\tilde{\mu}_x, \mu) \geq \text{ tr }(W^{-1}(I_p + W)^{-1}) \times \left[\frac{2cr}{n(n-r-p-2)} - \frac{c^2(n-r-2)(r+p+1)r}{n(n-r-p-1)(n-r-p-2)(n-r-p-4)} \right]. \tag{20.30}$$

Then we have that $R_2(\bar{X}, \mu) - R_2(\tilde{\mu}_x, \mu) \geq 0$ for $0 \leq c \leq \frac{2(n-r-p-1)(n-r-p-4)}{(n-r-2)(r+p+1)}$. This completes the proof. □

Proof of Theorem 20.3

Proof Consider the difference between $R_3(\bar{X}, \mu)$ and $R_3(\tilde{\mu}_x, \mu)$:

$$\begin{aligned} &R_3(\bar{X}, \mu) - R_3(\tilde{\mu}_x, \mu) \\ &= -2cE[\bar{X}'(\Sigma_x + \Sigma_y)^{-1}\bar{S}_x\bar{S}^{-1}(\bar{Y} - \bar{X})] \\ &\quad - c^2E[(\bar{Y} - \bar{X})'\bar{S}^{-1}\bar{S}_x(\Sigma_x + \Sigma_y)^{-1}\bar{S}_x\bar{S}^{-1}(\bar{Y} - \bar{X})] \\ &- -2c \text{ tr }[E((\Sigma_x + \Sigma_y)^{-1}\bar{S}_x\bar{S}^{-1}(\bar{Y} - \bar{X})\bar{X}')] \\ &\quad - c^2 \text{ tr }[E(\bar{S}^{-1}\bar{S}_x(\Sigma_x + \Sigma_y)^{-1}\bar{S}_x\bar{S}^{-1}(\bar{Y} - \bar{X})(\bar{Y} - \bar{X})')]. \end{aligned} \tag{20.31}$$

As $\bar{X}$, $\bar{Y}$, $\bar{S}_x$ and $\bar{S}$ are mutually independent, it follows from (20.31) that

$$\begin{aligned} &R_3(\bar{X}, \mu) - R_3(\tilde{\mu}_x, \mu) \\ &= \frac{2c}{n} \text{ tr }[(\Sigma_x + \Sigma_y)^{-1}E(\bar{S}_x)E(\bar{S}^{-1})\Sigma_x] \\ &\quad - \frac{c^2}{n} \text{ tr }[E(\bar{S}_x(\Sigma_x + \Sigma_y)^{-1}\bar{S}_x)E(\bar{S}^{-1}(\Sigma_x + \Sigma_y)\bar{S}^{-1})]. \end{aligned} \tag{20.32}$$

Using Lemma 20.1, after some manipulation, we get

$$\begin{aligned} R_3(\bar{X}, \mu) - R_3(\tilde{\mu}_x, \mu) &= \frac{2cr}{n(n-r-p-2)} \text{ tr }((\Sigma_x + \Sigma_y)^{-1}\Sigma_x(\Sigma_x + \Sigma_y)^{-1}\Sigma_x) \\ &\quad - \frac{c^2(n-r-2)}{n(n-r-p-1)(n-r-p-2)(n-r-p-4)} \\ &\quad \times \text{ tr }[(r \text{ tr }(\Sigma_x(\Sigma_x + \Sigma_y)^{-1})\Sigma_x + r(r+1)\Sigma_x(\Sigma_x + \Sigma_y)^{-1}\Sigma_x)(\Sigma_x + \Sigma_y)^{-1}] \\ &= \left[\frac{2cr}{n(n-r-p-2)} - \frac{c^2(n-r-2)r(r+1)}{n(n-r-p-1)(n-r-p-2)(n-r-p-4)} \right] \\ &\quad \times \text{ tr }((\Sigma_x + \Sigma_y)^{-1}\Sigma_x(\Sigma_x + \Sigma_y)^{-1}\Sigma_x) \end{aligned}$$

$$
\begin{aligned}
&- \frac{c^2(n-r-2)r}{n(n-r-p-1)(n-r-p-2)(n-r-p-4)} \\
&\times \operatorname{tr}(\Sigma_x(\Sigma_x+\Sigma_y)^{-1}) \operatorname{tr}(\Sigma_x(\Sigma_x+\Sigma_y)^{-1}) \\
&= \left[\frac{2cr}{n(n-r-p-2)} - \frac{c^2(n-r-2)r(r+1)}{n(n-r-p-1)(n-r-p-2)(n-r-p-4)}\right] \operatorname{tr}(T^2) \\
&- \frac{c^2(n-r-2)r}{n(n-r-p-1)(n-r-p-2)(n-r-p-4)} \operatorname{tr}^2(T) \\
&\geq \left[\frac{2cr}{n(n-r-p-2)} - \frac{c^2(n-r-2)r(p+r+1)}{n(n-r-p-1)(n-r-p-2)(n-r-p-4)}\right] \operatorname{tr}(T^2),
\end{aligned} \tag{20.33}
$$

where the inequality holds because of Lemma 20.2 and spectral decomposition (similar proof as in the Proof of Theorem 20.2).

It can be seen that $R_3(\bar{X}, \mu) - R_3(\tilde{\mu}_x, \mu) \geq 0$ for $0 \leq c \leq \frac{2(n-r-p-1)(n-r-p-4)}{(n-r-2)(p+r+1)}$. This completes the proof. □

Proof of Theorem 20.4

Proof The difference between $R_1(\tilde{\mu}_x, \mu)$ and $R_1(\bar{Y}, \mu)$ is given by

$$
\begin{aligned}
&R_1(\tilde{\mu}_x, \mu) - R_1(\bar{Y}, \mu) \\
&\quad = 2 \operatorname{tr}[E(\bar{Y}'\Sigma_x^{-1}(c\bar{S}_x\bar{S}^{-1} - I_p)(\bar{Y}-\bar{X}))] \\
&\quad + \operatorname{tr}[E((\bar{Y}-\bar{X})'(c\bar{S}^{-1}\bar{S}_x - I_p)\Sigma_x^{-1}(c\bar{S}_x\bar{S}^{-1} - I_p)(\bar{Y}-\bar{X}))] \\
&\quad = 2 \operatorname{tr}[\Sigma_x^{-1}(cE(\bar{S}_x)E(\bar{S}^{-1}) - I_p)E((\bar{Y}-\bar{X})\bar{Y}')] \\
&\quad + \operatorname{tr}[E((c\bar{S}^{-1}\bar{S}_x - I_p)\Sigma_x^{-1}(c\bar{S}_x\bar{S}^{-1} - I_p))E((\bar{Y}-\bar{X})(\bar{Y}-\bar{X})')] \\
&\quad = 2 \operatorname{tr}[\Sigma_x^{-1}(\frac{cr}{n-r-p-2}\Sigma_x(\Sigma_x+\Sigma_y)^{-1} - I_p)\Sigma_y/n] \\
&\quad + \operatorname{tr}\{[c^2E(\bar{S}^{-1}\bar{S}_x\Sigma_x^{-1}\bar{S}_x\bar{S}^{-1}) - 2cE(\bar{S}^{-1}\bar{S}_x)\Sigma_x^{-1} + \Sigma_x^{-1}](\Sigma_x+\Sigma_y)/n\}.
\end{aligned} \tag{20.34}
$$

Using Lemma 20.1, we get

$$
\begin{aligned}
&R_1(\tilde{\mu}_x, \mu) - R_1(\bar{Y}, \mu) \\
&\quad = \frac{2cr}{n(n-r-p-2)} \operatorname{tr}(\Sigma_y(\Sigma_x+\Sigma_y)^{-1}) - \frac{2}{n}\operatorname{tr}(\Sigma_y\Sigma_x^{-1}) \\
&\quad - \frac{2cpr}{n(n-r-p-2)} + \frac{1}{n}\operatorname{tr}(\Sigma_x^{-1}(\Sigma_x+\Sigma_y))
\end{aligned}
$$

$$+\frac{c^2r(n-r-2)(r+p+1)}{n(n-r-p-1)(n-r-p-2)(n-r-p-4)}\text{ tr }(\Sigma_x(\Sigma_x+\Sigma_y)^{-1})$$

$$=\left[\frac{c^2r(n-r-2)(r+p+1)}{n(n-r-p-1)(n-r-p-2)(n-r-p-4)}-\frac{2cr}{n(n-r-p-2)}\right]$$

$$\times\text{ tr }(\Sigma_x(\Sigma_x+\Sigma_y)^{-1})+\frac{1}{n}\text{ tr }(\Sigma_x^{-1}(\Sigma_x-\Sigma_y)). \qquad (20.35)$$

Obviously, the best selection of c is c_0. Substituting it in (20.35), we obtain

$$R_1(\tilde{\mu}_x,\mu)-R_1(\bar{Y},\mu)=\frac{1}{n}[p-\text{ tr }(\Sigma_x^{-1}\Sigma_y)]$$
$$-\frac{r(n-r-p-1)(n-r-p-4)}{n(n-r-p-2)(n-r-2)(r+p+1)}\text{ tr }(\Sigma_x(\Sigma_x+\Sigma_y)^{-1}). \qquad (20.36)$$

By Cauchy-Schwartz Inequality [9], we get

$$\text{tr }(\Sigma_x(\Sigma_x+\Sigma_y)^{-1})\geq\frac{p^2}{p+\text{ tr }(\Sigma_x^{-1}\Sigma_y)}. \qquad (20.37)$$

Using $c_m=\frac{r(n-r-p-1)(n-r-p-4)}{(n-r-p-2)(n-r-2)(r+p+1)}$, we get from (20.36) and (20.37) that

$$R_1(\tilde{\mu}_x,\mu)-R_1(\bar{Y},\mu)$$
$$\leq\frac{1}{n}\left[2p-\frac{p^2}{\text{tr }(\Sigma_x(\Sigma_x+\Sigma_y)^{-1})}-c_m\text{ tr }(\Sigma_x(\Sigma_x+\Sigma_y)^{-1})\right]. \qquad (20.38)$$

Then it can be seen that $R_1(\tilde{\mu}_x,\mu)-R_1(\bar{Y},\mu)\leq 0$ when $c_m\text{ tr }^2(T)-2p\text{ tr }(T)+p^2\geq 0$. A sufficient condition is found to be $\text{tr }(T)\leq\left(\frac{1}{c_m}-\sqrt{\frac{1}{c_m}-c_m}\right)p$. This completes the proof. □

Proof of Theorem 20.5
Proof Note that

$$R_2(\tilde{\mu}_{Ax},\mu)-R_2(\bar{Y},\mu)$$
$$=2\text{ tr }[\Sigma_y^{-1}(c(r-1)\Sigma_x(\Sigma_x+\Sigma_y)^{-1}/(n-r-p-2)-I_p)\Sigma_y/n]$$
$$+\text{ tr }\{[c^2E(\bar{S}^{-1}\bar{S}_x\Sigma_y^{-1}\bar{S}_x\bar{S}^{-1})-2cE(\bar{S}^{-1}\bar{S}_x)\Sigma_y^{-1}+\Sigma_y^{-1}](\Sigma_x+\Sigma_y)/n\}$$
$$=\frac{2}{n}\text{ tr }[c(r-1)\Sigma_x(\Sigma_x+\Sigma_y)^{-1}/(n-r-p-2)-I_p]$$

$$+\frac{c^2}{n}\operatorname{tr}[E(\bar{S}_x\Sigma_y^{-1}\bar{S}_x\bar{S}^{-1}(\Sigma_x+\Sigma_y)\bar{S}^{-1})]$$
$$-\frac{2c}{n}\operatorname{tr}[E(\bar{S}^{-1}\bar{S}_x)\Sigma_y^{-1}(\Sigma_x+\Sigma_y)]+\frac{1}{n}\operatorname{tr}[\Sigma_y^{-1}(\Sigma_x+\Sigma_y)]. \tag{20.39}$$

Using Lemma 20.1, we get

$$\begin{aligned}
&R_2(\tilde{\mu}_{Ax},\mu)-R_2(\bar{Y},\mu)\\
&=\frac{2cr}{n(n-r-p-2)}\operatorname{tr}(\Sigma_x(\Sigma_x+\Sigma_y)^{-1})-\frac{2p}{n}\\
&\quad-\frac{2cr}{n(n-r-p-2)}\operatorname{tr}(\Sigma_x\Sigma_y^{-1})+\frac{1}{n}\operatorname{tr}[\Sigma_y^{-1}(\Sigma_x+\Sigma_y)]\\
&\quad+\frac{c^2(n-r-2)}{n(n-r-p-1)(n-r-p-2)(n-r-p-4)}\operatorname{tr}[E(\bar{S}_x\Sigma_y^{-1}\bar{S}_x(\Sigma_x+\Sigma_y)^{-1})].
\end{aligned} \tag{20.40}$$

Using Lemma 20.1 again, we get

$$\begin{aligned}
&R_2(\tilde{\mu}_{Ax},\mu)-R_2(\bar{Y},\mu)\\
&=\frac{2cr}{n(n-r-p-2)}\operatorname{tr}(T)-\frac{2p}{n}-\frac{2cr}{n(n-r-p-2)}\operatorname{tr}(Q)+\frac{1}{n}(\operatorname{tr}(Q)+p)\\
&\quad+\frac{c^2(n-r-2)}{n(n-r-p-1)(n-r-p-2)(n-r-p-4)}[r\operatorname{tr}(Q)\operatorname{tr}(T)+r(r+1)\operatorname{tr}(TQ)]\\
&=\frac{2cr}{n(n-r-p-2)}\operatorname{tr}(T)-\frac{p}{n}-\frac{2cr}{n(n-r-p-2)}\operatorname{tr}(Q)+\frac{1}{n}\operatorname{tr}(Q)\\
&\quad+\frac{c^2(n-r-2)}{n(n-r-p-1)(n-r-p-2)(n-r-p-4)}\\
&\quad\times[r\operatorname{tr}(Q)\operatorname{tr}(T)+r(r+1)(\operatorname{tr}(Q)-\operatorname{tr}(T))]\\
&=\left[\frac{2cr}{n(n-r-p-2)}-\frac{c^2(n-r-2)r(r+1)}{n(n-r-p-1)(n-r-p-2)(n-r-p-4)}\right]\operatorname{tr}(T)-\frac{p}{n}\\
&\quad+\left[\frac{c^2(n-r-2)r(r+1)}{n(n-r-p-1)(n-r-p-2)(n-r-p-4)}-\frac{2cr}{n(n-r-p-2)}+\frac{1}{n}\right]\operatorname{tr}(Q)\\
&\quad+\frac{c^2(n-r-2)r}{n(n-r-p-1)(n-r-p-2)(n-r-p-4)}\operatorname{tr}(Q)\operatorname{tr}(T),
\end{aligned} \tag{20.41}$$

where $Q=\Sigma_x\Sigma_y^{-1}$ and $T=\Sigma_x(\Sigma_x+\Sigma_y)^{-1}=Q(I_p+Q)^{-1}$.

From (20.41), we have the following result

$$
\begin{aligned}
&R_2(\tilde{\mu}_{Ax}, \mu) - R_2(\bar{Y}, \mu) \\
&\le \left[\frac{2cr}{n(n-r-p-2)} - \frac{c^2(n-r-2)r(r+1)}{n(n-r-p-1)(n-r-p-2)(n-r-p-4)}\right] \mathrm{tr}\,(T) - \frac{p}{n} \\
&+ \left[\frac{c^2(n-r-2)r(r+1)}{n(n-r-p-1)(n-r-p-2)(n-r-p-4)} - \frac{2cr}{n(n-r-p-2)} + \frac{1}{n}\right] \mathrm{tr}\,(Q) \\
&+ \frac{c^2p(n-r-2)r}{n(n-r-p-1)(n-r-p-2)(n-r-p-4)}(\,\mathrm{tr}\,(Q) - \,\mathrm{tr}\,(T)) \\
&= \left[\frac{2cr}{n(n-r-p-2)} - \frac{2c^2(n-r-2)r(r+1)}{n(n-r-p-1)(n-r-p-2)(n-r-p-4)}\right] \mathrm{tr}\,(T) - \frac{p}{n} \\
&+ \left[\frac{2c^2(n-r-2)r(r-1)}{n(n-r-p-1)(n-r-p-2)(n-r-p-4)} - \frac{2cr}{n(n-r-p-2)} + \frac{1}{n}\right] \mathrm{tr}\,(Q) \\
&\le \left[\frac{2cr}{n(n-r-p-2)} - \frac{2c^2(n-r-2)r(r+1)}{n(n-r-p-1)(n-r-p-2)(n-r-p-4)}\right] (p - \frac{p^2}{p + \,\mathrm{tr}\,(Q)}) \\
&- \frac{p}{n} + \left[\frac{2c^2(n-r-2)r(r+1)}{n(n-r-p-1)(n-r-p-2)(n-r-p-4)} - \frac{2cr}{n(n-r-p-2)} + \frac{1}{n}\right] \mathrm{tr}\,(Q).
\end{aligned}
\tag{20.42}
$$

Obviously, a good value of c to select is $c = c_t = \frac{(n-r-p-1)(n-r-p-4)}{2(r+1)(n-r-2)}$. Using $c_s = \frac{r(n-r-p-1)(n-r-p-4)}{2(r+1)(n-r-2)(n-r-p-2)}$, it is easy to show that a sufficient condition is $\mathrm{tr}\,(Q) \le \frac{p}{\sqrt{1-c_s}}$. This completes the proof. □

Proof of Theorem 20.6

Proof Note that

$$
\begin{aligned}
&R_3(\tilde{\mu}_x, \mu) - R_3(\bar{Y}, \mu) \\
&= 2\,\mathrm{tr}\,[E(\bar{Y}'(\Sigma_x + \Sigma_y)^{-1}(c\bar{S}_x\bar{S}^{-1} - I_p)(\bar{Y} - \bar{X}))] \\
&\quad + \mathrm{tr}\,[E((\bar{Y} - \bar{X})'(c\bar{S}^{-1}\bar{S}_x - I_p)(\Sigma_x + \Sigma_y)(c\bar{S}_x\bar{S}^{-1} - I_p)(\bar{Y} - \bar{X}))] \\
&= 2\,\mathrm{tr}\,[(\Sigma_x + \Sigma_y)^{-1}(cE(\bar{S}_x)E(\bar{S}^{-1}) - I_p)E((\bar{Y} - \bar{X})\bar{Y}')] \\
&\quad + \mathrm{tr}\,[E((c\bar{S}^{-1}\bar{S}_x - I_p)(\Sigma_x + \Sigma_y)^{-1}(c\bar{S}_x\bar{S}^{-1} - I_p))E((\bar{Y} - \bar{X})(\bar{Y} - \bar{X})')] \\
&= 2\,\mathrm{tr}\,[(\Sigma_x + \Sigma_y)^{-1}(\frac{c(r-1)}{n-r-p-2}\Sigma_x(\Sigma_x + \Sigma_y)^{-1} - I_p)\Sigma_y/n] \\
&\quad + c^2\,\mathrm{tr}\,\{E(\bar{S}^{-1}\bar{S}_x(\Sigma_x + \Sigma_y)^{-1}\bar{S}_x\bar{S}^{-1})(\Sigma_x + \Sigma_y)/n\} \\
&\quad - 2c\,\mathrm{tr}\,\{E(\bar{S}^{-1}\bar{S}_x)/n\} + p/n.
\end{aligned}
\tag{20.43}
$$

Using Lemma 20.1, from (20.43) we have

$$
\begin{aligned}
& R_3(\tilde{\mu}_x, \mu) - R_3(\bar{Y}, \mu) \\
&= \frac{2cr}{n(n-r-p-2)} \operatorname{tr}\left(\Sigma_y(\Sigma_x+\Sigma_y)^{-1}\Sigma_x(\Sigma_x+\Sigma_y)^{-1}\right) \\
&\quad - \frac{2}{n}\operatorname{tr}\left(\Sigma_y(\Sigma_x+\Sigma_y)^{-1}\right) + \frac{p}{n} - \frac{2cr}{n(n-r-p-2)}\operatorname{tr}\left(\Sigma_x(\Sigma_x+\Sigma_y)^{-1}\right) \\
&\quad + \frac{c^2(n-r-2)}{n(n-r-p-1)(n-r-p-2)(n-r-p-4)} \\
&\quad \times \operatorname{tr}\{[r\operatorname{tr}(\Sigma_x(\Sigma_x+\Sigma_y)^{-1})\Sigma_x + r(r+1)\Sigma_x(\Sigma_x+\Sigma_y)^{-1}\Sigma_x](\Sigma_x+\Sigma_y)^{-1}\} \\
&= \frac{2cr}{n(n-r-p-2)}\operatorname{tr}(T-T^2) - \frac{2}{n}\operatorname{tr}(I_p - T) + \frac{p}{n} - \frac{2cr}{n(n-r-p-2)}\operatorname{tr}(T) \\
&\quad + \frac{c^2(n-r-2)}{n(n-r-p-1)(n-r-p-2)(n-r-p-4)}[r\operatorname{tr}^2(T) + r(r+1)\operatorname{tr}(T^2)] \\
&= -\frac{2cr}{n(n-r-p-2)}\operatorname{tr}(T^2) + \frac{2}{n}\operatorname{tr}(T) - \frac{p}{n} \\
&\quad + \frac{c^2(n-r-2)[r\operatorname{tr}^2(T) + r(r+1)\operatorname{tr}(T^2)]}{n(n-r-p-1)(n-r-p-2)(n-r-p-4)},
\end{aligned}
\tag{20.44}
$$

where $T = \Sigma_x(\Sigma_x+\Sigma_y)^{-1}$.

By Lemma 20.2 and spectral decomposition (similar proof as in the Proof of Theorem 20.2), we have $\operatorname{tr}(T^2) \geq \frac{\operatorname{tr}^2(T)}{p}$. From (20.44), we have

$$
\begin{aligned}
& R_3(\tilde{\mu}_x, \mu) - R_3(\bar{Y}, \mu) \leq \frac{2}{n}\operatorname{tr}(T) - \frac{p}{n} \\
&\quad + \left[\frac{c^2(n-r-2)(r+p+1)r}{(n-r-p-1)(n-r-p-2)(n-r-p-4)} - \frac{2cr}{(n-r-p-2)}\right]\frac{\operatorname{tr}^2(T)}{np}.
\end{aligned}
\tag{20.45}
$$

Obviously, a good value of c to select is $c = c_0$. Using $c_m = \frac{r(n-r-p-1)(n-r-p-4)}{(r+p+1)(n-r-2)(n-r-p-2)}$, we see from (20.45) that a sufficient condition can be $\operatorname{tr}(T) \leq \frac{1}{1+\sqrt{1-c_m}}p$. This completes the proof. □

References

1. Ahmed, S.E., Liu, S.: Asymptotic theory of simultaneous estimation of Poisson means. Linear Algebra Appl. **430**, 2734–2748 (2009)
2. Brown, L.D., Cohen, A.: Point and confidence estimation of a common mean and recovery of interblock information. Ann. Stat. **2**, 963–976 (1974)

3. Chiou, W.J., Cohen, A.: On estimating a common multivariate normal mean vector. Ann. Inst. Stat. Math. **37**, 499–506 (1985)
4. Graybill, F.A., Deal, R.B.: Combining unbiased estimators. Biometrics **15**, 543–550 (1959)
5. Hartung, J.: An alternative method for meta-analysis. Biomet. J. **41**, 901–916 (1999)
6. Hartung, J., Knapp, G., Sinha, B.K.: Statistical Meta-Analysis with Applications. Wiley, New York (2008)
7. Kollo, T., von Rosen, D.: Advanced Multivariate Statistics with Matrices. Mathematics and Its Applications, vol. 579. Springer, Dordrecht (2005)
8. Krishnamoorthy, K.: Estimation of a common multivariate normal mean vector. Ann. Inst. Stat. Math. **43**, 761–771 (1991)
9. Lütkepohl, H.: Handbook of Matrix. Wiley, New York (1996)
10. Ma, T.F., Liu, S.: Estimation of order-restricted means of two normal populations under the LINEX loss function. Metrika **76**, 409–425 (2013)
11. Ma, T.F., Ye, R.D., Jia, L.J.: Finite-sample properties of the Graybill-Deal estimator. J. Stat. Plan. Infer. **141**, 3675–3680 (2011)
12. Misra, N., Van der Meulen, E.C.: On estimation of the common mean of $k(\geq 2)$ normal populations with order restricted variances. Stat. Probab. Lett. **36**, 261–267 (1997)
13. Nair, K.A.: Variance and distribution of the Graybill-Deal estimator of the common mean of two normal populations. Ann. Stat. **8**, 212–216 (1980)
14. Nanayakkara, N., Cressie, N.: Combining two unbiased estimators of a common mean of two normal populations. Aust. J. Stat. **33**(1), 43–56 (1991)
15. Norwood, T.E., Hinkelmann, K.: Estimating the common mean of several normal populations. Ann. Stat. **5**, 1047–1050 (1977)
16. Pal, N., Lim, W.K.: A note on second order admissibility of the Graybill-Deal estimator of a common mean of several normal populations. J. Stat. Plan. Infer. **63**, 71–78 (1997)
17. Pal, N., Lin, J.J., Chang, C.H., Kumar, S.: A revisit to the common mean problem: comparing the maximum likelihood estimator with the Graybill-Deal estimator. Comput. Stat. Data Anal. **51**, 5673–5681 (2007)
18. Park, J., Park, D.H.: Testing the equality of a large number of normal population means. Comput. Stat. Data Anal. **56**, 1131–1149 (2012)
19. Sinha, B.K.: Unbiased estimation of the variance of the Graybill-Deal estimator of the common mean of several normal distributions. Can. J. Stat. **13**, 243–247 (1985)
20. Varian, H.R.: A Bayesian approach to real state assessment. In: Studies in Bayesian Econometric and Statistics in Honor of L.J. Savage, S.E. Fienberg and A. Zellner (eds.), pp. 195–208. North Holland, Amsterdam (1975)
21. von Rosen, D.: Bilinear Regression Analysis. Springer, Berlin (2018)
22. Ye, R.D., Ma, T.F., Wang, S.G.: Inferences on the common mean of several inverse Gaussian populations. Comput. Stat. Data Anal. **54**, 906–915 (2010)
23. Zellner, A.: Bayes estimation and prediction using asymmetric loss functions. J. Amer. Stat. Asso. **81**, 446–451 (1986)
24. Zou, G.H.: Admissible estimation for finite population under the Linex loss function. J. Stat. Plan. Infer. **61**, 373–384 (1997)

The manufacturer's authorised representative in the EU is Springer Nature Customer Service Centre GmbH, Europaplatz 3, 69115 Heidelberg, Germany. If you have any concerns regarding our products, please contact ProductSafety@springernature.com

Printed and bound by CPI Group (UK) Ltd, Croydon, CR0 4YY
30/09/2026
02225503-0012